CZECHOSLOVAK ACADEMY OF SCIENCES

DIFFERENTIAL AND INTEGRAL EQUATIONS

Boundary Value Problems and Adjoints

CZECHOSLOVAK ACADEMY OF SCIENCES

SCIENTIFIC EDITOR

Prof. Dr. Vlastimil Pták, DrSc.

REVIEWER

Dr. Ivo Vrkoč, CSc.

Differential and Integral Equations

Boundary Value Problems and Adjoints

ŠTEFAN SCHWABIK, MILAN TVRDÝ, OTTO VEJVODA
Mathematical Institute of the Czechoslovak Academy of Sciences, Prague

1979

D. REIDEL PUBLISHING COMPANY
DORDRECHT : HOLLAND / BOSTON : U.S.A. / LONDON : ENGLAND

Library of Congress Cataloging in Publication Data

Schwabik, Stefan, 1941 –
Differential and integral equations.

Bibliography: p.
Includes index.
1. Differential equations. 2. Integral equations.
3. Boundary value problems. I. Tvrdy, Milan,
1944– joint author. II. Vejvoda, Otto, 1922–, joint
author. III. Title.
QA 371.S 385 1979 515'.35 78-13194
ISBN 90-277-0802-9

Sold and distributed in the U.S.A., Canada and Mexico by D. Reidel Publishing Company, Inc.
Lincoln Building, 160 Old Derby Street, Hingham, Mass. 02043, U.S.A.
Distributed in Albania, Bulgaria, Chinese People's Republic, Czechoslovakia, Cuba, German Democratic Republic, Hungary, Korean People's Democratic Republic, Mongolia, Poland, Rumania, Vietnam, the USSR, and Yugoslavia by Academia, Prague

Sold and distributed in all other countries by D. Reidel Publishing Company, Dordrecht

Published by D. Reidel Publishing Company, P. O. Box 17, Dordrecht, Holland, in co-edition with ACADEMIA, Publishing House of the Czechoslovak Academy of Sciences, Prague

Printed in Czechoslovakia

Contents

Preface

The present book is devoted to certain problems which belong to the domain of integral equations and boundary value problems for differential equations. Its essential part is concerned with linear systems of integral and generalized differential equations having in general discontinuous solutions of bounded variation on an interval. For various types of boundary value problems we derive adjoint problems in order to provide solvability conditions based on the principles of functional analysis.

Our exposition starts with an introductory chapter on linear algebra, functional analysis, Perron-Stieltjes integral and functions of bounded variation. In this chapter we collect some results needed in the other parts of the book. The second chapter contains results on integral operators in the Banach space of functions of bounded variation on an interval and basic results concerning linear Fredholm-Stieltjes and Volterra-Stieltjes integral equations in this space. Generalized linear differential equations are studied in the third chapter. Chapters IV and V deal with linear boundary value problems for ordinary differential and integro-differential operators while the last chapter is devoted to the perturbation theory for nonlinear ordinary differential equations with nonlinear side conditions.

Our conventions on cross references are as follows: For example, III.2.1 refers to paragraph 1 in section 2 of the third chapter while 2.1 refers to paragraph 1 in section 2 of the current chapter. The same applies to formulas whose numbers are given in parentheses, i.e. (III.2,1) is the first formula in section 2 of the third chapter while (2,1) stands for the first formula in section 2 of the current chapter. Bibliographical references include the name of the author followed by a number in square brackets which refers to the list of the bibliography given at the end of the book.

We wish express our gratitude to Professor Jaroslav Kurzweil for his continuous support dating back to the beginning of our work in this field. His results on generalized differential equations, Perron-Stieltjes integral as well as his ideas concerning general boundary value problems underlie the results contained in this book.

A special acknowledgement is due to Dr. Ivo Vrkoč who read and critically examined all the manuscript and in many cases improved considerably our original version.

The authors

List of symbols

Matrices

Functions

Operators

Sets and spaces

I. *Introduction*

This chapter provides some auxiliary results and notations needed in the subsequent chapters. As most of them can be easily found in the plentiful literature on linear algebra, real functions, functional analysis etc. we give only the necessary references without including their proofs. More attention is paid only to the Perron-Stieltjes integral in sections 4, 5 and 6.

1. Preliminaries

1.1. Basic notations. By R we denote the set of all real numbers. For $a < b$ we denote by $[a, b]$ and (a, b) respectively the closed and the open interval with the endpoints a, b. Similarly $[a, b)$, $(a, b]$ means the corresponding halfopen intervals.

A matrix with m rows and n columns is called an $m \times n$-matrix, $n \times 1$-matrices are called column n-vectors and $1 \times m$-matrices are called row m-vectors.

Matrices which in general do not reduce to vectors are denoted by capitals while vectors are denoted by lower-case letters. Given an $m \times n$-matrix $\mathbf{A}$, its element in the j-th row and k-th column is usually denoted by $a_{j,k}$ ($\mathbf{A} = (a_{j,k})$, $j = 1, \ldots, m$, $k = 1, \ldots, n$). Furthermore, $\mathbf{A}^*$ denotes the transpose of $\mathbf{A}$ ($\mathbf{A}^* = (a_{k,j})$, $k = 1, \ldots, n$, $j = 1, \ldots, m$),

$$|\mathbf{A}| = \max_{j=1,\ldots,m} \sum_{k=1}^{n} |a_{j,k}|,$$

rank $(\mathbf{A})$ is the rank of $\mathbf{A}$ and det $(\mathbf{A})$ denotes the value of the determinant of $\mathbf{A}$. If $m = n$ and $\det(\mathbf{A}) \neq 0$, then $\mathbf{A}^{-1}$ denotes the inverse of $\mathbf{A}$. $\mathbf{I}_m$ is the identity $m \times m$-matrix and $\mathbf{0}_{m,n}$ is the zero $m \times m$-matrix ($\mathbf{I}_m = (\delta_{j,k})$ $j, k = 1, \ldots, m$, where $\delta_{j,k} = 1$ if $j = k$, $\delta_{j,k} = 0$ if $j \neq k$ and $\mathbf{0}_{m,n} = (n_{j,k})$ $j = 1, \ldots, m$, $k = 1, \ldots, n$, where $n_{j,k} = 0$ for all $j = 1, \ldots, m$ and $k = 1, \ldots, n$). Usually, if no confusion may arise, the indices are omitted. The addition and multiplication on the space of matrices are defined in the obvious way and the usual notation

$$\mathbf{A} + \mathbf{B}, \ \mathbf{AB}, \ \lambda\mathbf{A} \qquad (\lambda \in R)$$

is used. Let the matrices $\mathbf{A}$, $\mathbf{B}$, $\mathbf{C}$ be of the types $m \times n$, $m \times p$ and $q \times n$, respectively. Then $\mathbf{D} = [\mathbf{A}, \mathbf{B}]$ is the $m \times (n + p)$-matrix with $d_{j,k} = a_{j,k}$ for $j = 1, \dots, m$, $k = 1, \dots, n$ and $d_{j,k} = b_{j,k-n}$ for $j = 1, \dots, m$, $k = n+1, n+2, \dots, n+p$. Analogously

$$\mathbf{H} = \begin{bmatrix} \mathbf{A} \\ \mathbf{C} \end{bmatrix}$$

is the $(m + q) \times n$-matrix with $h_{j,k} = a_{j,k}$ if $j \le m$ and $h_{j,k} = c_{j-m,k}$ if $j > m$.

R_n is the space of all real column n-vectors and R_n^* is the space of all real row n-vectors, $R_1 = R_1^* = R$. For $\mathbf{x} \in R_n$, $\mathbf{x}^* \in R_n^*$ we write

$$|\mathbf{x}| = \max_{j=1,\dots,n} |x_j|$$

and

$$|\mathbf{x}^*| = \sum_{j=1}^{n} |x_j| .$$

Given an $m \times n$-matrix $\mathbf{A}$, $\mathbf{x} \in R_n$ and $\mathbf{y} \in R_m$, then $|\mathbf{A}\mathbf{x}| \le |\mathbf{A}|\,|\mathbf{x}|$ and $|\mathbf{y}^*\mathbf{A}| \le |\mathbf{y}^*|\,|\mathbf{A}|$. The Euclidean norm in R_n is denoted by $|.|_e$

$$\mathbf{x} \in R_n \to |\mathbf{x}|_e = (\mathbf{x}^*\mathbf{x})^{1/2} = \left(\sum_{j=1}^{n} x_j^2 \right)^{1/2} .$$

It is easy to see that any $\mathbf{x} \in R_n$ satisfies $|\mathbf{x}|_e = |\mathbf{x}^*|_e$ and $|\mathbf{x}| \le |\mathbf{x}|_e \le |\mathbf{x}^*| \le n|\mathbf{x}|$.

The space of all real $m \times n$-matrices is denoted by $L(R_n, R_m)$ $(L(R_n, R_n) = L(R_n))$.

If M, N are sets and f is a mapping defined on M with values in N then we write f: $M \to N$ or $x \in M \to f(x) \in N$. For example, if f is a real function defined on an interval $[a, b]$, we write simply f: $[a, b] \to R$.

The words "measure", "measurable" without specification stand always for Lebesgue measure in R_n and measurability with respect to Lebesgue measure.

1.2. Linear spaces. A nonempty set X is called a (real) linear space if for every $\mathbf{x}, \mathbf{y} \in X$ and $\lambda \in R$ the sum $\mathbf{x} + \mathbf{y} \in X$ and the product $\lambda \mathbf{x} \in X$ are defined and the operations satisfy the usual axioms of a linear space. The zero element in X is denoted by $\mathbf{0}$.

A subset $L \subset X$ is a linear subspace of X if L is a linear space with respect to the sum and product with a real number given in X.

The elements $\mathbf{x}_1, \dots, \mathbf{x}_n$ of X are called *linearly independent* if $\alpha_1 \mathbf{x}_1 + \dots + \alpha_n \mathbf{x}_n = \mathbf{0}$, $\alpha_i \in R$, $i = 1, \dots, n$ implies $\alpha_1 = \alpha_2 = \dots = \alpha_n = 0$. Otherwise the elements $\mathbf{x}_1, \dots, \mathbf{x}_n$ are *linearly dependent*.

If X is a linear space and a norm $\mathbf{x} \in X \to \|\mathbf{x}\| \in R$ is defined, X is called a *normed linear space*. If X is a normed linear space which is complete with respect to the metric induced by the norm, then X is called a *Banach space*.

A real linear space X is called an *inner product space* (or pre-Hilbert space) if on $X \times X$ a real function $(\boldsymbol{x}_1, \boldsymbol{x}_2)_X$ is defined $((\boldsymbol{x}_1, \boldsymbol{x}_2) \in X \times X \to (\boldsymbol{x}_1, \boldsymbol{x}_2)_X \in R)$ such that for all $\boldsymbol{x}, \boldsymbol{x}_1, \boldsymbol{x}_2, \boldsymbol{x}_3 \in X$

$$(\boldsymbol{x}_1 + \boldsymbol{x}_2, \boldsymbol{x}_3)_X = (\boldsymbol{x}_1, \boldsymbol{x}_3)_X + (\boldsymbol{x}_2, \boldsymbol{x}_3)_X ,$$
$$(\boldsymbol{x}_1, \boldsymbol{x}_2)_X = (\boldsymbol{x}_2, \boldsymbol{x}_1)_X ,$$
$$(\alpha\boldsymbol{x}_1, \boldsymbol{x}_2)_X = \alpha(\boldsymbol{x}_1, \boldsymbol{x}_2)_X ,$$
$$(\boldsymbol{x}, \boldsymbol{x})_X \geq 0 \quad \text{and} \quad (\boldsymbol{x}, \boldsymbol{x})_X \neq 0 \qquad \text{for} \quad \boldsymbol{x} \neq \boldsymbol{0} .$$

The real function $(.,.)_X$ is called an *inner product* on X.

If X is an inner product space then the relation

$$(*) \qquad \boldsymbol{x} \in X \to \|\boldsymbol{x}\|_X = (\boldsymbol{x}, \boldsymbol{x})_X^{1/2} \in R$$

defines a norm on X.

A real inner product space X which is complete with respect to the norm defined by $(*)$ is called a real Hilbert space. Consequently a Hilbert space is a Banach space whose norm is induced by an inner product on X.

1.3. Function spaces. We shall deal with some usual spaces of real functions on an interval $[a, b]$, $-\infty < a < b < +\infty$. The sum of two functions and the product of a scalar and a function is defined in the usual way. For more detailed information see e.g. Dunford, Schwartz [1].

(i) We denote by $C_n[a, b]$ the space of all continuous column n-vector functions $\boldsymbol{f}$: $[a, b] \to R_n$ and define

$$\boldsymbol{f} \in C_n[a, b] \to \|\boldsymbol{f}\|_{C_n[a,b]} = \sup_{t \in [a,b]} |\boldsymbol{f}(t)| .$$

$\|\cdot\|_{C_n[a,b]}$ is a norm on $C_n[a, b]$; $C_n[a, b]$ with respect to this norm forms a Banach space. The zero element in $C_n[a, b]$ is the function vanishing identically on $[a, b]$.

(ii) If $1 \leq p < \infty$ we denote by $L^p_n[a, b]$ the space of all measurable functions $\boldsymbol{f}$: $[a, b] \to R_n$ such that

$$\int_a^b |\boldsymbol{f}(t)|^p \, dt < \infty .$$

We set

$$\boldsymbol{f} \in L^p_n[a, b] \to \|\boldsymbol{f}\|_{L^p_n[a,b]} = \left(\int_a^b |\boldsymbol{f}(t)|^p \, dt\right)^{1/p} .$$

The elements of $L^p_n[a, b]$ are classes of functions which are equal to one another almost everywhere (a. e.)*) on $[a, b]$. For the purposes of this text it is not restrictive

*) If a statement is true except possibly on a set of measure zero then we say that the statement is true almost everywhere (a.e.).

if we consider functions instead of classes of functions which are equal a.e. on $[a, b]$.

$L_n^p[a, b]$ with respect to the norm $\|.\|_{L_n^p[a,b]}$ is a Banach space. By $L_n^\infty[a, b]$ we denote the space of all measurable essentially bounded functions $\boldsymbol{f}: [a, b] \to R_n$ with the norm defined by

$$\boldsymbol{f} \in L_n^\infty[a, b] \to \|\boldsymbol{f}\|_{L_n^\infty[a,b]} = \sup_{t \in [a,b]} \text{ess}\, |\boldsymbol{f}(t)| \,.$$

$L_n^\infty[a, b]$ is a Banach space with respect to the norm $\|.\|_{L_n^\infty[a,b]}$. The zero element in $L_n^p[a, b]$ $(1 \le p \le \infty)$ is the class of functions which vanish a.e. on $[a, b]$.

(iii) We denote by $BV_n[a, b]$ the space of all functions $\boldsymbol{f}: [a, b] \to R_n$ of bounded variation $var_a^b \boldsymbol{f} < \infty$ where

$$\text{var}_a^b \boldsymbol{f} = \sup \sum_{i=1}^{k} |\boldsymbol{f}(t_i) - \boldsymbol{f}(t_{i-1})|$$

and the supremum is taken over all finite subdivisions of $[a, b]$ of the form $a = t_0 < t_1 < \ldots < t_k = b$. Let $c \in [a, b]$ then

$$\text{var}_a^b \boldsymbol{f} = \text{var}_a^c \boldsymbol{f} + \text{var}_c^b \boldsymbol{f} .$$

If we define

$$\boldsymbol{f} \in BV_n[a, b] \to \|\boldsymbol{f}\|_{BV_n[a,b]} = |\boldsymbol{f}(a)| + \text{var}_a^b \boldsymbol{f}$$

then $\|.\|_{BV_n[a,b]}$ is a norm on $BV_n[a, b]$ and $BV_n[a, b]$ is a Banach space with respect to this norm.

By $NBV_n[a, b]$ the subspace of $BV_n[a, b]$ is denoted such that $\boldsymbol{f} \in NBV_n[a, b]$ if $\boldsymbol{f}$ is continuous from the right at every point of (a, b) and $\boldsymbol{f}(a) = \boldsymbol{0}$. The norm in $NBV_n[a, b]$ is defined by

$$\boldsymbol{f} \in NBV_n[a, b] \to \|\boldsymbol{f}\|_{NBV_n[a,b]} = \text{var}_a^b \boldsymbol{f} .$$

A function $\boldsymbol{f}: [a, b] \to R_n$ is called absolutely continuous if for every $\varepsilon > 0$ there exists $\delta > 0$ such that

$$\sum_{i=1}^{k} |\boldsymbol{f}(b_i) - \boldsymbol{f}(a_i)| < \varepsilon$$

where (a_i, b_i), $i = 1, \ldots, k$ are arbitrary pairwise disjoint subintervals in $[a, b]$ such that $\sum_{i=1}^{k} |b_i - a_i| < \delta$.

Let $AC_n[a, b]$ be the space of all absolutely continuous functions $\boldsymbol{f}: [a, b] \to R_n$. It is $AC_n[a, b] \subset BV_n[a, b]$ and $AC_n[a, b]$ is a Banach space with respect to the norm of $BV_n[a, b]$, i.e.

$$\boldsymbol{f} \in AC_n[a, b] \to \|\boldsymbol{f}\|_{AC_n[a,b]} = |\boldsymbol{f}(a)| + \text{var}_a^b \boldsymbol{f} .$$

The zero element in $AC_n[a, b]$ and $BV_n[a, b]$ is the function vanishing identically on $[a, b]$.

Given an interval $[a, b]$, we write simply C_n, L^p_n, L^∞_n, BV_n, NBV_n, AC_n instead of $C_n[a, b]$, $L^p_n[a, b]$, $L^\infty_n[a, b]$, $BV_n[a, b]$, $NBV_n[a, b]$, $AC_n[a, b]$ if no misunderstanding may arise. If $n = 1$ then the index n is omitted, e.g. $C_1[a, b] = C[a, b]$, $L^p_1[a, b] = L^p[a, b]$ etc. The index n is also sometimes omitted in symbols for the norms, i.e. instead of $\|.\|_{C_n}$, $\|\cdot\|_{BV_n}$, $\|\cdot\|_{L^p_n}$ we write $\|.\|_C$, $\|\cdot\|_{BV}$, $\|\cdot\|_{L^p}$, respectively.

A matrix valued function $\mathbf{F}: [a, b] \to L(R_n, R_m)$ is said to be measurable or continuous or of bounded variation or absolutely continuous or essentially bounded on $[a, b]$ if any of the functions

$$t \in [a, b] \to f_{i,j}(t) \in R \qquad (i = 1, 2, \dots, m, \quad j = 1, 2, \dots, n)$$

is measurable or continuous or of bounded variation or absolutely continuous or essentially bounded on $[a, b]$, respectively.

Let us mention that

$$\operatorname{var}_a^b \mathbf{F} = \sup \sum_{i=1}^{k} |\mathbf{F}(t_i) - \mathbf{F}(t_{i-1})|$$

where the supremum is taken over all finite subdivisions of $[a, b]$ of the form

$$a = t_0 < t_1 < \dots < t_k = b$$

and

$$\max_{\substack{j=1,2,\dots,m \\ l=1,2,\dots,n}} (\operatorname{var}_a^b f_{j,l}) \le \operatorname{var}_a^b \mathbf{F} \le \sum_{j=1}^{m} \sum_{l=1}^{n} \operatorname{var}_a^b f_{j,l}.$$

We denote $\|\mathbf{F}\|_{L^\infty} = \sup\operatorname{ess}_{t \in [a,b]} |\mathbf{F}(t)|$ and $\|\mathbf{F}\|_{L^p} = (\int_a^b |\mathbf{F}(t)|^p \, dt)^{1/p}$ for $1 \le p < \infty$. If $\mathbf{F}: [a, b] \to L(R_n, R_m)$ is measurable and $\|\mathbf{F}\|_{L^p} < \infty$ $(1 \le p \le \infty)$, then the matrix valued function $\mathbf{F}: [a, b] \to L(R_n, R_m)$ is said to be L^p-integrable on $[a, b]$. (Instead of L^1-integrable we write simply L-integrable.)

1.4. Properties of functions of bounded variation. If $f \in BV[a, b]$ then the limits $\lim_{t \to t_0+} f(t) = f(t_0+)$, $t_0 \in [a, b)$, $\lim_{t \to t_0-} f(t) = f(t_0-)$, $t_0 \in (a, b]$ exist and the set of discontinuity points of f in $[a, b]$ is at most countable.

If $f \in BV[a, b]$ then $f(t) = p(t) - n(t)$, $t \in [a, b]$ where $p, n: [a, b] \to R$ are nondecreasing functions on $[a, b]$. Let a sequence $t_1, t_2, \dots$ of points in $[a, b]$, $t_i \neq t_j$, $i \neq j$ and two sequences of real numbers $c_1, c_2, \dots, d_1, d_2, \dots$ be given such that $t_n = a$ implies $c_n = 0$ and $t_n = b$ implies $d_n = 0$. Assume that the series $\sum_n c_n$, $\sum_n d_n$ converge absolutely. Define on $[a, b]$ a function $s: [a, b] \to R$ by the relation

$$s(t) = \sum_{t_n \le t} c_n + \sum_{t_n < t} d_n.$$

Every function of this type is called a *break function* on $[a, b]$. Clearly $s(t_n+) - s(t_n) = d_n$ and $s(t_n) - s(t_n-) = c_n$, $n = 1, 2, \dots$ and $s(t+) = s(t) = s(t-)$ if $t \in [a, b]$, $t \neq t_n$, $n = 1, 2, \dots$. Further $s \in BV[a, b]$ and $\operatorname{var}_a^b s = \sum_n (|c_n| + |d_n|)$.

If $f \in BV[a,b]$ then there exist uniquely determined functions $f_c \in BV[a,b]$, $f_b \in BV[a,b]$ such that f_c is a continuous function on $[a,b]$, f_b is a break function on $[a,b]$ and $f = f_c + f_b$ (the Jordan decomposition of $f \in BV[a,b]$).

If $f \in BV[a,b]$ then the derivative f' of f exists a.e. on $[a,b]$.

If $f \in BV[a,b]$ then f is expressible in the form

$$f = f_{ac} + f_s + f_b$$

where $f_{ac} \in AC[a,b]$, f_b is a break function on $[a,b]$ and f_s; $[a,b] \to R$ is continuous on $[a,b]$ with the derivative $f_s' = 0$ a.e. on $[a,b]$ (the Lebesgue decomposition of $f \in BV[a,b]$).

If $f \in AC[a,b]$ then the derivative f' exists a.e. on $[a,b]$ and $f' \in L^1[a,b]$, i.e. $\int_a^b |f'(t)| \, dt < \infty$ and $\operatorname{var}_a^b f = \int_a^b |f'(t)| \, dt$.

The following statement is important:

Helly's Choice Theorem. *Let an infinite family F of real functions on $[a,b]$ be given. If there is $K \geq 0$ such that*

$$|f(t)| \leq K \quad \text{for} \quad t \in [a,b] \quad \text{and} \quad \operatorname{var}_a^b f \leq K \quad \text{for every} \quad f \in F$$

then the family F contains a sequence $\{f_n\}_{n=1}^\infty$ such that $\lim_{n\to\infty} f_n(t) = \varphi(t)$ for every $t \in [a,b]$ and $\varphi \in BV[a,b]$, i.e. the sequence $f_n(t)$ converges pointwise to a function φ: $[a,b] \to R$ which is also of bounded variation.

On functions of bounded variation see e.g. Natanson [1], Aumann [1].

2. Linear algebraical equations and generalized inverse matrices

Let us consider linear algebraical equations for $\mathbf{x} \in R_n$ and $\mathbf{y}^* \in R_m^*$

(2,1) $$\mathbf{A}\mathbf{x} = \mathbf{b},$$

(2,2) $$\mathbf{A}\mathbf{x} = \mathbf{0}$$

and

(2,3) $$\mathbf{y}^*\mathbf{A} = \mathbf{0},$$

where $\mathbf{A}$ is an $m \times n$-matrix $(\mathbf{A} \in L(R_n, R_m))$ and $\mathbf{b} \in R_m$.

By $N(\mathbf{A})$ we denote the set of all solutions to (2,2). Obviously, $N(\mathbf{A})$ is a linear subspace in R_n, i.e. if $\mathbf{x}_1, \mathbf{x}_2 \in N(\mathbf{A})$ and $\alpha_1, \alpha_2 \in R$, then $\mathbf{x}_1\alpha_1 + \mathbf{x}_2\alpha_2 \in N(\mathbf{A})$. It is well-known that

(2,4) $$\dim N(\mathbf{A}) = n - \operatorname{rank}(\mathbf{A}),$$

i.e. either (2,2) possesses only the trivial solution $\mathbf{x} = \mathbf{0}$ (if $\operatorname{rank}(\mathbf{A}) = n$) or $N(\mathbf{A})$ contains a subset of $k = n - \operatorname{rank}(\mathbf{A})$ elements $\mathbf{x}_1, \mathbf{x}_2, \ldots, \mathbf{x}_k$ which are linearly independent, while any subset of $k+1$ its elements is linearly dependent. (We say also that the homogeneous equation (2,2) has exactly $k = n - \operatorname{rank}(\mathbf{A})$ linearly independent solutions.) The set $\{\mathbf{x}_1, \mathbf{x}_2, \ldots, \mathbf{x}_k\}$ forms a basis of $N(\mathbf{A})$ and any $\mathbf{x} \in N(\mathbf{A})$ can be expressed as their linear combination

$$\mathbf{x} = \sum_{j=1}^{k} \mathbf{x}_j \alpha_j, \qquad \text{where} \quad \alpha_j \in R \quad (j = 1, 2, \ldots, k).$$

As (2,3) is equivalent to $\mathbf{A}^*\mathbf{y} = \mathbf{0}$, $N(\mathbf{A}^*)$ denotes the linear subspace in R_m^* of all solutions to (2,3) and

(2,5) $$\dim N(\mathbf{A}^*) = m - \operatorname{rank}(\mathbf{A}^*) = m - \operatorname{rank}(\mathbf{A}).$$

Furthermore, the equation (2,1) possesses a solution if and only if (2,3) implies $\mathbf{y}^*\mathbf{b} = 0$. In particular, (2,1) possesses a solution for any $\mathbf{b} \in R_m$ if and only if (2,3) implies $\mathbf{y}^* = \mathbf{0}$ ($\dim N(\mathbf{A}^*) = 0$).

The equation (2,4) is said to be an adjoint equation to (2,1).

The concept of a generalized inverse matrix introduced by R. Penrose (Penrose [1] and [2]) enables us to express the solutions to (2,1) if they exist.

The following assertion is helpful.

2.1. Lemma. $\mathbf{BAA}^* = \mathbf{CAA}^*$ *implies* $\mathbf{BA} = \mathbf{CA}$ *and* $\mathbf{BA}^*\mathbf{A} = \mathbf{CA}^*\mathbf{A}$ *implies* $\mathbf{BA}^* = \mathbf{CA}^*$.

Proof. If $\mathbf{BAA}^* = \mathbf{CAA}^*$, then $\mathbf{0} = (\mathbf{BAA}^* - \mathbf{CAA}^*)(\mathbf{B} - \mathbf{C})^* = (\mathbf{BA} - \mathbf{CA})(\mathbf{A}^*\mathbf{B}^* - \mathbf{A}^*\mathbf{C}^*)$, whence $\mathbf{BA} = \mathbf{CA}$ immediately follows. (Given a matrix $\mathbf{D}$, $\mathbf{DD}^* = \mathbf{0}$ if and only if $\mathbf{D} = \mathbf{0}$.) As $(\mathbf{A}^*)^* = \mathbf{A}$, the latter implication is a consequence of the former one.

2.2. Theorem. *Given* $\mathbf{A} \in L(R_n, R_m)$, *there exists a unique matrix* $\mathbf{X} \in L(R_m, R_n)$ *such that*

(2,6) $$\mathbf{AXA} = \mathbf{A},$$

(2,7) $$\mathbf{XAX} = \mathbf{X},$$

(2,8) $$\mathbf{X}^*\mathbf{A}^* = \mathbf{AX},$$

(2,9) $$\mathbf{A}^*\mathbf{X}^* = \mathbf{XA}.$$

Proof. (a) Putting (2,8) into (2,7) we obtain

(2,10) $$\mathbf{XX}^*\mathbf{A}^* = \mathbf{X}.$$

On the other hand, if (2,10) holds, then $\mathbf{AX} = \mathbf{AXX}^*\mathbf{A}^*$. Since $(\mathbf{AXX}^*\mathbf{A}^*)^* = \mathbf{AXX}^*\mathbf{A}^*$, this means that $(\mathbf{AX})^* = \mathbf{AX}$ and (2,8) holds. Moreover, (2,8) and (2,10) yields $\mathbf{X} = \mathbf{XX}^*\mathbf{A}^* = \mathbf{XAX}$, i.e. the couple of equations (2,7), (2,8) is equivalent to (2,10).

(b) Analogously, the system (2,6), (2,9) is equivalent to

(2,11) $$\mathbf{XAA}^* = \mathbf{A}^*.$$

(c) Furthermore, to find a solution $\mathbf{X}$ to the system (2,10), (2,11) it is sufficient to find a solution $\mathbf{B}$ to the equation

(2,12) $$\mathbf{B}\mathbf{A}^*\mathbf{A}\mathbf{A}^* = \mathbf{A}^* .$$

In fact, (2,12) implies immediately that $\mathbf{X} = \mathbf{B}\mathbf{A}^*$ satisfies (2,11) and consequently also (2,9). Hence

$$\mathbf{A}^*\mathbf{X}^*\mathbf{A}^* = \mathbf{X}\mathbf{A}\mathbf{A}^* = \mathbf{A}^* \quad \text{and} \quad \mathbf{X}\mathbf{X}^*\mathbf{A}^* = \mathbf{B}\mathbf{A}^*\mathbf{X}^*\mathbf{A}^* = \mathbf{B}\mathbf{A}^* = \mathbf{X} .$$

(d) Now, let us consider the set of $n \times n$-matrices $(\mathbf{A}^*\mathbf{A})^j$ $(j = 1, 2, \ldots)$. Since the dimension of the space of all real $n \times n$-matrices is finite (n^2), there exist a natural number k and real numbers $\lambda_1, \lambda_2, \ldots, \lambda_k$ such that $|\lambda_1| + |\lambda_2| + \ldots + |\lambda_k| > 0$ and

(2,13) $$\lambda_1 \mathbf{A}^*\mathbf{A} + \lambda_2 (\mathbf{A}^*\mathbf{A})^2 + \ldots + \lambda_k (\mathbf{A}^*\mathbf{A})^k = \mathbf{0} .$$

Let r be the smallest natural number such that $\lambda_r \neq 0$. If we put

(2,14) $$\mathbf{B} = -\lambda_r^{-1}\{\lambda_{r+1}\mathbf{I} + \lambda_{r+2}\mathbf{A}^*\mathbf{A} + \ldots + \lambda_k (\mathbf{A}^*\mathbf{A})^{k-r-1}\} ,$$

then according to (2,13)

$$\mathbf{B}(\mathbf{A}^*\mathbf{A})^{r+1} = (\mathbf{A}^*\mathbf{A})^r .$$

Hence if $r \geq 2$, $\mathbf{B}(\mathbf{A}^*\mathbf{A})^r \mathbf{A}^*\mathbf{A} = (\mathbf{A}^*\mathbf{A})^{r-1} \mathbf{A}^*\mathbf{A}$ and according to 2.1

$$\mathbf{B}(\mathbf{A}^*\mathbf{A})^r = (\mathbf{A}^*\mathbf{A})^{r-1} .$$

In this way we can successively obtain

$$\mathbf{B}(\mathbf{A}^*\mathbf{A})^j = (\mathbf{A}^*\mathbf{A})^{j-1} \qquad \text{for} \quad j = 2, 3, \ldots, r .$$

In particular, $\mathbf{B}(\mathbf{A}^*\mathbf{A})^2 = \mathbf{A}^*\mathbf{A}$ and by 2.1 $\mathbf{B}\mathbf{A}^*\mathbf{A}\mathbf{A}^* = \mathbf{A}^*$. The matrix $\mathbf{B}$ defined in (2,14) satisfies (2,12) and hence $\mathbf{X} = \mathbf{B}\mathbf{A}^*$ verifies the system (2,6)–(2,9).

(e) It remains to show that this $\mathbf{X}$ is unique. Let us notice that by (2,9) and (2,7)

$$\mathbf{A}^*\mathbf{X}^*\mathbf{X} = \mathbf{X}\mathbf{A}\mathbf{X} = \mathbf{X}$$

and by (2,8) and (2,6)

$$\mathbf{A}^*\mathbf{A}\mathbf{X} = \mathbf{A}^*\mathbf{X}^*\mathbf{A}^* = (\mathbf{A}\mathbf{X}\mathbf{A})^* = \mathbf{A}^*$$

Now, let us assume that $\mathbf{Y} \in L(R_m, R_n)$ is such that

(2,15) $$\mathbf{A}^*\mathbf{Y}^*\mathbf{Y} = \mathbf{Y} , \qquad \mathbf{A}^*\mathbf{A}\,\mathbf{Y} = \mathbf{A}^* .$$

Then, according to (2,10) and (2,11)

$$\mathbf{X} = \mathbf{X}\mathbf{X}^*\mathbf{A}^* = \mathbf{X}\mathbf{X}^*\mathbf{A}^*\mathbf{A}\mathbf{Y} = \mathbf{X}\mathbf{A}\mathbf{Y} = \mathbf{X}\mathbf{A}\mathbf{A}^*\mathbf{Y}^*\mathbf{Y} = \mathbf{A}^*\mathbf{Y}^*\mathbf{Y} = \mathbf{Y} .$$

2.3. Definition. The unique solution $\mathbf{X}$ of the system (2,6)–(2,9) will be called the *generalized inverse matrix* to $\mathbf{A}$ and written $\mathbf{X} = \mathbf{A}^{\#}$.

2.4. Remark. By the definition and by the proof of 2.2 $\mathbf{A}^{\#}$ fulfils the relations

$$(2,16)\quad \mathbf{A}\mathbf{A}^{\#}\mathbf{A} = \mathbf{A}\,,\quad \mathbf{A}^{\#}\mathbf{A}\mathbf{A}^{\#} = \mathbf{A}^{\#}\,,\quad (\mathbf{A}^{\#})^{*}\,\mathbf{A}^{*} = \mathbf{A}\mathbf{A}^{\#}\,,\quad \mathbf{A}^{*}(\mathbf{A}^{\#})^{*} = \mathbf{A}^{\#}\mathbf{A}$$

and

$$(2,17)\quad \mathbf{A}^{\#}(\mathbf{A}^{\#})^{*}\,\mathbf{A}^{*} = \mathbf{A}^{\#}\,,\quad \mathbf{A}^{\#}\mathbf{A}\mathbf{A}^{*} = \mathbf{A}^{*}\,,\quad \mathbf{A}^{*}(\mathbf{A}^{\#})^{*}\,\mathbf{A}^{\#} = \mathbf{A}^{\#}\,,\quad \mathbf{A}^{*}\mathbf{A}\mathbf{A}^{\#} = \mathbf{A}^{*}$$

(cf. (2,6)–(2,11) and (2,15)).

2.5. Remark. If $m = n$ and $\mathbf{A}$ possesses an inverse matrix $\mathbf{A}^{-1}$, then evidently $\mathbf{A}^{-1}$ is a generalized inverse matrix to $\mathbf{A}$.

2.6. Proposition. *Let $\mathbf{A} \in L(R_n, R_m)$, $\mathbf{B} \in L(R_p, R_m)$. Then the equation for $\mathbf{X} \in L(R_p, R_n)$*

$$(2,18)\qquad \mathbf{A}\mathbf{X} = \mathbf{B}$$

possesses a solution if and only if

$$(2,19)\qquad (\mathbf{I}_m - \mathbf{A}\mathbf{A}^{\#})\,\mathbf{B} = \mathbf{0}\,.$$

If this is true, any solution $\mathbf{X}$ of (2,18) is of the form

$$(2,20)\qquad \mathbf{X} = \mathbf{X}_0 + \mathbf{A}^{\#}\mathbf{B}\,,$$

where $\mathbf{X}_0$ is an arbitrary solution of the matrix equation

$$\mathbf{A}\mathbf{X}_0 = \mathbf{0}_{m,p}\,.$$

Proof. Let $\mathbf{A}\mathbf{X} = \mathbf{B}$, then by (2,6) $(\mathbf{I} - \mathbf{A}\mathbf{A}^{\#})\,\mathbf{B} = (\mathbf{A} - \mathbf{A}\mathbf{A}^{\#}\mathbf{A})\,\mathbf{X} = \mathbf{0}$. If (2,19) holds, then $\mathbf{B} = \mathbf{A}\mathbf{A}^{\#}\mathbf{B}$ and (2,18) is equivalent to $\mathbf{A}(\mathbf{X} - \mathbf{A}^{\#}\mathbf{B}) = \mathbf{0}$, i.e. to $\mathbf{X} = \mathbf{X}_0 + \mathbf{A}^{\#}\mathbf{B}$, where $\mathbf{A}\mathbf{X}_0 = \mathbf{0}$.

2.7. Proposition. *Let $\mathbf{A} \in L(R_n, R_m)$. Then $\mathbf{A}\mathbf{X}_0 = \mathbf{0}_{m,p}$ if and only if there exists $\mathbf{C} \in L(R_p, R_n)$ such that $\mathbf{X}_0 = (\mathbf{I}_n - \mathbf{A}^{\#}\mathbf{A})\,\mathbf{C}$.*

Proof. $\mathbf{A}(\mathbf{I}_n - \mathbf{A}^{\#}\mathbf{A})\,\mathbf{C} = (\mathbf{A} - \mathbf{A}\mathbf{A}^{\#}\mathbf{A})\,\mathbf{C} = \mathbf{0}$ for any $\mathbf{C} \in L(R_p, R_n)$. If $\mathbf{A}\mathbf{X}_0 = \mathbf{0}$, then $\mathbf{X}_0 = \mathbf{X}_0 - \mathbf{A}^{\#}\mathbf{A}\mathbf{X}_0 = (\mathbf{I} - \mathbf{A}^{\#}\mathbf{A})\,\mathbf{X}_0$.

Some further properties of generalized inverse matrices are listed in the following lemma.

2.8. Lemma. *Given $\mathbf{A} \in L(R_n, R_m)$,*

$$(2,21)\qquad \mathbf{A}^{\#\#} = (\mathbf{A}^{\#})^{\#} = \mathbf{A}\,,$$

$$(2,22)\qquad (\mathbf{A}^{*})^{\#} = (\mathbf{A}^{\#})^{*}\,,$$

$$(2,23)\qquad (\lambda\mathbf{A})^{\#} = \lambda^{-1}\mathbf{A}^{\#} \quad \textit{for any } \lambda \in R,\ \lambda \neq 0 \ \textit{ and } \ \mathbf{0}^{\#}_{m,n} = \mathbf{0}_{n,m}\,,$$

$$(2,24)\qquad (\mathbf{A}^{*}\mathbf{A})^{\#} = \mathbf{A}^{\#}(\mathbf{A}^{\#})^{*}\,,\quad (\mathbf{A}\mathbf{A}^{*})^{\#} = (\mathbf{A}^{\#})^{*}\,\mathbf{A}^{\#}\,.$$

(The relations (2,21)–(2,24) may be easily verified by substituting their right-hand sides in the defining relations for the required generalized inverse.)

2.9. Lemma. *Let* $\mathbf{A} \in L(R_n, R_m)$ *and let* $\mathbf{U} \in L(R_m, R_n)$ *and* $\mathbf{V} \in L(R_n, R_m)$ *be such that*

$$\mathbf{A^*AU} = \mathbf{A^*} \quad \text{and} \quad \mathbf{AA^*V} = \mathbf{A}.$$

Then

$$\mathbf{V^*AU} = \mathbf{A^\#}.$$

Proof. Let $\mathbf{A^*AU} = \mathbf{A^*}$ and $\mathbf{AA^*V} = \mathbf{A}$. Then by 2.6

$$\mathbf{U} = \mathbf{U}_0 + (\mathbf{A^*A})^\# \mathbf{A^*} \quad \text{and} \quad \mathbf{V} = \mathbf{V}_0 + (\mathbf{AA^*})^\# \mathbf{A},$$

where $\mathbf{A^*AU}_0 = \mathbf{0}$ and $\mathbf{AA^*V}_0 = \mathbf{0}$. It follows from 2.1 that $\mathbf{A^*AU}_0 = \mathbf{0}$ (i.e. $\mathbf{U}_0^*\mathbf{A^*A} = \mathbf{0A^*A}$) and $\mathbf{AA^*V}_0 = \mathbf{0}$ (i.e. $\mathbf{V}_0^*\mathbf{AA^*} = \mathbf{0AA^*}$) implies $\mathbf{AU}_0 = \mathbf{0}$ and $\mathbf{V}_0^*\mathbf{A} = \mathbf{0}$, respectively. Furthermore, by (2,22) and (2,24)

$$((\mathbf{AA^*})^\#)^* = (\mathbf{AA^*})^\# = (\mathbf{A}^\#)^* \mathbf{A}^\# \quad \text{and} \quad (\mathbf{A^*A})^\# = \mathbf{A}^\#(\mathbf{A}^\#)^*.$$

Hence by the definition of $\mathbf{A}^\#$ (cf. 2.4)

$$\mathbf{V^*AU} = [\mathbf{A}^*(\mathbf{A}^\#)^*][\mathbf{A}^\#\mathbf{AA}^\#][(\mathbf{A}^\#)^*\mathbf{A}^*] = \mathbf{A}^\#\mathbf{AA}^\#\mathbf{AA}^\# = \mathbf{A}^\#.$$

2.10. Lemma. *Given* $\mathbf{A} \in L(R_n, R_m)$, *there exist* $\mathbf{U} \in L(R_m, R_n)$ *and* $\mathbf{V} \in L(R_n, R_m)$ *such that*

$$\mathbf{A^*AU} = \mathbf{A^*}, \qquad \mathbf{AA^*V} = \mathbf{A}. \tag{2,25}$$

Proof. By (2,24) and (2,17)

$$(\mathbf{A^*A})^\# \mathbf{A^*} = \mathbf{A}^\#(\mathbf{A}^\#)^* \mathbf{A^*} = \mathbf{A}^\#$$

and by (2,16) and (2,22) $\mathbf{AA}^\# = (\mathbf{A}^\#)^* \mathbf{A}^* = (\mathbf{A}^*)^\# \mathbf{A}^*$. Thus

$$[\mathbf{I} - (\mathbf{A^*A})(\mathbf{A^*A})^\#]\mathbf{A^*} = \mathbf{A^*} - \mathbf{A^*AA}^\# = \mathbf{A^*} - \mathbf{A}^*(\mathbf{A}^*)^\# \mathbf{A}^* = \mathbf{0}.$$

Since $(\mathbf{A}^*)^* = \mathbf{A}$, this implies also

$$[\mathbf{I} - (\mathbf{AA^*})(\mathbf{AA^*})^\#]\mathbf{A} = \mathbf{0}.$$

The proof follows now from 2.6.

2.11. Remark. Let us notice that from the relations (2,16) defining the generalized inverse of $\mathbf{A}$, only $\mathbf{AA}^\#\mathbf{A} = \mathbf{A}$ was utilized in the proofs of 2.6 and 2.7. Some authors (see e.g. Reid [1]) define any matrix $\mathbf{X}$ fulfilling $\mathbf{AXA} = \mathbf{A}$ to be a generalized inverse of $\mathbf{A}$.

Let $\mathbf{A} \in L(R_n, R_m)$ and $h = \operatorname{rank}(\mathbf{A})$. If $h = n$, then $\mathbf{Ax} = \mathbf{0}$ if and only if $\mathbf{x} = \mathbf{0}$. Let us assume $h < n$. By (2,4) there exist an $n \times (n-h)$-matrix $\mathbf{X}_0$ such that its columns form a basis in $N(\mathbf{A})$, i.e. $\mathbf{Ax} = \mathbf{0}$ if and only if there exists $\mathbf{c} \in R_{n-h}$ such that $\mathbf{x} = \mathbf{X}_0\mathbf{c}$. Consequently $\mathbf{X} \in L(R_p, R_n)$ fulfils $\mathbf{AX} = \mathbf{0}_{m,p}$ if and only if there exists $\mathbf{C} \in L(R_p, R_{n-h})$ such that $\mathbf{X} = \mathbf{X}_0\mathbf{C}$. In particular, there exists $\mathbf{C}_0 \in L(R_n, R_{n-h})$ such that

$$\mathbf{I}_n - \mathbf{A}^{\#}\mathbf{A} = \mathbf{X}_0\mathbf{C}_0 . \tag{2,26}$$

Furthermore, let $h = \operatorname{rank}(\mathbf{A}) < m$. Then by (2,5) there exists $\mathbf{Y}_0 \in L(R_m, R_{m-h})$ such that its rows form a basis in $N(\mathbf{A}^*)$. Consequently $\mathbf{Y} \in L(R_m, R_p)$ fulfils $\mathbf{YA} = \mathbf{0}_{p,n}$ if and only if there exists $\mathbf{D} \in L(R_{m-h}, R_p)$ such that $\mathbf{Y} = \mathbf{DY}_0$. In particular, there exists $\mathbf{D}_0 \in L(R_{m-h}, R_m)$ such that

$$\mathbf{I}_m - \mathbf{AA}^{\#} = \mathbf{D}_0\mathbf{Y}_0 . \tag{2,27}$$

(If $h = m$, then $\mathbf{y}^*\mathbf{A} = \mathbf{0}$ if and only if $\mathbf{y}^* = \mathbf{0}$.)

2.12. Proposition. *Let* $\mathbf{A} \in L(R_n, R_m)$ *and* $\mathbf{X} = L(R_m, R_n)$. *Then* $\mathbf{AXA} = \mathbf{A}$ *if and only if there exist* $\mathbf{H}$ *and* $\mathbf{D} \in L(R_m, R_n)$ *such that*

$$\mathbf{X} = \mathbf{A}^{\#} + (\mathbf{I}_n - \mathbf{A}^{\#}\mathbf{A})\,\mathbf{H} + \mathbf{D}(\mathbf{I}_m - \mathbf{AA}^{\#})$$

or equivalently if and only if

$$\mathbf{X} = \mathbf{A}^{\#} + \mathbf{X}_0\mathbf{K} + \mathbf{LY}_0 ,$$

where $\mathbf{X}_0 \in L(R_{n-h}, R_n)$ *and* $\mathbf{Y}_0 \in L(R_m, R_{m-h})$ $(h = \operatorname{rank}(\mathbf{A}))$ *were defined above,* $\mathbf{K} \in L(R_m, R_{n-h})$ *and* $\mathbf{L} \in L(R_{m-h}, R_n)$ *are arbitrary, the term* $\mathbf{X}_0\mathbf{K}$ *vanishes if* $h = n$ *and the term* $\mathbf{LY}_0$ *vanishes if* $h = m$.

Proof. Let us assume $h < m$ and $h < n$. Let both $\mathbf{AX}_1\mathbf{A} = \mathbf{A}$ and $\mathbf{AX}_2\mathbf{A} = \mathbf{A}$. Then $\mathbf{A}(\mathbf{X}_1 - \mathbf{X}_2)\,\mathbf{A} = \mathbf{0}_{m,n}$ and hence $(\mathbf{X}_1 - \mathbf{X}_2)\,\mathbf{A} = (\mathbf{I}_n - \mathbf{A}^{\#}\mathbf{A})\,\mathbf{C}$ with some $\mathbf{C} \in L(R_n)$. By 2.6 and 2.7 this is possible if and only if

$$\mathbf{X}_1 - \mathbf{X}_2 = (\mathbf{I}_n - \mathbf{A}^{\#}\mathbf{A})\,\mathbf{CA}^{\#} + \mathbf{D}(\mathbf{I}_m - \mathbf{AA}^{\#})$$

or by (2,26) and (2,27) if and only if

$$\mathbf{X}_1 - \mathbf{X}_2 = \mathbf{X}_0[\mathbf{C}_0\mathbf{CA}^{\#}] + [\mathbf{DD}_0]\,\mathbf{Y}_0 .$$

Putting $\mathbf{CA}^{\#} = \mathbf{H}$, $\mathbf{C}_0\mathbf{CA}^{\#} = \mathbf{K}$ and $\mathbf{DD}_0 = \mathbf{L}$ we obtain the desired relations. The modification of the proof in the case that $h = m$ and/or $h = n$ is obvious.

2.13. Lemma. *Let* $\mathbf{A} \in L(R_n, R_m)$. *If* $\operatorname{rank}(\mathbf{A}) = m$, *then* $\det(\mathbf{AA}^*) \neq 0$. *If* $\operatorname{rank}(\mathbf{A}) = n$, *then* $\det(\mathbf{A}^*\mathbf{A}) \neq 0$.

Proof. Let $\operatorname{rank}(\mathbf{A}) = m$. Then by (2,5) $\mathbf{A}^*\mathbf{y} = \mathbf{0}$ if and only if $\mathbf{y} = \mathbf{0}$. Now, since $\mathbf{A}^* = \mathbf{A}^{\#}\mathbf{AA}^*$ (cf. (2,17)), $\mathbf{AA}^*\mathbf{y} = \mathbf{0}$ implies $\mathbf{A}^*\mathbf{y} = \mathbf{A}^{\#}\mathbf{AA}^*\mathbf{y} = \mathbf{0}$ and hence $\mathbf{y} = \mathbf{0}$. This implies that $\operatorname{rank}(\mathbf{AA}^*) = m$ (cf. (2,4)).

If $\operatorname{rank}(\mathbf{A}) = \operatorname{rank}(\mathbf{A}^*) = n$, then by the first assertion of the lemma $\operatorname{rank}(\mathbf{A}^*\mathbf{A}) = \operatorname{rank}(\mathbf{A}^*(\mathbf{A}^*)^*) = n$.

2.14. Remark. It is well known that rank $(\mathbf{AX}) = \min(\text{rank}(\mathbf{A}), \text{rank}(\mathbf{X}))$ whenever the product $\mathbf{AX}$ of the matrices $\mathbf{A}, \mathbf{X}$ is defined. Hence for a given $\mathbf{A} \in L(R_n, R_m)$ there exists $\mathbf{X} \in L(R_m, R_n)$ such that $\mathbf{AX} = \mathbf{I}_m$ only if rank $(\mathbf{A}) = m$. Analogously, there exists $\mathbf{X} \in L(R_m, R_n)$ such that $\mathbf{XA} = \mathbf{I}_n$ only if rank $(\mathbf{A}) = n$.

2.15. Lemma. *Let* $\mathbf{A} \in L(R_n, R_m)$. *If* rank $(\mathbf{A}) = m$, *then* $\mathbf{AA}^{\#} = \mathbf{I}_m$. *If* rank $(\mathbf{A}) = n$, *then* $\mathbf{A}^{\#}\mathbf{A} = \mathbf{I}_n$.

Proof. (a) Let rank $(\mathbf{A}) = m$. Then by 2.13 $(\mathbf{AA}^*)$ possesses an inverse $(\mathbf{AA}^*)^{-1}$ and according to the relation $\mathbf{A}^{\#}\mathbf{AA}^* = \mathbf{A}^*$ (cf. (2,17))

$$\mathbf{A}^{\#} = \mathbf{A}^*(\mathbf{AA}^*)^{-1} \tag{2,28}$$

and hence $\mathbf{AA}^{\#} = \mathbf{I}_m$.

(b) If rank $(\mathbf{A}) = n$, then the relation $\mathbf{A}^*\mathbf{AA}^{\#} = \mathbf{A}^*$ from (2,17) and 2.13 imply

$$\mathbf{A}^{\#} = (\mathbf{A}^*\mathbf{A})^{-1}\,\mathbf{A} \tag{2,29}$$

and hence $\mathbf{A}^{\#}\mathbf{A} = \mathbf{I}_n$.

2.16. Lemma. *Let* $\mathbf{A} \in L(R_m)$, $\mathbf{B} \in L(R_n, R_m)$ *and* $\mathbf{C} \in L(R_n)$. *If* rank $(\mathbf{A})$ = rank $(\mathbf{B})$ $= m$, *then* $(\mathbf{AB})^{\#} = \mathbf{B}^{\#}\mathbf{A}^{-1}$. *If* rank $(\mathbf{B})$ = rank $(\mathbf{C}) = n$, *then* $(\mathbf{BC})^{\#} = \mathbf{C}^{-1}\mathbf{B}^{\#}$.

Proof. Let rank $(\mathbf{A})$ = rank $(\mathbf{B}) = m$. Then by 2.15 $\mathbf{BB}^{\#} = \mathbf{I}$. Consequently $\mathbf{ABB}^{\#}\mathbf{A}^{-1} = \mathbf{I}$. Furthermore, $(\mathbf{B}^{\#}\mathbf{A}^{-1})(\mathbf{AB}) = \mathbf{B}^{\#}\mathbf{B} = \mathbf{B}^*(\mathbf{B}^{\#})^* = \mathbf{B}^*\mathbf{A}^*(\mathbf{A}^{-1})^*(\mathbf{B}^{\#})^*$ $= (\mathbf{AB})^*(\mathbf{B}^{\#}\mathbf{A}^{-1})^*$. This completes the proof of the former assertion. The latter one could be proved analogously.

For some more details about generalized inverse matrices see e.g. Reid [1] (Appendix B), Moore [1], Nashed [1] and "*Proceedings of Symposium on the Theory and Applications of Generalized Inverses of Matrices*" held at the Texas Technological College, Lubbock, Texas, March 1968, Texas Technological College Math. Series, No. 4.

3. Functional analysis

Here we review some concepts and results from linear functional analysis used in the subsequent chapters. For more information we mention e.g. Dunford, Schwartz [1], Heuser [1], Goldberg [1], Schechter [1].

Let X be a linear space over the real scalars R. If F, G are linear subspaces of X, then we set

$$F + G = \{\mathbf{z} \in X;\ \mathbf{z} = \mathbf{x} + \mathbf{y},\ \mathbf{x} \in F,\ \mathbf{y} \in G\}.$$

$F + G$ is evidently a linear subspace of X.

$F + G$ is called the direct sum of two linear subspaces F, G if $F \cap G = \{\mathbf{0}\}$. Let the direct sum of F and G be denoted by $F \oplus G$.

If $F \oplus G = X$ then G is called the complementary subspace to F in X.

It can be shown (see e.g. Heuser [1], II.4) that

(1) for any linear subspace $F \subset X$ there exists at least one complementary subspace $G \subset X$

(2) for any two complementary subspaces G_1, G_2 to a given subspace $F \subset X$ we have $\dim G_1 = \dim G_2$ where by dim the usual linear dimension of a linear set is denoted.

This enables us to define the codimension of a linear subspace $F \subset X$ as follows. Let $X = F \oplus G$; then we set

$$\operatorname{codim} F = \dim G .$$

(If $\dim G = \infty$ or $X = F$, we put $\operatorname{codim} F = \infty$ or $\operatorname{codim} F = 0$, respectively.)

If $F \subset X$ is a linear subspace, then we set $\mathbf{x} \sim \mathbf{y}$ for $\mathbf{x}, \mathbf{y} \in X$ if $\mathbf{x} - \mathbf{y} \in F$. By $\sim$ an equivalence relation on X is given. This equivalence relation decomposes X into disjoint classes of equivalent elements of X. If $\mathbf{x} \in X$ belongs to a given equivalence class with respect to the equivalence relation $\sim$ then all elements of this class belong to the set $\mathbf{x} + F$.

Let us denote by X/F the set of all equivalence classes with respect to the given equivalence relation. Let the equivalence class containing $\mathbf{x} \in X$ be denoted by $[\mathbf{x}]$, i.e.

$$[\mathbf{x}] = \mathbf{x} + F .$$

Then

$$X/F = \{[\mathbf{x}] = \mathbf{x} + F;\ \mathbf{x} \in X\} .$$

If we define $[\mathbf{x}] + [\mathbf{y}] = [\mathbf{x} + \mathbf{y}]$, $\alpha[\mathbf{x}] = [\alpha\mathbf{x}]$ where $\mathbf{x} \in [\mathbf{x}]$, $\mathbf{y} \in [\mathbf{y}]$, $\alpha \in R$ then X/F becomes a linear space over R called the quotient space. It can be shown that if $X = F \oplus G$, then there is a one-to-one correspondence between X/F and G (see e.g. Heuser [1], III.20). Hence

$$\operatorname{codim} F = \dim G = \dim (X/F) .$$

Let X and Y be linear spaces over R. We consider linear operators $\mathbf{A}$ which assign a unique element $\mathbf{Ax} = \mathbf{y} \in Y$ to every element $\mathbf{x} \in D(\mathbf{A}) \subset X$. The set $D(\mathbf{A})$ called the domain of $\mathbf{A}$ forms a linear subspace in X and the linearity relation

$$\mathbf{A}(\alpha\mathbf{x} + \beta\mathbf{z}) = \alpha\mathbf{Ax} + \beta\mathbf{Az}$$

holds for all $\mathbf{x}, \mathbf{z} \in X$, $\alpha, \beta \in R$.

The set of all linear operators $\mathbf{A}$ with values in Y such that $D(\mathbf{A}) \doteq X$ will be denoted by $L(X, Y)$. If $X = Y$, then we write simply $L(X)$ instead of $L(X, X)$. The identity operator $\mathbf{x} \in X \to \mathbf{x} \in X$ on X is usually denoted by $\mathbf{I}$. For an operator $\mathbf{A} \in L(X, Y)$ we use the following notations:

$$R(\mathbf{A}) = \{\mathbf{y} \in Y;\ \mathbf{y} = \mathbf{A}\mathbf{x},\ \mathbf{x} \in X\}$$

denotes the *range* of $\mathbf{A}$, the linear subspace of values of $\mathbf{A} \in L(X, Y)$ in Y.

$$N(\mathbf{A}) = \{\mathbf{x} \in X;\ \mathbf{A}\mathbf{x} = \mathbf{0} \in Y\}$$

denotes the *null-space* of $\mathbf{A} \in L(X, Y)$; $N(\mathbf{A}) \subset X$ is a linear subspace in X. Further we denote

$$\alpha(\mathbf{A}) = \dim N(\mathbf{A})$$

and

$$\beta(\mathbf{A}) = \operatorname{codim} R(\mathbf{A}) = \dim\left(Y/R(\mathbf{A})\right).$$

If $\alpha(\mathbf{A})$, $\beta(\mathbf{A})$ are not both infinite, then we define the index $\operatorname{ind} \mathbf{A}$ of $\mathbf{A} \in L(X, Y)$ by the relation

$$\operatorname{ind} \mathbf{A} = \beta(\mathbf{A}) - \alpha(\mathbf{A}).$$

The operator $\mathbf{A} \in L(X, Y)$ is called one-to-one if for $\mathbf{x}_1, \mathbf{x}_2 \in X$, $\mathbf{x}_1 \neq \mathbf{x}_2$ we have $\mathbf{A}\mathbf{x}_1 \neq \mathbf{A}\mathbf{x}_2$. Evidently $\mathbf{A} \in L(X, Y)$ is one-to-one if and only if $N(\mathbf{A}) = \{\mathbf{0}\}$ (or equivalently $\alpha(\mathbf{A}) = 0$).

The inverse operator $\mathbf{A}^{-1}$ for $\mathbf{A} \in L(X, Y)$ can be defined only if $\mathbf{A}$ is one-to-one. By definition $\mathbf{A}^{-1}$ is a linear operator from Y to X mapping $\mathbf{y} = \mathbf{A}\mathbf{x} \in Y$ to $\mathbf{x} \in X$. We have $D(\mathbf{A}^{-1}) = R(\mathbf{A})$, $R(\mathbf{A}^{-1}) = D(\mathbf{A}) = X$, $\mathbf{A}^{-1}(\mathbf{A}\mathbf{x}) = \mathbf{x}$ for $\mathbf{x} \in X$, $\mathbf{A}(\mathbf{A}^{-1}\mathbf{y}) = \mathbf{y}$ for $\mathbf{y} \in R(\mathbf{A})$. If $R(\mathbf{A}) = Y$ and $N(\mathbf{A}) = \{\mathbf{0}\}$ (i.e. $\alpha(\mathbf{A}) = \beta(\mathbf{A}) = 0$) then we can assign to any $\mathbf{y} \in Y$ the element $\mathbf{A}^{-1}\mathbf{y}$ which is the unique solution of the linear equation

$$\mathbf{A}\mathbf{x} = \mathbf{y}. \tag{3,1}$$

In this case we have $\mathbf{A}^{-1} \in L(Y, X)$. The linear equation (3,1) can be solved in general only for $\mathbf{y} \in R(\mathbf{A})$.

The linear equation (3,1) for $\mathbf{A} \in L(X, Y)$ is called uniquely solvable on $R(\mathbf{A})$ if for any $\mathbf{y} \in R(\mathbf{A})$ there is only one $\mathbf{x} \in X$ such that $\mathbf{A}\mathbf{x} = \mathbf{y}$. The equation (3,1) is uniquely solvable on $R(\mathbf{A})$ if and only if $\mathbf{A}$ is one-to-one (i.e. $N(\mathbf{A}) = \{\mathbf{0}\}$).

Let now X, X^+ be linear spaces. Assume that a bilinear form $\langle \mathbf{x}, \mathbf{x}^+ \rangle : X \times X^+ \to R$ is defined on $X \times X^+$ (i.e. $\langle \alpha\mathbf{x} + \beta\mathbf{y}, \mathbf{x}^+ \rangle = \alpha\langle \mathbf{x}, \mathbf{x}^+ \rangle + \beta\langle \mathbf{y}, \mathbf{x}^+ \rangle$, $\langle \mathbf{x}, \alpha\mathbf{x}^+ + \beta\mathbf{y}^+ \rangle = \alpha\langle \mathbf{x}, \mathbf{x}^+ \rangle + \beta\langle \mathbf{x}, \mathbf{y}^+ \rangle$ for every $\mathbf{x}, \mathbf{y} \in X$, $\mathbf{x}^+, \mathbf{y}^+ \in X^+$, $\alpha, \beta \in R$).

3.1. Definition. If X, X^+ are linear spaces, $\langle \mathbf{x}, \mathbf{x}^+ \rangle$ a bilinear form on $X \times X^+$ we say that the spaces X, X^+ form a *dual pair* (X, X^+) (*with respect to the bilinear form* $\langle .,. \rangle$) if

$$\langle \mathbf{x}, \mathbf{x}^+ \rangle = 0 \quad \text{for every} \quad \mathbf{x} \in X \quad \text{implies} \quad \mathbf{x}^+ = \mathbf{0} \in X^+$$

and

$$\langle \mathbf{x}, \mathbf{x}^+ \rangle = 0 \quad \text{for every} \quad \mathbf{x}^+ \in X^+ \quad \text{implies} \quad \mathbf{x} = \mathbf{0} \in X.$$

In Heuser [1], VI.40 the following important statement is proved.

3.2. Theorem. *Let (X, X^+) be a dual pair of linear spaces with respect to the bilinear form $\langle .,. \rangle$ defined on $X \times X^+$. Assume that $\mathbf{A} \in L(X)$ is such an operator that there is an operator $\mathbf{A}^+ \in L(X^+)$ such that*

$$\langle \mathbf{Ax}, \mathbf{x}^+ \rangle = \langle \mathbf{x}, \mathbf{A}^+\mathbf{x}^+ \rangle$$

for every $\mathbf{x} \in X$, $\mathbf{x}^+ \in X^+$.

If $\operatorname{ind} \mathbf{A} = \operatorname{ind} \mathbf{A}^+ = 0$, *then*

$$\alpha(\mathbf{A}) = \alpha(\mathbf{A}^+) = \beta(\mathbf{A}) = \beta(\mathbf{A}^+) < \infty$$

and moreover

$\mathbf{Ax} = \mathbf{y}$ *has a solution if and only if* $\langle \mathbf{y}, \mathbf{x}^+ \rangle = 0$ *for all* $\mathbf{x}^+ \in N(\mathbf{A}^+)$,

$\mathbf{A}^+\mathbf{x}^+ = \mathbf{y}^+$ *has a solution if and only if* $\langle \mathbf{x}, \mathbf{y}^+ \rangle = 0$ *for all* $\mathbf{x} \in N(\mathbf{A})$.

In the following we assume that X and Y are Banach spaces, i.e. normed linear spaces which are complete with respect to the norm given in X, Y respectively. The norm in a normed linear space X will be denoted by $\|.\|_X$ or simply $\|.\|$ when no misunderstanding may occur.

3.3. Definition. An operator $\mathbf{A} \in L(X, Y)$ is bounded if there exists a constant $M \in R$ such that

$$\|\mathbf{Ax}\| \le M \|\mathbf{x}\|$$

for all $\mathbf{x} \in X$.

The set of all bounded operators $\mathbf{A} \in L(X, Y)$ $(\mathbf{A} \in L(X))$ will be denoted by $B(X, Y)$ $(B(X))$.

It is well-known that $\mathbf{A} \in B(X, Y)$ if and only if $\mathbf{A}$ is continuous, i.e. for every sequence $\{\mathbf{x}_n\}_{n=1}^{\infty}$, $\lim_{n\to\infty} \mathbf{x}_n = \mathbf{x}$ we have $\lim_{n\to\infty} \mathbf{Ax}_n = \mathbf{Ax}$.

For $\mathbf{A} \in B(X, Y)$ we define

$$\|\mathbf{A}\|_{B(X,Y)} = \sup_{\|\mathbf{x}\|=1} \|\mathbf{Ax}\| = \sup_{\mathbf{x} \neq \mathbf{0}} \frac{\|\mathbf{Ax}\|}{\|\mathbf{x}\|}. \tag{3,2}$$

It can be proved that by the relation (3,2) a norm on $B(X, Y)$ is given and that $B(X, Y)$ with this norm is a Banach space (see e.g. Schechter [1], Chap. III.).

3.4. Theorem (Bounded Inverse Theorem). *If* $\mathbf{A} \in B(X, Y)$ *is such that* $R(\mathbf{A}) = Y$ *and* $N(\mathbf{A}) = \{\mathbf{0}\}$, *then* $\mathbf{A}^{-1}$ *exists and* $\mathbf{A}^{-1} \in B(Y, X)$.
(See Schechter [1], III. Theorem 4.1).

3.5. Definition. We denote $X^* = B(X, R)$, where R is the Banach space of real numbers with the norm given by $\alpha \in R \to |\alpha|$. The elements of X^* are called *linear bounded functionals* on X and X^* is the *dual space* to X. Given $\mathbf{f} \in X^*$, its value at $\mathbf{x} \in X$ is denoted also by

$$\mathbf{f}(\mathbf{x}) = \langle \mathbf{x}, \mathbf{f} \rangle_X .$$

If $f(x) = 0$ for any $x \in X$, f is said to be the zero functional on X and we write $f = 0$.

3.6. Remark. X^* equipped with the norm

$$\|f\|_{X^*} = \sup_{\|x\|_X = 1} |f(x)| = \sup_{x \neq 0} \frac{|f(x)|}{\|x\|_X} \qquad \text{for} \quad f \in X^*$$

(cf. (3,2)) is a Banach space. Furthermore,

$$x \in X,\ f \in X^* \to \langle x, f \rangle_X$$

is evidently a bilinear form on $X \times X^*$. Clearly, $\langle x,f \rangle_X = 0$ for any $x \in X$ if and only if f is the zero functional on X ($f = 0 \in X^*$). Moreover, it follows from the Hahn-Banach Theorem (see e.g. Schechter [1], II.3.2) that $\langle x, f \rangle_X = 0$ for any $f \in X^*$ if and only if $x = 0$. This means that the spaces X and its dual X^* form a dual pair (X, X^*) with respect to the bilinear form $\langle .,. \rangle_X$.

For some Banach spaces X there exist a Banach space E_X and a bilinear form $[.,.]_X$ on $X \times E_X$ such that $f \in X^*$ if and only if there exists $g \in E_X$ such that

$$\langle x, f \rangle_X = [x, g]_X \qquad \text{for any} \quad x \in X .$$

If this correspondence between E_X and X^* is an isometrical isomorphism*), we identify E_X with X^* and put

$$\langle x, g \rangle_X = [x, g]_X .$$

3.7. Definition. Let X, Y be Banach spaces. By $X \times Y$ we denote the space of all couples (x, y), where $x \in X$ and $y \in Y$. Given $(x, y), (u, v) \in X \times Y$ and $\lambda \in R$, we put $(x, y) + (u, v) = (x + u, y + v)$, $\lambda(x, y) = (\lambda x, \lambda y)$ and

$$\|(x, y)\|_{X \times Y} = \|x\|_X + \|y\|_Y .$$

(Clearly, $\|.\|_{X \times Y}$ is a norm on $X \times Y$ and $X \times Y$ equipped with this norm is a Banach space.)

3.8. Lemma. *If (X, X^+) and (Y, Y^+) are dual pairs with respect to the bilinear forms $[.,.]_X$ and $[.,.]_Y$, respectively, then $(X \times Y, X^+ \times Y^+)$ is a dual pair with respect to the bilinear form*

$$(x, y) \in X \times Y,\ (x^+, y^+) \in X^+ \times Y^+ \to$$
$$[(x, y), (x^+, y^+)]_{X \times Y} = [x, x^+]_X + [y, y^+]_Y .$$

---

*) A linear operator mapping a Banach space X into a Banach space Y is called an isomorphism if it is continuous and has a continuous inverse. An isomorphism Φ: $X \to Y$ is isometrical if $\|\Phi x\|_Y = \|x\|_X$ for any $x \in X$. Banach spaces X, Y are isometrically isomorphic if there exists an isometrical isomorphism mapping X onto Y.

Proof. $[.,.]_{X \times Y}$ is clearly a bilinear form. Furthermore, let us assume that

$$(3,3) \qquad [(\mathbf{x}, \mathbf{y}), (\mathbf{x}^+, \mathbf{y}^+)]_{X \times Y} = 0 \quad \text{for all} \quad (\mathbf{x}^+, \mathbf{y}^+) \in X^+ \times Y^+ .$$

In particular, we have

$$[(\mathbf{x}, \mathbf{y}), (\mathbf{x}^+, \mathbf{y}^+)]_{X \times Y} = [\mathbf{x}, \mathbf{x}^+]_X = 0$$

for all $(\mathbf{x}^+, \mathbf{y}^+) \in X^+ \times Y^+$ with $\mathbf{y}^+ = \mathbf{0}$. Since (X, X^+) is a dual pair this implies $\mathbf{x} = \mathbf{0}$ and (3,3) reduces to $[\mathbf{y}, \mathbf{y}^+]_Y = 0$ for all $\mathbf{y}^+ \in Y$, i.e. $\mathbf{y} = \mathbf{0}$. Analogously, we would show that $[(\mathbf{x}, \mathbf{y}), (\mathbf{x}^+, \mathbf{y}^+)]_{X \times Y} = 0$ for all $(\mathbf{x}, \mathbf{y}) \in X \times Y$ if and only if $\mathbf{x}^+ = \mathbf{0}$, $\mathbf{y}^+ = \mathbf{0}$.

3.9. Remark. In particular, $(X \times Y)^* = X^* \times Y^*$, where

$$\langle(\mathbf{x}, \mathbf{y}), (\xi, \boldsymbol{\eta})\rangle_{X \times Y} = \langle \mathbf{x}, \xi \rangle_X + \langle \mathbf{y}, \boldsymbol{\eta} \rangle_Y$$

for any $\mathbf{x} \in X$, $\mathbf{y} \in Y$, $\xi \in X^*$ and $\boldsymbol{\eta} \in Y^*$.

3.10. Examples. (i) It is well-known (cf. Dunford, Schwartz [1]) that $\mathbf{A}$ is a linear operator acting from R_n into R_m if and only if there exists a real $m \times n$-matrix $\mathbf{B}$ such that $\mathbf{A}\colon \mathbf{x} \in R_n \to \mathbf{B}\mathbf{x} \in R_m$. Thus the space of all linear operators acting from R_n into R_m and the space of all real $m \times n$-matrices may be identified. Clearly, $B(R_n, R_m) = L(R_n, R_m)$. In particular, $R_n^* = B(R_n, R) = L(R_n, R)$ is the space of all real row n-vectors, while

$$\langle \mathbf{x}, \mathbf{y}^* \rangle_{R_n} = \mathbf{y}^* \mathbf{x} \quad \text{for any} \quad \mathbf{y}^* \in R_n^* \quad \text{and} \quad \mathbf{x} \in R_n .$$

(ii) Let $-\infty < a < b < +\infty$. The dual space to $C_n[a, b]$ is isometrically isomorphic with the space $NBV_n[a, b]$ of column n-vector valued functions of bounded variation on $[a, b]$ which are right-continuous on (a, b) and vanishes at a. Given $\mathbf{y}^* \in NBV_n[a, b]$, the value of the corresponding functional on $\mathbf{x} \in C_n[a, b]$ is

$$(3,4) \qquad \langle \mathbf{x}, \mathbf{y}^* \rangle_C = \int_a^b \mathrm{d}[\mathbf{y}^*(t)]\, \mathbf{x}(t)$$

and

$$\|\mathbf{y}^*\|_C = \sup_{\|\mathbf{x}\|_C = 1} |\langle \mathbf{x}, \mathbf{y}^* \rangle_C| = \operatorname{var}_a^b \mathbf{y}^* = \|\mathbf{y}^*\|_{BV} .$$

(The integral in (3,4) is the usual Riemann-Stieltjes integral.) This result is called the Riesz Representation Theorem (see e.g. Dunford, Schwartz [1], IV.6.3). As a consequence $\mathbf{K} \in B(C_n[a, b], R_m)$ if and only if there exists a function $\mathbf{K}\colon [a, b] \to L(R_n, R_m)$ of bounded variation on $[a, b]$ and such that

$$\mathbf{K}\colon \mathbf{x} \in C_n[a, b] \to \int_a^b \mathrm{d}[\mathbf{K}(t)]\, \mathbf{x}(t) \in R_m .$$

Let us notice that the zero functional on $C_n[a, b]$ corresponds to the function $\mathbf{y}^* \in NBV_n[a, b]$ identically vanishing on $[a, b]$.

(iii) Let $-\infty < a < b < \infty$, $1 \le p < \infty$, $q = p/(p-1)$ if $p > 1$ and $q = \infty$ if $p = 1$. The dual space to $L_n^p[a, b]$ is isometrically isomorphic with $L_n^q[a, b]$ (whose elements are row n-vector valued functions). Given $\mathbf{y}^* \in L_n^q[a, b]$, the value of the corresponding functional on $\mathbf{x} \in L_n^p[a, b]$ is

$$\langle \mathbf{x}, \mathbf{y}^* \rangle_L = \int_a^b \mathbf{y}^*(t)\, \mathbf{x}(t)\, \mathrm{d}t \tag{3,5}$$

and

$$\|\mathbf{y}^*\|_{L^*} = \sup_{\|\mathbf{x}\|_{L^p}=1} |\langle \mathbf{x}, \mathbf{y}^* \rangle_L| = \|\mathbf{y}^*\|_{L^q}$$

(see e.g. Dunford, Schwartz [1], IV.8.1). (The integral in (3,5) is the usual Lebesgue integral.) The zero functional on $L_n^p[a, b]$ corresponds to any function $\mathbf{y}^* \in L_n^q[a, b]$ such that $\mathbf{y}^*(t) = \mathbf{0}$ a.e. on $[a, b]$.

(iv) Any Hilbert space H is isometrically isomorphic with its dual space. If $\mathbf{x}, \mathbf{y} \in H \to (\mathbf{x}, \mathbf{y})_H \in R$ is an inner product on H and $\mathbf{x} \in H \to \|\mathbf{x}\|_H = (\mathbf{x}, \mathbf{x})^{1/2}$ the corresponding norm on H, then given $\mathbf{h} \in H$, the value of the corresponding functional on $\mathbf{x} \in H$ is given by

$$\langle \mathbf{x}, \mathbf{h} \rangle_H = (\mathbf{x}, \mathbf{h})_H$$

and

$$\|\mathbf{h}\|_{H^*} = \sup_{\|\mathbf{x}\|_H=1} |\langle \mathbf{x}, \mathbf{h} \rangle_H| = \|\mathbf{h}\|_H .$$

If X, Y are Banach spaces and $\mathbf{A} \in B(X, Y)$, then for every $\mathbf{g} \in Y^*$ the mapping $\mathbf{x} \in X \to \langle \mathbf{Ax}, \mathbf{g} \rangle_Y$ is a linear bounded functional on X. (Given $\mathbf{x} \in X$ and $\mathbf{g} \in Y^*$, $|\langle \mathbf{Ax}, \mathbf{g} \rangle_Y| \le \|\mathbf{Ax}\|_Y \|\mathbf{g}\|_{Y^*} \le \|\mathbf{A}\|_{B(X,Y)} \|\mathbf{g}\|_{Y^*} \|\mathbf{x}\|_X$.) Thus there is an element of X^* denoted by $\mathbf{A}^*\mathbf{g}$ such that $\langle \mathbf{Ax}, \mathbf{g} \rangle_Y = \langle \mathbf{x}, \mathbf{A}^*\mathbf{g} \rangle_X$. This leads to the following

3.11. Definition. Given $\mathbf{A} \in B(X, Y)$, the operator $\mathbf{A}^*\colon Y^* \to X^*$ defined by

$$\langle \mathbf{Ax}, \mathbf{g} \rangle_Y = \langle \mathbf{x}, \mathbf{A}^*\mathbf{g} \rangle_X$$

for all $\mathbf{x} \in X$ and $\mathbf{g} \in Y^*$ is called the *adjoint operator to* $\mathbf{A}$.

Let us notice that $\mathbf{A}^* \in B(Y^*, X^*)$ and $\|\mathbf{A}^*\| = \|\mathbf{A}\|$ for any $\mathbf{A} \in B(X, Y)$. (See Schechter [1], III.2.)

3.12. Definition. For a given subset $M \subset X$ we define

$$M^\perp = \{\mathbf{f} \in X^*;\ \langle \mathbf{x}, \mathbf{f} \rangle_X = 0 \quad \text{for all} \quad \mathbf{x} \in M\}$$

and similarly for a subset $N \subset X^*$ we set

$${}^\perp N = \{\mathbf{x} \in X;\ \langle \mathbf{x}, \mathbf{f} \rangle_X = 0 \quad \text{for all} \quad \mathbf{f} \in N\} .$$

3.13. Definition. The operator $\mathbf{A} \in B(X, Y)$ is called *normally solvable* if the equation $\mathbf{Ax} = \mathbf{y}$ has a solution if and only if $\langle \mathbf{y}, \mathbf{f} \rangle_Y = 0$ for all solutions $\mathbf{f} \in Y^*$ of the adjoint equation $\mathbf{A}^*\mathbf{f} = \mathbf{0}$.

(In other words, $\mathbf{A} \in B(X, Y)$ is normally solvable if and only if the condition ${}^{\perp}N(\mathbf{A}^*) = R(\mathbf{A})$ is satisfied.)

3.14. Theorem. *If* $\mathbf{A} \in B(X, Y)$, *then the following statements are equivalent*

(i) $R(\mathbf{A})$ *is closed in* Y,
(ii) $R(\mathbf{A}^*)$ *is closed in* X^*,
(iii) $\mathbf{A}$ *is normally solvable* $(R(\mathbf{A}) = {}^{\perp}N(\mathbf{A}^*))$,
(iv) $R(\mathbf{A}^*) = N(\mathbf{A})^{\perp}$.

(See e.g. Goldberg [1], IV.1.2.)

3.15. Theorem. *Let* $\mathbf{A} \in B(X, Y)$ *have a closed range* $R(\mathbf{A})$ *in* Y. *Then*

$$\alpha(\mathbf{A}^*) = \beta(\mathbf{A}) \quad \text{and} \quad \alpha(\mathbf{A}) = \beta(\mathbf{A}^*) .$$

If ind $\mathbf{A}$ *is defined, then* ind $\mathbf{A}^*$ *is also defined and*

$$\operatorname{ind} \mathbf{A}^* = -\operatorname{ind} \mathbf{A} .$$

(See e.g. Goldberg [1], IV.2.3 or Schechter [1], V.4.)

3.16. Definition. If X, Y are Banach spaces then a linear operator $\mathbf{K} \in L(X, Y)$ is called *compact* (or *completely continuous*) if for every sequence $\{\mathbf{x}_n\}_{n=1}^{\infty}$, $\mathbf{x}_n \in X$ such that $\|\mathbf{x}_n\|_X \leq C = \text{const.}$ the sequence $\{\mathbf{Kx}_n\}_{n=1}^{\infty}$ in Y contains a subsequence which converges in Y.

Let the set of all compact operators in $L(X, Y)$ $(L(X))$ be denoted by $K(X, Y)$ $(K(X))$.

The set $K(X, Y) \subset L(X, Y)$ is evidently linear. Moreover every compact operator is bounded, i.e. $K(X, Y) \subset B(X, Y)$. Indeed, if $\mathbf{K} \in K(X, Y) \setminus B(X, Y)$, then there exists a sequence $\{\mathbf{x}_n\} \subset X$, $\|\mathbf{x}_n\|_X \leq C$ such that $\|\mathbf{Kx}_n\| \to \infty$ and the sequence $\{\mathbf{Kx}_n\} \subset Y$ cannot contain a subsequence which would be convergent in Y.

3.17. Theorem. *Suppose that* $\mathbf{K} \in B(X, Y)$ *and that there exists a sequence* $\{\mathbf{K}_n\} \subset K(X, Y)$ *such that* $\lim_{n \to \infty} \mathbf{K}_n = \mathbf{K}$ *in* $B(X, Y)$. *Then* $\mathbf{K} \in K(X, Y)$, *i.e.* $K(X, Y)$ *is a closed linear subspace in* $B(X, Y)$.
(See Schechter [1], IV.3.)

3.18. Proposition. *If* X, Y, Z *are Banach spaces,* $\mathbf{A} \in B(X, Y)$, $\mathbf{K} \in K(Y, Z)$, *then* $\mathbf{KA} \in K(X, Z)$. *Similarly* $\mathbf{BL} \in K(X, Z)$ *provided* $\mathbf{L} \in K(X, Y)$, $\mathbf{B} \in B(Y, Z)$.
(See Schechter [1], IV.3.)

For the adjoint of a compact operator we have

3.19. Theorem. $\mathbf{K} \in K(X, Y)$ *if and only if* $\mathbf{K}^* \in K(Y^*, X^*)$.
(See Goldberg [1], III.1.11 or Schechter [1], IV.4 for the "only if" part.)

3.20. Theorem. *Let* $\mathbf{K} \in K(X)$ *and let both the identity operator on* X *and the identity operator on* X^* *be denoted by* $\mathbf{I}$. *Then* $\mathbf{I} + \mathbf{K} \in B(X)$, $\mathbf{I} + \mathbf{K}^* \in B(X^*)$ *and*
(i) $R(\mathbf{I} + \mathbf{K})$ *is closed in* X *and* $R(\mathbf{I} + \mathbf{K}^*)$ *is closed in* X^*,
(ii) $\alpha(\mathbf{I} + \mathbf{K}) = \beta(\mathbf{I} + \mathbf{K}) = \alpha(\mathbf{I} + \mathbf{K}^*) = \beta(\mathbf{I} + \mathbf{K}^*) < \infty$.
(*In particular*, $\operatorname{ind}(\mathbf{I} + \mathbf{K}) = \operatorname{ind}(\mathbf{I} + \mathbf{K}^*) = 0$.)
(See Schechter [1], IV.3.)

3.21. Remark. It follows easily from the Bolzano-Weierstrass Theorem that any linear bounded operator with the range in a finite dimensional space is compact. ($B(X, R_n) = K(X, R_n)$ for any Banach space X.) Analogously $B(R_n, Y) = K(R_n, Y)$ for any Banach space Y.

3.22. Definition. Let E_X and E_Y be Banach spaces and let $\mathbf{J}_X \in B(X^*, E_X)$ and $\mathbf{J}_Y \in B(Y^*, E_Y)$ be isometrical isomorphisms of X^* onto E_X and Y^* onto E_Y, respectively. Let $[.,.]_X$ be a bilinear form on $X \times E_X$ such that $\langle \mathbf{x}, \xi \rangle_X = [\mathbf{x}, \mathbf{J}_X \xi]_X$ for any $\mathbf{x} \in X$ and $\xi \in X^*$ and let $[.,.]_Y$ be a bilinear form on $Y \times E_Y$ such that $\langle \mathbf{y}, \eta \rangle_Y = [\mathbf{y}, \mathbf{J}_Y \eta]_Y$ for any $\mathbf{y} \in Y$ and $\eta \in Y^*$. If $\mathbf{A} \in B(X, Y)$ and $\mathbf{B} \in L(E_Y, E_X)$ are such that

$$[\mathbf{A}\mathbf{x}, \varphi]_Y = [\mathbf{x}, \mathbf{B}\varphi]_X \qquad \text{for every} \quad \mathbf{x} \in X \quad \text{and} \quad \varphi \in E_Y,$$

then $\mathbf{B}$ is called a *representation of the adjoint operator to* $\mathbf{A}$.

3.23. Remark. If $\mathbf{A} \in B(X, Y)$ and $\mathbf{B} \in L(E_Y, E_X)$ is a representation of the adjoint operator $\mathbf{A}^* \in B(Y^*, X^*)$ to $\mathbf{A}$, then for any $\mathbf{x} \in X$ and $\varphi \in E_Y$ we have

$$[\mathbf{A}\mathbf{x}, \varphi]_Y = \langle \mathbf{A}\mathbf{x}, \mathbf{J}_Y^{-1}\varphi \rangle_Y = \langle \mathbf{x}, \mathbf{A}^*\mathbf{J}_Y^{-1}\varphi \rangle_X = [\mathbf{x}, \mathbf{J}_X\mathbf{A}^*\mathbf{J}_Y^{-1}\varphi]_X.$$

Thus $\mathbf{B} = \mathbf{J}_X\mathbf{A}^*\mathbf{J}_Y^{-1} \in B(E_Y, E_X)$. It follows easily that if we replace the dual spaces to X and Y respectively by the spaces E_X and E_Y isometrically isomorphic to them and the adjoint operators $\mathbf{A}^*$ and $\mathbf{K}^*$ to $\mathbf{A}$ and $\mathbf{K} \in B(X, Y)$, respectively, by its representations $\mathbf{B}$ and $\mathbf{C} \in B(E_Y, E_X)$ defined in 3.22, then Theorems 3.14, 3.15, 3.19 and 3.20 remain valid. This makes reasonable to use the notation $\mathbf{A}^*$ also for representations of the adjoint operator to $\mathbf{A}$.

In the rest of the section X stands for an inner product space endowed with the inner product $(.,.)_X$ and the corresponding norm $\mathbf{x} \in X \to \|\mathbf{x}\|_X = (\mathbf{x}, \mathbf{x})_X^{1/2}$. Furthermore, Y is a Hilbert space, $(.,.)_Y$ is the inner product defined on Y and $\|\mathbf{y}\|_Y = (\mathbf{y}, \mathbf{y})_Y^{1/2}$ for any $\mathbf{y} \in Y$.

3.24. Definition. Given $\mathbf{A} \in L(X, Y)$ and $\mathbf{y} \in Y$, $\mathbf{u} \in X$ is said to be a *least square solution to* (3,1) if

$$\|\mathbf{Au} - \mathbf{y}\|_Y \leq \|\mathbf{Ax} - \mathbf{y}\|_Y \qquad \text{for all} \quad \mathbf{x} \in X .$$

3.25. Proposition. *If* $\mathbf{A} \in L(X, Y)$ *and* $\mathbf{u}_0 \in X$ *is such that*

$$(3,6) \qquad (\mathbf{Ax}, \mathbf{Au}_0 - \mathbf{y})_Y = 0 \qquad \textit{for all} \quad \mathbf{x} \in X ,$$

then $\mathbf{u}_0$ *is a least square solution to* (3,1). *Furthermore,* $\mathbf{x} \in X$ *is a least square solution to* (3,1) *if and only if* $\mathbf{x} - \mathbf{u}_0 \in N(\mathbf{A})$.

Proof. Given $\mathbf{x} \in X$, $\mathbf{Ax} - \mathbf{y} = \mathbf{A}(\mathbf{x} - \mathbf{u}_0) + \mathbf{Au}_0 - \mathbf{y}$ and in virtue of (3,6)

$$\|\mathbf{Ax} - \mathbf{y}\|_Y^2 = \|\mathbf{A}(\mathbf{x} - \mathbf{u}_0)\|_Y^2 + 2(\mathbf{A}(\mathbf{x} - \mathbf{u}_0), \mathbf{Au}_0 - \mathbf{y})_Y + \|\mathbf{Au}_0 - \mathbf{y}\|_Y^2 =$$
$$= \|\mathbf{A}(\mathbf{x} - \mathbf{u}_0)\|_Y^2 + \|\mathbf{Au}_0 - \mathbf{y}\|_Y^2 \geq \|\mathbf{Au}_0 - \mathbf{y}\|_Y^2 .$$

Thus $\mathbf{u}_0$ is a least square solution to (3,1), while $\|\mathbf{Ax} - \mathbf{y}\|_Y = \|\mathbf{Au}_0 - \mathbf{y}\|_Y$ if and only if $\mathbf{A}(\mathbf{x} - \mathbf{u}_0) = \mathbf{0}$.

3.26. Remark. Let us notice that if $R(\mathbf{A})$ is closed in Y, then the Classical Projection Theorem (cf. e.g. Luenberger [1], p. 51) implies that the equation (3,1) possesses for any $\mathbf{y} \in Y$ a least square solution, while $\mathbf{u}_0 \in X$ is a least square solution to (3,1) if and only if (3,6) holds.

3.27. Definition. Given $\mathbf{A} \in L(X, Y)$ and $\mathbf{y} \in Y$, $\mathbf{u}_0 \in X$ is a best *approximate solution to* (3,1) if it is a least square solution to (3,1) of minimal norm (i.e. $\|\mathbf{u}_0\|_X \leq \|\mathbf{u}\|_X$ for any least square solution $\mathbf{u}$ of (3,1)).

3.28. Proposition. *Let* $\mathbf{A} \in L(X, Y)$ *and let* $\mathbf{u}_0 \in X$ *fulfil* (3,6). *If besides it*

$$(3,7) \qquad (\mathbf{v}, \mathbf{u}_0)_X = 0 \qquad \textit{for all} \quad \mathbf{v} \in N(\mathbf{A})$$

holds, then $\mathbf{u}_0$ *is a best approximate solution of* (3,1).

Proof. By 3.25 $\mathbf{u}_0$ is a least square solution to (3,1) and $\mathbf{u} - \mathbf{u}_0 \in N(\mathbf{A})$ for all least square solutions $\mathbf{u}$ of (3,1). Thus assuming (3,7) we have

$$\|\mathbf{u}\|_X^2 = \|\mathbf{u} - \mathbf{u}_0\|_X^2 + 2(\mathbf{u} - \mathbf{u}_0, \mathbf{u}_0)_X + \|\mathbf{u}_0\|_X^2 = \|\mathbf{u} - \mathbf{u}_0\|_X^2 + \|\mathbf{u}_0\|_X^2 \geq \|\mathbf{u}_0\|_X^2$$

for any least square solution $\mathbf{u}$ of (3,1). Let us notice that $\|\mathbf{u}_0\|_X = \|\mathbf{u}_0\|_X$ if and only if $\mathbf{u} = \mathbf{u}_0$.

3.29. Remark. Let $\mathbf{A} \in L(X, Y)$. If $k = \dim N(\mathbf{A}) < \infty$, then applying the Gramm-Schmidt orthogonalization process we may find a orthonormal basis $\mathbf{x}_1, \mathbf{x}_2, \ldots, \mathbf{x}_k$

in $N(\mathbf{A})$, i.e. $(\mathbf{x}_i, \mathbf{x}_j)_X = 0$ if $i \neq j$ and $(\mathbf{x}_i, \mathbf{x}_i)_X = 1$. Let us put

$$\mathbf{P}\colon \mathbf{x} \in X \to \sum_{i=1}^{k} (\mathbf{x}, \mathbf{x}_i)_X \, \mathbf{x}_i \,.$$

Then $\mathbf{P} \in B(X)$, $R(\mathbf{P}) = N(\mathbf{A})$ and $\mathbf{P}^2\mathbf{x} = \mathbf{P}(\mathbf{P}\mathbf{x}) = \mathbf{P}\mathbf{x}$ for every $\mathbf{x} \in X$. Moreover,

$$(3,8) \qquad (\mathbf{x} - \mathbf{P}\mathbf{x}, \mathbf{v})_X = 0 \qquad \text{for all} \quad \mathbf{x} \in X \quad \text{and} \quad \mathbf{v} \in N(\mathbf{A})\,.$$

If $R(\mathbf{A})$ is closed in Y, then there exists $\mathbf{Q} \in B(Y)$ such that $R(\mathbf{Q}) = R(\mathbf{A})$, $\mathbf{Q}^2 = \mathbf{Q}$ and

$$(3,9) \qquad (\mathbf{A}\mathbf{x}, \mathbf{Q}\mathbf{y} - \mathbf{y})_Y = 0 \qquad \text{for all} \quad \mathbf{y} \in Y \quad \text{and} \quad \mathbf{x} \in X$$

(cf. Luenberger [1]). $\mathbf{P}$ is said to be a *linear bounded orthogonal projection* of X onto $N(\mathbf{A})$ and $\mathbf{Q}$ is a linear bounded orthogonal projection of Y onto $R(\mathbf{A})$. Let us notice that since

$$R(\mathbf{I} - \mathbf{P}) = N(\mathbf{P}) \quad \text{and} \quad R(\mathbf{I} - \mathbf{Q}) = N(\mathbf{Q})\,,$$

$R(\mathbf{I} - \mathbf{P})$ and $R(\mathbf{I} - \mathbf{Q})$ are closed.

As a restriction $\mathbf{A}|_{R(\mathbf{I}-\mathbf{P})}$ of $\mathbf{A}$ onto $R(\mathbf{I} - \mathbf{P})$ is a one-to-one mapping of $R(\mathbf{I} - \mathbf{P})$ onto $R(\mathbf{A})$, it possesses a linear inverse operator $\mathbf{A}^+ \in L(R(\mathbf{A}), R(\mathbf{I} - \mathbf{P}))$, i.e.

$$(3,10) \qquad \mathbf{A}\mathbf{A}^+\mathbf{A} = \mathbf{A}\,.$$

As obviously $\mathbf{A}\mathbf{A}^+\mathbf{Q} = \mathbf{Q}$, it follows from (3,9) that $(\mathbf{A}\mathbf{x}, \mathbf{A}\mathbf{A}^+\mathbf{Q}\mathbf{y} - \mathbf{y})_Y = (\mathbf{A}\mathbf{x}, \mathbf{Q}\mathbf{y} - \mathbf{y})_Y = 0$ for every $\mathbf{y} \in Y$ and $\mathbf{x} \in X$. Hence by 3.25 $\mathbf{A}^+\mathbf{Q}\mathbf{y}$ is for any $\mathbf{y} \in Y$ a least square solution of (3,1).

Let us put

$$(3,11) \qquad \mathbf{A}^\# = (\mathbf{I} - \mathbf{P})\,\mathbf{A}^+\mathbf{Q}\,.$$

Evidently $\mathbf{A}(\mathbf{I} - \mathbf{P}) = \mathbf{A}$ and hence $(\mathbf{A}\mathbf{x}, \mathbf{A}\mathbf{A}^\#\mathbf{y} - \mathbf{y})_Y = (\mathbf{A}\mathbf{x}, \mathbf{A}\mathbf{A}^+\mathbf{Q}\mathbf{y} - \mathbf{y})_Y = 0$ for every $\mathbf{x} \in X$ and $\mathbf{y} \in Y$. Since according to (3,8) $(\mathbf{v}, \mathbf{A}^\#\mathbf{y})_X = (\mathbf{v}, (\mathbf{I} - \mathbf{P})\,\mathbf{A}^+\mathbf{Q}\mathbf{y})_X = 0$ for each $\mathbf{v} \in N(\mathbf{A})$ and $\mathbf{y} \in Y$, it follows from 3.28 that for every $\mathbf{y} \in Y$, $\mathbf{u}_0 = \mathbf{A}^\#\mathbf{y}$ is a best approximate solution to (3,1). Moreover, it is easy to verify that

$$(3,12) \qquad \mathbf{A}\mathbf{A}^\#\mathbf{A} = \mathbf{A}\,, \qquad \mathbf{A}^\#\mathbf{A}\mathbf{A}^\# = \mathbf{A}^\#\,, \qquad \mathbf{A}^\#\mathbf{A} = \mathbf{I} - \mathbf{P}\,, \qquad \mathbf{A}\mathbf{A}^\# = \mathbf{Q}\,.$$

3.30. Remark. If $\mathbf{A} \in B(X, Y)$, then the condition (3,6) becomes $\mathbf{A}^*\mathbf{A}\mathbf{u}_0 = \mathbf{A}^*\mathbf{y}$ or denoting $\mathbf{u}_0 = \mathbf{A}^\#\mathbf{y}$,

$$(3,13) \qquad \mathbf{A}^*\mathbf{A}\mathbf{A}^\# = \mathbf{A}^*\,.$$

Let us notice that if $R(\mathbf{A})$ is closed, then (3,12) implies (3,13). In fact, given $\mathbf{x} \in X$ and $\mathbf{y} \in Y$, we have by (3,9) $0 = (\mathbf{x}, \mathbf{A}^*\mathbf{Q}\mathbf{y} - \mathbf{A}^*\mathbf{y})_X$, i.e. $\mathbf{A}^*\mathbf{Q} = \mathbf{A}^*$. This together with the relation $\mathbf{A}\mathbf{A}^\# = \mathbf{Q}$ from (3,12) yields $\mathbf{A}^*\mathbf{A}\mathbf{A}^\# = \mathbf{A}^*\mathbf{Q} = \mathbf{A}^*$. Finally, as $\mathbf{A}^\# = \mathbf{A}^\#\mathbf{A}\mathbf{A}^\#$, $\mathbf{A}^\# = (\mathbf{I} - \mathbf{P})\,\mathbf{A}^\#$ and hence by (3,8) $(\mathbf{v}, \mathbf{A}^\#\mathbf{y})_X = (\mathbf{v}, (\mathbf{I} - \mathbf{P})\,\mathbf{A}^\#\mathbf{y})_X = 0$ for every $\mathbf{v} \in N(\mathbf{A})$ and $\mathbf{y} \in Y$. It means that (3,12) implies also (3,7).

Given $\mathbf{A} \in L(X, Y)$, any operator $\mathbf{A}^+ \in L(Y, X)$ satisfying (3,10) is called a *generalized inverse operator* to $\mathbf{A}$. If $\mathbf{A} \in B(X, Y)$, then the unique operator $\mathbf{A}^\# \in B(Y, X)$ satisfying (3,12) is called the *principal generalized inverse operator* of $\mathbf{A}$.

3.31. Remark. If $X = R_n$, $Y = R_m$ and $\mathbf{A}$ is an $m \times n$-matrix and $\mathbf{A}^\#$ its generalized inverse matrix defined by 2.2, then the vector $\mathbf{u}_0 = \mathbf{A}^\# \mathbf{b} \in R_n$ satisfies the conditions (3,13) and (3,7). In fact, as by 2.7 $\mathbf{v} \in N(\mathbf{A})$ if and only if $\mathbf{v} = (\mathbf{I} - \mathbf{A}^\# \mathbf{A})\,\mathbf{d}$ for some $\mathbf{d} \in R_n$, we have owing to (2,16) $(\mathbf{v}, \mathbf{A}^\# \mathbf{b}) = \mathbf{v}^* \mathbf{A}^\# \mathbf{b} = \mathbf{d}^*(\mathbf{I} - (\mathbf{A}^\# \mathbf{A})^*)\,\mathbf{A}^\# \mathbf{b} = \mathbf{d}^*(\mathbf{A}^\# - \mathbf{A}^\# \mathbf{A} \mathbf{A}^\#)\,\mathbf{b} = 0$. Furthermore, $\mathbf{A}^* \mathbf{A} \mathbf{A}^\# = \mathbf{A}^*$ by (2,17). Thus if R_n and R_m are equipped with the Euclidean norm $|.|_e$, $\mathbf{A}^\# \mathbf{b}$ is for any $\mathbf{b} \in R_m$ a unique best approximate solution of (2,1).

4. Perron-Stieltjes integral

This section contains the definition of the Perron-Stieltjes integral based on the work of J. Kurzweil [1], [2]. Some facts concerning this integral are collected here. These facts are necessary for the subsequent study of equations and problems involving the Perron-Stieltjes integrals.

Let a fixed interval $[a, b]$, $-\infty < a < b < +\infty$ be given. We denote by $\mathscr{S} = \mathscr{S}[a, b]$ the system of sets $S = R_2$ having the following property:

for every $\tau \in [a, b]$ there exists such a $\delta = \delta(\tau) > 0$ that $(\tau, t) \in S$ whenever $\tau \in [a, b]$ and $t \in [\tau - \delta(\tau), \tau + \delta(\tau)]$.

Evidently any set $S \in \mathscr{S}[a, b]$ is characterized by a real function $\delta\colon [a, b] \to (0, +\infty)$.

Let $f\colon [a, b] \to R$ and $g\colon [a, b] \to R$ be real functions, $-\infty < \alpha < a < b < \beta < +\infty$. If $g(t)$ is defined only for $t \in [a, b]$ then we assume automatically that $g(t) = g(a)$ for $t < a$ and $g(t) = g(b)$ for $t > b$. It is evident that if $\operatorname{var}_a^b g < \infty$, this arrangement yields $\operatorname{var}_\alpha^\beta g = \operatorname{var}_a^b g$ for any α, β such that $\alpha < a < b < \beta$.

4.1. Definition. A real valued finite function $M\colon [a, b] \to R$ is a *major function of f with respect to g* if there exists such a set $S \in \mathscr{S}[a, b]$ that

$$(\tau - \tau_0)(M(\tau) - M(\tau_0)) \geq (\tau - \tau_0)\, f(\tau_0)\,(g(\tau) - g(\tau_0))$$

for $(\tau_0, \tau) \in S$. The set of major functions of f with respect to g is denoted by $M(f, g)$.

A function $m\colon [a, b] \to R$ *is a minor function of f with respect to g* if $-m \in M(-f, g)$, i.e. if $-m$ is a major function of $-f$ with respect to g. The set of minor functions of f with respect to g is denoted by $m(f, g)$.

4.2. Definition. Let $M(f, g) \neq \emptyset$ and $m(f, g) \neq \emptyset$. The lower bound of the numbers $M(b) - M(a)$ where $M \in M(f, g)$ is called the *upper Perron-Stieltjes integral of f with respect to g from a to b* and is denoted by $\overline{\int_a^b} f\, dg$. Similarly the upper bound of the numbers $m(b) - m(a)$, $m \in m(f, g)$ is called the *lower Perron-Stieltjes integral of f with respect to g from a* to b and is denoted by $\underline{\int}_a^b f\,dg$.

4.3. Lemma. *If $M(f, g) \neq \emptyset$ and $m(f, g) \neq \emptyset$, then*

$$\underline{\int_a^b} f\,dg \leq \overline{\int_a^b} f\,dg\,.$$

For the proof of this lemma see Kurzweil [1], Lemma 1,1,1.

4.4. Definition. If $M(f, g) \neq \emptyset$, $m(f, g) \neq \emptyset$ and the equality

$$\underline{\int_a^b} f\,dg = \overline{\int_a^b} f\,dg$$

holds, then by the relation

$$\int_a^b f\,dg = \underline{\int_a^b} f\,dg$$

the *Perron-Stieltjes integral* $\int_a^b f\,dg$ *of the function f with respect to g from a* to b is defined. In this case f is called *integrable with respect to g* on $[a, b]$. If $a = b$, then we set $\int_a^b f\,dg = 0$ and if $b < a$, then we put $\int_a^b f\,dg = -\int_b^a f\,dg$.

Now we give a different definition of the Stieltjes integral which is also included in the paper Kurzweil [1] and is equivalent to Definition 4.4. This is a definition of the integral using integral sums which is close to the Riemann-Stieltjes definition.

For the given bounded interval $[a, b] \subset R$ we consider sequences of numbers $A = \{\alpha_0, \tau_1, \alpha_1, \ldots, \tau_k, \alpha_k\}$ such that

$$a = \alpha_0 < \alpha_1 < \ldots < \alpha_k = b\,, \tag{4,1}$$

$$\alpha_{j-1} \leq \tau_j \leq \alpha_j\,, \qquad j = 1, 2, \ldots, k\,. \tag{4,2}$$

For a given set $S \in \mathscr{S}[a, b]$, A satisfying (4,1) and (4,2) is called a subdivision of $[a, b]$ subordinate to S if

$$(\tau_j, t) \in S \qquad \text{for} \quad t \in [\alpha_{j-1}, \alpha_j]\,, \quad j = 1, 2, \ldots, k\,. \tag{4,3}$$

The set of all subdivisions A of the interval $[a, b]$ subordinate to S is denoted by $A(S)$.

In Kurzweil [1], Lemma 1.1.1 it is proved that for every $S \in \mathscr{S}[a, b]$ we have

$$A(S) \neq \emptyset\,. \tag{4,4}$$

If now the real functions $f\colon [a, b] \to R$, $g\colon [a, b] \to R$ are given and $A = \{\alpha_0, \tau_1, \alpha_1, \ldots, \tau_k, \alpha_k\}$ is a subdivision of $[a, b]$ which satisfies (4,1) and (4,2), we put

$$B_{f,g}(A) = \sum_{j=1}^{k} f(\tau_j)\left(g(\alpha_j) - g(\alpha_{j-1})\right). \tag{4,5}$$

If no misunderstanding may occur, we write simply $B(A)$ instead of $B_{f,g}(A)$.

4.5. Definition. Let $f\colon [a,b] \to R$ and $g\colon [a,b] \to R$. If there is a real number J such that to every $\varepsilon > 0$ there exists a set $S \in \mathscr{S}[a,b]$ such that

$$|B_{f,g}(A) - J| < \varepsilon \qquad \text{for any} \quad A \in A(S),$$

we define this number to be the *integral*

$$\int_a^b f\,\mathrm{d}g$$

of the function f with respect to g from a to b.

The completeness of the space R of all real numbers implies that *the integral $\int_a^b f\,\mathrm{d}g$ exists if and only if for any $\varepsilon > 0$ there exists a set $S \in \mathscr{S}[a,b]$ such that*

$$|B_{f,g}(A_1) - B_{f,g}(A_2)| < \varepsilon \qquad \textit{for all} \quad A_1, A_2 \in A(S).$$

In Kurzweil [1] (Theorem 1.2.1), the following statement is proved.

4.6. Theorem. *The integral $\int_a^b f\,\mathrm{d}g$ exists in the sense of Definition 4.4 if and only if $\int_a^b f\,\mathrm{d}g$ exists in the sense of Definition 4.5. If these integrals exist, then their values are equal.*

4.7. Remark. In Schwabik [3] it is shown that the integral introduced in 4.4 and 4.5 is equivalent to the usual Perron-Stieltjes integral defined e.g. in Saks [1]. Consequently the Riemann-Stieltjes, Lebesgue and Perron integrals are special cases of our integral. In particular, if one of the functions f, g is continuous and the other one is of bounded variation on $[a,b]$, then the integral $\int_a^b f\,\mathrm{d}g$ exists and is equal to the ordinary Riemann-Stieltjes integral of f with respect to g from a to b.

The σ-Young integral described in Hildebrandt [1] (II.19.3) is not included in the Perron-Stieltjes integral (see Example 2,1 in Schwabik [3]). However, if $f\colon [a,b] \to R$ is bounded and $g \in BV[a,b]$, then the existence of the σ-Young integral $Y\int_a^b f\,\mathrm{d}g$ implies the existence of the Perron-Stieltjes integral $\int_a^b f\,\mathrm{d}g$ and both integrals are then equal to one another (Schwabik [3], Theorem 3,2).

Now we give a survey of some fundamental properties of the Perron-Stieltjes integral. The proofs of Theorems 4.8 and 4.9 follow directly from Definition 4.5.

4.8. Theorem. *If $f\colon [a,b] \to R$, $g\colon [a,b] \to R$, $\lambda \in R$ and the integral $\int_a^b f\,\mathrm{d}g$ exists, then the integrals $\int_a^b \lambda f\,\mathrm{d}g$ and $\int_a^b f\,\mathrm{d}[\lambda g]$ exist and*

$$\int_a^b \lambda f\,\mathrm{d}g = \lambda \int_a^b f\,\mathrm{d}g, \qquad \int_a^b f\,\mathrm{d}[\lambda g] = \lambda \int_a^b f\,\mathrm{d}g.$$

4.9. Theorem. *If $f_1, f_2\colon [a,b] \to R$, $g\colon [a,b] \to R$ and the integrals $\int_a^b f_1\,\mathrm{d}g$ and*

$\int_a^b f_2\,\mathrm{d}g$ *exist, then the integral* $\int_a^b (f_1 + f_2)\,\mathrm{d}g$ *exists and*

$$\int_a^b (f_1 + f_2)\,\mathrm{d}g = \int_a^b f_1\,\mathrm{d}g + \int_a^b f_2\,\mathrm{d}g\,.$$

If $f\colon [a,b] \to R$, $g_1, g_2\colon [a,b] \to R$ *and the integrals* $\int_a^b f\,\mathrm{d}g_1$ *and* $\int_a^b f\,\mathrm{d}g_2$ *exist, then the integral* $\int_a^b f\,\mathrm{d}[g_1 + g_2]$ *exists and*

$$\int_a^b f\,\mathrm{d}[g_1 + g_2] = \int_a^b f\,\mathrm{d}g_1 + \int_a^b f\,\mathrm{d}g_2\,.$$

4.10. Theorem. *If* $f\colon [a,b] \to R$, $g\colon [a,b] \to R$ *and* $\int_a^b f\,\mathrm{d}g$ *exists, then for any* $c, d \in R$, $a \le c < d \le b$ *the integral* $\int_c^d f\,\mathrm{d}g$ *exists.*

4.11. Theorem. *If* $f\colon [a,b] \to R$, $g\colon [a,b] \to R$, $c \in [a,b]$ *and the integrals* $\int_a^c f\,\mathrm{d}g$, $\int_c^b f\,\mathrm{d}g$ *exist, then also the integral* $\int_a^b f\,\mathrm{d}g$ *exists and the equality*

$$\int_a^b f\,\mathrm{d}g = \int_a^c f\,\mathrm{d}g + \int_c^b f\,\mathrm{d}g$$

holds.

The statement 4.10 can be proved easily if 4.6 is taken into account. The proof of 4.11 is given in Kurzweil [1] (Theorem 1.3.4).

4.12. Theorem. *Let* $f\colon [a,b] \to R$, $g\colon [a,b] \to R$ *be given and let the integral* $\int_a^b f\,\mathrm{d}g$ *exist. If* $c \in [a,b]$, *then*

$$\lim_{\substack{t \to c\\ t \in [a,b]}} \left[\int_a^t f\,\mathrm{d}g - f(c)\,(g(t) - g(c))\right] = \int_a^c f\,\mathrm{d}g\,.$$

(See Kurzweil [1], Theorem 1.3.5.)

4.13. Corollary. *If the assumptions of* 4.12 *are satisfied, then*

$$\lim_{\substack{t \to c\\ t \in [a,b]}} \int_a^t f\,\mathrm{d}g = \int_a^c f\,\mathrm{d}g$$

if and only if $\lim_{\substack{t \to c\\ t \in [a,b]}} g(t) = g(c)$ *or* $f(c) = 0$.

If $g\colon [a,b] \to R$ *possesses the onesided limits* $g(c+)$, $g(c-)$ *at* $c \in [a,b]$ (*e.g. if* $g \in BV[a,b]$), *then*

$$\lim_{\substack{t \to c+\\ t \in [a,b]}} \int_a^t f\,\mathrm{d}g = \int_a^c f\,\mathrm{d}g + f(c)\,(g(c+) - g(c)) = \int_a^c f\,\mathrm{d}g + f(c)\,\Delta^+ g(c) \tag{4,6}$$

and

$$(4,7)\qquad \lim_{\substack{t\to c-\\ t\in[a,b]}} \int_a^t f\,dg = \int_a^c f\,dg - f(c)\,(g(c) - g(c-)) = \int_a^c f\,dg - f(c)\,\Delta^- g(c)$$

$$\text{for}\quad c \in (a, b]$$

where we have used the notation $\Delta^+ g(c) = g(c+) - g(c)$, $\Delta^- g(c) = g(c) - g(c-)$.

4.14. Lemma. *If* $f_i\colon [a, b] \to R$, $i = 1, 2$, $g \in BV[a, b]$ *and* $A = \{\alpha_0, \tau_1, \ldots, \tau_k, \alpha_k\}$ *is an arbitrary subdivision of* $[a, b]$ *satisfying* (4,1) *and* (4,2), *then*

$$(4,8)\qquad |B_{f_1,g}(A) - B_{f_2,g}(A)| \le \sup_{t\in[a,b]} |f_1(t) - f_2(t)| \operatorname{var}_a^b g\,.$$

Proof. Evidently

$$|B_{f_1,g}(A) - B_{f_2,g}(A)| = \left|\sum_{j=1}^{k} (f_1(\tau_j) - f_2(\tau_j))\,(g(\alpha_j) - g(\alpha_{j-1}))\right|$$

$$\le \sum_{j=1}^{k} |f_1(\tau_j) - f_2(\tau_j)|\,|g(\alpha_j) - g(\alpha_{j-1})|$$

$$\le \sup_{t\in[a,b]} |f_1(t) - f_2(t)| \sum_{j=1}^{k} |g(\alpha_j) - g(\alpha_{j-1})|$$

and (4,8) holds.

In the same trivial way the following lemma can be proved.

4.15. Lemma. *Let* $f\colon [a, b] \to R$, $|f(t)| \le M$ *for all* $t \in [a, b]$, $g_i \in BV[a, b]$, $i = 1, 2$. *Then for any subdivision* $A = \{\alpha_0, \tau_1, \ldots, \tau_k, \alpha_k\}$ *of the interval* $[a, b]$ *satisfying* (4,1) *and* (4,2) *we have*

$$(4,9)\qquad |B_{f,g_1}(A) - B_{f,g_2}(A)| \le M \operatorname{var}_a^b (g_1 - g_2)\,.$$

4.16. Lemma. *If* $f\colon [a, b] \to R$, $g \in BV[a, b]$ *and the integral* $\int_a^b f\,dg$ *exists, then the inequality*

$$\left|\int_a^b f\,dg\right| \le \sup_{t\in[a,b]} |f(t)| \operatorname{var}_a^b g$$

holds.

Proof. Since the integral $\int_a^b f\,dg$ exists, for every $\varepsilon > 0$ there exists $S \in \mathscr{S}[a, b]$ such that for any $A \in A(S)$ we have

$$\left|B_{f,g}(A) - \int_a^b f\,dg\right| < \varepsilon\,.$$

Let us set $f_1(t) = f(t)$, $f_2(t) = 0$ for $t \in [a, b]$. Then by 4.14 we have for any $A \in A(S)$

$$\left|\int_a^b f\,dg\right| \le \left|B_{f,g}(A) - \int_a^b f\,dg\right| + |B_{f,g}(A)| < \varepsilon + |B_{f_1,g}(A) - B_{f_2,g}(A)|$$

$$\le \varepsilon + \sup_{t\in[a,b]} |f(t)| \operatorname{var}_a^b g\,.$$

Hence the inequality is proved because $\varepsilon > 0$ is arbitrary.

4.17. Theorem. *If* $f_n\colon [a,b] \to R$, $n = 1, 2, \ldots$, $\lim_{n\to\infty} f_n = f$ *uniformly on* $[a,b]$, $g \in BV[a,b]$ *and* $\int_a^b f_n \,\mathrm{d}g$ *exists for all* $n = 1, 2, \ldots$, *then the limit* $\lim_{n\to\infty} \int_a^b f_n \,\mathrm{d}g$ *as well as the integral* $\int_a^b f \,\mathrm{d}g$ *exist and the equality*

(4,10) $$\lim_{n\to\infty} \int_a^b f_n \,\mathrm{d}g = \int_a^b f \,\mathrm{d}g$$

holds.

The proof of the existence of the limit $\lim_{n\to\infty} \int_a^b f_n \,\mathrm{d}g$ and of the integral $\int_a^b f \,\mathrm{d}g$ follows from 4.14. The equality of these quantities is an immediate consequence of 4.16.

4.18. Theorem. *Let* $g_n, g \in BV[a,b]$, $n = 1, 2, \ldots$ *and* $\lim_{n\to\infty} \operatorname{var}_a^b (g_n - g) = 0$. *Assume that* $f\colon [a,b] \to R$ *is bounded and* $\int_a^b f \,\mathrm{d}g_n$ *exists for all* $n = 1, 2, \ldots$. *Then the limit* $\lim_{n\to\infty} \int_a^b f \,\mathrm{d}g_n$ *as well as the integral* $\int_a^b f \,\mathrm{d}g$ *exist and*

(4,11) $$\lim_{n\to\infty} \int_a^b f \,\mathrm{d}g_n = \int_a^b f \,\mathrm{d}g .$$

(The proof follows from 4.15; cf. Schwabik [3], Proposition 2,3.)

If $f, g \in BV[a,b]$, then by Hildebrandt [1] (II.19.3.11) the σ-Young integral $Y\int_a^b f \,\mathrm{d}g$ exists. Taking into account the relationship of the σ-Young and the Perron-Stieltjes integrals (cf. 4.7) we obtain immediately the following.

4.19. Theorem. *If* $f, g \in BV[a,b]$, *then the integral* $\int_a^b f \,\mathrm{d}g$ *exists.*

4.20. Remark. For a given $\alpha \in [a,b]$ and for $t \in [a,b]$ we define

(4,12) $$\psi_\alpha^+(t) = 0 \quad \text{if} \quad t \le \alpha, \qquad \psi_\alpha^+(t) = 1 \quad \text{if} \quad \alpha < t$$

and

(4,13) $$\psi_\alpha^-(t) = 0 \quad \text{if} \quad t < \alpha, \qquad \psi_\alpha^-(t) = 1 \quad \text{if} \quad \alpha \le t .$$

The functions ψ_α^+, ψ_α^- are called *simple jump functions.*

A real function $f\colon [a,b] \to R$ is said to be a *finite-step function* on the interval $[a,b]$ if there is a finite sequence $a = d_0 < d_1 < \ldots < d_k = b$ of points in $[a,b]$ such that in every open interval (d_{i-1}, d_i) $(i = 1, 2, \ldots, k)$ the function f equals identically a constant $c_i \in R$. Let us put for $t \in [a,b]$ and $i = 1, 2, \ldots, k$

$$g_i(t) = c_i(\psi_{d_{i-1}}^+(t) - \psi_{d_i}^-(t)) + f(d_{i-1})(\psi_{d_{i-1}}^-(t) - \psi_{d_{i-1}}^+(t)) .$$

It is easy to see that $g_i(t) = f(t)$ if $t \in [d_{i-1}, d_i)$ and $g_i(t) = 0$ if $t \in [a,b] \setminus [d_{i-1}, d_i)$.

Hence, for any $t \in [a, b]$ we have

$$f(t) = \sum_{i=1}^{k} g_i(t) + f(b)\,\psi_b^-(t)$$

$$= \sum_{i=1}^{k} c_i(\psi_{d_{i-1}}^+(t) - \psi_{d_i}^-(t)) + f(d_{i-1})\,(\psi_{d_{i-1}}^-(t) - \psi_{d_{i-1}}^+(t)) + f(b)\,\psi_b^-(t),$$

i.e. any finite-step function can be expressed in the form of a finite linear combination of functions of the type ψ_α^+ and ψ_α^-.

Since any function $f\colon [a, b] \to R$ which possesses the onesided limits $f(c+)$ for any $c \in [a, b)$ and $f(c-)$ for any $c \in (a, b]$ can be approximated uniformly on $[a, b]$ by a sequence of finite-step functions (see e.g. assertion 7.3.2.1 (3) in Aumann [1]), it follows from 4.17 that to prove 4.19 it is sufficient to show that the integrals $\int_a^b \psi_\alpha^+\,dg$ and $\int_a^b \psi_\alpha^-\,dg$ exist for any $g \in BV[a, b]$ and any $\alpha \in [a, b]$.

4.21. Lemma. *Let $\alpha \in [a, b]$ and let $\psi_\alpha^+\colon [a, b] \to R$ and $\psi_\alpha^-\colon [a, b] \to R$ be the simple jump functions defined by (4,12) and (4,13) in 4.20.*

(a) *The integrals $\int_a^b g\,d\psi_\alpha^+$ and $\int_a^b g\,d\psi_\alpha^-$ exist for an arbitrary function $g\colon [a, b] \to R$ and*

$$\int_a^b g\,d\psi_\alpha^+ = \begin{cases} g(\alpha) & \text{if } \alpha < b, \\ 0 & \text{if } \alpha = b, \end{cases} \tag{4,14}$$

$$\int_a^b g\,d\psi_\alpha^- = \begin{cases} g(\alpha) & \text{if } \alpha > a, \\ 0 & \text{if } \alpha = a. \end{cases} \tag{4,15}$$

(b) *If $f \in BV[a, b]$ then the integrals $\int_a^b \psi_\alpha^+\,df$, $\int_a^b \psi_\alpha^-\,df$ exist and*

$$\int_a^b \psi_\alpha^+\,df = \begin{cases} f(b) - f(\alpha+) & \text{if } \alpha < b, \\ 0 & \text{if } \alpha = b, \end{cases} \tag{4,16}$$

$$\int_a^b \psi_\alpha^-\,df = \begin{cases} f(b) - f(\alpha-) & \text{if } a < \alpha, \\ 0 & \text{if } a = \alpha. \end{cases} \tag{4,17}$$

Proof. (a) If $\alpha = b$ then by definition $\psi_\alpha^+(t) = 0$ for every $t \in [a, b]$ and for any subdivision $A\colon a = \alpha_0 \le \tau_1 \le \alpha_1 \le \ldots \le \tau_k \le \alpha_k = b$ we have $B_{g,\psi_\alpha^+}(A) = 0$. Hence $\int_a^b g\,d\psi_\alpha^+ = 0$. If $\alpha < b$ let us define $\delta(t) = \frac{1}{4}|t - \alpha|$ for $t \in [a, b]$, $t \neq \alpha$, $\delta(\alpha) = 1$. Evidently $\delta\colon [a, b] \to (0, +\infty)$. We define

$$S = \{(\tau, t) \in R_2;\ \tau \in [a, b],\ t \in [\tau - \delta(\tau),\ \tau + \delta(\tau)]\},$$

by definition we have $S \in \mathscr{S}[a, b]$. For every subdivision $A \in A(S)$ we have $[\alpha_{j-1}, \alpha_j] \subset [\tau_j - \delta(\tau_j), \tau_j + \delta(\tau_j)]$, i.e.

$$0 < \alpha_j - \alpha_{j-1} \le 2\delta(\tau_j)$$

for any $j = 1, 2, \ldots, k$ (see (4,1), (4,2)). Moreover, there exists an index i, $1 \le i \le k$, such that $\alpha \in [\alpha_{i-1}, \alpha_i)$. If $\tau_i \neq \alpha$ then we obtain a contradictory inequality

$$0 < \alpha_i - \alpha_{i-1} \leq 2\delta(\tau_j) = \tfrac{1}{2}|\tau_i - \alpha| \leq \tfrac{1}{2}(\alpha_i - \alpha_{i-1}).$$

Hence $\tau_i = \alpha$. For every subdivision $A \in A(S)$ we have

$$B_{g,\psi_\alpha^+}(A) = \sum_{j=1}^{k} g(\tau_j)\,[\psi_\alpha^+(\alpha_j) - \psi_\alpha^+(\alpha_{j-1})] = g(\tau_i)\,[\psi_\alpha^+(\alpha_i) - \psi_\alpha^+(\alpha_{i-1})] = g(\tau_i) = g(\alpha).$$

Hence the integral $\int_a^b g\,\mathrm{d}\psi_\alpha^+$ exists and equals $g(\alpha)$ by Definition 4.5. The result for the integral $\int_a^b g\,\mathrm{d}\psi_\alpha^-$ can be proved similarly.

(b) The existence of the integrals $\int_a^b \psi_\alpha^+\,\mathrm{d}f$, $\int_a^b \psi_\alpha^-\,\mathrm{d}f$ follows imeediately from 4.19. It is not difficult to compute their values using 4.11 and 4.13. See also Schwabik [2], Proposition 2.1.

4.22. Lemma. *For $\alpha \in [a, b]$ define $\psi_\alpha(t) = 0$ if $t \in [a, b]$, $t \neq \alpha$, $\psi_\alpha(\alpha) = 1$. Then for any $g \in BV[a, b]$ the integrals $\int_a^b \psi_\alpha\,\mathrm{d}g$, $\int_a^b g\,\mathrm{d}\psi_\alpha$ exist and*

$$\int_a^b \psi_\alpha\,\mathrm{d}g = g(\alpha+) - g(\alpha-) = \Delta g(\alpha), \tag{4,18}$$

(recall that $g(a-) = g(a)$ and $g(b+) = g(b)$),

$$\int_a^b g\,\mathrm{d}\psi_\alpha = 0 \quad \text{if} \quad \alpha \in (a, b), \tag{4,19}$$

$$\int_a^b g\,\mathrm{d}\psi_a = -g(a), \qquad \int_a^b g\,\mathrm{d}\psi_b = g(b).$$

Proof. It is easy to see that $\psi_\alpha(t) = \psi_\alpha^-(t) - \psi_\alpha^+(t)$ where ψ_α^+, ψ_α^- are given by (4,12) (4,13) respectively. The existence of the integrals is clear by 4.19, the relations (4,18), (4,19) follow immediately from 4.21.

4.23. Lemma. *Let $g_B \in BV[a, b]$ be a break function, $f \in BV[a, b]$. Then the integral $\int_a^b f\,\mathrm{d}g_B$ exists and*

$$\int_a^b f\,\mathrm{d}g_B = f(a)\,\Delta^+ g_B(a) + \sum_{a<\tau<b} f(\tau)\,\Delta g_B(\tau) + f(b)\,\Delta^- g_B(b)$$

where $\Delta^+ g_B(t) = g_B(t+) - g_B(t)$, $\Delta^- g_B(t) = g_B(t) - g_B(t-)$, $\Delta g_B(t) = g_B(t+) - g_B(t-)$.

Proof. Since g_B is a break function, there exists an at most countable set $(t_1, t_2, \ldots)$ of points in $[a, b]$ and two sequences c_i^+, c_i^-, $i = 1, 2, \ldots$ such that

$$g_B(t) = \sum_{a \leq t_i < t} c_i^+ + \sum_{a < t_i \leq t} c_i^-$$

where $\operatorname{var}_a^b g_B = \sum_{a<t_i\leq b} |c_i^-| + \sum_{a\leq t_i<b} |c_i^+| < +\infty$. By definition it is $c_i^+ = \Delta^+ g_B(t_i)$,

$c_i^- = \Delta^- g_B(t_i)$. Using the functions ψ_α^+, ψ_α^- defined by (4,12), (4,13) we can write

$$g_B(t) = \sum_{i=1}^{\infty} [c_i^+ \psi_{t_i}^+(t) + c_i^- \psi_{t_i}^-(t)]$$
$$= \sum_{i=1}^{\infty} [\Delta^+ g_B(t_i) \psi_{t_i}^+(t) + \Delta^- g_B(t_i) \psi_{t_i}^-(t)] .$$

Let us define

$$g_B^N(t) = \sum_{i=1}^{N} [\Delta^+ g_B(t_i) \psi_{t_i}^+(t) + \Delta^- g_B(t_i) \psi_{t_i}^-(t)] ,$$

we have

$$\operatorname{var}_a^b (g_B - g_B^N) = \operatorname{var}_a^b \left(\sum_{i=N+1}^{\infty} [\Delta^+ g_B(t_i) \psi_{t_i}^+(t) + \Delta^- g_B(t_i) \psi_{t_i}^-(t)] \right) =$$
$$= \sum_{i=N+1}^{\infty} [|\Delta^+ g_B(t_i)| + |\Delta^- g_B(t_i)|] .$$

This yields

$$\lim_{N \to \infty} \operatorname{var}_a^b (g_B - g_B^N) = 0$$

since the series $\sum_{i=1}^{\infty} [|\Delta^+ g_B(t_i)| + |\Delta^- g_B(t_i)|] = \operatorname{var}_a^b g_B$ converges by the asumption.

Evaluating $\int_a^b f \, dg_B^N$ we obtain by the results of 4.21

$$\int_a^b f \, dg_B^N = \sum_{i=1}^{N} \left[\Delta^+ g_B(t_i) \int_a^b f \, d\psi_{t_i}^+ + \Delta^- g_B(t_i) \int_a^b f \, d\psi_{t_i}^- \right] =$$
$$= \sum_{i=1}^{N} [\Delta^+ g_B(t_i) f(t_i) + \Delta^- g_B(t_i) f(t_i)] .$$

Recall that we assume $g(a-) = g(a)$, $g(b) = g(b+)$. By 4.18 we have

$$\int_a^b f \, dg_B = \lim_{N \to \infty} \int_a^b f \, dg_B^N = \sum_{i=1}^{\infty} (\Delta^+ g_B(t_i) + \Delta^- g_B(t_i)) f(t_i)$$

and the proof is complete.

In Hildebrandt [1] (II. 19.3.14) the following result is proved for the Young integrals.

Osgood Convergence Theorem. *If f_n: $[a, b] \to R$, $n = 1, 2, \ldots$ are uniformly bounded on $[a, b]$, i.e. $|f_n(t)| \le M$ for all $t \in [a, b]$ and $n = 1, 2, \ldots$, $g \in BV[a, b]$, $\lim_{n \to \infty} f_n(t) = f(t)$ for all $t \in [a, b]$, and if $Y \int_a^b f_n \, dg$ and $Y \int_a^b f \, dg$ exist, then $\lim_{n \to \infty} Y \int_a^b f_n \, dg = Y \int_a^b f \, dg$.*

In virtue of the relations between the Young integral and the Perron-Stieltjes integral mentioned in 4.7 the following statement can be deduced.

4.24. Theorem. *If $f, g, f_n \in BV[a, b]$, $|f_n(t)| \leq M$ for all $t \in [a, b]$, $n = 1, 2, \ldots$ and $\lim_{n \to \infty} f_n(t) = f(t)$ for all $t \in [a, b]$, then the integrals $\int_a^b f_n \, dg$, $\int_a^b f \, dg$ exist and $\lim_{n \to \infty} \int_a^b f_n \, dg = \int_a^b f \, dg$.*

This statement follows from the above quoted Osgood Convergence Theorem in the following way: Since all functions in question belong to $BV[a, b]$, the integrals $\int_a^b f_n \, dg$, $\int_a^b f \, dg$, $Y \int_a^b f_n \, dg$ and $Y \int_a^b f \, dg$ exist and $\int_a^b f_n \, dg = Y \int_a^b f_n \, dg$, $\int_a^b f \, dg = Y \int_a^b f \, dg$ (see 4.7). Hence all the assumptions of the Osgood theorem are satisfied and our statement holds.

4.25. Theorem (Substitution Theorem). *If $h \in BV[a, b]$, $g\colon [a, b] \to R$ and $f\colon [a, b] \to R$, the integral $\int_a^b g \, dh$ exists and f is bounded on $[a, b]$, then the integral $\int_a^b f(t) \, d(\int_a^t g(\tau) \, dh(\tau))$ exists if and only if the integral $\int_a^b f(t) g(t) \, dh(t)$ exists and in this case the two integrals are equal.*

Proof. Let us show that the following statement holds. If $\int_a^b g \, dh$ exists then for every $\eta > 0$ there is an $S_1 \in \mathscr{S}[a, b]$ such that for every $A \in A(S_1)$, $A\colon a = \alpha_0 \leq \tau_1 \leq \ldots \leq \tau_k \leq \alpha_k = b$ we have

$$\sum_{j=1}^{k} \left| g(\tau_j) [h(\alpha_j) - h(\alpha_{j-1})] - \int_{\alpha_{j-1}}^{\alpha_j} g \, dh \right| < \eta . \tag{4,20}$$

Let $\eta > 0$ be given. By definition there exists $S_1 \in \mathscr{S}[a, b]$ such that if $A \in A(S_1)$ then

$$\left| B_{g,h}(A) - \int_a^b g \, dh \right| = \left| \sum_{j=1}^{k} \left\{ g(\tau_j) [h(\alpha_j) - h(\alpha_{j-1})] - \int_{\alpha_{j-1}}^{\alpha_j} g \, dh \right\} \right| < \frac{\eta}{8}$$

and if also $A' \in A(S_1)$ then

$$|B_{g,h}(A) - B_{g,h}(A')| < \frac{\eta}{4} .$$

Let $A\colon a = \alpha_0 \leq \tau_1 \leq \ldots \leq \tau_k \leq \alpha_k = b$, $A \in A(S_1)$ be fixed. Assume that $U_1 = \{j_1, j_2, \ldots, j_m\}$, $m \leq l$ is an arbitrary set of integers such that $1 \leq j_1 < j_2 < \ldots < j_m \leq k$. Since by 4.10 the integrals $\int_{\alpha_{j_i - 1}}^{\alpha_{j_i}} g \, dh$, $i = 1, 2, \ldots, m$ exist there is an $S_2 \in \mathscr{S}[a, b]$, $S_2 \subset S_1$ such that for any subdivision A_i of the interval $[\alpha_{j_i - 1}, \alpha_{j_i}]$ which is subordinate to S_1 we have

$$\left| B_{g,h}(A_i) - \int_{\alpha_{j_i - 1}}^{\alpha_{j_i}} g \, dh \right| < \frac{\eta}{4m} . \tag{4,21}$$

Let us refine the subdivision A in such a way that for $i = 1, \ldots, m$ the points $\alpha_{j_i - 1} \leq \tau_{j_i} \leq \alpha_{j_i}$ are replaced by the points of A_i and the points $\alpha_{j-1} \leq \tau_j \leq \alpha_j$, $j \notin U_1$

remain unchanged. Let us denote this refinement by A'; evidently $A' \in A(S_1)$.

We have

$$\left| \sum_{j \in U_1} \left(g(\tau_j) [h(\alpha_j) - h(\alpha_{j-1})] - \int_{\alpha_{j-1}}^{\alpha_j} g \, dh \right) \right|$$

$$\leq \left| \sum_{i=1}^{m} (g(\tau_{j_i}) [h(\alpha_{j_i}) - h(\alpha_{j_i - 1})] - B_{g,h}(A_i)) \right| + \left| \sum_{i=1}^{m} \left(B_{g,h}(A_i) - \int_{\alpha_{j_i - 1}}^{\alpha_{j_i}} g \, dh \right) \right|$$

$$= \left| \sum_{j=1}^{k} g(\tau_j) [h(\alpha_j) - h(\alpha_{j-1})] - \sum_{\substack{j=1 \\ j \notin U_1}}^{m} g(\tau_j) [h(\alpha_j) - h(\alpha_{j-1})] - \sum_{i=1}^{m} B_{g,h}(A_i) \right|$$

$$+ \sum_{i=1}^{m} \left| B_{g,h}(A_i) - \int_{\alpha_{j_i - 1}}^{\alpha_{j_i}} g \, dh \right| < |B_{g,h}(A) - B_{g,h}(A')| + m \frac{\eta}{4m} < \frac{\eta}{4} + \frac{\eta}{4} = \frac{\eta}{2}$$

because $A, A' \in A(S_1)$ and (4,21) holds.

Since the set $U_1 \subset \{1, \ldots, k\}$ of indices was arbitrary, we obtain that for a given $\eta > 0$ there exists $S_1 \in \mathscr{S}[a, b]$ such that for any $A \in A(S_1)$ and $U_1 \subset \{1, 2, \ldots, k\}$ the inequality

$$\left| \sum_{j = U_1} g(\tau_j) [h(\alpha_j) - h(\alpha_{j-1})] - \int_{\alpha_{j-1}}^{\alpha_j} g \, dh \right| < \frac{\eta}{2}$$

holds. Let us set

$$d_j = g(\tau_j) [h(\alpha_j) - h(\alpha_{j-1})] - \int_{\alpha_{j-1}}^{\alpha_j} g \, dh$$

and assume that U_1 is the set of all $j \in \{1, \ldots, k\}$ for which $d_j \geq 0$, $U_2 = \{1, \ldots, k\} \setminus U_1$. Then we have

$$\sum_{j=1}^{k} |d_j| = \sum_{j \in U_1} d_j - \sum_{j \in U_2} d_j \leq \left| \sum_{j \in U_1} d_j \right| + \left| \sum_{j \in U_2} d_j \right| \leq \eta,$$

i.e. (4,20) holds.

Now, let us prove the theorem. Assume that $\varepsilon > 0$ is given. If the integral $\int_a^b fg \, dh$ exists then by definition there exists $S_1 \in \mathscr{S}[a, b]$ such that for all $A \in A(S_1)$

(i) $$\left| \sum_{j=1}^{k} f(\tau_j) g(\tau_j) [h(\alpha_j) - h(\alpha_{j-1})] - \int_a^b fg \, dh \right| < \frac{\varepsilon}{2}.$$

Since the integral $\int_a^b g \, dh$ exists, by the above statement there is $S_2 \in \mathscr{S}[a, b]$ such that for any $A \in A(S_2)$ we have

(ii) $$\sum_{j=1}^{k} \left| g(\tau_j) [h(\alpha_j) - h(\alpha_{j-1})] - \int_{\alpha_{j-1}}^{\alpha_j} g \, dh \right| < \frac{\varepsilon}{2C}$$

where $C > 0$ is the bound for f, i.e. $|f(t)| \leq C$ for all $t \in [a, b]$. If we set $S = S_1 \cap S_2$ then $S \in \mathscr{S}[a, b]$ and for any $A \in A(S)$ the inequalities (i), (ii) are satisfied. Let us set $k(t) = \int_a^t g(\tau)\,\mathrm{d}h(\tau)$, $t \in [a, b]$. Then for $A \in A(S)$ we have by (i) and (ii)

$$\left|B_{f,k}(A) - \int_a^b fg\,\mathrm{d}h\right| = \left|\sum_{j=1}^k f(\tau_j)\int_{\alpha_{j-1}}^{\alpha_j} g\,\mathrm{d}h - \int_a^b fg\,\mathrm{d}h\right|$$

$$\leq \left|\sum_{j=1}^k f(\tau_j)\int_{\alpha_{j-1}}^{\alpha_j} g\,\mathrm{d}h - f(\tau_j)\,g(\tau_j)\left[h(\alpha_j) - h(\alpha_{j-1})\right]\right.$$

$$\left. + \sum_{j=1}^k f(\tau_j)\,g(\tau_j)\left[h(\alpha_j) - h(\alpha_{j-1})\right] - \int_a^b fg\,\mathrm{d}h\right|$$

$$< C\sum_{j=1}^k \left|g(\tau_j)\left[h(\alpha_j) - h(\alpha_{j-1})\right] - \int_{\alpha_{j-1}}^{\alpha_j} g\,\mathrm{d}h\right| + \frac{\varepsilon}{2} < \frac{C\varepsilon}{2C} + \frac{\varepsilon}{2} = \varepsilon\,.$$

Hence according to Definition 4.5 the integral $\int_a^b f\,\mathrm{d}k = \int_a^b f(t)\,\mathrm{d}\left(\int_a^t g\,\mathrm{d}h\right)$ exists and equals $\int_a^b fg\,\mathrm{d}h$. Using the same technique the second implication can be also proved.

4.26. Theorem. *Assume that for the functions* $g, h \in BV[a, b]$, $f\colon [a, b] \to R$, $\varphi\colon [a, b] \to R$ *the integrals* $\int_a^b f\,\mathrm{d}g$, $\int_a^b \varphi\,\mathrm{d}h$ *exist. If to every* $\tau \in [a, b]$ *there is a* $\delta^*(\tau) > 0$ *such that*

$$|t - \tau|\,|f(\tau)\,(g(t) - g(\tau))| \leq (t - \tau)\,\varphi(\tau)\,(h(t) - h(\tau)) \tag{4,22}$$

holds for every $\tau \in [a, b]$, $t \in [a, b] \cap [\tau - \delta^*(\tau), \tau + \delta^*(\tau)]$, *then*

$$\left|\int_a^b f\,\mathrm{d}g\right| \leq \int_a^b \varphi\,\mathrm{d}h\,.$$

This statement is proved in Kurzweil [2].

4.27. Corollary. *Assume that* $g \in BV[a, b]$. *If* $f\colon [a, b] \to R$, $|f(t)| \leq M = \text{const.}$ *for all* $t \in [a, b]$ *and* $\int_a^b f\,\mathrm{d}g$ *exists then for every* $[c, d] \subset [a, b]$ *we have*

$$\left|\int_c^d f\,\mathrm{d}g\right| \leq M\,\mathrm{var}_c^d\,g$$

and consequently $\mathrm{var}_a^b\left(\int_a^t f\,\mathrm{d}g\right) \leq M\,\mathrm{var}_a^b\,g < \infty$.

If $f \in BV[a, b]$ *then* $\int_a^b f\,\mathrm{d}g$ *exists and*

$$\left|\int_a^b f\,\mathrm{d}g\right| \leq \int_a^b |f(t)|\,\mathrm{d}(\mathrm{var}_a^t\,g) \leq \sup_{t\in[a,b]} |f(t)|\,\mathrm{var}_a^b\,g\,.$$

Proof. In the first case we have

$$|t - \tau|\,|f(\tau)\,(g(t) - g(\tau))| \leq (t - \tau)\,|f(\tau)|\,(\mathrm{var}_a^t\,g - \mathrm{var}_a^\tau\,g)$$
$$\leq (t - \tau)\,M(\mathrm{var}_a^t\,g - \mathrm{var}_a^\tau\,g)$$

for every $\tau \in [a, b]$, $t \in [a, b]$. Since the integral $\int_c^d M \, d(\text{var}_a^t g)$ exists and equals $M \, \text{var}_c^d g$ we obtain the result by 4.26. The second statement can be derived in a similar way, when 4.19 and the fact that $|f| \in BV[a, b]$ are taken into account.

4.28. Theorem. *Let us assume that* $g\colon [a, b] \to R$ *is nondecreasing,* $f_1, f_2\colon [a, b] \to R$, $f_1(t) \leq f_2(t)$ *for all* $t \in [a, b]$ *and* $\int_a^b f_i \, dg$ *exists for* $i = 1, 2$. *Then*

$$\int_a^b f_1 \, dg \leq \int_a^b f_2 \, dg .$$

This statement follows from 4.26.

4.29. Theorem. *If* $h\colon [a, b] \to R$ *is nonnegative, nondecreasing and continuous from the left in* $[a, b]$ *(i.e.* $h(t-) = h(t)$ *for every* $t \in (a, b]$*), then*

$$\int_a^b h^k(t) \, dh(t) \leq \frac{1}{k+1} [h^{k+1}(b) - h^{k+1}(a)] \tag{4,23}$$

for any $k = 0, 1, 2, \ldots$. *If* $h\colon [a, b] \to R$ *is assumed to be nonnegative, nonincreasing and continuous from the right (i.e.* $(h(t+) = h(t)$ *for every* $t \in [a, b)$*), then*

$$\int_a^b h^k(t) \, dh(t) \geq \frac{1}{k+1} [h^{k+1}(b) - h^{k+1}(a)] \tag{4,24}$$

for any $k = 0, 1, 2, \ldots$

The proof of the first part is given in Kurzweil [2]. The second part can be proved similarly.

4.30. Theorem. *Assume that* $g\colon [a, b] \to R$ *is a nonnegative nondecreasing function,* $\varphi\colon [a, b] \to R$ *nonnegative and bounded, i.e.* $\varphi(t) \leq C = \text{const.}$ *for all* $t \in [a, b]$.

(a) *If* g *is continuous from the right on* $[a, b)$ *and if there exist nonnegative constants* K_1, K_2 *such that*

$$\varphi(\xi) \leq K_1 + K_2 \int_\xi^b \varphi(\tau) \, dg(\tau) \tag{4,25}$$

for every $\xi \in [a, b]$, *then*

$$\varphi(\tau) \leq K_1 e^{K_2(g(b) - g(\tau))} \tag{4,26}$$

for any $\tau \in [a, b]$.

(b) *If* g *is continuous from the left on* $(a, b]$ *and if there exist nonnegative constants* K_1, K_2 *such that*

$$\varphi(\xi) \leq K_1 + K_2 \int_a^\xi \varphi(\tau) \, dg(\tau) \tag{4,27}$$

for every $\xi \in [a, b]$, *then*

(4,28) $$\varphi(\tau) \le K_1 e^{K_2(g(\tau) - g(a))}$$

for any $\tau \in [a, b]$.

Proof. We prove only (a). The statement (b) can be proved in the same way. Let us define

$$w(t) = L e^{K_2(g(b) - g(t))}, \qquad t \in [a, b]$$

where $L \ge 0$ is a constant.

For any $\xi \in [a, b]$ we have

$$L + K_2 \int_\xi^b w(\tau)\, dg(\tau) = L + K_2 L \int_\xi^b e^{K_2(g(b) - g(\tau))}\, dg(\tau)$$

$$= L\left(1 + K_2 \int_\xi^b \sum_{i=1}^{\infty} \frac{K_2}{i!} (g(b) - g(\tau))^i\, dg(\tau)\right).$$

Since the series $\sum_{i=0}^{\infty} K_2(g(b) - g(\tau))^i / i!$ evidently converges uniformly on $[a, b]$, 4.17 ensures that in the last term the integration and summation are interchangeable. Hence by (4,24) from 4.29 we obtain

$$L + K_2 \int_\xi^b w(\tau)\, dg(\tau) = L\left(1 + K_2 \sum_{i=0}^{\infty} \frac{K_2^i}{i!} \int_\xi^b (g(b) - g(\tau))^i\, dg(\tau)\right) =$$

$$= L\left(1 - K_2 \sum_{i=0}^{\infty} \frac{K_2^i}{i!} \int_\xi^b (g(b) - g(\tau))^i\, d(g(b) - g(\tau))\right) \le$$

$$\le L\left(1 + \sum_{i=0}^{\infty} \frac{K_2^i}{i!} (g(b) - g(\xi))^i\right) = L e^{K_2(g(b) - g(\xi))}.$$

Let $\varepsilon > 0$ be arbitrary. We set

$$w_\varepsilon(t) = (K_1 + \varepsilon)\, e^{K_2(g(b) - g(t))}, \qquad t \in [a, b].$$

Then

(4,29) $$K_1 + \varepsilon + K_2 \int_\xi^b w_\varepsilon(\tau)\, dg(\tau) \le w_\varepsilon(\xi), \qquad \xi \in [a, b].$$

For the difference $m_\varepsilon(\xi) = \varphi(\xi) - w_\varepsilon(\xi)$ we have by (4,25), (4,29)

(4,30) $$m_\varepsilon(\xi) \le -\varepsilon + K_2 \int_\xi^b m_\varepsilon(\tau)\, dg(\tau), \qquad \xi \in [a, b]$$

and, in particular, $m_\varepsilon(b) \le -\varepsilon < 0$. Moreover, it is easy to see that $|m_\varepsilon(\xi)| \le C_1 = \text{const.}$ for $\xi \in [a,b]$. By 4.12 we have

$$\begin{aligned} m_\varepsilon(\xi) &\le -\varepsilon + K_2\, m_\varepsilon(b)\,[g(b) - g(b-)] + \lim_{\delta\to 0+} K_2 \int_\xi^{b-\delta} m_\varepsilon(\tau)\, \mathrm{d}g(\tau) \\ &\le -\varepsilon + K_2\, m_\varepsilon(b)\,[g(b) - g(b-)] + C_2[g(b-) - g(\xi)]\,, \qquad C_2 = K_2 C_1\,. \end{aligned}$$

Since $g \in BV$, there exists $\eta > 0$ such that if $0 \le b - \xi \le \eta$ then $C_2(g(b-) - g(\xi)) < \varepsilon/2$. Hence for $\xi \in [b-\eta, b]$ we have $m_\varepsilon(\xi) < 0$. Let us set

$$T = \inf\{t \in [a,b];\ m_\varepsilon(\xi) < 0 \quad \text{for} \quad \xi \in [t,b]\}\,. \tag{4,31}$$

We have shown that $T < b$ and we have evidently $m_\varepsilon(t) < 0$ for $t \in (T, b]$. Further by (4,30) and 4.12

$$\begin{aligned} m_\varepsilon(T) &\le -\varepsilon + K_2 \int_T^b m_\varepsilon(\tau)\, \mathrm{d}g(\tau) \\ &= -\varepsilon + K_2\, m_\varepsilon(T)\,(g(T+) - g(T)) + \lim_{\delta\to 0+} K_2 \int_{T+\delta}^b m_\varepsilon(\tau)\, \mathrm{d}g(\tau) \le -\varepsilon < 0 \end{aligned}$$

since $g(T+) - g(T) = 0$ and $\int_{T+\delta}^b m_\varepsilon(\tau)\, \mathrm{d}g(\tau) \le 0$ for every $\delta > 0$.

If $T > a$ then we repeat the above procedure and show in the same way that there exists an $\eta > 0$ such that $m_\varepsilon(\xi) < 0$ for all $\xi \in [T-\eta, T]$. This contradicts (4,31). Hence $T = a$ and $m_\varepsilon(\xi) < 0$ for all $\xi \in [a,b]$, i.e.

$$\varphi(\xi) < K_1 e^{K_2(g(b) - g(\xi))} + \varepsilon e^{K_2(g(b) - g(a))}$$

for all $\xi \in [a,b]$ and $\varepsilon > 0$. This yields (4,26).

4.31. Theorem. *Let* $h\colon [a,b] \times [c,d] \to R$ *be such that* $|h(s,t)| \le M < \infty$ *and* $\operatorname{var}_a^b h(.,t) + \operatorname{var}_c^d h(s,.) < \infty$ *for every* $(t,s) \in [a,b] \times [c,d]$. *Then for any* $f \in BV[a,b]$ *and any* $g \in BV[c,d]$ *both the iterated integrals*

$$\int_a^b \mathrm{d}f(s) \left(\int_c^d h(s,t)\, \mathrm{d}g(t) \right) \quad \text{and} \quad \int_c^d \left(\int_a^b \mathrm{d}f(s)\, h(s,t) \right) \mathrm{d}g(t)$$

exist and are equal.

(See Hildebrandt [2], p. 356 and [1], II.19.)

4.32. Theorem (Dirichlet formula). *If* $h\colon [a,b] \times [a,b] \to R$ *is bounded on* $[a,b] \times [a,b]$ *and* $\operatorname{var}_a^b h(s,.) < \infty$ *for every* $s \in [a,b]$, $\operatorname{var}_a^b h(.,t) < \infty$ *for every* $t \in [a,b]$, *then for any* $f, g \in BV[a,b]$ *we have*

(4,32)
$$\int_a^b \mathrm{d}g(t)\left(\int_a^t h(s,t)\,\mathrm{d}f(s)\right)$$
$$= \int_a^b \left(\int_s^b \mathrm{d}g(t)\,h(s,t)\right)\mathrm{d}f(s) + \sum_{t\in(a,b]} \Delta^- g(t)\,h(t,t)\,\Delta^- f(t) - \sum_{t\in[a,b)} \Delta^+ g(t)\,h(t,t)\,\Delta^+ f(t)$$

where $\Delta^- g(t) = g(t) - g(t-)$, $\Delta^+ g(t) = g(t+) - g(t)$.

Proof. Let us define $k(s,t) = h(s,t)$ for $a \le s \le t \le b$ and $k(s,t) = 0$ for $a \le t < s \le b$. Then $k: [a,b] \times [a,b] \to R$ evidently satisfies the assumptions of 4.31 and this theorem gives

(4,33)
$$\int_a^b \mathrm{d}g(t)\left(\int_a^b k(s,t)\,\mathrm{d}f(s)\right) = \int_a^b \left(\int_a^b \mathrm{d}g(t)\,k(s,t)\right)\mathrm{d}f(s)\,.$$

Moreover for $t \in [a,b)$ it is

$$\int_a^b k(s,t)\,\mathrm{d}f(s) = \int_a^t h(s,t)\,\mathrm{d}f(s) + \int_t^b k(s,t)\,\mathrm{d}f(s)$$
$$= \int_a^t h(s,t)\,\mathrm{d}f(s) + h(t,t)\,\Delta^+ f(t)\,,$$

since from 4.13 and from the definition of $k(s,t)$ we have by (4,6)

$$\int_t^b k(s,t)\,\mathrm{d}f(s) = \lim_{\tau\to t+}\left[\int_\tau^b k(s,t)\,\mathrm{d}f(s) + k(t,t)\,(f(t+) - f(t))\right]$$
$$= k(t,t)\,\Delta^+ f(t) = h(t,t)\,\Delta^+ f(t)\,.$$

If $t = b$, then $\int_a^b k(s,b)\,\mathrm{d}f(s) = \int_a^b h(s,b)\,\mathrm{d}f(s)$. Hence for an arbitrary $t \in [a,b]$ we can write

(4,34)
$$\int_a^b k(s,t)\,\mathrm{d}f(s) = \int_a^t h(s,t)\,\mathrm{d}f(s) + h(t,t)\,\Delta^+ f(t)$$

if we set $\Delta^+ f(b) = 0$.

A similar argument gives

(4,35)
$$\int_a^b \mathrm{d}g(t)\,k(s,t) = \int_s^b \mathrm{d}g(t)\,h(s,t) + \Delta^- g(s)\,h(s,s)$$

for every $s \in [a,b]$ if the convention $\Delta^- g(a) = 0$ is used. Setting (4,34) and (4,35) into (4,33) we obtain

(4,36)
$$\int_a^b \mathrm{d}g(t)\left(\int_a^t h(s,t)\,\mathrm{d}f(s)\right)$$
$$= \int_a^b \mathrm{d}f(s)\left(\int_s^b h(s,t)\,\mathrm{d}g(t)\right) + \int_a^b \Delta^- g(s)\,h(s,s)\,\mathrm{d}f(s) - \int_a^b \mathrm{d}g(t)\,h(t,t)\,\Delta^+ f(t)\,.$$

Since $g \in BV[a, b]$, there is an at most countable set of points $\alpha_1, \alpha_2, \ldots$ in $[a, b]$ such that $\Delta^- g(s) = 0$ for all $s \in [a, b]$, $s \neq \alpha_i$ and $\sum_{i=1}^{\infty} |\Delta^- g(\alpha_i)| \leq \operatorname{var}_a^b g < +\infty$. Let us set $H(s) = \Delta^- g(s)\, h(s, s)$ for any $s \in [a, b]$. Then $H(s) = 0$ for all $s \in [a, b]$, $s \neq \alpha_i$, $i = 1, 2, \ldots$ and

$$\int_a^b \Delta^- g(s)\, h(s, s)\, \mathrm{d}f(s) = \int_a^b H(s)\, \mathrm{d}f(s).$$

Let us define for $N = 1, 2, \ldots$ and $s \in [a, b]$

$$H_N(s) = \sum_{i=1}^{N} \Delta^- g(\alpha_i)\, h(\alpha_i, \alpha_i)\, \psi_{\alpha_i}(s)$$

where $\psi_\alpha(s) = 0$ if $s \neq \alpha$ and $\psi_\alpha(\alpha) = 1$.

Evidently $H_N(s) = 0$ for all $s \in [a, b]$, $s \neq \alpha_1, \alpha_2, \ldots, \alpha_N$ and $H_N(\alpha_i) = H(\alpha_i)$ for $i = 1, 2, \ldots, N$. For $s \in [a, b]$, $s \notin \alpha_1, \alpha_2, \ldots, \alpha_N$ we have

$$|H_N(s) - H(s)| = |H(s)| \leq \sup_{i=N+1,\ldots} |H(\alpha_i)| < \sum_{i=N+1}^{\infty} |\Delta^- g(\alpha_i)\, h(\alpha_i, \alpha_i)|$$
$$\leq M \sum_{i=N+1}^{\infty} |\Delta^- g(\alpha_i)|$$

where M is the bound of $|h(s, t)|$.

Since the series $\sum_{i=1}^{\infty} |\Delta^- g(\alpha_i)|$ is convergent, we obtain that for any $\varepsilon > 0$ there is a natural N such that $M \sum_{i=N+1}^{\infty} |\Delta^- g(\alpha_i)| < \varepsilon$ and also

$$|H_N(s) - H(s)| < \varepsilon$$

for all $s \in [a, b]$, i.e. $\lim_{N\to\infty} H_N(s) = H(s)$ uniformly in $[a, b]$. Using (4,18) we conclude

$$\int_a^b H_N(s)\, \mathrm{d}f(s) = \sum_{i=1}^{N} \Delta^- g(\alpha_i)\, h(\alpha_i, \alpha_i)\, \Delta f(\alpha_i)$$

and by 4.17 we obtain

$$\int_a^b \Delta^- g(s)\, h(s, s)\, \mathrm{d}f(s) = \int_a^b H(s)\, \mathrm{d}f(s) = \lim_{N\to\infty} \int_a^b H_N(s)\, \mathrm{d}f(s) =$$
$$= \sum_{i=1}^{\infty} \Delta^- g(\alpha_i)\, h(\alpha_i, \alpha_i)\, \Delta f(\alpha_i) = \sum_{s\in(a,b]} \Delta^- g(s)\, h(s, s)\, \Delta f(s).$$

Similarly it can be proved that

$$\int_a^b \mathrm{d}g(t)\, (h(t, t)\, \Delta^+ f(t)) = \sum_{t\in[a,b)} \Delta g(t)\, h(t, t)\, \Delta^+ f(t).$$

If we set these expressions into (4,36) we obtain

$$\int_a^b dg(t)\left(\int_a^t h(s,t)\,df(s)\right)$$

$$= \int_a^b\left(\int_s^b dg(t)\,h(s,t)\right)df(s) + \sum_{s\in(a,b)}\left[\Delta^- g(s)\,h(s,s)\,\Delta f(s) - \Delta g(s)\,h(s,s)\,\Delta^+ f(s)\right]$$
$$+ \Delta^- g(b)\,h(b,b)\,\Delta f(b) - \Delta g(a)\,h(a,a)\,\Delta^+ f(a)$$

and this yields the result.

4.33. Theorem (integration-by-parts). *Let $f, g \in BV[a,b]$; then for any interval $[c,d] \subset [a,b]$ we have*

$$\int_c^d f\,dg + \int_c^d g\,df = f(d)\,g(d) - f(c)\,g(c) - \sum_{c\le\tau<d}\Delta^+ f(\tau)\,\Delta^+ g(\tau) + \sum_{c<\tau\le d}\Delta^- f(\tau)\,\Delta^- g(\tau)$$

where $\Delta^+ f(\tau) = f(\tau+) - f(\tau)$, $\Delta^- f(\tau) = f(\tau) - f(\tau-)$ and similarly for $\Delta^+ g(\tau)$, $\Delta^- g(\tau)$.

Proof. If we set $h(s,t) \equiv 1$ on $[a,b]\times[a,b]$ then for every $f, g \in BV[a,b]$ we have by 4.32

(4,37)
$$\int_c^d\left(\int_c^t df(s)\right)dg(t)$$
$$= \int_c^d\left(\int_s^d dg(t)\right)df(s) + \sum_{t\in(c,d]}\Delta^- g(t)\,\Delta^- f(t) - \sum_{t\in[c,d)}\Delta^+ g(t)\,\Delta^+ f(t).$$

Moreover,

$$\int_c^d\left(\int_c^t df(s)\right)dg(t) = \int_c^d (f(t) - f(c))\,dg(t) = \int_c^d f(t)\,dg(t) - f(c)\,(g(d) - g(c))$$

and similarly

$$\int_c^d\left(\int_s^d dg(t)\right)df(s) = -\int_c^d g(t)\,df(t) + g(d)\,(f(d) - f(c)).$$

Inserting this into (4,37) we obtain the result. (A direct proof of the integration-by-parts theorem 4.33 is given in Kurzweil [3].)

The Lebesgue-Stieltjes integral has been defined and studied in many monographs on integration theory. (See e.g. Saks [1], Hildebrandt [1], Dunford, Schwartz [1] etc.) In the next theorem its relationship with the Perron-Stieltjes integral is cleared up. The proof follows e.g. from Theorem VI (8.1) in Saks [1].

4.34. Theorem. *Let $g \in BV[a,b]$ and $f\colon [a,b] \to R$ be such that the Lebesgue-Stieltjes integral (L-S) $\int_{(a,b)} f\,dg$ over the open interval (a,b) exists. Then the Perron-*

Stieltjes integral $\int_a^b f\,\mathrm{d}g$ *also exists and*

$$\int_a^b f\,\mathrm{d}g = (L - S)\int_{(a,b)} f\,\mathrm{d}g + f(a)\,\Delta^+ g(a) + f(b)\,\Delta^- g(b)\,.$$

4.35. Remark. If f: $[a,b] \to R$ is bounded, h: $[a,b] \to R$ is Lebesgue integrable on $[a,b]$ $(h \in L^1[a,b])$ and $g(t) = g(a) + \int_a^t h(\tau)\,\mathrm{d}\tau$ on $[a,b]$ $(g \in AC[a,b])$, then in virtue of 4.25 and 4.34

$$\int_a^b f(t)\,\mathrm{d}g(t) = \int_a^b f(t)\,h(t)\,\mathrm{d}t\,,$$

where the right-hand side integral is the Lebesgue one.

For the proof of the following assertion see e.g. Natanson [1] (Corollary of Theorem XII.4.2). It is also included as a special case in the "symmetrical Fubini theorem" for Lebesgue-Stieltjes integrals (cf. Hildebrandt [1], X.3.2).

4.36. Theorem (Tonelli, Hobson). *If* h: $D = [a,b] \times [c,d] \to R$ *is measurable and if any one of the three Lebesgue integrals*

$$\iint_D |h(t,s)|\,\mathrm{d}t\,\mathrm{d}s\,, \qquad \int_a^b \left(\int_c^d |h(t,s)|\,\mathrm{d}s\right)\mathrm{d}t\,, \qquad \int_c^d \left(\int_a^b |h(t,s)|\,\mathrm{d}t\right)\mathrm{d}s$$

exists, then the Lebesgue integrals

$$\iint_D h(t,s)\,\mathrm{d}t\,\mathrm{d}s\,, \qquad \int_a^b \left(\int_c^d h(t,s)\,\mathrm{d}s\right)\mathrm{d}t\,, \qquad \int_c^d \left(\int_a^b h(t,s)\,\mathrm{d}t\right)\mathrm{d}s$$

all exist and are equal to one another.

One of the most helpful tools for the investigation of integro-differential and functional-differential equations is the "*unsymmetrical Fubini theorem*" 4.38. For its proof the following lemma is needed.

4.37. Lemma. *Let* h: $[a,b] \times [c,d] \to R$ *be such that* $h(.,s)$ *is measurable on* $[a,b]$ *for any* $s \in [c,d]$, $\chi(t) = |h(t,c)| + \mathrm{var}_c^d\, h(t,.) < \infty$ *for a.e.* $t \in [a,b]$ *and* $\chi \in L^p[a,b]$, $1 \le p < \infty$. *Then*

(a) *given* $f \in L^q[a,b]$ *with* $q = p/(p-1)$ *if* $p > 1$ *and* $q = \infty$ *if* $p = 1$, *the function*

$$\varphi\colon\ s \in [c,d] \to \int_a^b f(t)\,h(t,s)\,\mathrm{d}s$$

is defined for any $s \in [c,d]$, *belongs to* $BV[c,d]$ *and*

$$\text{(4,38)} \qquad \varphi(s+) = \int_a^b f(t)\,h(t,s+)\,\mathrm{d}t \qquad \textit{for any} \quad s \in [c,d)\,,$$

$$\varphi(s-) = \int_a^b f(t)\,h(t,s-)\,\mathrm{d}t \qquad \textit{for any} \quad s \in (c,d]\,;$$

(b) *given* $g \in C[c,d]$ (*or* $g \in BV[c,d]$), *the function*

$$\eta\colon t \in [a,b] \to \int_c^d \mathrm{d}_s[h(t,s)]\, g(s)$$

is defined a.e. on $[a,b]$ *and belongs to* $L^p[a,b]$.

Proof. Clearly, $\varphi(s)$ is defined for any $s \in [c,d]$. For an arbitrary subdivision $c = s_0 < s_1 < \ldots < s_k = d$ of $[c,d]$ we have

$$\sum_{j=1}^{k} |\varphi(s_j) - \varphi(s_{j-1})| \le \int_a^b |f(t)| \sum_{j=1}^{k} |h(t,s_j) - h(t,s_{j-1})|\, \mathrm{d}t$$
$$\le \int_a^b |f(t)|\, \chi(t)\, \mathrm{d}t < \infty\,,$$

i.e. $\varphi \in BV[c,d]$. Furthermore,

$$|f(t)\, h(t,\sigma)| \le |f(t)|\, \chi(t) \qquad \text{for a.e.} \quad t \in [a,b] \quad \text{and any} \quad \sigma \in [c,d]\,.$$

Applying the Lebesgue Dominated Convergence Theorem we obtain immediately (4,38).

(b) Under our assumptions $\eta(t)$ is defined a.e. on $[a,b]$. If $g\colon [c,d] \to R$ is a finite step function with jumps at $s_j \in [c,d]$ $(j = 1, 2, \ldots, k)$ (cf. 4.20), then according to 4.21 $\eta(t)$ is a.e. on $[a,b]$ equal to a linear combination of the values $h(t,b)$, $h(t,a)$, $h(t,s_j+)$ and $h(t,s_j-)$ $(j = 1, 2, \ldots, k)$. In particular, in this case η is measurable on $[a,b]$. Making use of the fact that any function g which is continuous on $[a,b]$ or of bounded variation on $[a,b]$ can be approximated uniformly on $[a,b]$ by finite step functions (Aumann [1]) and applying 4.17 we complete the proof of the measurability of η on $[a,b]$. By 4.16

$$|\eta(t)| \le \chi(t)\, (\sup_{s \in [c,d]} |g(s)|) \qquad \text{a.e. on} \quad [a,b]$$

and hence $\eta \in L^p[a,b]$ for any $g \in C[c,d]$ (or $g \in BV[c,d]$).

4.38. Theorem (Cameron, Martin). *Let* $h\colon [a,b] \times [c,d] \to R$ *fulfil the assumptions of* 4.37. *Then for any* $f \in L^q[a,b]$, *where* $q = p/(p-1)$ *if* $p > 1$ *and* $q = \infty$ *if* $p = 1$, *and any* $g \in C[c,d]$ (*or* $g \in BV[c,d]$) *the integrals*

$$\int_a^b f(t) \left(\int_c^d \mathrm{d}_s[h(t,s)]\, g(s) \right) \mathrm{d}t \quad \textit{and} \quad \int_c^d \mathrm{d}_s \left[\int_a^b f(t)\, h(t,s)\, \mathrm{d}t \right] g(s)$$

both exist and are equal to one another.

Proof. Let the functions $\varphi\colon [c,d] \to R$ and $\eta\colon [a,b] \to R$ be defined as in 4.37. By 4.19 and 4.37 both the integrals

$$\int_a^b f(t)\, \eta(t)\, \mathrm{d}t \quad \text{and} \quad \int_c^d \mathrm{d}[\varphi(s)]\, g(s)$$

exist. Let $g_n\colon [c,d] \to R$ $(n = 1, 2, \ldots)$ be a sequence of finite step functions such that $\lim_{n\to\infty} g_n(t) = g(t)$ uniformly on $[c,d]$. (Such a sequence exists according to 7.3.2.1 (3) in Aumann [1].) To prove the theorem it is sufficient by 4.17 and 4.20 to show that

$$\int_a^b f(t)\,\eta(t)\,\mathrm{d}t = \int_c^d \mathrm{d}[\varphi(s)]\,g(s) \tag{4,39}$$

holds for all simple jump functions $g(s) = \psi_\alpha^+(s)$ or $g(s) = \psi_\alpha^-(s)$ $(\alpha \in [c,d])$ defined by (4,12) and (4,13). Let $\alpha \in [c,d]$ and $g(s) = \psi_\alpha^+(s)$ on $[c,d]$, then in virtue of 4.21

$$\eta(t) = \begin{cases} h(t,d) - h(t,\alpha+) & \text{if} \quad \alpha < d \\ 0 & \text{if} \quad \alpha = d \end{cases}$$

and

$$\int_c^d \mathrm{d}[\varphi(s)]\,g(s) = \begin{cases} \varphi(d) - \varphi(\alpha+) & \text{if} \quad \alpha < d \\ 0 & \text{if} \quad \alpha = d \end{cases}$$

and (4,39) follows from (4,38). Analogously we can show that (4,39) holds also if $g(s) = \psi_\alpha^-(s)$ on $[c,d]$.

4.39. Integrals of matrix valued functions. If $\mathbf{F} = (f_{i,j})$, $i = 1, 2, \ldots, p$; $j = 1, 2, \ldots, r$; $\mathbf{G} = (g_{j,k})$, $j = 1, 2, \ldots, r$, $k = 1, 2, \ldots, q$ are matrix valued functions defined on the interval $[a,b]$ $(f_{i,j}\colon [a,b] \to R,\ g_{j,k}\colon [a,b] \to R)$, then we use the following symbols

$$\int_a^b \mathbf{F}\,\mathrm{d}\mathbf{G} = (\alpha_{i,k}), \qquad i = 1, 2, \ldots, p, \quad k = 1, 2, \ldots, q,$$

and

$$\int_a^b \mathrm{d}[\mathbf{F}]\,\mathbf{G} = (\beta_{i,k}), \qquad i = 1, 2, \ldots, p, \quad k = 1, 2, \ldots, q$$

where

$$\alpha_{i,k} = \sum_{j=1}^r \int_a^b f_{i,j}\,\mathrm{d}g_{j,k} \quad \text{and} \quad \beta_{i,k} = \sum_{j=1}^r \int_a^b g_{j,k}\,\mathrm{d}f_{i,j},$$

whenever the integrals appearing in these sums exist. In the same way it is possible to define also integrals of the type $\int_a^b \mathbf{F}\,\mathrm{d}[\mathbf{G}]\,\mathbf{H}$ etc. if the products of matrices occurring in the expressions are well defined.

Since the integral of a matrix valued function with respect to a matrix valued function is a matrix whose elements are sums of Perron-Stieltjes integrals of real scalar functions with respect to real scalar functions, all statements from this section can be used also for such integrals.

5. The space BV_n

In this section we recall some basic properties of the linear space of functions with a bounded variation from the functional analytic point of view.

Let us consider the linear set of all functions $x: [0, 1] \to R$ with a bounded variation $\mathrm{var}_0^1 x$. Let this linear set with the norm

$$(5,1) \qquad x \in BV \to \|x\|_{BV} = |x(0)| + \mathrm{var}_0^1 x$$

be denoted by $BV[0, 1]$ or simply BV.

It is easy to check that (5,1) satisfies all the axioms of a norm.

If $x \in BV$, then evidently

$$(5,2) \quad |x(t)| \le |x(t) - x(0)| + |x(0)| \le |x(0)| + \mathrm{var}_0^1 x \le \|x\|_{BV} \quad \text{for any} \quad t \in [0, 1].$$

5.1. Proposition. *The normed linear space BV is a Banach space (i.e. BV is complete).*

(See Dunford, Schwartz [1] or Hildebrandt [1], II.8.6.)

Further it can be easily shown that BV is not separable. Indeed, if we set $x_\alpha(t) = 0$ for $0 \le t \le \alpha$, $x_\alpha(t) = 1$ for $\alpha < t \le 1$ for any $\alpha \in (0, 1)$, then evidently $x_\alpha \in BV$ for any $\alpha \in (0, 1)$ and

$$\|x_\alpha - x_\beta\|_{BV} = 2$$

provided $\alpha, \beta \in (0, 1)$, $\alpha \neq \beta$. Hence BV cannot contain a countable subset which would be dense in BV. This implies that BV is not separable.

In the same way we can introduce the Banach space BV_n of all column n-vector functions $\boldsymbol{x} = (x_1, \ldots, x_n)^*$: $[0, 1] \to R_n$ of bounded variation if for the definition of $\mathrm{var}_0^1 \boldsymbol{x}$ some norm in R_n is used. The norm in BV_n is given by

$$\boldsymbol{x} \in BV_n \to \|\boldsymbol{x}\|_{BV_n} = |\boldsymbol{x}(0)| + \mathrm{var}_0^1 \boldsymbol{x}\,.$$

It is evident that $\boldsymbol{x}$: $[0, 1] \to R_n$ belongs to BV_n if and only if any component x_i, $i = 1, 2, \ldots, n$ belongs to BV. Hence it is sufficient to consider only the space BV instead of BV_n. All essential properties of BV are transferable to BV_n.

Let us consider some subspaces of BV which are of interest for the subsequent investigations.

By NBV we denote the set of all functions $\varphi \in BV$ for which $\varphi(t+) = \varphi(t)$ if $t \in (0, 1)$ and $\varphi(0) = 0$.

Similarly NBV^- denotes the set of all functions $\varphi \in BV$ such that $\varphi(t-) = \varphi(t)$ for $t \in (0, 1)$ and $\varphi(0) = 0$. Further we denote by S the linear set of all functions $w \in BV$ such that $w(t+) = w(t-) = c = \text{const.}$ for every $t \in (0, 1)$, $w(0) = w(0+) = c$, $w(1) = w(1-) = c$.

5.2. Proposition. *The linear sets NBV, NBV⁻, S are closed in BV.*

Proof. Let $\{\varphi_l\}$, $l = 1, 2, \ldots$ be a sequence with $\varphi_l \in NBV$, such that $\lim_{l\to\infty} \|\varphi_l - \varphi\|_{BV} = 0$ for some $\varphi \in BV$. For $t \in (0, 1)$ we have

$$|\varphi(t+) - \varphi(t)| = |\varphi(t+) - \varphi_l(t+) - (\varphi(t) - \varphi_l(t))| \leq \|\varphi_l - \varphi\|_{BV}$$

for any natural l since $\varphi_l \in NBV$. Hence $\varphi(t+) = \varphi(t)$. Similarly for any l we have

$$|\varphi(0)| = |\varphi_l(0) - \varphi(0)| \leq \|\varphi_l - \varphi\|_{BV}$$

and consequently $\varphi(0) = 0$ and $\varphi \in NBV$. The closedness of NBV^- and S can be proved by the same reasoning.

We denote by AC the linear set of all absolutely continuous functions on $[0, 1]$. If $x \in AC$ then by definition there exists $\delta > 0$ such that for every system $[a_i, b_i]$, $i = 1, \ldots, k$ of nonoverlapping intervals on $[0, 1]$ with

$$\sum_{i=1}^{k} (b_i - a_i) < \delta$$

we have

$$\sum_{i=1}^{k} |x(b_i) - x(a_i)| < 1\,.$$

If we subdivide the interval $[0, 1]$ into m intervals by the division points $0 = c_0 < c_1 < \ldots < c_m = 1$ such that $c_i - c_{i-1} < \delta$, $i = 1, 2, \ldots, m$, then $\operatorname{var}_{c_{i-1}}^{c_i} x < 1$ for $i = 1, 2, \ldots, m$ and consequently $\operatorname{var}_0^1 x = \sum_{i=1}^{m} \operatorname{var}_{c_{i-1}}^{c_i} x < m$. Hence $x \in BV$ and the inclusion $AC \subset BV$ holds.

5.3. Proposition. *The linear set AC is closed in BV.*

Proof. Let $\lim_{k\to\infty} \|\varphi_k - \varphi\|_{BV} = 0$ for $\varphi \in BV$ and $\varphi_k \in AC$, $k = 1, 2, \ldots$. For an arbitrary system $[a_i, b_i]$, $i = 1, \ldots, k$ of nonoverlapping intervals in $[0, 1]$ we have

$$\sum_{i=1}^{k} |\varphi(b_i) - \varphi(a_i)| \leq \sum_{i=1}^{k} |\varphi_l(b_i) - \varphi(b_i) - (\varphi_l(a_i) - \varphi(a_i))| + \sum_{i=1}^{k} |\varphi_l(b_i) - \varphi_l(a_i)|$$
$$\leq \|\varphi_l - \varphi\|_{BV} + \sum_{i=1}^{k} |\varphi_l(b_i) - \varphi_l(a_i)|$$

for any $l = 1, 2, \ldots$. Let $\varepsilon > 0$ be given. Let us choose an integer $l_0 \geq 1$ such that $\|\varphi_l - \varphi\|_{BV} < \varepsilon/2$ for $l > l_0$. For any fixed $l > l_0$ there is $\delta > 0$ such that if

$$\sum_{i=1}^{k} (b_i - a_i) < \delta$$

then

$$\sum_{i=1}^{k} |\varphi_l(b_i) - \varphi_l(a_i)| < \varepsilon\,.$$

Hence by the inequality given above we have $\sum_{i=1}^{k} |\varphi(b_i) - \varphi(a_i)| < \varepsilon$ and $\varphi \in AC$.

5.4. Remark. From 5.2 and 5.3 it is evident that if the closed linear sets NBV, NBV^-, S, AC in BV are equipped with the norm (5,1) of BV, then they are Banach spaces.

By NBV_n, NBV_n^-, S_n, AC_n we denote the closed linear subsets in BV_n which are defined similarly as NBV, NBV^-, S, AC for n-vector functions. For the same reason as above NBV_n, NBV_n^-, S_n, AC_n equipped with the norm of BV_n are Banach spaces.

Let us now assume that $x \in BV$ and define $w(0) = w(1) = x(0)$, $w(t) = x(t) - x(t+) + x(0)$ for $t \in (0, 1)$. Then evidently $w \in S$, since the difference $x(t) - x(t+)$ is nonzero only on an at most countable set $A \subset (0, 1)$ and

$$\operatorname{var}_0^1 w = 2\sum_{t \in A} |x(t+) - x(t)| \leq 2 \operatorname{var}_0^1 x < \infty .$$

Further let us set $\varphi(t) = x(t) - w(t)$ for $t \in [0, 1]$. It is $\varphi(0) = x(0) - w(0) = 0$, $\varphi(t) = x(t+) - x(0)$ for $t \in (0, 1)$, $\varphi(1) = x(1) - x(0)$, i.e. $\varphi \in NBV$.

In this way we have obtained

$$x = \varphi + w$$

for any $x \in BV$ where $\varphi \in NBV$ and $w \in S$. Since evidently $NBV \cap S = \{0\}$, this decomposition is unique. Hence the Banach space BV can be written in the form of the direct sum of closed subspaces NBV and S, i.e.

$$BV = NBV \oplus S . \tag{5,3}$$

Similarly it can be shown that also the decomposition

$$BV = NBV^- \oplus S$$

holds.

For any $x \in BV$ and $\psi \in BV$ we can define the expression

$$f(x) = \int_0^1 x(t) \, d\psi(t) . \tag{5,4}$$

By 4.19 the integral on the right-hand side in (5,4) exists. The functional f is evidently linear. Further it is

$$|f(x)| = \left| \int_0^1 x(t) \, d\psi(t) \right| \leq \sup_{t \in [0,1]} |x(t)| \operatorname{var}_0^1 \psi \leq \|x\|_{BV} \|\psi\|_{BV}$$

(see 4.27). Hence if f is given by (5,4) with $\psi \in BV$, then $f \in BV^*$.

5.5. Proposition. *Assume that* $w \in BV$. *Then*

$$\int_0^1 x(t) \, dw(t) = 0 \tag{5,5}$$

for any $x \in BV$ *if and only if* $w \in S$.

Proof. Let us suppose that $\int_0^1 x(t)\,\mathrm{d}w(t) = 0$ for any $x \in BV$. For a given $\alpha \in [0, 1]$ we define $x_\alpha(t) = 0$ if $t \in [0, 1] \setminus \{\alpha\}$, $x_\alpha(\alpha) = 1$. Then evidently $x_\alpha \in BV$ and we obtain by the assumption

$$\int_0^1 x_\alpha(t)\,\mathrm{d}w(t) = w(\alpha+) - w(\alpha-) = 0\,,$$

i.e. $w(\alpha+) = w(\alpha-)$ for any $\alpha \in (0, 1)$ and $\int_0^1 x_1(t)\,\mathrm{d}w(t) = w(1) - w(1-) = 0$, $\int_0^1 x_0(t)\,\mathrm{d}w(t) = w(0+) - w(0) = 0$ (cf. 4.22). This means that w differs from a continuous function only on an at most countable subset in $(0, 1)$.

Assume that $w \notin S$. Then there exist two points $\alpha, \beta \in [0, 1]$, $\alpha < \beta$ such that α, β are points of continuity for w and $w(\alpha) \neq w(\beta)$. We define $x_{[\alpha,\beta]}(t) = 1$ for $t \in [\alpha, \beta]$ and $x_{[\alpha,\beta]}(t) = 0$ for $t \in [0, 1] \setminus [\alpha, \beta]$. Evidently $x_{[\alpha,\beta]} \in BV$. Using the properties of the integral we obtain the relation

$$\int_0^1 x_{[\alpha,\beta]}(t)\,\mathrm{d}w(t) = w(\alpha) - w(\alpha-) + \int_\alpha^\beta \mathrm{d}w(t) + w(\beta+) - w(\beta)$$

$$= \int_\alpha^\beta \mathrm{d}w(t) = w(\beta) - w(\alpha) \neq 0$$

which contradicts the assumption. Hence $w \in S$. Let us assume that $w \in S$; w is evidently a break function with $\Delta w(t) = w(t+) - w(t-) = 0$ for every $t \in (0, 1)$ and $\Delta^+ w(0) = w(0+) - w(0) = 0$, $\Delta^- w(1) = w(1) - w(1-) = 0$. Hence by 4.23 we have $\int_0^1 x(t)\,\mathrm{d}w(t) = 0$ for every $x \in BV$.

5.6. Corollary. *Let $\psi \in BV$ be given. Using (5,3) ψ can be uniquely written in the form $\psi = \varphi + w$ where $\varphi \in NBV$, $w \in S$ and*

$$\int_0^1 x(t)\,\mathrm{d}\psi(t) = \int_0^1 x(t)\,\mathrm{d}\varphi(t)$$

for every $x \in BV$.

Let us define for $x \in BV$, $\varphi \in NBV$ the relation

$$\langle x, \varphi \rangle = \int_0^1 x(t)\,\mathrm{d}\varphi(t)\,. \tag{5,6}$$

This relation evidently defines a bilinear form on $BV \times NBV$.

5.7. Lemma. *Let $\varphi \in NBV$. If $\langle x, \varphi \rangle = 0$ for every $x \in BV$, then $\varphi = 0$.*

Let $x \in BV$. If $\langle x, \varphi \rangle = 0$ for every $\varphi \in NBV$, then $x = 0$.

Proof. (1) If $\langle x, \varphi \rangle = 0$ for every $x \in BV$, then $\varphi \in S$ by 5.5. Hence $\varphi \in NBV \cap S$ and by (5,3) we obtain $\varphi = 0$.

(2) Assume that $\langle x, \varphi \rangle = 0$ for every $\varphi \in NBV$ but $x \neq 0$. Then either there exists $a \in (0, 1]$ such that $x(a) \neq 0$ or $x(t) = 0$ for all $t \in (0, 1]$ and $x(0) \neq 0$. In the

first case we set $\varphi(t) = 0$ for $t \in [0, a)$, $\varphi(t) = 1$ for $t \in [a, 1]$. Evidently $\varphi \in NBV$ and φ is a simple jump function (see 4.20). By 4.21 we have $\langle x, \varphi \rangle = \int_0^1 x(t)\,d\varphi(t) = x(a) \neq 0$ and this contradicts the assumption. For the second case we set $\varphi(t) = 1$ for $t \in (0, 1]$, $\varphi(0) = 0$, then $\varphi \in NBV$ is also a simple jump function ($\varphi = \psi_0$) and by 4.21 we have $\langle x, \varphi \rangle = \int_0^1 x(t)\,d\varphi(t) = x(0) \neq 0$. Again we have obtained a contradiction and our lemma is proved.

5.8. Proposition. *The pair of spaces BV, NBV forms a dual pair (BV, NBV) with respect to the bilinear form $\langle .,. \rangle$ given by* (5,6).

Proof follows immediately from 5.7 and from the definition of a dual pair given in 3.1.

5.9. Remark. It follows easily from 5.8 that (BV_n, NBV_n) is a dual pair with respect to the bilinear form

$$\mathbf{x} \in BV_n, \boldsymbol{\varphi} \in NBV_n \rightarrow \langle \mathbf{x}, \boldsymbol{\varphi} \rangle = \int_0^1 \mathbf{x}^*(t)\,d\boldsymbol{\varphi}(t) = \sum_{j=1}^n \int_0^1 x_j(t)\,d\varphi_j(t).$$

Let us mention that for every fixed $\boldsymbol{\varphi} \in NBV_n$ by $\langle \mathbf{x}, \boldsymbol{\varphi} \rangle$ a bounded linear functional on BV_n is defined. In fact, we have by 4.27

$$|\langle \mathbf{x}, \boldsymbol{\varphi} \rangle| \leq \left| \int_0^1 \mathbf{x}^*(t)\,d\boldsymbol{\varphi}(t) \right| \leq \big(\sup_{t \in [0,1]} |\mathbf{x}(t)| \big) (\operatorname{var}_0^1 \boldsymbol{\varphi}) = (\operatorname{var}_0^1 \boldsymbol{\varphi})\, \|\mathbf{x}\|_{BV_n}$$

for every $\mathbf{x} \in BV_n$ and $\boldsymbol{\varphi} \in NBV_n$.

The space BV_n has important subspaces called the *Sobolev spaces* W_n^p $(1 \leq p < \infty)$ including in particular the space AC_n of absolutely continuous functions on $[0,1]$.

5.10. Definition. Given a real number p, $1 \leq p < \infty$, W_n^p denotes the space of all absolutely continuous functions $\mathbf{x}\colon [0, 1] \rightarrow R_n$ whose derivatives $\mathbf{x}'$ are L^p-integrable on $[0, 1]$. Furthermore,

$$\|\mathbf{x}\|_{W_n^p} = |\mathbf{x}(0)| + \left(\int_0^1 |\mathbf{x}'(t)|^p\,dt \right)^{1/p} = |\mathbf{x}(0)| + \|\mathbf{x}'\|_{L^p} \qquad \text{for any} \quad \mathbf{x} \in W_n^p.$$

($W_1^p = W^p$ and instead of $\|.\|_{W_n^p}$ we write $\|.\|_{W^p}$.)

5.11. Remark. Evidently, any W_n^p $(p \in R, p \geq 1)$ equipped with the norm $\|.\|_{W^p}$ is a linear normed space.

5.12. Remark. It is well-known that any $\mathbf{x} \in BV_n$ possesses a.e. on $[0, 1]$ a derivative $\mathbf{x}'(t)$ which is L-integrable on $[0, 1]$ $(\mathbf{x}' \in L_n^1)$. Furthermore, $\mathbf{x} \in AC_n$ if and only if there is $\mathbf{z} \in L_n^1$ such that

$$\mathbf{x}(t) = \mathbf{x}(0) + \int_0^t \mathbf{z}(\tau)\,d\tau \qquad \text{on } [0, 1],$$

i.e. $W_n^1 = AC_n$. Given $\mathbf{x} \in AC_n$, we have $\operatorname{var}_0^1 \mathbf{x} = \|\mathbf{x}'\|_{L^1}$ and therefore also the norms $\|.\|_{AC}$ and $\|.\|_{W^1}$ are identical (cf. e.g. Natanson [1]).

5.13. Proposition. *Given $p \in R$, $p \geq 1$, the space W_n^p is isometrically isomorphic with the product space $L_n^q \times R_n$ and its dual space is isometrically isomorphic with $L_n^q \times R_n^*$, where $q = p/(p-1)$ if $p > 1$ and $q = \infty$ if $p = 1$.*

Proof. (a) The mapping $\mathbf{x} \in W_n^p \to (\mathbf{x}', \mathbf{x}(0)) \in L_n^p \times R_n$ and its inverse $(\mathbf{z}, \mathbf{c}) \in L_n^p \times R_n \to \mathbf{x}(t) = \mathbf{c} + \int_0^t \mathbf{z}(\tau)\,d\tau \in W_n^p$ establish an isometrical isomorphism between W_n^p and $L_n^p \times R_n$.

(b) Let f be an arbitrary linear bounded functional on W_n^p and let us put for any $\mathbf{c} \in R_n$ and $\mathbf{z} \in L_n^p$ $f_1(\mathbf{z}) = f(\Psi \mathbf{z})$ and $f_2(\mathbf{c}) = f(\Phi \mathbf{c})$, where

$$\Psi\colon \mathbf{z} \in L_n^p \to \int_0^t \mathbf{z}(\tau)\,d\tau \in W_n^p, \qquad \Phi\colon \mathbf{c} \in R_n \to \mathbf{u}(t) \equiv \mathbf{c} \in W_n^p.$$

Then f_1 and f_2 are linear bounded functionals on L_n^p and R_n, respectively, while $f(\mathbf{x}) = f(\Psi \mathbf{x}' + \Phi \mathbf{x}(0)) = f_1(\mathbf{x}') + f_2(\mathbf{x}(0))$ for any $\mathbf{x} \in W_n^p$. Consequently, given $f \in (W_n^p)^*$, there exist uniquely determined $\mathbf{y}^* \in L_n^q$ ($q = p/(p-1)$ if $p > 1$, $q = \infty$ if $p = 1$) and $\boldsymbol{\lambda}^* \in R_n^*$ such that (cf. 3.10)

$$f(\mathbf{x}) = \int_0^1 \mathbf{y}^*(t)\,\mathbf{x}'(t)\,dt + \boldsymbol{\lambda}^*\,\mathbf{x}(0) \qquad \text{for any} \quad \mathbf{x} \in W_n^p.$$

Furthermore,

$$\|f_1\| = \sup_{\|\mathbf{z}\|_{L^p}=1} |f_1(\mathbf{z})| = \|\mathbf{y}^*\|_{L^q}, \qquad \|f_2\| = \sup_{|\mathbf{c}|=1} |f_2(\mathbf{c})| = |\boldsymbol{\lambda}^*|$$

and hence

$$\|f\| = \sup_{\|\mathbf{x}\|_{W^p}=1} |f(\mathbf{x})| = \|(\mathbf{y}^*, \boldsymbol{\lambda}^*)\|_{L^q \times R} = \|\mathbf{y}^*\|_{L^q} + |\boldsymbol{\lambda}^*|.$$

5.14. Remark. In accordance with 3.6 we denote for $\mathbf{x} \in W_n^p$, $\mathbf{y}^* \in L_n^q$ and $\boldsymbol{\lambda}^* \in R_n^*$

$$\langle \mathbf{x}, (\mathbf{y}^*, \boldsymbol{\lambda}^*) \rangle_W = \langle \mathbf{x}', \mathbf{y}^* \rangle_L + \boldsymbol{\lambda}^*\,\mathbf{x}(0) = \int_0^1 \mathbf{y}^*(t)\,\mathbf{x}'(t)\,dt + \boldsymbol{\lambda}^*\,\mathbf{x}(0).$$

Let us notice that $\mathbf{x} \in W_n^p \to \langle \mathbf{x}, (\mathbf{y}^*, \boldsymbol{\lambda}^*) \rangle_W$ is the zero functional on W_n^p if and only if $\mathbf{y}^*(t) = \mathbf{0}$ a.e. on $[0,1]$ and $\boldsymbol{\lambda}^* = \mathbf{0}$. As a consequence we have

5.15. Proposition. *If $\mathbf{y}^* \in L_n^q$ and $\boldsymbol{\lambda}^* \in R_n^*$, then*

$$\int_0^1 \mathbf{y}^*(t)\,\mathbf{x}'(t)\,dt + \boldsymbol{\lambda}^*\,\mathbf{x}(0) = 0 \qquad \text{for any} \quad \mathbf{x} \in W_n^p$$

or

$$\int_0^1 \mathbf{y}^*(t)\,\mathbf{z}(t)\,dt + \boldsymbol{\lambda}^*\mathbf{c} = 0 \qquad \text{for any} \quad \mathbf{z} \in L_n^p \quad \text{and} \quad \mathbf{c} \in R_n$$

if and only if $\mathbf{y}^*(t) = \mathbf{0}$ a.e. on $[0,1]$ and $\boldsymbol{\lambda}^* = \mathbf{0}$.

5.16. Proposition. $\mathbf{S} \in B(W_n^p, R_m)$ *if and only if there exist an* $m \times n$*-matrix* $\mathbf{M}$ *and an* $m \times n$*-matrix valued function* $\mathbf{K}$ *with* $\|\mathbf{K}\|_{L^q} < \infty$ $(q = p/(p-1)$ *if* $p > 1$, $q = \infty$ *if* $p = 1)$ *such that*

$$\mathbf{Sx} = \mathbf{M}\,\mathbf{x}(0) + \int_0^1 \mathbf{K}(t)\,\mathbf{x}'(t)\,\mathrm{d}t \qquad \text{for any} \quad \mathbf{x} \in W_n^p .$$

5.17 Lemma. *Let* $f \in BV$ *be right-continuous on* $[0, 1)$ *and left-continuous at* 1 *and* $f(1) = 0$. *Then*

$$\int_0^1 x(s)\,\mathrm{d}f(s) = 0 \qquad \text{for any} \quad x \in W_n^p \quad \text{with} \quad x(0) = x(1) = 0$$

if and only if $f(t) \equiv 0$ *on* $[0, 1]$.

Proof. Let us assume that $f(t) \not\equiv 0$ on $[0, 1]$, e.g. let $f(t_0) \neq 0$. Then $\operatorname{var}_0^1 f \geq |f(1) - f(t_0)| = |f(t_0)| > 0$. Let $\varepsilon > 0$ be such that $\alpha = \operatorname{var}_0^1 f \geq 3\varepsilon > 0$. By the definition of a variation there exists a subdivision $\{0 = t_0 < t_1 < \ldots < t_m = 1\}$ of $[0, 1]$ such that

$$\sum_\sigma |\Delta f| = \sum_{j=1}^q |f(s_j) - f(s_{j-1})| > \alpha - \varepsilon$$

for any of its refinements $\sigma = \{0 = s_0 < s_1 < \ldots < s_q = 1\}$. In virtue of the one-sided continuity of f there exist $\tau_j \in (0, 1)$ $(j = 1, 2, \ldots, m)$ such that $0 < \tau_0 < < \tau_1 < \ldots < \tau_m < 1$, $t_{j-1} < \tau_{j-1} < t_j$ $(j = 1, 2, \ldots, m-1)$, $t_{m-1} < \tau_{m-1} < \tau_m < t_m = 1$ and

$$\sum_{j=0}^{m-1} |f(t_j) - f(\tau_j)| + |f(1) - f(\tau_m)| \leq \sum_{j=0}^{m-1} \operatorname{var}_{t_j}^{\tau_j} f + \operatorname{var}_{\tau_m}^1 f < \varepsilon .$$

Putting $x(0) = 0$, $x(t) = \operatorname{sign}(f(t_j) - f(\tau_{j-1}))$ for $t \in [\tau_{j-1}, t_j]$ $(j = 1, 2, \ldots, m-1)$, $x(t) = \operatorname{sign}(f(\tau_m) - f(\tau_{m-1}))$ for $t \in [\tau_{m-1}, \tau_m]$, $x(1) = 0$ and extending the definition of x to the whole $[0, 1]$ in such a way that x is linear on the rest of $[0, 1]$, we obtain

$$\left|\sum_{j=0}^{m-1} \int_{t_j}^{\tau_j} x(s)\,\mathrm{d}[f(s)] + \int_{\tau_m}^1 x(s)\,\mathrm{d}[f(s)]\right| \leq \sum_{j=0}^{m-1} \operatorname{var}_{t_j}^{\tau_j} f + \operatorname{var}_{\tau_m}^1 f < \varepsilon .$$

Hence

$$\left|\int_0^1 x(s)\,\mathrm{d}[f(s)]\right|$$

$$= \left|\sum_{j=1}^{m-1} |f(t_j) - f(\tau_{j-1})| + |f(\tau_m) - f(\tau_{m-1})| + \sum_{j=0}^{m-1} \int_{t_j}^{\tau_j} x(s)\,\mathrm{d}[f(s)] + \int_{\tau_m}^1 x(s)\,\mathrm{d}[f(s)]\right|$$

$$> \sum_{j=1}^{m-1} |f(t_j) - f(\tau_{j-1})| + |f(\tau_m) - f(\tau_{m-1})| - \varepsilon > \sum_\sigma |\Delta f| - 2\varepsilon > \alpha - 3\varepsilon > 0 ,$$

where $\sigma = \{0 = t_0 < \tau_0 < t_1 < \tau_1 < \ldots < t_{m-1} < \tau_{m-1} < \tau_m < t_m = 1\}$. Since obviously $x \in W^p$ and $x(0) = x(1) = 0$, this completes the proof.

6. Variation of functions of two variables

Various definitions of the variation of functions of two or more variables are known. In our considerations we use one of them, the so called *Vitali variation*. This section is devoted to the definition of this sort of variation for functions of two variables and to the fundamental properties of functions with finite variation in this sense.

Let a nondegenerate interval $I = [a, b] \times [c, d] \subset R_2$ be given. We consider a real function $k\colon I \to R$ defined on I.

For a given subinterval $J = [a', b'] \times [c', d'] \subset I$, $a \leq a' \leq b' \leq b$, $c \leq c' \leq d' \leq d$ we set

$$m_k(J) = k(b', d') - k(b', c') - k(a', d') + k(a', c'). \tag{6,1}$$

Let us define

$$\mathrm{v}_I(k) = \sup \sum_i |m_k(J_i)|, \tag{6,2}$$

where the supremum is taken over all finite systems of nonoverlapping intervals $J_i \subset I$ (i.e. for the interiors J_i^0 of the intervals J_i we assume that $J_i^0 \cap J_j^0 = \emptyset$ whenever $i \neq j$).

6.1. Definition (Vitali). The real function $k\colon I \to R$ is of bounded variation on I if $\mathrm{v}_I(k) < +\infty$.

6.2. Remark. If on the interval $I = [a, b] \times [c, d]$ an $n \times n$-matrix $\mathbf{K}(s, t) = (k_{ij}(s, t))$ $(i, j = 1, \ldots, r)$ is given, i.e. $\mathbf{K}\colon I \to L(R_n)$, then we can set

$$m_{\mathbf{K}}(J) = \mathbf{K}(b', d') - \mathbf{K}(b', c') - \mathbf{K}(a', d') + \mathbf{K}(a', c')$$

as above and define the number $\mathrm{v}_I(\mathbf{K}) = \sup \sum_i |m_{\mathbf{K}}(J_i)|$ in the same way as in (6,2) where the norm in the sum on the right-hand side is some norm of an $n \times n$-matrix (cf. 1.1). For the case of the norm defined in 1.1 we have evidently $\mathrm{v}_I(k_{ij}) \leq \mathrm{v}_I(\mathbf{K})$ for all $i, j = 1, 2, \ldots, n$.

6.3. Remark. Assume that $a = \alpha_0 < \alpha_1 < \ldots < \alpha_k = b$, $c = \gamma_0 < \gamma_1 < \ldots < \gamma_l = d$ are some finite subdivisions of the intervals $[a, b], [c, d]$ respectively. The finite system of subintervals

$$J_{ij} = [\alpha_{i-1}, \alpha_i] \times [\gamma_{j-1}, \gamma_j], \qquad i = 1, \ldots, k, \quad j = 1, \ldots, l$$

is called a net-type subdivision of the interval $I = [a, b] \times [c, d]$. Evidently every net-type subdivision of I is a finite system of nonoverlapping intervals.

It is easy to see that for every finite system of nonoverlapping intervals $J_i \subset I$ there is a net-type subdivision of I such that every J_i is the union of some of its elements. Using this fact it is not difficult to show that for the definition of $\mathrm{v}_I(k)$

from (6,2) the supremum can be taken over all finite net-type subdivisions and the number $v_I(k)$ remains unchanged.

6.4. Examples. Assume that $f \in BV[a, b]$, $g \in BV[c, d]$. Then for $k(s, t) = f(s)\,g(t)$: $[a, b] \times [c, d] \to R$ we have by definition

$$v_{[a,b]\times[c,d]}(k) = \operatorname{var}_a^b f \operatorname{var}_c^d g < \infty .$$

Let us set

$$h(s, t) = 0 \quad \text{for} \quad 0 \le t < s \le 1 , \qquad h(s, t) = 1 \quad \text{for} \quad 0 \le s \le t \le 1 .$$

Then for every net-type subdivision $J_{ij} = [\alpha_{i-1}, \alpha_i] \times [\alpha_{j-1}, \alpha_j]$, $i, j = 1, \ldots, k$, $0 = \alpha_0 < \alpha_1 < \ldots < \alpha_k = 1$ we have

$$\sum_{i,j=1}^{k} |m_h(J_{i,j})| = \sum_{i=1}^{k} |m_h(J_{i,i})| + \sum_{i=2}^{k} |m_h(J_{i,j-1})| = 2k - 1$$

since $m_h(J_{i,i}) = 1$, $m_h(J_{i,i-1}) = 1$ and $m_h(J_{i,j}) = 0$ if $j \neq i, i-1$. Hence $v_{[0,1]\times[0,1]}(h)$ cannot be finite.

The following lemma can be easily verified.

6.5. Lemma. *If $I_j \subset I \subset R_2$, $j = 1, \ldots, m$ is a finite system of nonoverlapping intervals in I and k: $I \to R$, then*

(6,3)
$$\sum_{j=1}^{m} v_{I_j}(k) \le v_I(k) .$$

6.6. Lemma. *Let k: $I = [a, b] \times [c, d] \to R$ be given such that $v_I(k) < \infty$, $\operatorname{var}_a^b k(., \gamma_0) < \infty$ for some $\gamma_0 \in [c, d]$, i.e. $k(., \gamma_0) \in BV[a, b]$ for some $\gamma_0 \in [c, d]$. Then $k(., \gamma) \in BV[a, b]$ for all $\gamma \in [c, d]$ and*

(6,4)
$$\operatorname{var}_a^b k(., \gamma) \le v_I(k) + \operatorname{var}_a^b k(., \gamma_0) .$$

If k: $[a, b] \times [c, d] \to R$ and $\gamma \in [c, d]$ is fixed, then we denote the usual variation of the function $k(s, \gamma)$ in the interval $[a, b]$ by $\operatorname{var}_a^b k(., \gamma)$. Similarly for $\operatorname{var}_c^d k(\alpha, .)$ where $\alpha \in [a, b]$ is fixed.

Proof. For any $\gamma, \gamma_0 \in [c, d]$, $\alpha_{j-1}, \alpha_j \in [a, b]$ we have

$$|k(\alpha_j, \gamma) - k(\alpha_{j-1}, \gamma)| \le |m_{J_j}(k)| + |k(\alpha_j, \gamma_0) - k(\alpha_{j-1}, \gamma_0)|$$

where $J_j = [\alpha_{j-1}, \alpha_j] \times [\gamma_0, \gamma]$. Hence for each finite decomposition $a = \alpha_0 < \alpha_1 < \ldots < \alpha_k = b$ we have

$$\sum_{j=1}^{k} |k(\alpha_j, \gamma) - k(\alpha_{j-1}, \gamma)|$$

$$\le \sum_{j=1}^{k} |m_{J_j}(k)| + \sum_{j=1}^{k} |k(\alpha_j, \gamma_0) - k(\alpha_{j-1}, \gamma_0)| \le v_I(k) + \operatorname{var}_a^b k(., \gamma_0)$$

and this inequality implies (6,4).

For a given $k: I \to R$, $I = [a,b] \times [c,d]$ we put

$$(6{,}5) \qquad \omega_1(a) = 0, \qquad \omega_1(\sigma) = \mathrm{v}_{[a,\sigma]\times[c,d]}(k) \qquad \text{for} \quad \sigma \in (a,b]$$

and similarly

$$(6{,}6) \qquad \omega_2(c) = 0, \qquad \omega_2(\tau) = \mathrm{v}_{[a,b]\times[c,\tau]}(k) \qquad \text{for} \quad \tau \in (c,d].$$

6.7. Lemma. *The function $\omega_1: [a,b] \to R$ from (6,5) is nondecreasing on $[a,b]$, $\omega_1(b) = \mathrm{v}_I(k)$; hence $\omega_1 \in BV[a,b]$ if $\mathrm{v}_I(k) < +\infty$. Similarly for the function $\omega_2: [c,d] \to R$ from (6,6).*

The proof follows easily from the definitions.

6.8. Lemma. *If $k: I \to R$, $I = [a,b] \times [c,d]$, $\mathrm{v}_I(k) < \infty$ and $\mathrm{var}_a^b\, k(.,c) < \infty$, then the set of discontinuity points of k in the first variable s lies on a denumerable system of lines in I, which are parallel to the t-axis.*

Proof. For any $s, s_0 \in [a,b]$, $t \in [c,d]$ we have

$$\begin{aligned} |k(s,t) - k(s_0,t)| &\le |k(s,t) - k(s,c) - k(s_0,t) + k(s_0,c)| + |k(s,c) - k(s_0,c)| \\ &\le |\omega_1(s) - \omega_1(s_0)| + |\mathrm{var}_a^s\, k(.,c) - \mathrm{var}_a^{s_0}\, k(.,c)| \end{aligned}$$

where $\omega_1: [a,b] \to R$ is given by (6,5). Since $\omega_1 \in BV[a,b]$ by 6.7 and the function $\mathrm{var}_a^s\, k(.,c)$ is also of bounded variation on $[a,b]$, the above inequality gives that there exists an at most denumerable set of points $M \subset [a,b]$ such that $\lim_{s \to s_0} k(s,t) = k(s_0,t)$ whenever $s_0 \in [a,b] \setminus M$ and $t \in [c,d]$ are arbitrary. This yields our proposition.

6.9. Lemma. *If $k: I \to R$, $\mathrm{v}_I(k) < \infty$, $\mathrm{var}_a^b\, k(.,c) < \infty$, $\mathrm{var}_c^d\, k(a,.) < \infty$, then the set of discontinuities of k in $I = [a,b] \times [c,d]$ lies on a denumerable set of lines in I parallel to the coordinate axes.*

This proposition is proved in Hildebrandt [1], III.5.4. If $k(s,t)$ satisfies the assumptions of 6.8 then $h(s,t) = k(s,t) - k(a,t)$ satisfies the assumptions of 6.9 and 6.8 is a corollary of 6.9.

6.10. Lemma. *If $k: I \to R$, $I = [a,b] \times [c,d]$, $\mathrm{v}_I(k) < +\infty$, then for an arbitrary subdivision $c = \gamma_0 < \gamma_1 < \ldots < \gamma_l = d$ and any two points $s_1, s_2 \in [a,b]$ we have*

$$\left| \sum_{j=1}^{l} [\mathrm{var}_a^{s_2} (k(.,\gamma_j) - k(.,\gamma_{j-1})) - \mathrm{var}_a^{s_1} (k(.,\gamma_j) - k(.,\gamma_{j-1}))] \right| \le |\omega_1(s_2) - \omega_1(s_1)|$$

where $\omega_1: [a,b] \to R$ is defined by (6,5).

Proof. Let us set $h(s,t) = k(s,t) - k(s,c)$ for $(s,t) \in I$. Then $h(s,c) = 0$ for any $s \in [a,b]$ and by 6.6 $\mathrm{var}_a^b\, h(.,t) < \infty$ for any $t \in [c,d]$ because evidently $\mathrm{v}_I(h) < \infty$. Hence $\mathrm{var}_a^s\, h(.,t)$ is finite for any $s \in [a,b]$, $t \in [c,d]$. For any $j = 1, \ldots, l$ we have

$h(s, \gamma_j) - h(s, \gamma_{j-1}) = k(s, \gamma_j) - k(s, \gamma_{j-1})$ and $\operatorname{var}_a^s (k(., \gamma_j) - k(., \gamma_{j-1}))$ is also finite for every $s \in [0, 1]$. This implies that for any $j = 1, \ldots, l$ we have

$$\begin{aligned} &|\operatorname{var}_a^{s_2} (k(., \gamma_j) - k(., \gamma_{j-1})) - \operatorname{var}_a^{s_1} (k(., \gamma_j) - k(., \gamma_{j-1}))| \\ &\leq |\operatorname{var}_{s_1}^{s_2} (k(., \gamma_j) - k(., \gamma_{j-1}))| \leq \mathrm{v}_{[s_1,s_2] \times [\gamma_{j-1},\gamma_j]}(k). \end{aligned}$$

By 6.5 we obtain the inequality

$$\begin{aligned} &\sum_{j=1}^{l} \mathrm{v}_{[s_1,s_2] \times [\gamma_{j-1},\gamma_j]}(k) \leq \mathrm{v}_{[s_1,s_2] \times [c,d]}(k) \\ &\leq |\mathrm{v}_{[a,s_2] \times [c,d]}(k) - \mathrm{v}_{[a,s_1] \times [c,d]}(k)| = |\omega_1(s_2) - \omega_1(s_1)| \end{aligned}$$

which yields our result.

6.11. Lemma. *If* $k: I \to R$, $I = [a, b] \times [c, d]$, $\mathrm{v}_I(k) < \infty$ *and for some* $s_0 \in [a, b]$ *the relation*

(6,7) $$\lim_{s \to s_0 \pm} |k(s, t) - k(s_0, t)| = 0$$

holds for all $t \in [c, d]$, *then*

(6,8) $$\lim_{s \to s_0 \pm} \omega_1(s) = \omega_1(s_0)$$

where $\omega_1: [a, b] \to R$ *is defined by* (6.5).

This is proved in Schwabik [2], Lemma 2.1.

6.12. Remark. If for $k: I \to R$ we have $\mathrm{v}_I(k) < \infty$ and $\operatorname{var}_a^b k(., c) < \infty$, then by 6.8 the relation (6,7) is satisfied for all $s_0 \in [a, b]$ except for a denumerable set of points in $[a, b]$. Moreover, in this case $k(., t) \in BV[a, b]$ for every $t \in [c, d]$ (cf. 6.6). Hence by the elementary properties of functions of bounded variation the onesided limits $\lim_{\sigma \to s_0+} k(\sigma, t) = k(s_0+, t)$, $\lim_{\sigma \to s_0-} k(\sigma, t) = k(s_0-, t)$ exist for every $s_0 \in [a, b)$, $s_0 \in (a, b]$, respectively, and for every $t \in [c, d]$.

6.13. Lemma. *If* $k: I \to R$ $(I = [a, b] \times [c, d])$ *is given, then for every* $s_1, s_2 \in [a, b]$ *we have*

(6,9) $$\operatorname{var}_c^d (k(s_2, .) - k(s_1, .)) \leq |\omega_1(s_2) - \omega_1(s_1)|$$

where $\omega_1: [a, b] \to R$ *is defined in* (6,5).

Proof. For an arbitrary subdivision $c = \gamma_0 < \gamma_1 < \ldots < \gamma_l = d$ we have by 6.5

$$\begin{aligned} &\sum_{j=1}^{l} |k(s_2, \gamma_j) - k(s_1, \gamma_j) - k(s_2, \gamma_{j-1}) + k(s_1, \gamma_{j-1})| \\ &\leq \mathrm{v}_{[s_1,s_2] \times [c,d]}(k) \leq |\mathrm{v}_{[a,s_2] \times [c,d]}(k) - \mathrm{v}_{[a,s_1] \times [c,d]}(k)| = |\omega_1(s_2) - \omega_1(s_1)| \end{aligned}$$

and proceeding to the supremum for all finite subdivisions of $[c, d]$ we obtain (6,9).

6.14. Lemma. *Assume that* $k\colon I \to R$ $(I = [a,b] \times [c,d])$ *is given with* $\mathrm{v}_I(k) < \infty$ *and for some* $s_0 \in [a,b)$ *the limit*

$$\lim_{s \to s_0+} k(s,t) = k(s_0+, t) \tag{6,10}$$

exists for every $t \in [c,d]$. *Then*

$$\lim_{\delta \to 0+} \operatorname{var}_c^d (k(s_0+\delta, .) - k(s_0+, .)) = 0 .$$

Proof. Define $k^0\colon I \to R$ such that $k^0(s,t) = k(s,t)$ if $(s,t) \in I$, $s \neq s_0$ and $k^0(s_0,t) = k(s_0+, t)$. Since $\operatorname{var}_c^d (k(s_0+, .) - k(s_0, .)) < \infty$ we obtain $\mathrm{v}_I(k^0) < \infty$. Let $\omega_1^0\colon [a,b] \to R$, $\omega_1^0(a) = 0$, $\omega_1^0(\sigma) = \mathrm{v}_{[a,\sigma] \times [c,d]}(k^0)$ for $\sigma \in (a,b]$. Since $\lim_{s \to s_0+} (k^0(s,t) - k^0(s_0,t)) = 0$ for every $t \in [c,d]$, we have by 6.11 $\lim_{s \to s_0+} \omega_1^0(s) = \omega_1^0(s_0)$. For every $\delta > 0$ such that $s_0 + \delta \in [a,b]$ we have by 6.13

$$\begin{aligned}\operatorname{var}_c^d (k^0(s_0+\delta, .) - k^0(s_0, .)) &= \operatorname{var}_c^d (k(s_0+\delta, .) - k(s_0+, .)) \\ &\leq |\omega_1^0(s_0+\delta) - \omega_1^0(s_0)| .\end{aligned}$$

The limitation process $\delta \to 0+$ yields our result.

6.15. Corollary. *If* $k\colon I \to R$ $(I = [a,b] \times [c,d])$ *is such that* $\mathrm{v}_I(k) < \infty$ *and* $\operatorname{var}_a^b k(., c) < \infty$, *then for any* $s_0 \in [a,b)$ *we have*

$$\operatorname{var}_c^d (k(s_0+, .) - k(s_0, .)) \leq \omega_1(s_0+) - \omega_1(s_0)$$

where $\omega_1\colon [a,b] \to R$ *is given by* (6,5).

Proof. The assumptions assure by 6.6 that $\operatorname{var}_a^b k(., t) < \infty$ for every $t \in [c,d]$ and consequently the limit $\lim_{s \to s_0+} k(s,t) = k(s_0+, t)$ exist for every $t \in [c,d]$. The statement follows immediately from 6.13.

6.16. Corollary. *If* $k\colon I \to R$, $\mathrm{v}_I(k) < \infty$, $\operatorname{var}_a^b k(., c) < \infty$, *then for any* $s_0 \in [a,b)$ *we have*

$$\lim_{\delta \to 0+} \sup_{t \in [c,d]} |k(s_0+\delta, t) - k(s_0+, t)| = 0 ,$$

i.e.

$$\lim_{\delta \to 0+} k(s_0+\delta, t) = k(s_0+, t) \quad \textit{uniformly in } [c,d] .$$

Proof. For any $t \in [c,d]$ we have evidently

$$|k(s_0+\delta, t) - k(s_0+, t)| \leq |k(s_0+\delta, c) - k(s_0+, c)| + \operatorname{var}_c^d (k(s_0+\delta, .) - k(s_0+, .))$$

and our result follows immediately from the fact that $\lim_{s \to s_0+} k(s,c) = k(s_0+, c)$ exists and from 6.14.

6.17. Remark. It is easy to see that the statements from 6.14, 6.15 and 6.16 are also reformulable for the case of left-hand limits.

Further it is clear that 6.4–6.16 are also valid if the real function $k\colon I \to R$ is replaced by a matrix valued function $\boldsymbol{K}(s,t) = (k_{ij}(s,t))$. If some continuity properties are needed, then the usual norm of a matrix is used. Compare also 6.2.

6.18. Theorem. *Let* $k\colon I \to R$, $I = [a,b] \times [c,d]$ *be given. Let us suppose that* $v_I(k) < +\infty$ *and* $\mathrm{var}_c^d\, k(a,.) < \infty$.

If $g \in BV[c,d]$, *then the integral*

$$\int_c^d g(t)\, \mathrm{d}_t[k(s,t)] \tag{6,12}$$

exists for every $s \in [a,b]$. *For any* $s \in [a,b]$ *the inequality*

$$\left|\int_c^d g(t)\, \mathrm{d}_t[k(s,t)]\right| \le \int_c^d |g(t)|\, \mathrm{d}_t[\mathrm{var}_c^t\, k(s,.)] \le \sup_{t\in[c,d]} |g(t)|\, \mathrm{var}_c^d\, k(s,.) \tag{6,13}$$

holds and moreover

$$\mathrm{var}_a^b\left(\int_c^d g(t)\, \mathrm{d}_t[k(.,t)]\right) \le \int_c^d |g(t)|\, \mathrm{d}\omega_2(t) \le \sup_{t\in[c,d]} |g(t)|\, v_I(k) \tag{6,14}$$

where $\omega_2\colon [c,d] \to R$ *is defined by* (6,6). *Thus the integral* (6,12) *as a function of the variable s belongs to* $BV[a,b]$.

Proof. By 6.6 $k(s,.) \in BV[c,d]$ for every $s \in [a,b]$. Hence by 4.19 the integral (6,12) exists for every $s \in [a,b]$. The inequality (6,13) follows immediately from 4.27. In order to prove (6,14) we assume that an arbitrary subdivision $a = \alpha_0 < \alpha_1 < \ldots < \alpha_k = b$ of the interval $[a,b]$ is given. By 4.27 we have

$$\left|\int_c^d g(t)\, \mathrm{d}_t[k(\alpha_i,t) - k(\alpha_{i-1},t)]\right| \le \int_c^d |g(t)|\, \mathrm{d}(\mathrm{var}_c^t\,(k(\alpha_i,.) - k(\alpha_{i-1},.))).$$

Consequently

$$\sum_{i=1}^k \left|\int_c^d g(t)\, \mathrm{d}_t[k(\alpha_i,t) - k(\alpha_{i-1},t)]\right| \tag{6,15}$$

$$\le \int_c^d |g(t)|\, \mathrm{d}\left(\sum_{i=1}^k \mathrm{var}_c^t\,(k(\alpha_i,.) - k(\alpha_{i-1},.))\right).$$

Using 6.10 we obtain for all $t, \tau \in [a,b]$

$$|t-\tau|\, |g(\tau)| \left|\sum_{i=1}^k \mathrm{var}_c^t\,(k(\alpha_i,.) - k(\alpha_{i-1},.)) - \sum_{i=1}^k \mathrm{var}_c^\tau\,(k(\alpha_i,.) - k(\alpha_{i-1},.))\right|$$

$$\le (t-\tau)\, |g(\tau)|\, (\omega_2(t) - \omega_2(\tau))$$

since $\omega_2\colon [c,d] \to R$ is nondecreasing and consequently 4.26 gives the estimate

$$\int_c^d |g(t)| \,\mathrm{d}\Big(\sum_{i=1}^k \operatorname{var}_c^t (k(\alpha_i, .) - k(\alpha_{i-1}, .))\Big) \le \int_c^d |g(t)| \,\mathrm{d}\omega_2(t).$$

Since this holds for every subdivision of $[a,b]$ we get by (6,15) the inequality

$$\operatorname{var}_a^b \Big(\int_c^d g(t) \,\mathrm{d}_t[k(.,t)]\Big) \le \int_c^d |g(t)| \,\mathrm{d}\omega_2(t).$$

By 4.27 we have

$$\int_c^d |g(t)| \,\mathrm{d}\omega_2(t) \le \sup_{t\in[c,d]} |g(t)| \operatorname{var}_c^d \omega_2 = \sup_{t\in[c,d]} |g(t)| \,\mathrm{v}_I(k).$$

6.19. Corollary. *If the assumptions of* 6.18 *are satisfied, then*

$$\sup_{s\in[a,b]} \Big|\int_c^d g(t) \,\mathrm{d}_t[k(s,t)]\Big| \le \sup_{t\in[c,d]} |g(t)| (\operatorname{var}_c^d k(a, .) + \mathrm{v}_I(k)). \tag{6,16}$$

Proof. For any $s\in[a,b]$ we have by 4.27

$$\Big|\int_c^d g(t) \,\mathrm{d}_t[k(s,t)]\Big| \le \Big|\int_c^d g(t) \,\mathrm{d}_t[k(a,t)]\Big| + \operatorname{var}_a^b \Big(\int_c^d g(t) \,\mathrm{d}_t[k(.,t)]\Big)$$

$$\le \sup_{t\in[c,d]} |g(t)| \operatorname{var}_c^d k(a, .) + \operatorname{var}_a^b \Big(\int_c^d g(t) \,\mathrm{d}_t[k(.,t)]\Big).$$

(6,16) follows now easily from (6,14).

6.20. Theorem. *Let* $k\colon I = [a,b] \times [c,d] \to R$ *be given. Suppose that* $\mathrm{v}_I(k) < \infty$, $\operatorname{var}_c^d k(a, .) < \infty$ *and* $\operatorname{var}_a^b k(., c) < \infty$. *If* $f\in BV[a,b]$, $g\in BV[c,d]$, *then*

$$\int_c^d g(t) \,\mathrm{d}_t\Big(\int_a^b k(s,t) \,\mathrm{d}f(s)\Big) = \int_a^b \Big(\int_c^d g(t) \,\mathrm{d}_t[k(s,t)]\Big) \,\mathrm{d}f(s) \tag{6,17}$$

holds and the integrals on both sides of (6,17) *exist.*

Proof. By 6.18 $\int_c^d g(t) \,\mathrm{d}_t[k(.,t)] \in BV[a,b]$ and 4.19 yields the existence of the integral on the right-hand side of (6,17). By 6.6 we obtain $k(.,t)\in BV[a,b]$ for every $t\in[c,d]$ and by 4.19 also the existence of the integral $\int_a^b k(s,t)\,\mathrm{d}f(s)$ for any $t\in[c,d]$. Let $c = \gamma_0 < \gamma_1 < \ldots < \gamma_l = d$ be an arbitrary subdivision of $[c,d]$. For any $s\in[a,b]$ and $i = 1,\ldots,l$ we have

$$|k(s,\gamma_i) - k(s,\gamma_{i-1})|$$

$$\le |k(s,\gamma_i) - k(a,\gamma_i) - k(s,\gamma_{i-1}) + k(a,\gamma_{i-1})| + |k(a,\gamma_i) - k(a,\gamma_{i-1})|$$

$$\le \mathrm{v}_{[a,c]\times[\gamma_{i-1},\gamma_i]}(k) + |k(a,\gamma_i) - k(a,\gamma_{i-1})|.$$

Hence by 4.27 and 6.5

$$\sum_{i=1}^{l} \left| \int_a^b (k(s,\gamma_i) - k(s,\gamma_{i-1}))\, \mathrm{d}f(s) \right|$$

$$\leq \sum_{i=1}^{l} \left[\mathrm{v}_{[a,b]\times[\gamma_{i-1},\gamma_i]}(k) + |k(a,\gamma_i) - k(a,\gamma_{i-1})| \right] \mathrm{var}_a^b f$$

$$\leq (\mathrm{v}_I(k) + \mathrm{var}_c^d\, k(a,.))\, \mathrm{var}_a^b f < \infty\,.$$

Taking the supremum over all finite subdivisions of $[c, d]$ on the left-hand side of this inequality we obtain

$$\mathrm{var}_c^d \left(\int_a^b k(s,.)\, \mathrm{d}f(s) \right) \leq (\mathrm{v}_I(k) + \mathrm{var}_c^d\, k(a,.))\, \mathrm{var}_a^b f < \infty\,. \tag{6,18}$$

From 4.27 the existence of the integral on the left-hand side of (6,17) follows.

Let now $\alpha \in [c, d]$ and let $\psi_\alpha^+(t)$ be the simple jump function defined for $t \in [c, d]$ (see 4.20). By 4.21 we have

$$\int_c^d \psi_\alpha^+(t)\, \mathrm{d}_t[k(s,t)] = k(s,d) - k(s,\alpha+)$$

and

$$\int_a^b \left(\int_c^d \psi_\alpha^+(t)\, \mathrm{d}_t[k(s,t)] \right) \mathrm{d}f(s) = \int_a^b (k(s,d) - k(s,\alpha+))\, \mathrm{d}f(s)\,. \tag{6,19}$$

On the other hand, we have by 4.21

$$\int_c^d \psi_\alpha^+(t)\, \mathrm{d}_t \left[\int_a^b k(s,t)\, \mathrm{d}f(s) \right] = \int_a^b k(s,d)\, \mathrm{d}f(s) - \lim_{\delta\to 0+} \int_a^b k(s,\alpha+\delta)\, \mathrm{d}f(s) \tag{6,20}$$

$$= \lim_{\delta\to 0+} \int_a^b (k(s,d) - k(s,\alpha+\delta))\, \mathrm{d}f(s)\,.$$

By 4.27 we have

$$\left| \int_a^b (k(s,\alpha+) - k(s,\alpha+\delta))\, \mathrm{d}f(s) \right| \leq \sup_{s\in[a,b]} |k(s,\alpha+) - k(s,\alpha+\delta)|\, \mathrm{var}_a^b f$$

and by 6.16 we obtain

$$\lim_{\delta\to 0+} \int_a^b (k(s,\alpha+) - k(s,\alpha+\delta))\, \mathrm{d}f(s) = 0\,.$$

Hence by (6,20)

$$\int_c^d \psi_\alpha^+(t)\, \mathrm{d}_t \left[\int_a^b k(s,t)\, \mathrm{d}f(s) \right] = \int_a^b (k(s,d) - k(s,\alpha+))\, \mathrm{d}f(s)$$

and this together with (6,19) yields that for $g = \psi_\alpha^+$ the equality (6,17) is satisfied.

In the same way it can be proved that (6,17) holds if we set $g(t) = \psi_\alpha^-(t)$, where ψ_α^- is the simple jump function given by (4,13). From these facts and from the linearity of the integral it is now clear that (6,17) holds whenever $g \in BV[c, d]$ is a finite step function (cf. 4.20).

Let now $g \in BV[c, d]$. There is a sequence $g_l \in BV[c, d]$, $l = 1, 2, \ldots$ of finite step functions such that $\lim_{l \to \infty} g_l(t) = g(t)$ uniformly on $[c, d]$ (see Aumann [1], 7.3.2.1).

Since by (6,18) it is $\int_a^b k(s, .)\, \mathrm{d}f(s) \in BV[c, d]$, we have by 4.17

$$\lim_{l \to \infty} \int_c^d g_l(t)\, \mathrm{d}_t \left(\int_a^b k(s, t)\, \mathrm{d}f(s) \right) = \int_c^d g(t)\, \mathrm{d}_t \left(\int_a^b k(s, t)\, \mathrm{d}f(s) \right). \tag{6,21}$$

Further by 6.19 we obtain

$$\sup_{s \in [a,b]} \left| \int_c^d (g(t) - g_l(t))\, \mathrm{d}_t[k(s, t)] \right| \leq \sup_{t \in [c,d]} |g(t) - g_l(t)| \, (\mathrm{var}_a^d\, k(a, .) + \mathrm{v}_I(k)).$$

Hence

$$\lim_{l \to \infty} \int_c^d g_l(t)\, \mathrm{d}_t[k(s, t)] = \int_c^d g(t)\, \mathrm{d}_t[k(s, t)]$$

uniformly on $[a, b]$ and by 4.17 the relation

$$\lim_{l \to \infty} \int_a^b \left(\int_c^d g_l(t)\, \mathrm{d}_t[k(s, t)] \right) \mathrm{d}f(s) = \int_a^b \left(\int_c^d g(t)\, \mathrm{d}_t[k(s, t)] \right) \mathrm{d}f(s) \tag{6,22}$$

holds. Since g_l are finite step functions we have for any $l = 1, 2, \ldots$

$$\int_a^b \left(\int_c^d g_l(t)\, \mathrm{d}_t[k(s, t)] \right) \mathrm{d}f(s) = \int_c^d g_l(t)\, \mathrm{d}_t \left(\int_a^b k(s, t)\, \mathrm{d}f(s) \right)$$

as was shown above. Consequently, by (6,21) and (6,22) we obtain the desired equality (6,17) and the proof is complete.

6.21. Remark. If all assumptions of 6.20 are satisfied, then it can be proved that the equality

$$\int_c^d g(t)\, \mathrm{d}_t \left(\int_a^b f(s)\, \mathrm{d}_s[k(s, t)] \right) = \int_a^b f(s)\, \mathrm{d}_s \left(\int_c^d g(t)\, \mathrm{d}_t[k(s, t)] \right) \tag{6,23}$$

also holds (see Schwabik [2]).

6.22. Theorem. *Let* $\mathbf{K}(s, t)\colon I = [a, b] \times [c, d] \to L(R_n)$ *be given*, $\mathbf{K}(s, t) = (k_{ij}(s, t))$, $i, j = 1, \ldots, n$. *Suppose that* $\mathrm{v}_I(\mathbf{K}) < \infty$, $\mathrm{var}_c^d\, \mathbf{K}(a, .) < \infty$, $\mathrm{var}_a^b\, \mathbf{K}(., c) < \infty$. *If* $\mathbf{x} \in BV_n[c, d]$, $\mathbf{y} \in BV_n[a, b]$, *then the equality*

$$\int_a^b \left(\int_c^d \mathrm{d}_t[\mathbf{K}(s, t)]\, \mathbf{x}(t) \right)^* \mathrm{d}\mathbf{y}(s) = \int_c^d \mathbf{x}^*(t)\, \mathrm{d}_t \left(\int_a^b \mathbf{K}^*(s, t)\, \mathrm{d}\mathbf{y}(s) \right) \tag{6,24}$$

holds and the integrals on both sides of (6,24) *exist.*

Proof. By definition we have

$$(6,25)\qquad \int_a^b \left(\int_c^d \mathrm{d}_t[\mathbf{K}(s,t)]\,\mathbf{x}(t)\right)^* \mathrm{d}\mathbf{y}(s)$$

$$= \sum_{i=1}^n \int_a^b \left(\sum_{j=1}^n \int_c^d x_j(t)\,\mathrm{d}_t[k_{ij}(s,t)]\right) \mathrm{d}y_i(s) = \sum_{i=1}^n \sum_{j=1}^n \int_a^b \left(\int_c^d x_j(t)\,\mathrm{d}_t[k_{ij}(s,t)]\right) \mathrm{d}y_i(s).$$

Since all x_j, y_i, k_{ij}, $i,j = 1,\dots,n$ satisfy the assumptions of 6.20 we can use this theorem for the interchanging of the order of integrations in the expression (6,25). If we do this we obtain

$$\int_a^b \left(\int_c^d \mathrm{d}_t[\mathbf{K}(s,t)]\,\mathbf{x}(t)\right)^* \mathrm{d}\mathbf{y}(s) = \sum_{i=1}^n \sum_{j=1}^n \int_c^d x_j(t)\,\mathrm{d}_t\left(\int_a^b k_{ij}(s,t)\,\mathrm{d}y_i(s)\right)$$

$$= \sum_{j=1}^n \int_c^d x_j(t)\,\mathrm{d}_t\left(\int_a^b \sum_{i=1}^n k_{ij}(s,t)\,\mathrm{d}y_i(s)\right) = \int_c^d \mathbf{x}^*(t)\,\mathrm{d}_t\left(\int_a^b \mathbf{K}^*(s,t)\,\mathrm{d}\mathbf{y}(s)\right)$$

and (6,24) is proved.

6.23. Remark. A similar formulation in terms of a matrix valued function $\mathbf{K}$ and vectors $\mathbf{x}, \mathbf{y}$ can be given for the equality (6,23) from 6.21.

6.24. Remark. In this paragraph only such results on functions of bounded variation in two variables are presented which are in some manner used in the forthcomming investigations of integral equations in the space BV_n. For the reader interested in this topic we refer to further results contained in the book Hildebrandt [1], III.4. (for example Helly's Choice Theorem, Jordan decomposition, etc.).

6.25. Remark. Let $I = [a,b] \times [c,d]$ be given. Let us denote by $SBV(I)$ the set of all functions $k\colon I \to R$ such that $\mathrm{v}_I(K) < \infty$, $\mathrm{var}_a^b\, k(.,c) < \infty$, $\mathrm{var}_c^d\, k(a,.) < \infty$. $SBV(I)$ is evidently a linear set. $SBV(I)$ can be normed by setting

$$\|k\| = |k(a,c)| + \mathrm{var}_a^b\, k(.,c) + \mathrm{var}_c^d\, k(a,.) + \mathrm{v}_I(k).$$

Evidently

$$|k(s,t)| \le \|k\| \qquad \text{for every} \quad (s,t) \in I.$$

The same holds even if the functions on I are matrix valued.

7. Nonlinear operators and nonlinear operator equations in Banach spaces

This section provides the basic tools for the investigation of nonlinear boundary value problems for ordinary differential equations contained in Chapter V. The reader interested in more details concerning differential and integral calculus on

Banach spaces is referred to the monographs on functional analysis (e.g. Kantorovič, Akilov [1]).

Throughout the paragraph, X, Y and Z are Banach spaces.

7.1. Preliminaries. Given a Banach space X with the norm $\|.\|_X$, $\varrho_0 > 0$ and $\mathbf{x}_0 \in X$, $\mathfrak{B}(\mathbf{x}_0, \varrho_0; X)$ denotes the set of all $\mathbf{x} \in X$ such that $\|\mathbf{x} - \mathbf{x}_0\|_X \le \varrho_0$.

Let $\mathbf{F}$ be an operator acting from X into Y and defined on $D \subset X$ ($\mathbf{F}$: $D \to Y$). $\mathbf{F}$ is *lipschitzian* on $D_0 \subset D$ if there exists a real number λ, $0 \le \lambda < \infty$, such that

$$\|\mathbf{F}(\mathbf{x}') - \mathbf{F}(\mathbf{x}'')\|_Y \le \lambda \|\mathbf{x}' - \mathbf{x}''\|_X$$

for all $\mathbf{x}', \mathbf{x}'' \in D_0$. If $\lambda < 1$, $\mathbf{F}$ is said to be *contractive* on D_0.

The operator $\mathbf{F}$: $D \subset X \times Z \to Y$ is said to be *locally lipschitzian* on $D_0 \subset D$ near $\mathbf{z} = \mathbf{z}_0$ if for any $\mathbf{x}_0 \in D_0$ there exist $\varrho_0 > 0$, $\sigma_0 > 0$ and $\lambda \ge 0$ such that $\mathbf{x}', \mathbf{x}'' \in \mathfrak{B}(\mathbf{x}_0, \varrho_0; X)$ and $\mathbf{z} \in \mathfrak{B}(\mathbf{z}_0, \sigma_0; Z)$ implies $(\mathbf{x}', \mathbf{z}) \in D$, $(\mathbf{x}'', \mathbf{z}) \in D$ and

$$\|\mathbf{F}(\mathbf{x}', \mathbf{z}) - \mathbf{F}(\mathbf{x}'', \mathbf{z})\|_Y \le \lambda \|\mathbf{x}' - \mathbf{x}''\|_X .$$

7.2. Gâteaux derivative. The operator $\mathbf{F}$ acting from X into Y and defined on $D \subset X$ is *Gâteaux differentiable at* $\mathbf{x}_0 \in D$ if there exists a bounded linear operator $\mathbf{G} \in B(X, Y)$ such that for any $\xi \in X$

$$\lim_{\vartheta \to 0} \left\| \frac{\mathbf{F}(\mathbf{x}_0 + \vartheta\xi) - \mathbf{F}(\mathbf{x}_0)}{\vartheta} - \mathbf{G}\xi \right\|_Y = 0 .$$

$\mathbf{G}$ is the *Gâteaux derivative* of $\mathbf{F}$ at $\mathbf{x} = \mathbf{x}_0$ and is denoted by $\mathbf{G} = \mathbf{F}'(\mathbf{x}_0)$. If $\mathbf{F}'(\mathbf{x})$ exists for all $\mathbf{x} \in D'$, where $D' \subset D$ is an open subset in X, and the mapping

$$\mathbf{F}': \mathbf{x} \in D_0 \to \mathbf{F}'(\mathbf{x}) \in B(X, Y)$$

possesses the Gâteaux derivative $\mathbf{H} \in B(X, B(X, Y))$ at $\mathbf{x} = \mathbf{x}_0 \in D_0$, $\mathbf{H}$ is said to be the *second order Gâteaux derivative* of $\mathbf{F}$ at $\mathbf{x} = \mathbf{x}_0$ and $\mathbf{H} = \mathbf{F}''(\mathbf{x}_0)$.

In general, if $\mathbf{H}$ is the *k-th order Gâteaux derivative* of $\mathbf{F}$ on $D_0 \subset D \subset X$ and $\mathbf{L}$ is the Gâteaux derivative of $\mathbf{H}$ at $\mathbf{x} = \mathbf{x}_0 \in D_0$, then $\mathbf{L}$ is the $(k+1)$-th order Gâteaux derivative of $\mathbf{F}$ at $\mathbf{x}_0$ and $\mathbf{L} = \mathbf{F}^{(k+1)}(\mathbf{x}_0)$.

Let $X_1, X_2, \ldots, X_n$ be Banach spaces. Let $\mathbf{F}$: $(\mathbf{x}_1, \mathbf{x}_2, \ldots, \mathbf{x}_n) \to \mathbf{F}(\mathbf{x}_1, \mathbf{x}_2, \ldots, \mathbf{x}_n) \in Y$ be an operator from the product space $\Xi = X_1 \times X_2 \times \ldots \times X_n$ into a Banach space Y. The derivative of $\mathbf{F}$ at a point $\mathbf{x} = (\mathbf{x}_1, \mathbf{x}_2, \ldots, \mathbf{x}_n)$ with respect to the j-th variable (i.e. if we fix the other variables and $\mathbf{F}$ is considered as an operator from X_j into Y) is denoted by $\mathbf{F}_j'(\mathbf{x})$ or $\mathbf{F}_{\mathbf{x}_j}'(\mathbf{x})$. ($\mathbf{F}'(\mathbf{x})$ is defined and continuous on the open subset $D \subset \Xi$ if and only if $\mathbf{F}_j'(\mathbf{x})$ $(j = 1, 2, \ldots, n)$ are defined and continuous on D. Then for any $\mathbf{x} \in \Xi$ and $\xi = (\xi_1, \xi_2, \ldots, \xi_n) \in \Xi$

$$[\mathbf{F}'(\mathbf{x})]\, \xi = \sum_{j=1}^{n} [\mathbf{F}_j'(\mathbf{x})]\, \xi_j .) .$$

If on $D \subset \Xi$ $\mathbf{F}$ possesses all the derivatives $\mathbf{F}_j^{(p_j)}(\mathbf{x})$ $(j = 1, 2, \ldots, n)$ which are continuous in $\mathbf{x}$ on D, we shall write $\mathbf{F} \in C^{p_1, p_2, \ldots, p_n}(D)$. If $\mathbf{F}$ is continuous on D, we shall write $\mathbf{F} \in C(D)$.

Let us summarize some basic properties of the Gâteaux derivative.

(i) *Any linear mapping* $\mathbf{A} \in B(X, Y)$ *is Gâteaux differentiable on* X *and* $\mathbf{A}'(\mathbf{x}) = \mathbf{A}$ *for any* $\mathbf{x} \in X$.

(ii) *If the operators* $\mathbf{F}_1, \mathbf{F}_2 \colon X \to Y$ *are Gâteaux differentiable at* $\mathbf{x}_0 \in X$ *and* $\alpha_1, \alpha_2 \in R$, *then also* $\alpha_1 \mathbf{F}_1 + \alpha_2 \mathbf{F}_2$ *is Gâteaux differentiable at* $\mathbf{x}_0$ *and*

$$(\alpha_1 \mathbf{F}_1 + \alpha_2 \mathbf{F}_2)'(\mathbf{x}_0) = \alpha_1 \mathbf{F}_1'(\mathbf{x}_0) + \alpha_2 \mathbf{F}_2'(\mathbf{x}_0).$$

(iii) *Let the operators* $\mathbf{F} \colon X \to Y$ *and* $\mathbf{G} \colon Y \to Z$ *be Gâteaux differentiable on open subsets* $D_{\mathbf{F}} \subset X$ *and* $D_{\mathbf{G}} \subset Y$ $(D_{\mathbf{F}} \supset \mathbf{F}(D_{\mathbf{F}}))$, *respectively. Then, if the mapping*

$$\mathbf{y} \in D_{\mathbf{G}} \subset Y \to \mathbf{G}'(\mathbf{y}) \in B(Y, B(Y, Z))$$

is continuous $(\mathbf{G} \in C^1(D_{\mathbf{G}}))$, *then the composed operator* $\mathbf{T} = \mathbf{GF} \colon X \to Z$ *is Gâteaux differentiable on* D. *If, moreover,* $\mathbf{F} \in C^1(D)$, *then also* $\mathbf{T} \in C^1(D)$.

(iv) *If the operator* $\mathbf{F} \colon X \to Y$ *is Gâteaux differentiable at any point* $\mathbf{x}$ *of the domain* D *in* X *and* $\|\mathbf{F}'(\mathbf{x})\|_{B(X,Y)} \leq M < \infty$ *for any* $\mathbf{x} \in D$, *then* $\mathbf{F}$ *is lipschitzian on* D *(with the Lipschitz constant* M*).*

7.3. Abstract functions. *The operators acting from* R *into a Banach space* Y *are called abstract functions.*

The derivative $\mathbf{f}'$ of the abstract function $\mathbf{f} \colon R \to Y$ at the point $t_0 \in R$ is defined by

$$\lim_{t \to t_0} \left\| \frac{\mathbf{f}(t) - \mathbf{f}(t_0)}{t - t_0} - \mathbf{f}'(t_0) \right\|_Y = 0.$$

Let the abstract function $\mathbf{f} \colon R \to Y$ be defined and continuous on the interval $[a, b]$ $(-\infty < a < b < \infty)$. Then there exists $\mathbf{y} \in Y$ such that given $\varepsilon > 0$, there is a $\delta > 0$ such that for any subdivision $\sigma = \{a = t_0 < t_1 < \ldots < t_{m_\sigma} = b\}$ of the interval $[a, b]$ with $(t_j - t_{j-1}) \leq \delta$ $(j = 1, 2, \ldots, m_\sigma)$ and for an arbitrary choice of $t_j' \in (t_{j-1}, t_j)$ $(j = 1, 2, \ldots, m_\sigma)$ it holds

$$\left\| \sum_{j=1}^{m_\sigma} \mathbf{f}(t_j')(t_j - t_{j-1}) - \mathbf{y} \right\|_Y < \varepsilon.$$

We denote

$$\mathbf{y} = \int_a^b \mathbf{f}(t) \, dt$$

and $\mathbf{y}$ is said to be the abstract Riemann integral of $\mathbf{f}(t)$ over the interval $[a, b]$.

The abstract Riemann integral possesses analogous properties as the usual

Riemann integral of functions $[a, b] \to R$. In particular, if $\|\boldsymbol{f}(t)\|_Y \leq M < \infty$ on $[a, b]$, then

$$\left\| \int_a^b \boldsymbol{f}(t)\, \mathrm{d}t \right\|_Y \leq \int_a^b \|\boldsymbol{f}(t)\|_Y\, \mathrm{d}t \leq M(b - a).$$

Furthermore, if $\boldsymbol{f}'$ exists and is continuous on $(\alpha, \beta) \supset [a, b]$, then

$$\int_a^b \boldsymbol{f}'(t)\, \mathrm{d}t = \boldsymbol{f}(b) - \boldsymbol{f}(a).$$

7.4. Lemma (Mean Value Theorem). *Let X, Y, Z be Banach spaces, and $\mathbf{x}_0 \in X$, $\mathbf{z}_0 \in Z$. Let the operator $\mathbf{F}$: $X \times Z \to Y$ be defined and Gâteaux differentiable on $\mathfrak{B}(\mathbf{x}_0, \varrho_0; X) \times \mathfrak{B}(\mathbf{z}_0, \sigma_0; Z)$ $(\varrho_0 > 0,\ \sigma_0 > 0)$. Then for any $\mathbf{x}_1, \mathbf{x}_2 \in \mathfrak{B}(\mathbf{x}_0, \varrho_0; X)$ and $\mathbf{z}_1, \mathbf{z}_2 \in \mathfrak{B}(\mathbf{z}_0, \sigma_0; Z)$*

$$\mathbf{F}(\mathbf{x}_2, \mathbf{z}_2) - \mathbf{F}(\mathbf{x}_1, \mathbf{z}_1) = \int_0^1 [\mathbf{F}'_x(\mathbf{x}_1 + \vartheta(\mathbf{x}_2 - \mathbf{x}_1), \mathbf{z}_1 + \vartheta(\mathbf{z}_2 - \mathbf{z}_1))]\, (\mathbf{x}_2 - \mathbf{x}_1)\, \mathrm{d}\vartheta$$
$$+ \int_0^1 [\mathbf{F}'_z(\mathbf{x}_1 + \vartheta(\mathbf{x}_2 - \mathbf{x}_1), \mathbf{z}_1 + \vartheta(\mathbf{z}_2 - \mathbf{z}_1))]\, (\mathbf{z}_2 - \mathbf{z}_1)\, \mathrm{d}\vartheta.$$

(*The mapping*

$$\begin{aligned} \vartheta \in [0, 1] \to & [\mathbf{F}'(\mathbf{x}_1 + \vartheta(\mathbf{x}_2 - \mathbf{x}_1), \mathbf{z}_1 + \vartheta(\mathbf{z}_2 - \mathbf{z}_1))]\, [(\mathbf{x}_2, \mathbf{z}_2) - (\mathbf{x}_1, \mathbf{z}_1)] \\ = & [\mathbf{F}'_x(\mathbf{x}_1 + \vartheta(\mathbf{x}_2 - \mathbf{x}_1), \mathbf{z}_1 + \vartheta(\mathbf{z}_2 - \mathbf{z}_1))]\, (\mathbf{x}_2 - \mathbf{x}_1) \\ & + [\mathbf{F}'_z(\mathbf{x}_1 + \vartheta(\mathbf{x}_2 - \mathbf{x}_1), \mathbf{z}_1 + \vartheta(\mathbf{z}_2 - \mathbf{z}_1))]\, (\mathbf{z}_2 - \mathbf{z}_1) \in Y \end{aligned}$$

is an abstract function.)

7.5. Theorem (Implicit Function Theorem). *Let X, Y and Z be Banach spaces, $\mathbf{x}_0 \in X$, $\mathbf{z}_0 \in Z$, $\varrho_0 > 0$, $\sigma_0 > 0$. Let the operator $\mathbf{F}$: $X \times Z \to Y$ be defined and continuous on $\mathfrak{B}(\mathbf{x}_0, \varrho_0; X) \times \mathfrak{B}(\mathbf{z}_0, \sigma_0; Z)$, while*

(i) $\mathbf{F}(\mathbf{x}_0, \mathbf{z}_0) = \mathbf{0}$;
(ii) $\mathbf{F} \in C^{1,0}(\mathfrak{B}(\mathbf{x}_0, \varrho_0; X) \times \mathfrak{B}(\mathbf{z}_0, \sigma_0; Z))$ (cf. 7.2);
(iii) $\mathbf{F}'_x(\mathbf{x}_0, \mathbf{z}_0)$ *possesses a bounded inverse operator.*

Then there exist $\varrho > 0$ and $\sigma > 0$ such that for any $\mathbf{z} \in \mathfrak{B}(\mathbf{z}_0, \sigma; Z)$ there exists a unique solution $\mathbf{x} = \varphi(\mathbf{z}) \in \mathfrak{B}(\mathbf{x}_0, \varrho; X)$ to the equation

$$\mathbf{F}(\mathbf{x}, \mathbf{z}) = \mathbf{0}. \tag{7,1}$$

Moreover, the mapping $\mathbf{z} \in \mathfrak{B}(\mathbf{z}_0, \sigma; Z) \to \varphi(\mathbf{z}) \in \mathfrak{B}(\mathbf{x}_0, \varrho; X)$ is continuous.

(Proof follows easily by applying Corollary 7.7 of the Contraction Mapping Principle 7.6 to the equation

$$\mathbf{x} = \mathbf{x} - [\mathbf{F}'_x(\mathbf{x}_0, \mathbf{z}_0)]^{-1}\, \mathbf{F}(\mathbf{x}, \mathbf{z}).)$$

7.6. Theorem (Contraction Mapping Principle). *Let X be a Banach space and let $D \subset X$ be closed. Let the operator $\mathbf{T}\colon X \to X$ be contractive on D and $\mathbf{T}(D) \subset D$. Then there exists a unique $\mathbf{x} \in D$ such that $\mathbf{x} = \mathbf{T}(\mathbf{x})$.*
(The sought solution is the limit of successive approximations

$$\mathbf{x}_n = \mathbf{T}(\mathbf{x}_{n-1}) \quad (n = 1, 2, \ldots),$$

where $\mathbf{x}_0$ may be an arbitrary element of D.)

7.7. Corollary. *Let X and Z be Banach spaces. Let $\mathbf{x}_0 \in X$, $\mathbf{z}_0 \in Z$, $\varrho_0 > 0$, $\sigma_0 > 0$, $0 \le \lambda < 1$ and let $\mathbf{T}$ be a continuous mapping of $\mathfrak{B}(\mathbf{x}_0, \varrho_0; X) \times \mathfrak{B}(\mathbf{z}_0, \sigma_0; Z)$ into X such that*

(i) $$\|\mathbf{T}(\mathbf{x}_1, \mathbf{z}) - \mathbf{T}(\mathbf{x}_2, \mathbf{z})\|_X \le \lambda \|\mathbf{x}_1 - \mathbf{x}_2\|_X$$

for all $\mathbf{x}_1, \mathbf{x}_2 \in \mathfrak{B}(\mathbf{x}_0, \varrho_0; X)$ and $\mathbf{z} \in \mathfrak{B}(\mathbf{z}_0, \sigma_0; Z)$;

(ii) $$\|\mathbf{T}(\mathbf{x}_0, \mathbf{z}) - \mathbf{x}_0\|_X < \varrho_0(1 - \lambda)$$

for all $\mathbf{z} \in \mathfrak{B}(\mathbf{z}_0, \sigma_0; Z)$.

Then, given $\mathbf{z} \in \mathfrak{B}(\mathbf{z}_0, \sigma_0; Z)$, there exists a unique element $\mathbf{x} = \boldsymbol{\varphi}(\mathbf{z}) \in \mathfrak{B}(\mathbf{x}_0, \varrho_0; X)$ such that $\mathbf{x} = \mathbf{T}(\mathbf{x}, \mathbf{z})$.

The mapping $\mathbf{z} \in \mathfrak{B}(\mathbf{z}_0, \sigma_0; Z) \to \boldsymbol{\varphi}(\mathbf{z}) \in \mathfrak{B}(\mathbf{x}_0, \varrho_0; X)$ is continuous.

Another version of the Implicit Function Theorem which is of interest for our purposes is the following theorem which also follows from the Contraction Mapping Principle.

7.8. Theorem. *Let X and Y be Banach spaces. Let $\mathbf{x}_0 \in X$, $\varrho_0 > 0$ and $\varkappa_0 > 0$. Let the operators $\mathbf{F}\colon X \to Y$ and $\mathbf{G}\colon X \times [0, \varkappa_0] \to Y$ satisfy the assumptions*

(i) $\mathbf{F}(\mathbf{x}_0) = \mathbf{0}$;
(ii) $\mathbf{F} \in C^1(\mathfrak{B}(\mathbf{x}_0, \varrho_0; X))$;
(iii) $\mathbf{F}'(\mathbf{x}_0)$ *possesses a bounded inverse operator;*
(iv) $\mathbf{G}$ *is locally lipschitzian on $\mathfrak{B}(\mathbf{x}_0, \varrho_0; X)$ near $\varepsilon = 0$.*

Then there exist $\varrho > 0$ and $\varkappa > 0$ such that for any $\varepsilon \in [0, \varkappa]$ there is a unique solution $\mathbf{x} = \boldsymbol{\varphi}(\varepsilon) \in \mathfrak{B}(\mathbf{x}_0, \varrho; X)$ of the equation

(7,2) $$\mathbf{F}(\mathbf{x}) + \varepsilon\, \mathbf{G}(\mathbf{x}, \varepsilon) = \mathbf{0}\,.$$

Moreover, the mapping $\varepsilon \in [0, \varkappa] \to \boldsymbol{\varphi}(\varepsilon) \in \mathfrak{B}(\mathbf{x}_0, \varrho; X)$ is continuous.

7.9. Quasilinear equation — noncritical case. Of special interest are quasilinear (weakly nonlinear) equations of the form

(7,3) $$\mathbf{L}\mathbf{x} - \varepsilon\, \mathbf{N}(\mathbf{x}, \varepsilon) = \mathbf{0}\,,$$

where $\mathbf{L}$ is a linear bounded operator acting from a Banach space X into a Banach space Y with the definition domain $D(\mathbf{L}) = X$ $(\mathbf{L} \in B(X, Y))$ and $\mathbf{N}$ is in general a nonlinear operator acting from $X \times R_1$ into Y.

The case when $\mathbf{L}$ possesses a bounded inverse operator is called *noncritical case.* In such a case the equation (7,3) is reduced to the equivalent equation

$$\mathbf{x} = \varepsilon \mathbf{L}^{-1} \mathbf{N}(\mathbf{x}, \varepsilon) . \tag{7,4}$$

For $\varepsilon = 0$ (7,4) has the unique solution $\mathbf{x}_0 = \mathbf{0}$. To solve it for $\varepsilon > 0$ we may apply Theorem 7.8, where $\mathbf{F} = \mathbf{L}$ and $\mathbf{G} = -\mathbf{N}$.

7.10. Quasilinear equation — critical case. A linear bounded operator $\mathbf{L} \in B(X, Y)$ possesses a bounded inverse if and only if $N(\mathbf{L}) = \{\mathbf{0}\}$ and $R(\mathbf{L}) = Y$ (cf. Bounded Inverse Theorem 3.4).

In a general case when either $\dim N(\mathbf{L}) > 0$ or $R(\mathbf{L}) \subsetneq Y$ the projection method may sometimes be used to consider the equation (7,3).

Let $\mathbf{L} \in B(X, Y)$ be such that

$$\begin{aligned} R(\mathbf{L}) \quad \text{is closed}, \quad \alpha(\mathbf{L}) = \dim N(\mathbf{L}) < \infty\,, \\ \beta(\mathbf{L}) = \operatorname{codim} R(\mathbf{L}) < \infty \end{aligned} \tag{7,5}$$

($\mathbf{L}$ is said to be *noetherian*). Then there exist linear bounded projections $\mathbf{P}$ of X onto $N(\mathbf{L})$ $(\mathbf{P} \in B(X), R(\mathbf{P}) = N(\mathbf{L}), \mathbf{P}^2 = \mathbf{P})$ and $\mathbf{Q}$ of Y onto $R(\mathbf{L})$ $(\mathbf{Q} \in B(Y),$ $R(\mathbf{Q}) = R(\mathbf{L}), \mathbf{Q}^2 = \mathbf{Q})$ such that $R(\mathbf{I} - \mathbf{P})$ is closed in X, $\dim R(\mathbf{I} - \mathbf{Q}) = \beta(\mathbf{L})$ and

$$X = N(\mathbf{L}) \oplus R(\mathbf{I} - \mathbf{P}), \qquad Y = R(\mathbf{L}) \oplus R(\mathbf{I} - \mathbf{Q}) \tag{7,6}$$

(cf. Goldberg [1] II.1.14 and II.1.16). Thus $\mathbf{L}\mathbf{x} = \varepsilon \mathbf{N}(\mathbf{x}, \varepsilon)$ if and only if both

$$\mathbf{Q}(\mathbf{L}\mathbf{x} - \varepsilon \mathbf{N}(\mathbf{x}, \varepsilon)) = \mathbf{L}\mathbf{x} - \varepsilon \mathbf{Q} \mathbf{N}(\mathbf{x}, \varepsilon) = \mathbf{0} \tag{7,7}$$

and

$$(\mathbf{I} - \mathbf{Q})(\mathbf{L}\mathbf{x} - \varepsilon \mathbf{N}(\mathbf{x}, \varepsilon)) = -\varepsilon(\mathbf{I} - \mathbf{Q}) \mathbf{N}(\mathbf{x}, \varepsilon) = \mathbf{0} . \tag{7,8}$$

Any $\mathbf{x} \in X$ may be written in the form $\mathbf{x} = \mathbf{P}\mathbf{x} + (\mathbf{I} - \mathbf{P}) \mathbf{x}$. For $\mathbf{x} \in X$ let us denote $\mathbf{u} = (\mathbf{I} - \mathbf{P}) \mathbf{x}$ and $\mathbf{v} = \mathbf{P}\mathbf{x}$. Then the system (7,7), (7,8) becomes

$$\mathbf{L}_1 \mathbf{u} - \varepsilon \mathbf{Q} \mathbf{N}_1(\mathbf{u}, \mathbf{v}, \varepsilon) = \mathbf{0} , \qquad (\mathbf{I} - \mathbf{Q}) \mathbf{N}_1(\mathbf{u}, \mathbf{v}, \varepsilon) = \mathbf{0} ,$$

where

$$\mathbf{L}_1 \colon \mathbf{u} \in R(\mathbf{I} - \mathbf{P}) \to \mathbf{L}\mathbf{u} \in R(\mathbf{L}) = R(\mathbf{Q})$$

and

$$\mathbf{N}_1(\mathbf{u}, \mathbf{v}, \varepsilon) = \mathbf{N}(\mathbf{u} + \mathbf{v}, \varepsilon)$$

for $\mathbf{u} \in R(\mathbf{I} - \mathbf{P})$, $\mathbf{v} \in N(\mathbf{L})$ and $\varepsilon \in [0, \varkappa_0]$. Clearly, $\mathbf{L}_1 \in B(R(\mathbf{I} - \mathbf{P}), R(\mathbf{L}))$ is a one-to-one mapping of $R(\mathbf{I} - \mathbf{P})$ onto $R(\mathbf{L})$. ($\mathbf{L}_1 \mathbf{u} = \mathbf{0}$ implies $\mathbf{u} \in R(\mathbf{P})$ and since $R(\mathbf{P}) \cap R(\mathbf{I} - \mathbf{P}) = \{\mathbf{0}\}$, $\mathbf{u} = \mathbf{0}$).

7.11. Theorem. *Let* $\mathbf{L} \in B(X, Y)$ *fulfil* (7,5) *and let* $\mathbf{P} \in B(X)$ *and* $\mathbf{Q} \in B(Y)$ *be the corresponding projections of* X *onto* $N(\mathbf{L})$ *and of* Y *onto* $R(\mathbf{L})$, *respectively. Let* $\mathbf{h} \in R(\mathbf{L})$ *and* $\mathbf{L}\mathbf{x}_0 = \mathbf{h}$.

Let $\varrho_0 > 0$, $\varkappa_0 > 0$ *and* $D = \mathfrak{B}(\mathbf{x}_0, \varrho_0; X) \times [0, \varkappa_0]$, *Let* $\mathbf{N} \in C^{1,0}(D)$, $\mathbf{N}(\mathbf{x}_0, 0) \in R(\mathbf{L})$ *and* $(\mathbf{I} - \mathbf{Q})\, \mathbf{N}'_{\mathbf{x}}(\mathbf{x}_0, 0)$ *possesses a bounded inverse.*

Then there are $\varkappa > 0$ *and* $\varrho > 0$ *such that for any* $\varepsilon \in [0, \varkappa]$ *there exists a unique solution* $\mathbf{x} = \boldsymbol{\varphi}(\varepsilon) \in \mathfrak{B}(\mathbf{x}_0, \varrho; X)$ *of the equation*

$$\mathbf{L}\mathbf{x} = \mathbf{h} + \varepsilon\, \mathbf{N}(\mathbf{x}, \varepsilon)\,. \tag{7,9}$$

The mapping $\boldsymbol{\varphi}\colon \varepsilon \in [0, \varkappa] \to \boldsymbol{\varphi}(\varepsilon) \in \mathfrak{B}(\mathbf{x}_0, \varrho; X)$ *is continuous.*

Proof. Let us denote $U = R(\mathbf{I} - \mathbf{P})$, $V = R(\mathbf{P}) = N(\mathbf{L})$. Then U and V are Banach spaces with the norms induced by $\|.\|_X$. Given $\mathbf{x} \in X$, let us put $\mathbf{u} = (\mathbf{I} - \mathbf{P})\,\mathbf{x}$ and $\mathbf{v} = \mathbf{P}\mathbf{x}$. In particular, $\mathbf{u}_0 = (\mathbf{I} - \mathbf{P})\,\mathbf{x}_0$, $\mathbf{v}_0 = \mathbf{P}\mathbf{x}_0$. Since $\mathbf{h} \in R(\mathbf{L})$, $(\mathbf{I} - \mathbf{Q})\,\mathbf{h} = \mathbf{0}$ and (7,9) becomes

$$\mathbf{L}_1\mathbf{u} - \mathbf{h} - \varepsilon\mathbf{Q}\, \mathbf{N}(\mathbf{u} + \mathbf{v}, \varepsilon) = \mathbf{0}\,, \qquad (\mathbf{I} - \mathbf{Q})\, \mathbf{N}(\mathbf{u} + \mathbf{v}, \varepsilon) = \mathbf{0}\,,$$

where $\mathbf{L}_1 = \mathbf{L}|_U \in B(U, R(\mathbf{L}))$ possesses a bounded inverse. Let $D_1 \subset U \times V \times [0, \varkappa_0]$ denote the set of all $(\mathbf{u}, \mathbf{v}, \varepsilon) \in U \times V \times [0, \varkappa_0]$ such that $\|\mathbf{u} - \mathbf{u}_0\|_X \leq \frac{1}{2}\varrho_0$ and $\|\mathbf{v} - \mathbf{v}_0\|_X \leq \frac{1}{2}\varrho_0$. Given $(\mathbf{u}, \mathbf{v}, \varepsilon) \in D_1$, $(\mathbf{u} + \mathbf{v}, \varepsilon) \in D$ and we may define

$$\mathbf{T}(\mathbf{u}, \mathbf{v}, \varepsilon) = \begin{pmatrix} \mathbf{L}_1\mathbf{u} - \mathbf{h} - \varepsilon\mathbf{Q}\, \mathbf{N}(\mathbf{u} + \mathbf{v}, \varepsilon) \\ (\mathbf{I} - \mathbf{Q})\, \mathbf{N}(\mathbf{u} + \mathbf{v}, \varepsilon) \end{pmatrix} \in R(\mathbf{L}) \times R(\mathbf{I} - \mathbf{Q})\,.$$

Clearly, $\mathbf{T}$ is a continuous mapping of $D_1 \subset U \times V \times [0, \varkappa_0]$ into $Y \times Y$. Moreover, for any $(\mathbf{u}, \mathbf{v}, \varepsilon) \in D_1$ and $(\boldsymbol{\xi}, \boldsymbol{\eta}) \in U \times V$

$$[\mathbf{T}'_{(\mathbf{u},\mathbf{v})}(\mathbf{u}, \mathbf{v}, \varepsilon)]\,(\boldsymbol{\xi}, \boldsymbol{\eta}) = \begin{pmatrix} \mathbf{L}_1\boldsymbol{\xi} - \varepsilon\mathbf{Q}[\mathbf{N}'_{\mathbf{x}}(\mathbf{u} + \mathbf{v}, \varepsilon)]\,(\boldsymbol{\xi} + \boldsymbol{\eta}) \\ (\mathbf{I} - \mathbf{Q})\,[\mathbf{N}'_{\mathbf{x}}(\mathbf{u} + \mathbf{v}, \varepsilon)]\,(\boldsymbol{\xi} + \boldsymbol{\eta}) \end{pmatrix},$$

the mapping $(\mathbf{u}, \mathbf{v}, \varepsilon) \in D_1 \to \mathbf{T}'_{(\mathbf{u},\mathbf{v})}(\mathbf{u}, \mathbf{v}, \varepsilon) \in B(U \times V, Y \times Y)$ being continuous.

Since $\mathbf{N}(\mathbf{u}_0 + \mathbf{v}_0, 0) \in R(\mathbf{L})$ and $\mathbf{L}_1\mathbf{u}_0 = \mathbf{h}$, $\mathbf{T}(\mathbf{u}_0, \mathbf{v}_0, 0) = \mathbf{0}$. Moreover,

$$[\mathbf{T}'_{(\mathbf{u},\mathbf{v})}(\mathbf{u}_0, \mathbf{v}_0, 0)]\,(\boldsymbol{\xi}, \boldsymbol{\eta}) = \begin{pmatrix} \mathbf{L}_1\boldsymbol{\xi} \\ (\mathbf{I} - \mathbf{Q})\, \mathbf{N}'_{\mathbf{x}}(\mathbf{x}_0, 0)\,(\boldsymbol{\xi} + \boldsymbol{\eta}) \end{pmatrix}$$

for any $(\boldsymbol{\xi}, \boldsymbol{\eta}) \in U \times V$. It is easy to see that for any $\mathbf{p} \in R(\mathbf{L})$ and $\mathbf{q} \in R(\mathbf{I} - \mathbf{Q})$

$$[\mathbf{T}'_{(\mathbf{u},\mathbf{v})}(\mathbf{u}_0, \mathbf{v}_0, 0)]\,(\boldsymbol{\xi}, \boldsymbol{\eta}) = \begin{pmatrix} \mathbf{p} \\ \mathbf{q} \end{pmatrix}$$

if and only if $\boldsymbol{\xi} = \mathbf{L}_1^{-1}\mathbf{p}$ and $\boldsymbol{\eta} = [(\mathbf{I} - \mathbf{Q})\, \mathbf{N}'_{\mathbf{x}}(\mathbf{x}_0, 0)]^{-1}\, \mathbf{q} - \boldsymbol{\xi}$. Applying the Implicit Function Theorem 7.5 we complete the proof.

II. *Integral equations in the space* $BV_n[0, 1]$

1. Some integral operators in the space $BV_n[0, 1]$

In this paragraph we assume that on the twodimensional interval $I = [0, 1] \times [0, 1] \subset R_2$ an $n \times n$-matrix valued function $\mathbf{K}(s, t) = k_{ij}(s, t))$, $i, j = 1, 2, \ldots, n$ is given, i.e. $\mathbf{K}: I \to L(R_n)$. Moreover let the twodimensional variation of $\mathbf{K}: I \to L(R_n)$ be finite, i.e. (cf. I.6.1)

(1,1)
$$\mathrm{v}_I(\mathbf{K}) < \infty .$$

The operator $\int_0^1 \mathrm{d}_t[\mathbf{K}(s, t)]\, \mathbf{x}(t)$

Let us assume that $\mathbf{x} \in BV_n[0, 1] = BV_n$ is given, i.e. $\mathbf{x}(t) = (x_1(t), x_2(t), \ldots, x_n(t))^*$; $t \in [0, 1]$. If it is assumed that

(1,2)
$$\operatorname{var}_0^1 \mathbf{K}(0, .) < \infty ,$$

then by I.6.6 we obtain $\operatorname{var}_0^1 \mathbf{K}(s, .) \leq \mathrm{v}_I(\mathbf{K}) + \operatorname{var}_0^1 \mathbf{K}(0, .) < +\infty$ for every $s \in [0, 1]$. This yields by I.4.19 the existence of the Perron-Stieltjes integral

(1,3)
$$\int_0^1 \mathrm{d}_t[\mathbf{K}(s, t)]\, \mathbf{x}(t) = \mathbf{y}(s)$$

for any $s \in [0, 1]$. The integral (1,3) evidently defines a function $\mathbf{y}: [0, 1] \to R_n$. By I.6.18 we have

(1,4)
$$\operatorname{var}_0^1 \mathbf{y} \leq \sup_{t \in [0,1]} |\mathbf{x}(t)|\, \mathrm{v}_I(\mathbf{K})$$

and consequently $\mathbf{y} \in BV_n$. Hence the integral (1,3) defines an operator acting in the Banach space BV_n. Let us denote this operator by

(1,5)
$$\mathbf{Kx} = \int_0^1 \mathrm{d}_t[\mathbf{K}(s, t)]\, \mathbf{x}(t), \qquad \mathbf{x} \in BV_n .$$

1.1. Theorem. *If* $\mathbf{K}\colon I \to L(R_n)$ *satisfies* (1,1) *and* (1,2) *then the operator* $\mathbf{K}$ *defined by* (1,5) *is a bounded linear operator on* BV_n ($\mathbf{K} \in B(BV_n)$) *and*

$$\|\mathbf{K}\|_{B(BV_n)} \leq \operatorname{var}_0^1 \mathbf{K}(0, .) + v_I(\mathbf{K}) . \tag{1,6}$$

Proof. The linearity of the operator $\mathbf{K}$ is evident. Further for any $\mathbf{x} \in BV_n$ it is

$$\|\mathbf{K}\mathbf{x}\|_{BV_n} = \left|\int_0^1 d_t[\mathbf{K}(0, t)]\, \mathbf{x}(t)\right| + \operatorname{var}_0^1\left(\int_0^1 d_t[\mathbf{K}(., t)]\, \mathbf{x}(t)\right)$$

$$\leq \sup_{t\in[0,1]} |\mathbf{x}(t)| \left(\operatorname{var}_0^1 \mathbf{K}(0, .) + v_I(\mathbf{K})\right) \leq \left(\operatorname{var}_0^1 \mathbf{K}(0, .) + v_I(\mathbf{K})\right) \|\mathbf{x}\|_{BV_n}$$

where (I.6,13) and (I.6,14) from I.6.18 was used. This implies the boundedness of $\mathbf{K}$ and the inequality (1,6).

1.2. Lemma. *If* $\mathbf{K}\colon I \to L(R_n)$ *and* $\tilde{\mathbf{K}}\colon I \to L(R_n)$ *satisfy* (1,1) *and* (1,2), *then*

$$\int_0^1 d_t[\mathbf{K}(s, t)]\, \mathbf{x}(t) = \int_0^1 d_t[\tilde{\mathbf{K}}(s, t)]\, \mathbf{x}(t) \tag{1,7}$$

for every $\mathbf{x} \in BV_n$ *and* $s \in [0, 1]$ *if and only if the difference*

$$\mathbf{W}(s, t) = \mathbf{K}(s, t) - \tilde{\mathbf{K}}(s, t)$$

satisfies

$$\mathbf{W}(s, t+) = \mathbf{W}(s, t-) = \mathbf{W}(s, 1-) = \mathbf{W}(s, 0+) = \mathbf{W}(s, 1) = \mathbf{W}(s, 0) \tag{1,8}$$

for every $s \in [0, 1]$ *and* $t \in (0, 1)$.

Proof. The assumptions on $\mathbf{K}, \tilde{\mathbf{K}}$ guarantee that for $\mathbf{W}\colon I \to L(R_n)$ we have $v_I(\mathbf{W}) < \infty$ and $\operatorname{var}_0^1 \mathbf{W}(0, .) < \infty$. Hence by I.6.6 also $\operatorname{var}_0^1 \mathbf{W}(s, .) < \infty$ for every $s \in [0, 1]$. The equality (1,7) can be written in the form $\int_0^1 d_t[\mathbf{W}(s, t)]\, \mathbf{x}(t) = 0$. The assertion of our lemma follows now immediately from I.6.5 since (1,8) is equivalent to the fact that for every $s \in [0, 1]$ the elements of the matrix $\mathbf{W}(s, .)$ belong to $S[0, 1]$.

1.3. Corollary. *If* $\mathbf{K}, \tilde{\mathbf{K}}\colon I \to L(R_n)$ *satisfies* (1,1) *and* (1,2) *where for the difference* $\mathbf{W}(s, t) = \mathbf{K}(s, t) - \tilde{\mathbf{K}}(s, t)$ *the chain of equalities* (1,8) *holds for any* $s \in [0, 1]$ *and* $t \in (0, 1)$, *then the operator* $\tilde{\mathbf{K}} \in B(BV_n)$ *defined by the relation*

$$\tilde{\mathbf{K}}\mathbf{x} = \int_0^1 d_t[\tilde{\mathbf{K}}(s, t)]\, \mathbf{x}(t), \qquad \mathbf{x} \in BV_n$$

is identical with the operator $\mathbf{K} \in B(BV_n)$ *defined by* (1,5).

If we define for any $s \in [0, 1]$

$$\hat{\mathbf{K}}(s, t) = \mathbf{K}(s, t+) - \mathbf{K}(s, 0) \quad \text{for} \quad t \in (0, 1), \tag{1,9}$$
$$\hat{\mathbf{K}}(s, 0) = 0, \qquad \hat{\mathbf{K}}(s, 1) = \mathbf{K}(s, 1) - \mathbf{K}(s, 0),$$

then $\mathrm{v}_I(\hat{\mathbf{K}}) < \infty$, $\mathrm{var}_0^1 \hat{\mathbf{K}}(0,.) < \infty$ *and the difference* $\hat{\mathbf{W}}(s,t) = \mathbf{K}(s,t) - \hat{\mathbf{K}}(s,t)$ *satisfies* (1,8) *for any* $s \in [0,1]$ *and* $t \in (0,1)$. *Hence the operator*

$$\hat{\mathbf{K}}\mathbf{x} = \int_0^1 \mathrm{d}_t[\hat{\mathbf{K}}(s,t)]\,\mathbf{x}(t), \qquad \mathbf{x} \in BV_n$$

is the same as the operator $\mathbf{K} \in B(BV_n)$ *defined by* (1,5), *i.e.* $\mathbf{K} = \hat{\mathbf{K}}$.

Proof. The first part of this corollary simply follows from 1.2. For the second part it is necessary to show that $\hat{\mathbf{K}}: I \to L(R_n)$ from (1,9) satisfies (1,1) and (1,2).

Assume that $0 = \alpha_0 < \alpha_1 < \ldots < \alpha_k = 1$ is an arbitrary subdivision of $[0,1]$ and $J_{ij} = [\alpha_{i-1}, \alpha_i] \times [\alpha_{j-1}, \alpha_j]$, $i,j = 1,\ldots,k$ is the corresponding net-type subdivision of I (see I.6.3). We have for any given $\delta > 0$

$$\sum_{i=1}^k |\mathbf{K}(\alpha_i, \alpha_1 + \delta) - \mathbf{K}(\alpha_i, \alpha_0) - \mathbf{K}(\alpha_{i-1}, \alpha_1 + \delta) + \mathbf{K}(\alpha_{i-1}, \alpha_0)|$$

$$+ \sum_{j=2}^k \sum_{i=1}^k |\mathbf{K}(\alpha_i, \alpha_j + \delta) - \mathbf{K}(\alpha_i, \alpha_{j-1} + \delta) - \mathbf{K}(\alpha_{i-1}, \alpha_j + \delta) + \mathbf{K}(\alpha_{i-1}, \alpha_{j-1} + \delta)| \le \mathrm{v}_I(\mathbf{K})$$

where we assume that $\mathbf{K}(s,t) = \mathbf{K}(s,1)$ if $t > 1$. Since for $\mathbf{K}: I \to L(R_n)$ (1,1) and (1,2) hold, the limit $\lim_{\delta \to 0+} \mathbf{K}(s, t+\delta) = \mathbf{K}(s, t+)$ exists for every $s \in [0,1]$, $t \in [0,1]$. Passing to the limit $\delta \to 0+$ in the above inequality we obtain for $\hat{\mathbf{K}}$ the inequality

$$\sum_{j=1}^k \sum_{i=1}^k |m_{\hat{\mathbf{K}}}(J_{ij})|$$

$$\sum_{j=1}^k \sum_{i=1}^k |\hat{\mathbf{K}}(\alpha_i, \alpha_j) - \hat{\mathbf{K}}(\alpha_i, \alpha_{j-1}) - \hat{\mathbf{K}}(\alpha_{i-1}, \alpha_j) + \hat{\mathbf{K}}(\alpha_{i-1}, \alpha_{j-1})| \le \mathrm{v}_I(\mathbf{K})$$

which holds for every net-type subdivision J_{ij} of I. Hence (see I.6.3) we obtain $\mathrm{v}_I(\hat{\mathbf{K}}) \le \mathrm{v}_I(\mathbf{K}) < \infty$. Since $\mathrm{var}_0^1 \mathbf{K}(0,.) < \infty$ and $\hat{\mathbf{K}}(0,t) = \mathbf{K}(0,t+) - \mathbf{K}(0,0)$ differs from $\mathbf{K}(0,t) - \mathbf{K}(0,0)$ only on an at most countable set of points in $[0,1]$, the variation $\mathrm{var}_0^1 \hat{\mathbf{K}}(0,.)$ is finite. For $\hat{\mathbf{W}}(s,t) = \mathbf{K}(s,t) - \hat{\mathbf{K}}(s,t)$ we have evidently

$$\hat{\mathbf{W}}(s,t-) = \mathbf{K}(s,t-) - \hat{\mathbf{K}}(s,t-) = \mathbf{K}(s,t-) - \lim_{\tau \to t-} \mathbf{K}(s,\tau+) + \mathbf{K}(s,0) = \mathbf{K}(s,0)$$

if $s \in [0,1]$, $t \in (0,1)$. Similarly also $\hat{\mathbf{W}}(s,t+) = \hat{\mathbf{W}}(s,1-) = \hat{\mathbf{W}}(s,1) = \hat{\mathbf{W}}(s,0+) = \hat{\mathbf{W}}(s,0) = \mathbf{K}(s,0)$ holds and the assertion of the second part of the corollary is valid.

1.4. Remark. The corollary 1.3 states that we can assume without any loss of generality that the kernel $\mathbf{K}: I \to L(R_n)$, which defines by (1,5) the operator $\mathbf{K} \in B(BV_n)$, satisfies

(1,10) $$\mathbf{K}(s,t+) = \mathbf{K}(s,t) \quad \text{for any} \quad s \in [0,1], \quad t \in (0,1)$$

and

(1,11) $$\mathbf{K}(s,0) = \mathbf{0} \quad \text{for any} \quad s \in [0,1].$$

It is clear that if in (1,9) the right-hand limit $\mathbf{K}(s, t+)$ is replaced by the left-hand limit $\mathbf{K}(s, t-)$, then 1.3 holds too. This justifies the possibility of replacing the condition (1,10) by

$$(1,10') \qquad \mathbf{K}(s, t-) = \mathbf{K}(s, t) \quad \text{for any} \quad s \in [0, 1], \quad t \in (0, 1).$$

Hence without any restriction it can be assumed that the kernel $\mathbf{K}: I \to L(R_n)$ defining the operator $\mathbf{K} \in B(BV_n)$ by (1,5) satisfies (1,10') and (1,11), $\mathbf{K}$ remaining unchanged also in this case.

Moreover, any operator $\mathbf{K} \in B(BV_n)$ given by (1,5) with $\mathbf{K}: I \to L(R_n)$ satisfying (1,1) and (1,2) can be represented by a kernel $\mathbf{K}: I \to L(R_n)$ satisfying the additional assumptions (1,10), (1,11) (or (1,10'), (1,11)). Using the notations from I.5 the additional assumptions (1,10), (1,11) ((1,10'), (1,11)) state that the elements $k_{ij}(s, t)$ of $\mathbf{K}: I \to L(R_n)$ as functions of the second variable t belong to the class NBV (NBV^-).

1.5. Theorem. *If* $\mathbf{K}: I \to L(R_n)$ *satisfies* (1,1) *and* (1,2), *then the operator* $\mathbf{K} \in B(BV_n)$ *defined by* (1,5) *is compact, i.e.* $\mathbf{K} \in K(BV_n)$.

Proof. For proving $\mathbf{K} \in K(BV_n)$ we use I.3.16. Let $\{\mathbf{x}_k\}$, $\mathbf{x}_k \in BV_n$, $k = 1, 2, \ldots$ be an arbitrary sequence with

$$\|\mathbf{x}_k\|_{BV_n} = |\mathbf{x}_k(0)| + \operatorname{var}_0^1 \mathbf{x}_k \le C = \text{const.}, \qquad k = 1, 2, \ldots.$$

By Helly's Choice Theorem (cf. I.1.4) there exists a function $\tilde{\mathbf{x}} \in BV_n$ and a subsequence $\mathbf{x}_{k_l}$, $l = 1, 2, \ldots$ of $\{\mathbf{x}_k\}$ such that $\lim_{l \to \infty} \mathbf{x}_{k_l}(t) = \tilde{\mathbf{x}}(t)$ for any $t \in [0, 1]$.

Let us put

$$\mathbf{z}_l(t) = \mathbf{x}_{k_l}(t) - \tilde{\mathbf{x}}(t), \qquad t \in [0, 1], \quad l = 1, 2, \ldots.$$

Then $\|\mathbf{z}_l\|_{BV_n} \le C + \|\tilde{\mathbf{x}}\|_{BV_n} < \infty$, $\mathbf{z}_l \in BV_n$, $l = 1, 2, \ldots$ and

$$(1,12) \qquad \lim_{l \to \infty} \mathbf{z}_l(t) = \mathbf{0} \quad \text{for any} \quad t \in [0, 1].$$

Using I.6.18 (see (I.6,14)) we have

$$(1,13) \qquad \operatorname{var}_0^1 \left(\int_0^1 \mathrm{d}_t[\mathbf{K}(.,t)] (\mathbf{x}_{k_l}(t) - \tilde{\mathbf{x}}(t)) \right) = \operatorname{var}_0^1 \left(\int_0^1 \mathrm{d}_t[\mathbf{K}(.,t)] \, \mathbf{z}_l(t) \right) \le \int_0^1 |\mathbf{z}_l(t)| \, \mathrm{d}\omega_2(t)$$

where $\omega_2: [0, 1] \to R$ is nondecreasing, $\omega_2(0) = 0$, $\omega_2(1) = \mathrm{v}_I(\mathbf{K})$, (see I.6.7). For every $t \in [0, 1]$ and $l = 1, 2, \ldots$ we have evidently $0 \le |\mathbf{z}_l(t)| \le \|\mathbf{z}_l\|_{BV_n} \le C + \|\tilde{\mathbf{x}}\|_{BV_n}$ and the real valued function $|\mathbf{z}_l(t)|: [0, 1] \to R$ belongs to $BV[0, 1]$ for every $l = 1, 2, \ldots$,. Hence by I.4.19 the integral $\int_0^1 |\mathbf{z}_l(t)| \, \mathrm{d}\omega_2(t)$ exists for every $l = 1, 2, \ldots$. I.4.24 implies by (1,12)

$$\lim_{l \to \infty} \int_0^1 |\mathbf{z}_l(t)| \, \mathrm{d}\omega_2(t) = 0$$

and this together with (1,13) leads to the relation

$$(1,14) \qquad \lim_{l\to\infty} \operatorname{var}_0^1 \left(\int_0^1 \mathrm{d}_t[\mathbf{K}(.,t)]\, \mathbf{x}_{k_l}(t) - \int_0^1 \mathrm{d}_t[\mathbf{K}(.,t)]\, \tilde{\mathbf{x}}(t) \right) = 0 .$$

By (I.6.13) we have further

$$\left| \int_0^1 \mathrm{d}_t[\mathbf{K}(0,t)]\, \mathbf{z}_l(t) \right| \le \int_0^1 |\mathbf{z}_l(t)|\, \mathrm{d}[\operatorname{var}_0^t \mathbf{K}(0,.)]$$

and the same argument as above gives by (1,12)

$$(1,15) \qquad \lim_{l\to\infty} \left| \int_0^1 \mathrm{d}_t[\mathbf{K}(0,t)]\, \mathbf{x}_{k_l}(t) - \int_0^1 \mathrm{d}_t[\mathbf{K}(0,t)]\, \tilde{\mathbf{x}}(t) \right| = 0 .$$

Let us now denote $\tilde{\mathbf{y}}(s) = \int_0^1 \mathrm{d}_t[\mathbf{K}(s,t)]\, \tilde{\mathbf{x}}(t)$. By 1,1 evidently $\tilde{\mathbf{y}} \in BV_n$ and by (1,14) and (1,15) we obtain

$$\lim_{l\to\infty} \|\mathbf{K}\mathbf{x}_{k_l} - \tilde{\mathbf{y}}\|_{BV_n} = \lim_{l\to\infty} \{|\mathbf{K}\mathbf{x}_{k_l}(0) - \tilde{\mathbf{y}}(0)| + \operatorname{var}_0^1 (\mathbf{K}\mathbf{x}_{k_l} - \tilde{\mathbf{y}})\} = 0 ,$$

i.e. the sequence $\{\mathbf{K}\mathbf{x}_k\}$ contains a subsequence which converges in BV_n. Hence $\mathbf{K} \in K(BV_n)$.

The operator $\int_0^1 \mathbf{K}(s,t)\, \mathrm{d}\boldsymbol{\varphi}(s)$

Let us assume that $\boldsymbol{\varphi} \in BV_n$ is given, $\boldsymbol{\varphi}(t) = (\varphi_1(t), \varphi_2(t), \ldots, \varphi_n(t))^*$, $t \in [0,1]$. If

$$(1,16) \qquad \operatorname{var}_0^1 \mathbf{K}(.,0) < \infty ,$$

then by I.6.6 we obtain $\operatorname{var}_0^1 \mathbf{K}(.,t) \le \mathrm{v}_I(\mathbf{K}) + \operatorname{var}_0^1 \mathbf{K}(.,0) < \infty$ for every $t \in [0,1]$ provided (1,1) is fulfilled. In this case by I.4.19 the Perron-Stieltjes integral

$$(1,17) \qquad \int_0^1 \mathbf{K}(s,t)\, \mathrm{d}\boldsymbol{\varphi}(s) = \boldsymbol{\psi}(t)$$

exists for every $t \in [0,1]$.

Let us show that the function $\boldsymbol{\psi}: [0,1] \to R_n$ defined by (1,17) is of bounded variation on $[0,1]$ if (1,16), (1,1), (1,2) are assumed.

Let $0 = \gamma_0 < \gamma_1 < \ldots < \gamma_l = 1$ be an arbitrary subdivision of $[0,1]$. By I.4.27 we have

$$|\boldsymbol{\psi}(\gamma_i) - \boldsymbol{\psi}(\gamma_{i-1})| = \left| \int_0^1 (\mathbf{K}(s,\gamma_i) - \mathbf{K}(s,\gamma_{i-1}))\, \mathrm{d}\boldsymbol{\varphi}(s) \right|$$

$$\le \sup_{s\in[0,1]} |\mathbf{K}(s,\gamma_i) - \mathbf{K}(s,\gamma_{i-1})| \operatorname{var}_0^1 \boldsymbol{\varphi} \le (\mathrm{v}_{[0,1]\times[\gamma_{i-1},\gamma_i]}(\mathbf{K}) + |\mathbf{K}(0,\gamma_i) - \mathbf{K}(0,\gamma_{i-1})|) \operatorname{var}_0^1 \boldsymbol{\varphi}$$

because for every $s \in [0,1]$

$$\begin{aligned}
&|\mathbf{K}(s,\gamma_i) - \mathbf{K}(s,\gamma_{i-1})| \\
&\le |\mathbf{K}(s,\gamma_i) - \mathbf{K}(s,\gamma_{i-1}) - \mathbf{K}(0,\gamma_i) + \mathbf{K}(0,\gamma_{i-1})| + |\mathbf{K}(0,\gamma_i) - \mathbf{K}(0,\gamma_{i-1})| \\
&\le \mathrm{v}_{[0,1]\times[\gamma_{i-1},\gamma_i]}(\mathbf{K}) + |\mathbf{K}(0,\gamma_i) - \mathbf{K}(0,\gamma_{i-1})|
\end{aligned}$$

(cf. I.6). Hence by I.6.5

$$\begin{aligned}
&\sum_{i=1}^{l} |\psi(\gamma_i) - \psi(\gamma_{i-1})| \\
&\le \sum_{i=1}^{l} \left(\mathrm{v}_{[0,1]\times[\gamma_{i-1},\gamma_i]}(\mathbf{K}) + |\mathbf{K}(0,\gamma_i) - \mathbf{K}(0,\gamma_{i-1})|\right) \mathrm{var}_0^1 \varphi \\
&\le [\mathrm{v}_I(\mathbf{K}) + \mathrm{var}_0^1 \mathbf{K}(0,.)] \,\mathrm{var}_0^1 \varphi \le [\mathrm{v}_I(\mathbf{K}) + \mathrm{var}_0^1 \mathbf{K}(0,.)] \,\|\varphi\|_{BV_n}
\end{aligned} \tag{1,18}$$

for all subdivisions $0 = \gamma_0 < \gamma_1 < \ldots < \gamma_l = 1$ and so $\mathrm{var}_0^1 \psi < \infty$. In this way the integral (1,17) defines an operator acting on BV_n; we set

$$\tilde{\mathbf{K}}\varphi = \int_0^1 \mathbf{K}(s,t)\,\mathrm{d}\varphi(s), \qquad \varphi \in BV_n. \tag{1,19}$$

1.6. Theorem. *If* $\mathbf{K}: I \to L(R_n)$ *satisfies* (1,1), (1,2) *and* (1,16), *then the operator* $\tilde{\mathbf{K}}$ *defined by* (1,19) *is a bounded linear operator on* BV_n; *i.e.* $\tilde{\mathbf{K}} \in B(BV_n)$ *and*

$$\|\tilde{\mathbf{K}}\|_{B(BV_n)} \le |\mathbf{K}(0,0)| + \mathrm{var}_0^1 \mathbf{K}(.,0) + \mathrm{var}_0^1 \mathbf{K}(0,.) + \mathrm{v}_I(\mathbf{K}). \tag{1,20}$$

Proof. The linearity of $\tilde{\mathbf{K}}$ is obvious. For any $\varphi \in BV_n$ by I.4.27 we have

$$\left|\int_0^1 \mathbf{K}(s,0)\,\mathrm{d}\varphi(s)\right| \le \sup_{s\in[0,1]} |\mathbf{K}(s,0)| \,\mathrm{var}_0^1 \varphi \le (|\mathbf{K}(0,0)| + \mathrm{var}_0^1 \mathbf{K}(.,0))\, \|\varphi\|_{BV_n}.$$

Using (1,18) we obtain

$$\begin{aligned}
\|\tilde{\mathbf{K}}\varphi\|_{BV_n} &= \left|\int_0^1 \mathbf{K}(s,0)\,\mathrm{d}\varphi(s)\right| + \mathrm{var}_0^1 \int_0^1 \mathbf{K}(s,.)\,\mathrm{d}\varphi(s) \\
&\le [|\mathbf{K}(0,0)| + \mathrm{var}_0^1 \mathbf{K}(0,.) + \mathrm{var}_0^1 \mathbf{K}(.,0) + \mathrm{v}_I(\mathbf{K})]\, \|\varphi\|_{BV_n}.
\end{aligned}$$

Hence $\tilde{\mathbf{K}} \in B(BV_n)$ and (1,20) holds.

1.7. Lemma. *Let* $\mathbf{M}: [0,1] \to L(R_n)$ *be an* $n \times n$*-matrix valued function such that*

$$\mathrm{var}_0^1 \mathbf{M} < \infty. \tag{1,21}$$

Assume that a fixed $\sigma \in [a,b]$ *is given. Define for* $\mathbf{x} \in BV_n$ *the operators*

$$\mathbf{Mx} = \mathbf{M}(t)\,\mathbf{x}(\sigma), \qquad \mathbf{M}^+\mathbf{x} = \mathbf{M}(t)\,\Delta^+\mathbf{x}(\sigma), \qquad \mathbf{M}^-\mathbf{x} = \mathbf{M}(t)\,\Delta^-\mathbf{x}(\sigma)$$

where $\Delta^+\mathbf{x}(\sigma) = \mathbf{x}(\sigma+) - \mathbf{x}(\sigma)$, $\Delta^-\mathbf{x}(\sigma) = \mathbf{x}(\sigma) - \mathbf{x}(\sigma-)$. *The operators* $\mathbf{M}, \mathbf{M}^+, \mathbf{M}^-$ *are compact linear operators on* BV_n, *i.e.* $\mathbf{M}, \mathbf{M}^+, \mathbf{M}^- \in K(BV_n)$.

Proof. Since evidently

$$\|\mathbf{M}\mathbf{x}\|_{BV_n} = |\mathbf{M}(0)\,\mathbf{x}(\sigma)| + \operatorname{var}_0^1(\mathbf{M}(.)\,\mathbf{x}(\sigma)) \leq [|\mathbf{M}(0)| + \operatorname{var}_0^1 \mathbf{M}]\,|\mathbf{x}(\sigma)|$$
$$\leq [|\mathbf{M}(0)| + \operatorname{var}_0^1 \mathbf{M}]\,\|\mathbf{x}\|_{BV_n}$$

we have $\mathbf{M} \in B(BV_n)$ and

$$\|\mathbf{M}\|_{B(BV_n)} \leq [|\mathbf{M}(0)| + \operatorname{var}_0^1 \mathbf{M}].$$

The same argument gives also $\mathbf{M}^+, \mathbf{M}^- \in B(BV_n)$ and the inequalities

$$\|\mathbf{M}^+\|_{B(BV_n)} \leq [|\mathbf{M}(0)| + \operatorname{var}_0^1 \mathbf{M}], \qquad \|\mathbf{M}^-\|_{B(BV_n)} \leq [|\mathbf{M}(0)| + \operatorname{var}_0^1 \mathbf{M}].$$

Let us denote by $B = \{\mathbf{x} \in BV_n;\ \|\mathbf{x}\|_{BV_n} \leq 1\}$ the unit ball in BV_n. $\mathbf{M}^+(B) = \{\mathbf{y} \in BV_n;\ \mathbf{y} = \mathbf{M}^+\mathbf{x},\ \mathbf{x} \in B\}$ is the image of B under the map $\mathbf{M}^+$. Let $\mathbf{y}_k \in \mathbf{M}^+(B)$, $k = 1, 2, \ldots$ be an arbitrary sequence in $\mathbf{M}^+(B)$, i.e. there is a sequence $\mathbf{x}_k \in B$ such that $\mathbf{y}_k = \mathbf{M}^+\mathbf{x}_k$. Since $\mathbf{x}_k \in B$, $k = 1, 2, \ldots$ we have

$$|\Delta^+\mathbf{x}_k(\sigma)| \leq \operatorname{var}_0^1 \mathbf{x}_k \leq \|\mathbf{x}_k\|_{BV_n} \leq 1$$

and there is a subsequence $\{\mathbf{x}_{k_l}\}$, $l = 1, 2, \ldots$ such that $\lim_{l\to\infty} \Delta^+\mathbf{x}_{k_l}(\sigma) = \mathbf{z} \in R_n$ and $\mathbf{M}(t)\,\mathbf{z} \in BV_n$. Since evidently

$$\|\mathbf{M}^+\mathbf{x}_{k_l} - \mathbf{M}(t)\,\mathbf{z}\|_{BV_n} \leq (|\mathbf{M}(0)| + \operatorname{var}_0^1 \mathbf{M})\,|\Delta^+\mathbf{x}_{k_l} - \mathbf{z}|$$

we obtain that

$$\lim_{l\to\infty} \mathbf{y}_{k_l} = \lim_{l\to\infty} \mathbf{M}^+\mathbf{x}_{k_l} = \mathbf{M}(t)\,\mathbf{z} \quad \text{in} \quad BV_n$$

and $\mathbf{M}^+ \in K(BV_n)$.

For an analogous reason the results $\mathbf{M} \in K(BV_n)$, $\mathbf{M}^- \in K(BV_n)$ are derivable.

1.8. Lemma. *Let $\{\sigma_l\}_{l=1}^{\infty}$ be an arbitrary sequence of real numbers in $[0, 1]$. Suppose that $\mathbf{M}_l$: $[0, 1] \to L(R_n)$, $l = 1, 2, \ldots$ is a sequence of $n \times n$-matrix valued functions satisfying*

(1,22)
$$\sum_{l=1}^{\infty} (|\mathbf{M}_l(0)| + \operatorname{var}_0^1 \mathbf{M}_l) < \infty.$$

Define for $\mathbf{x} \in BV_n$ the series

(1,23)
$$\mathbf{R}\mathbf{x} = \sum_{l=1}^{\infty} \mathbf{M}_l(t)\,\Delta^+\mathbf{x}(\sigma_l),$$

(1,24)
$$\mathbf{L}\mathbf{x} = \sum_{l=1}^{\infty} \mathbf{M}_l(t)\,\Delta^-\mathbf{x}(\sigma_l)$$

where $\Delta^+\mathbf{x}(\sigma) = \mathbf{x}(\sigma+) - \mathbf{x}(\sigma)$, for $\sigma \in [0, 1)$ $\Delta^-\mathbf{x}(\sigma) = \mathbf{x}(\sigma) - \mathbf{x}(\sigma-)$, for $\sigma \in (0, 1]$ $\Delta^+\mathbf{x}(1) = \mathbf{0}$, $\Delta^-\mathbf{x}(0) = \mathbf{0}$.
Both expressions (1,23) and (1,24) define compact operators on BV_n, i.e. $\mathbf{R}, \mathbf{L} \in K(BV_n)$.

Proof. We prove this lemma only for $\mathbf{R}$; the proof for $\mathbf{L}$ is similar. First let us prove that $\mathbf{R} \in B(BV_n)$. The linearity of the operator $\mathbf{R}$ is evident. Let $0 = \alpha_0 < \alpha_1 < \ldots < \alpha_k = 1$ be an arbitrary subdivision of $[0, 1]$. We have

$$\sum_{j=1}^{k} \left| \sum_{l=1}^{\infty} (\mathbf{M}_l(\alpha_j) - \mathbf{M}_l(\alpha_{j-1}))\, \Delta^+ \mathbf{x}(\sigma_l) \right| \leq \sum_{j=1}^{k} \sum_{l=1}^{\infty} |\mathbf{M}_l(\alpha_j) - \mathbf{M}_l(\alpha_{j-1})| \operatorname{var}_0^1 \mathbf{x}$$

$$= \sum_{l=1}^{\infty} \left(\sum_{j=1}^{k} |\mathbf{M}_l(\alpha_j) - \mathbf{M}_l(\alpha_{j-1})| \right) \operatorname{var}_0^1 \mathbf{x} \leq \sum_{l=1}^{\infty} \operatorname{var}_0^1 \mathbf{M}_l \operatorname{var}_0^1 \mathbf{x}\,.$$

Hence

$$\operatorname{var}_0^1 \mathbf{Rx} \leq \left(\sum_{l=1}^{\infty} \operatorname{var}_0^1 \mathbf{M}_l \right) \operatorname{var}_0^1 \mathbf{x} \leq \left(\sum_{l=1}^{\infty} \operatorname{var}_0^1 \mathbf{M}_l \right) \|\mathbf{x}\|_{BV_n}\,.$$

Further

$$\left| \sum_{l=1}^{\infty} \mathbf{M}_l(0)\, \Delta^+ \mathbf{x}(\sigma_l) \right| \leq \sum_{l=1}^{\infty} |\mathbf{M}_l(0)| \operatorname{var}_0^1 \mathbf{x} \leq \left(\sum_{l=1}^{\infty} |\mathbf{M}_l(0)| \right) \|\mathbf{x}\|_{BV_n}$$

and consequently

$$\|\mathbf{Rx}\|_{BV_n} \leq \left[\sum_{l=1}^{\infty} (|\mathbf{M}_l(0)| + \operatorname{var}_0^1 \mathbf{M}_l) \right] \|\mathbf{x}\|_{BV_n}, \quad \text{i.e.} \quad \mathbf{R} \in B(BV_n)\,.$$

Let us now define for every $N = 1, 2, \ldots$ the operator

$$R_N \mathbf{x} = \sum_{l=1}^{N} \mathbf{M}_l(t)\, \Delta^+ \mathbf{x}(\sigma_l)\,, \qquad \mathbf{x} \in BV_n\,.$$

1.7 implies that $\mathbf{R}_N$ is compact for every $N = 1, 2, \ldots$ because $\mathbf{R}_N$ is a finite sum of compact operators. Further for every $\mathbf{x} \in BV_n$ we have

$$\mathbf{Rx} - \mathbf{R}_N \mathbf{x} = \sum_{l=N+1}^{\infty} \mathbf{M}_l(t)\, \Delta^+ \mathbf{x}(\sigma_l)$$

and as above also

$$\|\mathbf{Rx} - \mathbf{R}_N \mathbf{x}\|_{BV_n} \leq \left[\sum_{l=N+1}^{\infty} (|\mathbf{M}_l(0)| + \operatorname{var}_0^1 \mathbf{M}_l) \right] \|\mathbf{x}\|_{BV_n}\,.$$

Hence by the assumption (1,22) we obtain that $\lim_{N \to \infty} \mathbf{R}_N = \mathbf{R}$ in $B(BV_n)$ and therefore by I.3.17 we get $\mathbf{R} \in K(BV_n)$.

1.9. Theorem. *If* $\mathbf{K}\colon I \to L(R_n)$ *satisfies* (1,1), (1,2) *and* (1,16), *then the operator* $\tilde{\mathbf{K}} \in B(BV_n)$ *defined by* (1,19) *is compact, i.e.* $\tilde{\mathbf{K}} \in K(BV_n)$.

Proof. In 1.6 we have proved that $\tilde{\mathbf{K}} \in B(BV_n)$. The assumptions guarantee by I.6.5 that $\operatorname{var}_0^1 \mathbf{K}(.\,, t) < \infty$ for every $t \in [0, 1]$. Hence by the integration-by-parts formula I.4.33 we get

$$(1,25) \qquad \int_0^1 \mathbf{K}(s, t)\, \mathrm{d}\varphi(s) = -\int_0^1 \mathrm{d}_s[\mathbf{K}(s, t)]\, \varphi(s) + \mathbf{K}(1, t)\, \varphi(1) - \mathbf{K}(0, t)\, \varphi(0)$$

$$- \sum_{0 \leq \sigma < 1} \Delta_s^+ \mathbf{K}(\sigma, t)\, \Delta^+ \varphi(\sigma) + \sum_{0 < \sigma \leq 1} \Delta_s^- \mathbf{K}(\sigma, t)\, \Delta^- \varphi(\sigma)$$

for any $t \in [0, 1]$, where $\Delta_s^+ \mathbf{K}(\sigma, t) = \mathbf{K}(\sigma +, t) - \mathbf{K}(\sigma, t)$, $\Delta_s^- \mathbf{K}(\sigma, t) = \mathbf{K}(\sigma, t) - \mathbf{K}(\sigma -, t)$, $\Delta^+ \varphi(\sigma) = \varphi(\sigma +) - \varphi(\sigma)$, $\Delta^- \varphi(\sigma) = \varphi(\sigma) - \varphi(\sigma -)$.

By 1.5 the integral $\int_0^1 \mathrm{d}_s[\mathbf{K}(s, t)]\, \varphi(s)$ defines a compact operator on BV_n. Further by (1,1) and (1,2) we have $\operatorname{var}_0^1 \mathbf{K}(s, .) < \infty$ for any $s \in [0, 1]$ (cf. I.6.6). Hence by 1.7 the expressions $\mathbf{K}(1, t)\, \varphi(1)$, $\mathbf{K}(0, t)\, \varphi(0)$ determine compact operators on BV_n. If we prove that the last two terms on the right-hand side in (1,25) define compact operators on BV_n; then $\tilde{\mathbf{K}} \in B(BV_n)$ is expressed by (1,25) in the form of the finite sum of compact operators and is therefore also compact.

Let us consider the term

$$\sum_{0 \le \sigma < 1} \Delta_s^+ \mathbf{K}(\sigma, t)\, \Delta^+ \varphi(\sigma) = \mathbf{R}\varphi \tag{1,26}$$

from the expression (1,25). Since (1,1) and (1,16) are assumed, the set of discontinuity points of $\mathbf{K}(s, t)$ in the first variable lies on an at most denumerable system of lines parallel to the t-axis (see I.6.8) i.e. there is a sequence σ_l, $l = 1, 2, \ldots$, $\sigma_l \in [0, 1]$ such that $\Delta_s^+ \mathbf{K}(\sigma, t) = \mathbf{0}$ whenever $\sigma \neq \sigma_l$, $l = 1, 2, \ldots, \sigma \in [0, 1)$, and $t \in [0, 1]$ is arbitrary. Hence the sum $\mathbf{R}\varphi$ from (1,26) can be written in the form

$$\mathbf{R}\varphi = \sum_{l=1}^{\infty} \Delta_s^+ \mathbf{K}(\sigma_l, t)\, \Delta^+ \varphi(\sigma_l).$$

By I.6.15 we have

$$\operatorname{var}_0^1 \Delta_s^+ \mathbf{K}(\sigma_l, .) \le \omega_1(\sigma_l +) - \omega_1(\sigma_l),$$

where $\omega_1 \colon [0, 1] \to R$ is defined by (I.6,5) for $\mathbf{K} \colon I \to L(R_n)$. Hence (see I.6.7)

$$\sum_{l=1}^{\infty} \operatorname{var}_0^1 \Delta_s^+ \mathbf{K}(\sigma_l, .) \le \sum_{l=1}^{\infty} (\omega_1(\sigma_l +) - \omega_1(\sigma_l)) \le \operatorname{var}_0^1 \omega_1 = v_I(\mathbf{K}).$$

Further evidently

$$\sum_{l=1}^{\infty} |\Delta_s^+ \mathbf{K}(\sigma_l, 0)| \le \operatorname{var}_0^1 \mathbf{K}(., 0) < \infty$$

by (1,16). Hence

$$\sum_{l=1}^{\infty} (|\Delta_s^+ \mathbf{K}(\sigma_l, 0)| + \operatorname{var}_0^1 \Delta_s^+ \mathbf{K}(\sigma_l, .)) < \infty.$$

All assumptions of 1.8 being satisfied we obtain that $\mathbf{R}\varphi$ is a compact operator acting on BV_n. In a similar way we can show that the expression $\sum_{0 < \sigma \le 1} \Delta_s^- \mathbf{K}(\sigma, t)\, \Delta^- \varphi(\sigma)$ from (1,25) also defines a compact operator on BV_n and this yields our theorem.

From 1.9 the following can easily be deduced.

1.10. Theorem. *If* $\mathbf{K} \colon I \to L(R_n)$ *satisfies* (1,1), (1,2) *and* (1,16) *and moreover*

$$\begin{aligned} &\mathbf{K}(s, t+) = \mathbf{K}(s, t) \quad \text{for any} \quad s \in [0, 1],\ t \in (0, 1),\\ &\mathbf{K}(s, 0) = \mathbf{0} \quad \text{for any} \quad s \in [0, 1] \end{aligned} \tag{1,27}$$

then the expression

$$K'\varphi = \int_0^1 K^*(s,t)\,d\varphi(s), \qquad \varphi \in NBV_n \tag{1,28}$$

defines a compact linear operator acting on NBV_n, *i.e.* $K' \in K(NBV_n)$. *(By* $K^*(s,t)$ *the transposition of the matrix* $K(s,t)$ *is denoted.)*

Proof. Using the properties of the norm of a matrix (see I.1.1) we easily obtain that for $K^*: I \to L(R_n)$ we have $\operatorname{var}_0^1 K^*(0,.) < \infty$, $\operatorname{var}_0^1 K^*(.,0) < \infty$, $v_I(K^*) < \infty$, $K^*(s,t+) = K^*(s,t)$ for any $s \in [0,1]$, $t \in (0,1)$ and $K^*(s,0) = 0$ for any $s \in [0,1]$ whenever the assumptions of the theorem are satisfied. By 1.9 the operator $\tilde{K}\psi = \int_0^1 K^*(s,t)\,d\psi(s)$, $\psi \in BV_n$ belongs to $K(BV_n)$. The operator K' given by (1,28) is evidently a restriction of $\tilde{K}$ to the closed subspace $NBV_n \subset BV_n$ (cf. I.5.2). For an arbitrary $\psi \in BV_n$ we have by (1,27)

$$\int_0^1 K^*(s,0)\,d\psi(s) = 0 \qquad \text{and for any} \quad t \in (0,1)$$

$$\lim_{\delta \to 0+} \int_0^1 K^*(s,t+\delta)\,d\psi(s) = \int_0^1 K^*(s,t)\,d\psi(s)$$

since by I.4.27 we have

$$\left| \int_0^1 (K^*(s,t+\delta) - K^*(s,t))\,d\psi(s) \right| \le \sup_{s \in [0,1]} |K^*(s,t+\delta) - K^*(s,t)|\; \|\psi\|_{BV_n}$$

and by I.6.16

$$\lim_{\delta \to 0+} \sup_{s \in [0,1]} |K^*(s,t+\delta) - K^*(s,t)| = 0\,.$$

Hence the above mentioned operator $\tilde{K} \in K(BV_n)$ maps BV_n into NBV_n when (1,27) is satisfied and its restriction K' to the closed subspace $NBV_n \subset BV_n$ consequently belongs to $K(NBV_n)$.

Let us now consider the pair of Banach spaces BV_n, NBV_n which form a dual pair (BV_n, NBV_n) with respect to the bilinear form

$$\langle x, \varphi \rangle = \int_0^1 x^*(t)\,d\varphi(t), \qquad x \in BV_n\,, \quad \varphi \in NBV_n \tag{1,29}$$

(see I.5.9). By the results from 1.3 we have

$$Kx = \int_0^1 d_t[K(s,t)]\,x(t) = \int_0^1 d_t[\hat{K}(s,t)]\,x(t)\,, \qquad s \in [0,1]$$

for every $x \in BV_n$, where $\hat{K}(s,t)$ is defined by (1,9) and $\hat{K}(s,t)$ evidently satisfies (1,1), (1,2), (1,16) and (1,27) (i.e. the assumptions of 1.10). Hence

$$\langle Kx, \varphi \rangle = \left\langle \int_0^1 d_t[\hat{K}(.,t)]\,x(t), \varphi \right\rangle$$

for every $\mathbf{x} \in BV_n$, $\varphi \in NBV_n$. Using I.6.22 we obtain

$$\langle \mathbf{Kx}, \varphi \rangle = \left\langle \mathbf{x}, \int_0^1 \hat{\mathbf{K}}^*(s, .)\, d\varphi(s) \right\rangle \qquad \text{for every} \quad \mathbf{x} \in BV_n, \quad \varphi \in NBV_n,$$

i.e.

$$\langle \mathbf{Kx}, \varphi \rangle = \langle \mathbf{x}, \mathbf{K}'\varphi \rangle$$

where

$$\mathbf{K}'\varphi = \int_0^1 \hat{\mathbf{K}}^*(s, t)\, d\varphi(s), \qquad t \in [0, 1], \quad \varphi \in NBV_n \tag{1,30}$$

and $\mathbf{K}'$ is a compact operator acting on the space NBV_n. Resuming these results we have

1.11. Theorem. *If $\mathbf{K}$: $I \to L(R_n)$ satisfies (1,1), (1,2), then for the operator $\mathbf{K} \in K(BV_n)$ given by (1,3) we have*

$$\langle \mathbf{Kx}, \varphi \rangle = \langle \mathbf{x}, \mathbf{K}'\varphi \rangle$$

for every $\mathbf{x} \in BV_n$, $\varphi \in NBV_n$ where $\mathbf{K}' \in K(NBV_n)$ is given by (1,30) and the bilinear form $\langle \mathbf{x}, \varphi \rangle$ on $BV_n \times NBV_n$ is given by (1,29).

2. Fredholm-Stieltjes integral equations

In this section we consider the Fredholm-Stieltjes integral equation

$$\mathbf{x}(t) - \int_0^1 d_s[\mathbf{K}(t, s)]\, \mathbf{x}(s) = \mathbf{f}(t)$$

in the Banach space $BV_n[0, 1] = BV_n$.

The fundamental results concerning equations of this kind are contained in the following

2.1. Theorem. *If $\mathbf{K}$: $I \to L(R_n)$ ($I = [0, 1] \times [0, 1] \subset R_2$)* satisfies

$$v_I(\mathbf{K}) < \infty, \tag{2,1}$$

$$\mathrm{var}_0^1\, \mathbf{K}(0, .) < \infty, \tag{2,2}$$

then either

I. *the Fredholm-Stieltjes integral equation*

$$\mathbf{x}(t) - \int_0^1 d_s[\mathbf{K}(t, s)]\, \mathbf{x}(s) = \mathbf{f}(t), \qquad t \in [0, 1] \tag{2,3}$$

admits a unique solution in BV_n for any $\mathbf{f} \in BV_n$ or

II. *the homogeneous Fredholm-Stieltjes integral equation*

$$(2,4) \qquad \mathbf{x}(t) - \int_0^1 \mathrm{d}_s[\mathbf{K}(t,s)]\,\mathbf{x}(s) = \mathbf{0}$$

admits r linearly independent solutions $\mathbf{x}_1, \ldots, \mathbf{x}_r \in BV_n$ *where r is a positive integer.*

If moreover it is assumed that

$$(2,5) \qquad \operatorname{var}_0^1 \mathbf{K}(.,0) < \infty\,,$$

$$(2,6) \qquad \mathbf{K}(t,s+) = \mathbf{K}(t,s) \quad \text{for any} \quad t \in [0,1], \quad s \in (0,1)$$

and

$$(2,7) \qquad \mathbf{K}(t,0) \;= \mathbf{0} \qquad \text{for any} \quad t \in [0,1]\,,$$

then in the case I. *the equation*

$$(2,8) \qquad \boldsymbol{\varphi}(s) - \int_0^1 \mathbf{K}^*(t,s)\,\mathrm{d}\boldsymbol{\varphi}(t) = \boldsymbol{\psi}(s)$$

admits a unique solution in NBV_n *for any* $\boldsymbol{\psi} \in NBV_n$ *and in the case* II. *the corresponding homogeneous equation*

$$(2,9) \qquad \boldsymbol{\varphi}(s) - \int_0^1 \mathbf{K}^*(t,s)\,\mathrm{d}\boldsymbol{\varphi}(t) = \mathbf{0}$$

admits also r linearly independent solutions $\boldsymbol{\varphi}_1, \boldsymbol{\varphi}_2, \ldots, \boldsymbol{\varphi}_r \in NBV_n$.

Proof. Let us denote by

$$\mathbf{A}\mathbf{x} = (\mathbf{I} - \mathbf{K})\,\mathbf{x} = \mathbf{x}(t) - \int_0^1 \mathrm{d}_s[\mathbf{K}(t,s)]\,\mathbf{x}(s)\,, \qquad \mathbf{x} \in BV_n$$

the linear operator corresponding to the Fredholm-Stieltjes integral equation (2,3). By $\mathbf{I}$ we denote the identity operator on BV_n and $\mathbf{K}$ is the operator defined by (1,5). Since 1.5 implies $\mathbf{K} \in K(BV_n)$, we have by I.3.20 $\operatorname{ind} \mathbf{A} = \operatorname{ind}(\mathbf{I} - \mathbf{K}) = 0$ and this implies the first part of our theorem immediately.

Under the assumptions of the second part we have by 1.11 $\langle \mathbf{K}\mathbf{x}, \boldsymbol{\varphi} \rangle = \langle \mathbf{x}, \mathbf{K}'\boldsymbol{\varphi} \rangle$ for every $\mathbf{x} \in BV_n$, $\boldsymbol{\varphi} \in NBV_n$ where $\mathbf{K}'\boldsymbol{\varphi} = \int_0^1 \mathbf{K}^*(t,s)\,\mathrm{d}\boldsymbol{\varphi}(t)$ is a compact operator acting on NBV_n (see 1,10). Hence $\operatorname{ind}(\mathbf{I} - \mathbf{K}') = 0$ and by I.3.20 we have $\alpha(\mathbf{I} - \mathbf{K}) = \alpha(\mathbf{I} - \mathbf{K}') = \beta(\mathbf{I} - \mathbf{K}) = \beta(\mathbf{I} - \mathbf{K}')$. This completes the proof.

2.2. Theorem. *If* $\mathbf{K}\colon I \to L(R_n)$ *satisfies* (2,1), (2,2), (2,5), (2,6) *and* (2,7), *then the equation* (2,3) *has a solution in* BV_n *if and only if*

$$(2,10) \qquad \int_0^1 \mathbf{f}^*(t)\,\mathrm{d}\boldsymbol{\varphi}(t) = 0$$

for any solution $\varphi \in NBV_n$ *of the homogeneous equation* (2,9) *and symmetrically the equation* (2,8) *has a solution in* NBV_n *if and only if*

$$\int_0^1 \mathbf{x}^*(t) \, \mathrm{d}\psi(t) = 0 \tag{2,11}$$

for any solution $\mathbf{x} \in BV_n$ *of the homogeneous equation* (2,4).

Proof. In the proof of 2.1 it was shown that all assumptions of Theorem I.3.2 are satisfied. Hence this statement is only a reformulation of the results from I.3.2.

2.3. Remark. 2.1 and 2.3 represent Fredholm theorems for the Stieltjes integral equations (2,3) and (2,8). It is of interest that the corresponding integral operators occuring in these equations are not connected with one another by the usual concept of adjointness. In this concrete situation the difficulties with the analytic description of the dual BV_n^* obstruct the analytic description of the adjoint $\mathbf{K}^*$. Fortunately the concept of the conjugate operator $\mathbf{K}'$ with respect to suitably described total subspace NBV_n works in our case and the results are given in an acceptable form.

2.4. Remark. Let us mention that in accordance with 1.4 in the same way the conjugate equation (2,8) in NBV_n can be replaced by the same equation working in NBV_n^- when instead of (2,6) we assume that $\mathbf{K}(t, s-) = \mathbf{K}(t, s)$ for any $t \in [0, 1]$, $s \in (0, 1)$.

2.5. Theorem. *Let* $\mathbf{K}$: $I = [0, 1] \times [0, 1] \to L(R_n)$ *satisfy* (2,1), (2,2) *and* (2,5). *If the homogeneous Fredholm-Stieltjes integral equation*

$$\mathbf{x}(t) - \int_0^1 \mathrm{d}_s[\mathbf{K}(t, s)] \, \mathbf{x}(s) = \mathbf{0}, \qquad t \in [0, 1] \tag{2,4}$$

has only the trivial solution $\mathbf{x} = \mathbf{0}$ *in* BV_n, *then there exists a unique* $n \times n$*-matrix valued function* $\boldsymbol{\Gamma}(t, s)$: $I \to L(R_n)$ *such that*

$$\mathrm{v}_I(\boldsymbol{\Gamma}) < \infty, \tag{2,12}$$

$$\mathrm{var}_0^1 \boldsymbol{\Gamma}(.,0) < \infty, \tag{2,13}$$

$$\mathrm{var}_0^1 \boldsymbol{\Gamma}(0, .) < \infty, \tag{2,14}$$

and for all $(t, s) \in I$ *the equation*

$$\boldsymbol{\Gamma}(t, s) = \mathbf{K}(t, s) + \int_0^1 \mathrm{d}_r[\mathbf{K}(t, r)] \, \boldsymbol{\Gamma}(r, s) \tag{2,15}$$

is satisfied.

Moreover for any $\mathbf{f} \in BV_n$ *the unique solution* $\mathbf{x} \in BV_n$ *of the Fredholm-Stieltjes integral equation* (2,3) *is given by the formula*

$$\mathbf{x}(t) = \mathbf{f}(t) + \int_0^1 \mathrm{d}_s[\boldsymbol{\Gamma}(t, s)]\, \mathbf{f}(s), \qquad t \in [0, 1]. \tag{2,16}$$

Proof. Let us set $\mathbf{A} = \mathbf{I} - \mathbf{K} \in B(BV_n)$ where $\mathbf{K}\mathbf{x} = \int_0^1 \mathrm{d}_s[\mathbf{K}(t, s)]\, \mathbf{x}(s)$, $\mathbf{x} \in BV_n$ and $\mathbf{I}$ is the identity operator on BV_n. By assumption we have $N(\mathbf{A}) = \{\mathbf{0}\}$. Since $\mathbf{K} \in B(BV_n)$ is compact by 1.5, we have $0 = \alpha(\mathbf{A}) = \beta(\mathbf{A}) = \dim(BV_n/R(\mathbf{A}))$ by I.3.20. Since $R(\mathbf{A})$ is closed, we obtain $R(\mathbf{A}) = BV_n$. Hence the Bounded Inverse Theorem I.3.4 implies that the inverse operator $\mathbf{A}^{-1} \in B(BV_n)$ exists and for any $\mathbf{f} \in BV_n$ the unique solution of (2,3) is given by $\mathbf{A}^{-1}\mathbf{f}$ and for this solution the estimate

$$\|\mathbf{x}\|_{BV_n} \leq C\|\mathbf{f}\|_{BV_n} \tag{2,17}$$

holds where $C = \|\mathbf{A}^{-1}\|_{B(BV_n)}$ is a constant.

Let us consider the matrix equation (2,15). Evidently the l-th column $\boldsymbol{\Gamma}_l(t, s)$ of $\boldsymbol{\Gamma}(t, s)\colon I \to L(R_n)$, $l = 1, 2, \ldots, n$ satisfies the equation

$$\boldsymbol{\Gamma}_l(t, s) = \mathbf{K}_l(t, s) + \int_0^1 \mathrm{d}_r[\mathbf{K}(t, r)]\, \boldsymbol{\Gamma}_l(r, s), \tag{2,18}$$

i.e. $\boldsymbol{\Gamma}_l(t, s)$ satisfies in the first variable the equation (2,3) with $\mathbf{f}(t) = \mathbf{K}_l(t, s)$ for any $s \in [0, 1]$. We have $\mathbf{f} \in BV_n$ since by I.6.6 $\mathrm{var}_0^1 \mathbf{K}(., s) \leq \mathrm{v}_I(\mathbf{K}) + \mathrm{var}_0^1 \mathbf{K}(., 0)$ and (2,1), (2,5) are assumed. By 2.1 the equation (2,18) has exactly one solution for any fixed $s \in [0, 1]$ and consequently the same holds also for the matrix equation (2,15).

Let us now consider the properties of the matrix $\boldsymbol{\Gamma}(t, s)$ defined by (2,15). By (2,17) the inequality

$$\|\boldsymbol{\Gamma}_l(., s)\|_{BV_n} \leq C\|\mathbf{K}_l(., s)\|_{BV_n}$$

holds for every $s \in [0, 1]$, $l = 1, 2, \ldots, n$. Hence (from the definition of the norm in BV_n) we obtain for any $s \in [0, 1]$ the inequality

$$|\boldsymbol{\Gamma}(0, s)| + \mathrm{var}_0^1 \boldsymbol{\Gamma}(., s) \leq C(|\mathbf{K}(0, s)| + \mathrm{var}_0^1 \mathbf{K}(., s))$$

which yields (2,13).

Let $0 = \alpha_0 < \alpha_1 < \ldots < \alpha_k = 1$ be an arbitrary decomposition of $[0, 1]$. If $\boldsymbol{\Gamma}(t, s)$ satisfies (2,15), then for any $j = 1, \ldots, k$ and $t \in [0, 1]$ we have

$$\boldsymbol{\Gamma}(t, \alpha_j) - \boldsymbol{\Gamma}(t, \alpha_{j-1}) = \mathbf{K}(t, \alpha_j) - \mathbf{K}(t, \alpha_{j-1}) + \int_0^1 \mathrm{d}_r[\mathbf{K}(t, r)]\, (\boldsymbol{\Gamma}(r, \alpha_j) - \boldsymbol{\Gamma}(r, \alpha_{j-1})),$$

i.e. the difference $\boldsymbol{\Gamma}(t, \alpha_j) - \boldsymbol{\Gamma}(t, \alpha_{j-1})$ satisfies a matrix equation of the type (2,15) and consequently by the Bounded Inverse Theorem I.3.4 we have as above

$$\begin{aligned} &|\boldsymbol{\Gamma}(0, \alpha_j) - \boldsymbol{\Gamma}(0, \alpha_{j-1})| + \mathrm{var}_0^1 (\boldsymbol{\Gamma}(., \alpha_j) - \boldsymbol{\Gamma}(., \alpha_{j-1})) \\ &\quad \leq C(|\mathbf{K}(0, \alpha_j) - \mathbf{K}(0, \alpha_{j-1})| + \mathrm{var}_0^1 (\mathbf{K}(., \alpha_j) - \mathbf{K}(., \alpha_{j-1})). \end{aligned} \tag{2,19}$$

Hence

$$\sum_{j=1}^{k} |\boldsymbol{\Gamma}(0,\alpha_j) - \boldsymbol{\Gamma}(0,\alpha_{j-1})| \le C(\mathrm{var}_0^1 \mathbf{K}(0,.) + \sum_{j=1}^{k} \mathrm{var}_0^1 (\mathbf{K}(.,\alpha_j) - \mathbf{K}(.,\alpha_{j-1}))$$
$$\le C(\mathrm{var}_0^1 \mathbf{K}(0,.) + \mathrm{v}_I(\mathbf{K})),$$

and since the subdivision $0 = \alpha_0 < \alpha_1 < \ldots < \alpha_k = 1$ was arbitrary we get (2,14) by passing to the supremum over all finite subdivisions of $[0, 1]$.

Let now $J_{ij} = [\alpha_{i-1}, \alpha_i] \times [\alpha_{j-1}, \alpha_j]$, $i, j = 1, 2, \ldots, k$ be the net-type subdivision of I corresponding to the arbitrary subdivision $0 = \alpha_0 < \alpha_1 < \ldots < \alpha_k = 1$ of $[0, 1]$. For $\boldsymbol{\Gamma}: I \to L(R_n)$ satisfying (2,15) we obtain by (2,19) the following inequality

$$\sum_{i,j=1} |m_{\Gamma}(J_{ij})| = \sum_{i,j=1}^{k} |\boldsymbol{\Gamma}(\alpha_i, \alpha_j) - \boldsymbol{\Gamma}(\alpha_i, \alpha_{j-1}) - \boldsymbol{\Gamma}(\alpha_{i-1}, \alpha_j) + \boldsymbol{\Gamma}(\alpha_{i-1}, \alpha_{j-1})|$$
$$\le \sum_{i,j=1}^{k} \mathrm{var}_{\alpha_{i-1}}^{\alpha_i} (\boldsymbol{\Gamma}(.,\alpha_j) - \boldsymbol{\Gamma}(.,\alpha_{j-1})) = \sum_{j=1}^{k} \mathrm{var}_0^1 (\boldsymbol{\Gamma}(.,\alpha_j) - \boldsymbol{\Gamma}(.,\alpha_{j-1}))$$
$$\le C \sum_{j=1}^{k} (|\mathbf{K}(0,\alpha_j) - \mathbf{K}(0,\alpha_{j-1})| + \mathrm{var}_0^1 (\mathbf{K}(.,\alpha_j) - \mathbf{K}(.,\alpha_{j-1})))$$
$$\le C(\mathrm{var}_0^1 \mathbf{K}(0,.) + \mathrm{v}_I(\mathbf{K})).$$

This inequality yields evidently (2,12) and the first part of the theorem is proved.

Now we prove that by (2,16) really the unique solution of (2,3) is given. Since $\boldsymbol{\Gamma}: I \to L(R_n)$ satisfies (2,12) and (2,14), by I.6.18 the integral $\int_0^1 \mathrm{d}_s[\boldsymbol{\Gamma}(t,s)]\, \boldsymbol{f}(s)$ exists for any $\boldsymbol{f} \in BV_n$ and $t \in [0, 1]$. Putting (2,16) into the left-hand side of (2,3) we obtain the expression

$$\boldsymbol{f}(t) + \int_0^1 \mathrm{d}_s[\boldsymbol{\Gamma}(t,s)]\, \boldsymbol{f}(s) - \int_0^1 \mathrm{d}_r[\mathbf{K}(t,r)] \left(\boldsymbol{f}(r) + \int_0^1 \mathrm{d}_s[\boldsymbol{\Gamma}(r,s)]\, \boldsymbol{f}(s)\right) = \boldsymbol{l}(t).$$

Hence

$$\boldsymbol{l}(t) = \boldsymbol{f}(t) + \int_0^1 \mathrm{d}_s[\boldsymbol{\Gamma}(t,s) - \mathbf{K}(t,s)]\, \boldsymbol{f}(s) - \int_0^1 \mathrm{d}_r[\mathbf{K}(t,r)] \int_0^1 \mathrm{d}_s[\boldsymbol{\Gamma}(r,s)]\, \boldsymbol{f}(s).$$

Using I.6.20 we obtain

$$\int_0^1 \mathrm{d}_r[\mathbf{K}(t,r)] \int_0^1 \mathrm{d}_s[\boldsymbol{\Gamma}(r,s)]\, \boldsymbol{f}(s) = \int_0^1 \mathrm{d}_s \left(\int_0^1 \mathrm{d}_r[\mathbf{K}(t,r)]\, \boldsymbol{\Gamma}(r,s)\right) \boldsymbol{f}(s), \qquad \text{i.e.}$$

$$\boldsymbol{l}(t) = \boldsymbol{f}(t) + \int_0^1 \mathrm{d}_s \left[\boldsymbol{\Gamma}(t,s) - \mathbf{K}(t,s) + \int_0^1 \mathrm{d}_r[\mathbf{K}(t,r)]\, \boldsymbol{\Gamma}(r,s)\right] \boldsymbol{f}(s) = \boldsymbol{f}(t)$$

since $\boldsymbol{\Gamma}: I \to L(R_n)$ satisfies (2,15) and consequently (2,16) gives the solution of (2,3). This concludes the proof of our theorem.

2.6. Remark. The matrix valued function $\boldsymbol{\Gamma}(t,s): I \to L(R_n)$ given in 2.6 is the resolvent of the Fredholm-Stieltjes integral equation (2,3). This resolvent gives by (2,16) the unique solution of (2,3) for every $\boldsymbol{f} \in BV_n$. For the existence of the resolvent $\boldsymbol{\Gamma}(t, s)$ the assumption $\alpha(\mathbf{A}) = \dim(\mathbf{I} - \mathbf{K}) = 0$ is essential.

Further let us investigate the equation (2,3) when $r = \alpha(\mathbf{A}) = \dim(\mathbf{I} - \mathbf{K}) \neq 0$. By assertion II. from 2.1 the homogeneous equation (2,4) admits in this case r linearly independent solutions $\boldsymbol{x}_1, \ldots, \boldsymbol{x}_r \in BV_n$ and $R(\mathbf{I} - \mathbf{K}) \neq BV_n$, i.e. (2,3) has no solutions for all $\boldsymbol{f} \in BV_n$

The following theorem holds in this situation. Let $\mathbf{K}\colon I = [0,1] \times [0,1] \to L(R_n)$ satisfy (2,1), (2,2), (2,5) and $\mathbf{K}(t, s+) = \mathbf{K}(t, s)$ for any $t \in [0,1]$, $s \in (0,1)$, $\mathbf{K}(t, 0) = \mathbf{0}$ for any $t \in [0,1]$. Then there exists an $n \times n$-matrix valued function $\hat{\boldsymbol{\Gamma}}(t, s)\colon I \to L(R_n)$ such that $\mathrm{v}_I(\hat{\boldsymbol{\Gamma}}) < \infty$, $\mathrm{var}_0^1 \hat{\boldsymbol{\Gamma}}(.,0) < \infty$, $\mathrm{var}_0^1 \hat{\boldsymbol{\Gamma}}(0,.) < \infty$ and if the Fredholm-Stieltjes integral equation (2,3) has solutions for $\boldsymbol{f} \in BV_n$ (i.e. if $\boldsymbol{f} \in R(\mathbf{I} - \mathbf{K})$, see also 2.3), then one of them is given by the formula

(2,20) $$\boldsymbol{x}(t) = \boldsymbol{f}(t) + \int_0^1 \mathrm{d}_s[\hat{\boldsymbol{\Gamma}}(t,s)]\, \boldsymbol{f}(s), \qquad t \in [0,1].$$

The general form of solutions of (2,3) *is*

$$\boldsymbol{x}(t) = \boldsymbol{f}(t) + \int_0^1 \mathrm{d}_s[\hat{\boldsymbol{\Gamma}}(t,s)]\, \boldsymbol{f}(s) + \sum_{i=1}^{r} \alpha_i \boldsymbol{x}^i(t),$$

where $\boldsymbol{x}^i$, $i = 1, \ldots, r$ *are linearly independent solutions of the homogeneous equation* (2,4) (cf. 2.1) *and* $\alpha_1, \ldots, \alpha_r$ *are arbitrary constants.*

The proof of this assertion is based on some pseudoresolvent technique using projections in BV_n. The theorem is completely proved in Schwabik [6].

3. Volterra-Stieltjes integral equations

In this section we consider integral equations of the form

(3,1) $$\boldsymbol{x}(t) - \int_0^t \mathrm{d}_s[\mathbf{K}(t,s)]\, \boldsymbol{x}(s) = \boldsymbol{f}(t), \qquad t \in [0,1]$$

in the Banach space $BV_n[0,1] = BV_n$ with $\boldsymbol{f} \in BV_n$. Equations of the form (3,1) are called *Volterra-Stieltjes integral equations.*

Throughout this paragraph it will be assumed that $\mathbf{K}\colon I = [0,1] \times [0,1] \to L(R_n)$ satisfies

(3,2) $$\mathrm{v}_I(\mathbf{K}) < \infty$$

and

(3,3) $$\mathrm{var}_0^1 \mathbf{K}(0,.) < \infty.$$

Let us mention that (3,3) can be replaced by $\mathrm{var}_0^1 \mathbf{K}(t_0,.) < \infty$, where $t_0 \in [0,1]$ is arbitrary. This follows from I.6.6.

Since (3,2) and (3,3) are assumed, for every fixed $t \in [0,1]$ we have $\operatorname{var}_0^1 \mathbf{K}(t,.) < \infty$ by I.6.6. Hence for any $\mathbf{x} \in BV_n$ and $t \in [0,1]$ the integral $\int_0^t \mathrm{d}_s[\mathbf{K}(t,s)]\,\mathbf{x}(s)$ exists by I.4.19.

Let us show that the equation (3,1) is a special case of the Fredholm-Stieltjes integral equation considered in the previous Section II.2.

To any given kernel $\mathbf{K}\colon I \to L(R_n)$ satisfying (3,2) and (3,3) we define a new "triangular" kernel $\mathbf{K}^\Delta\colon I \to L(R_n)$ as follows:

$$(3,4)\qquad \begin{aligned} &\mathbf{K}^\Delta(t,s) = \mathbf{K}(t,s) - \mathbf{K}(t,0) && \text{if } 0 \le s \le t \le 1\,,\\ &\mathbf{K}^\Delta(t,s) = \mathbf{K}(t,t) - \mathbf{K}(t,0) = \mathbf{K}^\Delta(t,t) && \text{if } 0 \le t < s \le 1\,. \end{aligned}$$

It is obvious that $\mathbf{K}^\Delta(t,0) = \mathbf{0}$ for any $t \in [0,1]$ and $\mathbf{K}^\Delta(0,s) = \mathbf{K}^\Delta(0,0) = \mathbf{0}$ for any $s \in [0,1]$. Let

$$J_{ij} = [\alpha_{i-1}, \alpha_i] \times [\alpha_{j-1}, \alpha_j]\,, \qquad i,j = 1,\dots,k$$

be an arbitrary net-type subdivision of the interval I corresponding to the subdivision $0 = \alpha_0 < \alpha_1 < \dots < \alpha_k = 1$ of $[0,1]$. By definition (3,4) of $\mathbf{K}^\Delta$ we have

$$\begin{aligned} &m_{\mathbf{K}^\Delta}(J_{ij}) = m_{\mathbf{K}}(J_{ij}) && \text{if } 0 \le j < i \le k\,,\\ &m_{\mathbf{K}^\Delta}(J_{ij}) = \mathbf{0} && \text{if } 0 \le i < j \le k \end{aligned}$$

and

$$m_{\mathbf{K}^\Delta}(J_{ii}) = \mathbf{K}(\alpha_i, \alpha_i) - \mathbf{K}(\alpha_i, \alpha_{i-1}) \qquad \text{for } i = 1,2,\dots,k\,.$$

(For $m_{\mathbf{K}}(J)$ see I.6.2.) Hence

$$\begin{aligned} \sum_{i,j=1}^{k} |m_{\mathbf{K}^\Delta}(J_{ij})| &= \sum_{i=1}^{k} \sum_{j=1}^{i-1} |m_{\mathbf{K}}(J_{ij})| + \sum_{i=1}^{k} |\mathbf{K}(\alpha_i,\alpha_i) - \mathbf{K}(\alpha_i,\alpha_{i-1})| \\ \le \sum_{i=1}^{k} \sum_{j=1}^{i-1} |m_{\mathbf{K}}(J_{ij})| &+ \sum_{i=1}^{k} |\mathbf{K}(\alpha_i,\alpha_i) - \mathbf{K}(\alpha_i,\alpha_{i-1}) - \mathbf{K}(0,\alpha_i) + \mathbf{K}(0,\alpha_{i-1})| \\ &+ \sum_{i=1}^{k} |\mathbf{K}(0,\alpha_i) - \mathbf{K}(0,\alpha_{i-1})| \le \mathrm{v}_I(\mathbf{K}) + \operatorname{var}_0^1 \mathbf{K}(0,.)\,. \end{aligned}$$

Consequently we obtain by definition (cf. I.6.1, I.6.3)

$$(3,5)\qquad \mathrm{v}_I(\mathbf{K}^\Delta) \le \mathrm{v}_I(\mathbf{K}) + \operatorname{var}_0^1 \mathbf{K}(0,.) < \infty$$

Since $\mathbf{K}^\Delta(t,s)$ is by definition constant on the interval $[t,1]$ for every $t \in [0,1]$, we have

$$(3,6)\qquad \int_t^1 \mathrm{d}_s[\mathbf{K}^\Delta(t,s)]\,\mathbf{x}(s) = \mathbf{0}$$

for every $\mathbf{x} \in BV_n$. Further

$$\int_0^t \mathrm{d}_s[\mathbf{K}^\Delta(t,s) - \mathbf{K}(t,s)]\,\mathbf{x}(s) = -\int_0^t \mathrm{d}_s[\mathbf{K}(t,0)]\,\mathbf{x}(s) = \mathbf{0}\,,$$

i.e.

$$\int_0^t \mathrm{d}_s[\mathbf{K}^\Delta(t,s)]\,\mathbf{x}(s) = \int_0^t \mathrm{d}_s[\mathbf{K}(t,s)]\,\mathbf{x}(s)\,.$$

Using (3,6) we obtain for an arbitrary $T \in [0,1]$ the equality

$$\int_0^t \mathrm{d}_s[\mathbf{K}(t,s)]\,\mathbf{x}(s) = \int_0^T \mathrm{d}_s[\mathbf{K}^\Delta(t,s)]\,\mathbf{x}(s) \tag{3,7}$$

for any $\mathbf{x} \in BV_n$ and $t \in [0, T]$.

Let us summarize these results.

3.1. Proposition. *Let* $\mathbf{K}\colon I \to L(R_n)$ *satisfy* (3,2), (3,3). *Then for the triangular kernel* $\mathbf{K}^\Delta\colon I \to L(R_n)$ *defined by the relations* (3, 4) *the following is valid.*

(a) $\mathrm{var}_0^1\, \mathbf{K}^\Delta(.,0) = 0,\quad \mathrm{var}_0^1\, \mathbf{K}^\Delta(0,.) = 0,\quad \mathrm{v}_I(\mathbf{K}^\Delta) < \infty,$

(b) *for every* $\mathbf{x} \in BV_n$, $T \in [0,1]$ *and* $t \in [0, T]$ *the equality* (3,7) *holds, i.e. by the relation*

$$\mathbf{K}\mathbf{x} = \int_0^t \mathrm{d}_s[\mathbf{K}(t,s)]\,\mathbf{x}(s) = \int_0^1 \mathrm{d}_s[\mathbf{K}^\Delta(t,s)]\,\mathbf{x}(s)\,, \qquad \mathbf{x} \in BV_n \tag{3,8}$$

an operator on BV_n *is defined and by* 1.5 *we have* $\mathbf{K} \in K(BV_n)$.

3.2. Remark. Proposition 3.1 states that the Volterra-Stieltjes integral equation (3,1) is equivalent to the Fredholm-Stieltjes integral equation

$$\mathbf{x}(t) - \int_0^1 \mathrm{d}_s[\mathbf{K}^\Delta(t,s)]\,\mathbf{x}(s) = \mathbf{f}(t)\,, \qquad t \in [0,1]\,. \tag{3,9}$$

Hence by Theorem 2.1 either the equation (3,1) admits a unique solution in BV_n for every $\mathbf{f} \in BV_n$ or the corresponding homogeneous equation

$$\mathbf{x}(t) - \int_0^t \mathrm{d}_s[\mathbf{K}(t,s)]\,\mathbf{x}(s) = \mathbf{0}\,, \qquad t \in [0,1] \tag{3,10}$$

has a finite number of linearly independent solutions in BV_n. Our aim is to give conditions under which the equation (3,1) is really of Volterra type, i.e. when the equation (3,10) admits only the trivial solution $\mathbf{x} = \mathbf{0}$ in BV_n.

3.3. Theorem. *Let the kernel* $\mathbf{K}\colon I \to L(R_n)$ *satisfy* (3,2), (3,3) *and*

$$\lim_{\sigma \to s-} |\mathbf{K}(t,\sigma) - \mathbf{K}(t,s)| = 0 \tag{3,11}$$

for any $t \in [0,1]$, $s \in (0,1]$. *Then the homogeneous Volterra-Stieltjes integral equation* (3,10) *has only the trivial solution* $\mathbf{x} = \mathbf{0}$ *in* BV_n.

Proof. Let $\mathbf{K}^\Delta: I \to L(R_n)$ be the triangular kernel corresponding to $\mathbf{K}$ by the relations (3,4). Since (3,11) holds we have also

$$(3,12) \qquad \lim_{\sigma \to s-} |\mathbf{K}^\Delta(t, \sigma) - \mathbf{K}^\Delta(t, s)| = 0$$

for any $t \in [0, 1]$, $s \in (0, 1]$. Let us set

$$\omega_2^\Delta(0) = 0, \qquad \omega_2^\Delta(s) = \mathrm{v}_{[0,1]\times[0,s]}(\mathbf{K}^\Delta) \qquad \text{for} \quad s \in (0, 1].$$

The function ω_2^Δ: $[0, 1] \to R$ is evidently nonnegative and nondecreasing (see I.6.7). Since (3,12) holds we have $\omega_2^\Delta(s-) = \omega_2^\Delta(s)$ for every $s \in (0, 1]$ by I.6.11, i.e. ω_2^Δ is left continuous on $[0, 1]$.

Assume that $\mathbf{x} \in BV_n$ is a solution of (3,10). Then evidently $\mathbf{x}(0) = \mathbf{0}$ and

$$|\mathbf{x}(s)| \leq |\mathbf{x}(0)| + \mathrm{var}_0^s \, \mathbf{x} = \mathrm{var}_0^s \, \mathbf{x}$$

for every $s \in [0, 1]$. Using (b) from 3.1 we get

$$\mathrm{var}_0^\xi \, \mathbf{x} = \mathrm{var}_0^\xi \left(\int_0^t \mathrm{d}_s[\mathbf{K}(t, s)] \, \mathbf{x}(s) \right) = \mathrm{var}_0^\xi \left(\int_0^\xi \mathrm{d}_s[\mathbf{K}^\Delta(\,.\,, s)] \, \mathbf{x}(s) \right)$$

for every $\xi \in [0, 1]$. If (I.6,14) from I.6.18 is used then we obtain

$$\mathrm{var}_0^\xi \, \mathbf{x} = \mathrm{var}_0^\xi \left(\int_0^\xi \mathrm{d}_s[\mathbf{K}^\Delta(\,.\,, s)] \, \mathbf{x}(s) \right) \leq \int_0^s |\mathbf{x}(s)| \, \mathrm{d}\omega_2^\Delta(s) \leq \int_0^\xi \mathrm{var}_0^s \, \mathbf{x} \, \mathrm{d}\omega_2^\Delta(s)$$

and I.4.30 yields the inequality $\mathrm{var}_0^\xi \, \mathbf{x} \leq 0$ for every $\xi \in [0, 1]$. Hence $\mathbf{x}(s) \equiv \mathbf{0}$ on $[0, 1]$, i.e. $\mathbf{x} = \mathbf{0} \in BV_n$.

3.4. Example. Let us define $h(t) = 0$ if $0 \leq t < \frac{1}{2}$, $h(t) = 1/t$ if $\frac{1}{2} \leq t \leq 1$, $g(s) = 0$ if $0 \leq s < \frac{1}{2}$, $g(s) = s$ if $\frac{1}{2} \leq s \leq 1$. Evidently $h, g \in BV$. If we set $k(t, s) = h(t) \, g(s)$ for $(t, s) \in I = [0, 1] \times [0, 1]$, then clearly $\mathrm{v}_I(k) < \infty$ (cf. I.6.4), $\mathrm{var}_0^1 \, k(0, .) < \infty$ and $\mathrm{var}_0^1 \, k(\,.\,, 0) < \infty$. Let us consider the homogeneous Volterra-Stieltjes integral equation

$$x(t) = \int_0^t \mathrm{d}_s[k(t, s)] \, x(s) = h(t) \int_0^t x(s) \, \mathrm{d}g(s), \qquad t \in [0, 1].$$

Let us set $y(s) = 0$ for $0 \leq s < \frac{1}{2}$, $y(s) = 1$ for $\frac{1}{2} \leq s \leq 1$. By easy computations using I.4,21 we obtain

$$\int_0^t y(s) \, \mathrm{d}g(s) = 0 \qquad \text{if} \quad 0 < t < \tfrac{1}{2},$$

$$\int_0^t y(s) \, \mathrm{d}g(s) = \tfrac{1}{2} \, y(\tfrac{1}{2}) + \int_{1/2}^t y(s) \, \mathrm{d}s = t \qquad \text{if} \quad \tfrac{1}{2} \leq t \leq 1$$

and consequently

$$h(t) \int_0^t y(s) \, \mathrm{d}g(s) = y(t)$$

for every $t \in [0,1]$. Hence $y \in BV$ is a solution of the homogeneous Volterra-Stieltjes integral equation and $y \neq 0$. The condition (3,11) is in this case affected. In fact, $\lim_{\sigma \to 1/2-} (k(t,\sigma) - k(t,\frac{1}{2})) = \frac{1}{2}h(t)$ and $h(t) \neq 0$ e.g. for $t = \frac{1}{2}$.

3.5. Remark. Example 3.4 shows that for $\mathbf{K}: I \to L(R_n)$ satisfying (3,2) and (3,3) the corresponding homogeneous Volterra-Stieltjes integral equation (3,10) need not have in general only trivial solutions, i.e. for the corresponding operator $\mathbf{K} \in K(BV_n)$ we can obtain in general a nontrivial null space $N(\mathbf{I} - \mathbf{K})$. If (3,11) is assumed, then this situation cannot occur. The condition (3,11) is too restrictive as will be shown in the following. We shall give necessary and sufficient conditions on $\mathbf{K}: I \to L(R_n)$ satisfying (3,2) and (3,3) such that the equation (3,10) has only the trivial solution in BV_n.

3.6. Proposition. *Let* $\mathbf{M}: I = [0,1] \times [0,1] \to L(R_n)$ *satisfy* $\mathrm{v}_I(\mathbf{M}) < \infty$, $\mathrm{var}_0^1 \mathbf{M}(0,.) < \infty$. *Then for any* $a \in [0,1]$ *there exists a nondecreasing bounded function* $\xi: [a,1] \to [0,+\infty)$ *such that for every* $b \in [a,1]$ *and* $\mathbf{x} \in BV_n$ *we have*

$$\mathrm{var}_a^b \left(\int_a^t \mathrm{d}_s[\mathbf{M}(t,s)]\, \mathbf{x}(s) \right) \leq |\mathbf{x}(a)|\, (\xi(a+) - \xi(a)) + \|\mathbf{x}\|_{BV_n[a,b]} (\xi(b) - \xi(a+)). \tag{3,13}$$

Proof. Let $\mathbf{M}^{\Delta}: I \to L(R_n)$ be the triangular kernel which corresponds to $\mathbf{M}: I \to L(R_n)$ (see 3.1). For any $t \in [a,b]$ we have (see (3,7))

$$\int_a^t \mathrm{d}_s[\mathbf{M}(t,s)]\, \mathbf{x}(s) = \int_a^b \mathrm{d}_s[\mathbf{M}^{\Delta}(t,s)]\, \mathbf{x}(s). \tag{3,14}$$

Let us define the function

$$\xi(t) = \mathrm{v}_{[a,1] \times [a,t]}(\mathbf{M}^{\Delta}) \quad \text{for} \quad t \in (a,1], \quad \xi(a) = 0.$$

$\xi: [a,1] \to R$ is evidently nondecreasing and bounded on $[a,1]$ (cf. I.6.7).

From (I.6,14) in I.6.18 we obtain

$$\mathrm{var}_a^b \left(\int_a^t \mathrm{d}_s[\mathbf{M}(t,s)]\, \mathbf{x}(s) \right) \leq \int_a^b |\mathbf{x}(s)|\, \mathrm{d}\xi(s). \tag{3,15}$$

Using I.4.13 we have

$$\int_a^b |\mathbf{x}(s)|\, \mathrm{d}\xi(s) = |\mathbf{x}(a)|\, (\xi(a+) - \xi(a)) + \lim_{\delta \to 0+} \int_{a+\delta}^b |\mathbf{x}(s)|\, \mathrm{d}\xi(s) \tag{3,16}$$

and for any $0 < \delta < b - a$ by I.4.27

$$\int_{a+\delta}^b |\mathbf{x}(s)|\, \mathrm{d}\xi(s) \leq \sup_{s \in [a+\delta, b]} |\mathbf{x}(s)|\, (\xi(b) - \xi(a+\delta)) \leq \|\mathbf{x}\|_{BV_n[a,b]} (\xi(b) - \xi(a+)).$$

Hence (3,15) and (3,16) imply (3,13).

3.7. Proposition. *Let* $\boldsymbol{H}: [0,1] \to L(R_n)$ *be such that*

(i) $\operatorname{var}_0^1 \boldsymbol{H} < \infty$,

(ii) *there is an at most countable set of points* $t_i \in [0,1]$, $i = 1, 2, \dots$ *such that*

$$\boldsymbol{H}(t) = \boldsymbol{0} \quad \text{for} \quad t \in [0,1], \quad t \neq t_i, \quad i = 1, 2, \dots,$$

(iii) *the matrix* $\boldsymbol{I} - \boldsymbol{H}(t)$ *is regular for all* $t \in [0,1]$. *Let us define the linear operator*

$$\boldsymbol{Tz} = [\boldsymbol{I} - \boldsymbol{H}(t)]^{-1} \boldsymbol{z}(t), \qquad t \in [0,1] \quad \text{for} \quad \boldsymbol{z} \in BV_n. \tag{3,17}$$

Then there exists a constant $C \geq 0$ *such that*

$$\|\boldsymbol{Tz}\|_{BV_n} \leq C \|\boldsymbol{z}\|_{BV_n} \qquad \text{for every} \quad \boldsymbol{z} \in BV_n, \tag{3,18}$$

i.e. $\boldsymbol{T} \in B(BV_n)$.

Proof. By (iii) the inverse matrix $[\boldsymbol{I} - \boldsymbol{H}(t)]^{-1}$ exists for every $t \in [0,1]$ and the operator $\boldsymbol{T}$ from (3,17) is well-defined.

Since $\boldsymbol{I} = (\boldsymbol{I} - \boldsymbol{H}(t))[\boldsymbol{I} - \boldsymbol{H}(t)]^{-1} = [\boldsymbol{I} - \boldsymbol{H}(t)]^{-1} - \boldsymbol{H}(t)[\boldsymbol{I} - \boldsymbol{H}(t)]^{-1}$ for any $t \in [0,1]$, we have $[\boldsymbol{I} - \boldsymbol{H}(t)]^{-1} = \boldsymbol{I} + \boldsymbol{H}(t)[\boldsymbol{I} - \boldsymbol{H}(t)]^{-1}$ and for any $\boldsymbol{z} \in BV_n$ we have

$$\boldsymbol{Tz} = \boldsymbol{z} + \boldsymbol{u} \tag{3,19}$$

where

$$\boldsymbol{u}(t) = \boldsymbol{H}(t)[\boldsymbol{I} - \boldsymbol{H}(t)]^{-1} \boldsymbol{z}(t), \qquad t \in [0,1]. \tag{3,20}$$

The assumption (ii) implies $\boldsymbol{u}(t) = \boldsymbol{0}$ for any $t \in [0,1]$, $t \neq t_i$, $i = 1, 2, \dots$. Hence evidently

$$\begin{aligned} \operatorname{var}_0^1 \boldsymbol{u} &= 2 \sum_{i=1}^{\infty} |\boldsymbol{u}(t_i)| = 2 \sum_{i=1}^{\infty} |\boldsymbol{H}(t_i)[\boldsymbol{I} - \boldsymbol{H}(t_i)]^{-1} \boldsymbol{z}(t_i)| \\ &\leq 2 \|\boldsymbol{z}\|_{BV_n} \sum_{i=1}^{\infty} |\boldsymbol{H}(t_i)| \, |[\boldsymbol{I} - \boldsymbol{H}(t_i)]^{-1}|. \end{aligned} \tag{3,21}$$

By (i) and (ii) we have

$$\sum_{i=1}^{\infty} |\boldsymbol{H}(t_i)| \leq |\boldsymbol{H}(0)| + |\boldsymbol{H}(1)| + 2 \sum_{t_i \in (0,1)} |\boldsymbol{H}(t_i)| = \operatorname{var}_0^1 \boldsymbol{H} < \infty;$$

hence there exists an integer $i_0 > 0$ such that $|\boldsymbol{H}(t_i)| < \frac{1}{2}$ for any $i > i_0$. This implies

$$|[\boldsymbol{I} - \boldsymbol{H}(t_i)]^{-1}| \leq 1 + |\boldsymbol{H}(t_i)| + \dots + |\boldsymbol{H}(t_i)|^k + \dots = \frac{1}{1 - |\boldsymbol{H}(t_i)|} < 2$$

for $i > i_0$ and immediately also the inequality

$$\begin{aligned} \sum_{i=1}^{\infty} |\boldsymbol{H}(t_i)| \, |[\boldsymbol{I} - \boldsymbol{H}(t_i)]^{-1}| &\leq \sum_{i=1}^{i_0} |\boldsymbol{H}(t_i)| \, |[\boldsymbol{I} - \boldsymbol{H}(t_i)]^{-1}| + 2 \sum_{i=i_0+1}^{\infty} |\boldsymbol{H}(t_i)| \\ &\leq \Big(\max_{i=1,2,\dots,i_0} |[\boldsymbol{I} - \boldsymbol{H}(t_i)]^{-1}| + 2 \Big) \sum_{i=1}^{\infty} |\boldsymbol{H}(t_i)| = C_0 < \infty \end{aligned}$$

which yields by (3,21)

$$\operatorname{var}_0^1 \mathbf{u} \leq 2C_0 \|\mathbf{z}\|_{BV_n}.$$

Hence (see (3,19))

$$\|\mathbf{Tz}\|_{BV_n} = |[\mathbf{I} - \mathbf{H}(0)]^{-1} \mathbf{z}(0)| + \operatorname{var}_0^1 \mathbf{Tz} \leq (|[\mathbf{I} - \mathbf{H}(0)]^{-1}| + 1 + 2C_0) \|\mathbf{z}\|_{BV_n}$$

and (3,18) is satisfied with $C = 1 + 2C_0 + |[\mathbf{I} - \mathbf{H}(0)]^{-1}|$.

3.8. Proposition. *Let us assume that* $\mathbf{K}: I = [0,1] \times [0,1] \to L(R_n)$ *satisfies* (3,2) *and* (3,3). *Define* $\mathbf{M}: I \to L(R_n)$ *as follows*:

(3,22)
$$\begin{aligned} \mathbf{M}(t,s) &= \mathbf{K}(t,s) \quad && \text{if } (t,s) \in I, \ t \neq s, \\ \mathbf{M}(t,t) &= \mathbf{K}(t,t-) \quad && \text{if } t \in (0,1], \\ \mathbf{M}(0,0) &= \mathbf{K}(0,0). \end{aligned}$$

Then

(i) $v_I(\mathbf{M}) < \infty$, $\operatorname{var}_0^1 \mathbf{M}(0,.) < \infty$,

(ii) *if* $\mathbf{x} \in BV_n$, *then for any fixed* $t \in [0,1]$ *we have*

$$\lim_{\tau \to t-} \int_0^\tau d_s[\mathbf{M}(t,s)]\, \mathbf{x}(s) = \int_0^t d_s[\mathbf{M}(t,s)]\, \mathbf{x}(s),$$

i.e. the integral $\int_0^t d_s[\mathbf{M}(t,s)]\, \mathbf{x}(s)$ *does not depend on the value* $\mathbf{x}(t) \in R_n$,

(iii) *for every* $\mathbf{x} \in BV_n$ *and* $t \in [0,1]$ *we have*

(3,23)
$$\int_0^t d_s[\mathbf{K}(t,s)]\, \mathbf{x}(s) = \int_0^t d_s[\mathbf{M}(t,s)]\, \mathbf{x}(s) + \mathbf{H}(t)\, \mathbf{x}(t)$$

where

(3,24)
$$\begin{aligned} \mathbf{H}(t) &= \mathbf{K}(t,t) - \mathbf{K}(t,t-) \quad \text{for } t \in (0,1], \\ \mathbf{H}(0) &= \mathbf{0}, \end{aligned}$$

(iv) *for* $\mathbf{H}: [0,1] \to L(R_n)$ *given by* (3,24) *there exists an at most countable set of points* $t_i \in [0,1]$, $i = 1, 2, \ldots$ *such that* $\mathbf{H}(t) = \mathbf{0}$ *for* $t \in [0,1]$, $t \neq t_i$, $i = 1, 2, \ldots$ *and* $\operatorname{var}_0^1 \mathbf{H} < \infty$.

Proof. In order to prove (i) let us mention that $\mathbf{M}(0,s) = \mathbf{K}(0,s)$ for all $s \in [0,1]$ and consequently $\operatorname{var}_0^1 \mathbf{M}(0,.) = \operatorname{var}_0^1 \mathbf{K}(0,.) < \infty$. Further let $0 = \alpha_0 < \alpha_1 < \ldots < \alpha_k = 1$ be an arbitrary subdivision of $[0,1]$ and let

$$J_{ij} = [\alpha_{i-1}, \alpha_i] \times [\alpha_{j-1}, \alpha_j], \qquad i, j = 1, \ldots, k$$

be the corresponding net-type subdivision of I. We consider the sum $\sum_{i,j=1}^k |m_{\mathbf{M}}(J_{ij})|$ where

$$m_{\mathbf{M}}(J_{ij}) = \mathbf{M}(\alpha_i, \alpha_j) - \mathbf{M}(\alpha_i, \alpha_{j-1}) - \mathbf{M}(\alpha_{i-1}, \alpha_j) + \mathbf{M}(\alpha_{i-1}, \alpha_{j-1})$$

for $i, j = 1, \ldots, k$. Usual considerations using the definition of $\boldsymbol{M}$ in (3,22) give

$$\sum_{i,j=1}^{k} |m_{\boldsymbol{M}}(J_{ij})| \leq \mathrm{v}_I(\boldsymbol{K}) + 4 \sum_{j=1}^{k} |\boldsymbol{K}(\alpha_j, \alpha_j) - \boldsymbol{K}(\alpha_j, \alpha_j -)|.$$

Since

$$\sum_{j=1}^{k} |\boldsymbol{K}(\alpha_j, \alpha_j) - \boldsymbol{K}(\alpha_j, \alpha_j -)|$$

$$\leq \sum_{j=1}^{k} |\boldsymbol{K}(\alpha_j, \alpha_j) - \boldsymbol{K}(\alpha_j, \alpha_j -) - \boldsymbol{K}(0, \alpha_j) + \boldsymbol{K}(0, \alpha_j -)| + \sum_{j=1}^{k} |\boldsymbol{K}(0, \alpha_j) - \boldsymbol{K}(0, \alpha_j -)|$$

$$\leq \mathrm{v}_I(\boldsymbol{K}) + \mathrm{var}_0^1 \boldsymbol{K}(0, .),$$

we obtain

$$\sum_{i,j=1}^{k} |m_{\boldsymbol{M}}(J_{ij})| \leq 5\mathrm{v}_I(\boldsymbol{K}) + 4 \,\mathrm{var}_0^1 \boldsymbol{K}(0, .) < \infty$$

and (i) holds since J_{ij} was an arbitrary net-type subdivision.

Let $t \in (0, 1]$ be fixed, $\boldsymbol{x}, \boldsymbol{y} \in BV_n$, $\boldsymbol{x}(s) = \boldsymbol{y}(s)$ for $s \in [0, t)$. By I.4.21 and from the definition of $\boldsymbol{M}$ we obtain

$$\int_0^t \mathrm{d}_s[\boldsymbol{M}(t, s)] (\boldsymbol{x}(s) - \boldsymbol{y}(s)) = (\boldsymbol{M}(t, t) - \boldsymbol{M}(t, t-)) (\boldsymbol{x}(t) - \boldsymbol{y}(t)) = \boldsymbol{0}.$$

Hence

$$\int_0^t \mathrm{d}_s[\boldsymbol{M}(t, s)] \, \boldsymbol{x}(s) = \int_0^t \mathrm{d}_s[\boldsymbol{M}(t, s)] \, \boldsymbol{y}(s)$$

or in other words: for all $\boldsymbol{x} \in BV_n$ we have

$$\lim_{\tau \to t-} \int_0^\tau \mathrm{d}_s[\boldsymbol{M}(t, s)] \, \boldsymbol{x}(s) = \int_0^t \mathrm{d}_s[\boldsymbol{M}(t, s)] \, \boldsymbol{x}(s).$$

For $t = 0$ the statement is trivial. Hence (ii) is proved.

Further for any $t \in (0, 1]$ and $\boldsymbol{x} \in BV_n$ we have

$$\int_0^t \mathrm{d}_s[\boldsymbol{K}(t, s) - \boldsymbol{M}(t, s)] \, \boldsymbol{x}(s) = [\boldsymbol{K}(t, t) - \boldsymbol{M}(t, t) - \boldsymbol{K}(t, t-) + \boldsymbol{M}(t, t-)] \, \boldsymbol{x}(t)$$

$$= [\boldsymbol{K}(t, t) - \boldsymbol{K}(t, t-)] \, \boldsymbol{x}(t) = \boldsymbol{H}(t) \, \boldsymbol{x}(t),$$

and (3,23) holds. For $t = 0$ the equality (3,24) is evident. Hence (iii) is valid.

By I.6.8 the set of discontinuity points of $\boldsymbol{K}(t, s)$ in the second variable s lies on an at most countable system of lines in I which are parallel to the t axis, i.e. there is an at most countable system t_i, $i = 1, 2, \ldots$ of points in $[0, 1]$ such that $\boldsymbol{H}(t) = \boldsymbol{K}(t, t) - \boldsymbol{K}(t, t-) = \boldsymbol{0}$ for all $t \in [0, 1]$, $t \neq t_i$, $i = 1, 2, \ldots$.

For any $t \in [0, 1]$ we have evidently

$$|\boldsymbol{H}(t)| = |\boldsymbol{K}(t, t) - \boldsymbol{K}(t, t-)|$$

$$\leq |\boldsymbol{K}(t, t) - \boldsymbol{K}(t, t-) - \boldsymbol{K}(0, t) + \boldsymbol{K}(0, t-)| + |\boldsymbol{K}(0, t) - \boldsymbol{K}(0, t-)|.$$

Let $0 = \alpha_0 < \alpha_1 < \ldots < \alpha_k = b$ be an arbitrary finite subdivision of $[0, 1]$. Then

$$\sum_{i=1}^{k} |\mathbf{H}(\alpha_i) - \mathbf{H}(\alpha_{i-1})| \le 2 \sum_{i=1}^{k} |\mathbf{H}(\alpha_i)|$$

$$\le 2 \sum [|\mathbf{K}(\alpha_i, \alpha_i) - \mathbf{K}(\alpha_i, \alpha_i -) - \mathbf{K}(0, \alpha_i) + \mathbf{K}(0, \alpha_i -)| + |\mathbf{K}(0, \alpha_i) - \mathbf{K}(0, \alpha_{i-1})|]$$

$$\le 2(\mathrm{v}_I(\mathbf{K}) + \mathrm{var}_0^1 \mathbf{K}(0, .)) < \infty$$

and (iv) is also proved.

3.9. Theorem. *Let the kernel* $\mathbf{K}: I \to L(R_n)$ $(I = [0, 1] \times [0, 1])$ *satisfy* (3,2) *and* (3,3). *Then the homogeneous Volterra-Stieltjes integral equation* (3,10) *has only the trivial solution* $\mathbf{x} = \mathbf{0}$ *in* BV_n *if and only if the matrix* $\mathbf{I} - (\mathbf{K}(t, t) - \mathbf{K}(t, t-))$ *is regular for any* $t \in (0, 1]$ *).

Proof. By (iii) from 3.8 the equation (3,10) can be written in the equivalent form

$$\mathbf{x}(t) = \int_0^t \mathrm{d}_s[\mathbf{M}(t, s)]\, \mathbf{x}(s) + \mathbf{H}(t)\, \mathbf{x}(t), \qquad t \in [0, 1] \tag{3,25}$$

where $\mathbf{M}: I \to L(R_n)$, $\mathbf{H}: [0, 1] \to L(R_n)$ are defined by (3,22), (3,24) respectively. Hence if we assume that for any $t \in [0, 1]$ the matrix $\mathbf{I} - \mathbf{H}(t) = \mathbf{I} - (\mathbf{K}(t, t) - \mathbf{K}(t, t-))$ is regular, then the inverse $[\mathbf{I} - \mathbf{H}(t)]^{-1}$ exists for any $t \in [0, 1]$ and (3,25) can be rewritten in the equivalent form

$$\mathbf{x}(t) = [\mathbf{I} - \mathbf{H}(t)]^{-1} \int_0^t \mathrm{d}_s[\mathbf{M}(t, s)]\, \mathbf{x}(s), \qquad t \in [0, 1]. \tag{3,26}$$

This equality can be formally written in the form $\mathbf{x} = \mathbf{TMx}$ where

$$\mathbf{Tz} = [\mathbf{I} - \mathbf{H}(t)]^{-1}\, \mathbf{z}(t) \qquad \text{for} \quad \mathbf{z} \in BV_n$$

and

$$\mathbf{Mx} = \int_0^t \mathrm{d}_s[\mathbf{M}(t, s)]\, \mathbf{x}(s) \qquad \text{for} \quad \mathbf{x} \in BV_n.$$

Assume that $\mathbf{x} \in BV_n$ is a solution of (3,10). Then evidently $\mathbf{x}(0) = \mathbf{0}$ and by 3.7, 3.6 we have for any δ $(0 < \delta \le 1)$

$$\|\mathbf{x}\|_{BV_n[0,\delta]} = |\mathbf{x}(0)| + \mathrm{var}_0^\delta\, \mathbf{x} = \|\mathbf{TMx}\|_{BV_n[0,\delta]} \le C \|\mathbf{Mx}\|_{BV_n[0,\delta]} \tag{3,27}$$

$$= C\, \mathrm{var}_0^\delta \left(\int_0^t \mathrm{d}_s[\mathbf{M}(t, s)]\, \mathbf{x}(s) \right) \le C(|\mathbf{x}(0)|\, (\xi(0+) - \xi(0)) + \|\mathbf{x}\|_{BV_n[0,\delta]} (\xi(\delta) - \xi(0+))$$

$$= C(\xi(\delta) - \xi(0+))\, \|\mathbf{x}\|_{BV_n[0,\delta]}$$

*) In this case we have $\mathbf{K}(0, 0) = \mathbf{K}(0, 0-)$ if we use the agreement $\mathbf{K}(0, s) = \mathbf{K}(0, 0)$ for $s < 0$, i.e. in fact $\mathbf{I} - (\mathbf{K}(0, 0) - \mathbf{K}(0, 0-)) = \mathbf{I}$ is also regular. Nevertheless this is not used in the proof of the theorem and the result does not depend on the behaviour of $\mathbf{K}(0, 0) - \mathbf{K}(0, 0-)$.

where $\xi: [0,1] \to [0,+\infty)$ is bounded and nondecreasing by 3.6 and $C \geq 0$ is a constant (cf. 3.7). The function ξ is of bounded variation and has consequently onesided limits at all points of $[0,1]$. Hence we can find a $\delta > 0$ such that $C(\xi(\delta) - \xi(0+)) < \frac{1}{2}$ and by (3,27) we obtain

$$\|\mathbf{x}\|_{BV_n[0,\delta]} < \tfrac{1}{2}\|\mathbf{x}\|_{BV_n[0,\delta]},$$

i.e. $\mathbf{x}(t) = \mathbf{0}$ for all $t \in [0,\delta]$.

Let us now assume that $t^* \in [0,1]$ is the supremum of all such positive δ that the solution $\mathbf{x} \in BV_n$ of the equation (3,10) equals zero on $[0,\delta]$. Evidently $\mathbf{x}(t) = \mathbf{0}$ for all $t \in [0,t^*)$. Since by (ii) from 3.8 we have

$$\int_0^{t^*} \mathrm{d}_s[\mathbf{M}(t^*,s)]\,\mathbf{x}(s) = \lim_{\tau \to t^*-} \int_0^{\tau} \mathrm{d}_s[\mathbf{M}(t^*,s)]\,\mathbf{x}(s) = \mathbf{0}$$

and $[\mathbf{I} - \mathbf{H}(t^*)]^{-1}$ exists, we have by (3,26) $\mathbf{x}(t^*) = \mathbf{0}$, i.e. $\mathbf{x}(t) = \mathbf{0}$ for $t \in [0,t^*]$. Now assuming $t^* < 1$ we have

$$\mathbf{x}(t) = [\mathbf{I} - \mathbf{H}(t)]^{-1} \int_0^t \mathrm{d}_s[\mathbf{M}(t,s)]\,\mathbf{x}(s) = [\mathbf{I} - \mathbf{H}(t)]^{-1} \int_{t^*}^t \mathrm{d}_s[\mathbf{M}(t,s)]\,\mathbf{x}(s)$$

for all $t \in [t^*,1]$. Using the same procedure as above we can determine a $\delta > 0$ such that the inequality

$$\|\mathbf{x}\|_{BV_n[t^*,t^*+\delta]} \leq \tfrac{1}{2}\|\mathbf{x}\|_{BV_n[t^*,t^*+\delta]}$$

holds and consequently $\mathbf{x}(t) = \mathbf{0}$ for $t \in [t^*, t^* + \delta]$. Hence we obtain a contradiction to the property of t^*. In this way we have $t^* = 1$, i.e. $\mathbf{x}(t) = \mathbf{0}$ for all $t \in [0,1]$ and the "if" part of the theorem is proved.

For the proof of the "only if" part of the theorem we refer to the Fredholm alternative included in 2.1. (cf. also 3.2) which states that either (3,10) has only the trivial solution $\mathbf{x} = \mathbf{0}$ in BV_n or there exists $\mathbf{f} \in BV_n$ such that the equation (3,1) has no solutions in BV_n.

Let us now assume that the matrix $\mathbf{I} - (\mathbf{K}(t,t) - \mathbf{K}(t,t-)) = \mathbf{I} - \mathbf{H}(t)$ is not regular for all $t \in (0,1]$. This may occur only for a finite set of points $0 < t_1 < \ldots < t_k$ in $(0,1]$ because $\operatorname{var}_0^1 \mathbf{H} < \infty$ by (iv) from 3.8 and consequently $|\mathbf{H}(t)| < \frac{1}{2}$ for all $t \in [0,1]$ except for a finite set of points in $(0,1]$. Hence $[\mathbf{I} - \mathbf{H}(t)]^{-1}$ exists for all $t \in [0,1]$, $t \neq t_i$, $i = 1,2,\ldots,k$, and $\mathbf{I} - \mathbf{H}(t_i)$, $i = 1,2,\ldots,k$ is not regular. Evidently there exists $\mathbf{y} \in R_n$ such that $\mathbf{y} \notin R(\mathbf{I} - \mathbf{H}(t_1))$, i.e. the linear algebraic equation

$$(\mathbf{I} - \mathbf{H}(t_1))\,\mathbf{x} = \mathbf{y}$$

has no solutions in R_n. Let us define

$$\mathbf{f}(t) = \mathbf{0} \quad \text{for} \quad t \in [0,1], \quad t \neq t_1, \quad \mathbf{f}(t_1) = \mathbf{y}$$

and consider the nonhomogeneous equation (3,1) with this right-hand side. Let us assume that $\mathbf{x} \in BV_n$ is a solution of this equation. In the same way as in the proof of the "if" part we can show that $\mathbf{x}(t) = \mathbf{0}$ for all $t \in [0, t_1)$ since $[\mathbf{I} - \mathbf{H}(t)]^{-1}$ exists for all $t \in [0, t_1)$. Using the expression (3,23) and (ii) from 3.8 for $\int_0^{t_1} \mathrm{d}_s[\mathbf{K}(t_1, s)]\, \mathbf{x}(s)$ we obtain

$$[\mathbf{I} - \mathbf{H}(t_1)]\, \mathbf{x}(t_1) = \int_0^{t_1} \mathrm{d}_s[\mathbf{M}(t_1, s)]\, \mathbf{x}(s) + \mathbf{f}(t_1) = \mathbf{y}$$

and $\mathbf{x}(t_1)$ cannot be determined since $\mathbf{y} \notin R(\mathbf{I} - \mathbf{H}(t_1))$ and consequently there is no $\mathbf{x} \in BV_n$ satisfying (3,1) with the given $\mathbf{f} \in BV_n$. By the above quoted Fredholm alternative the equation (3,10) possesses nontrivial solutions and our theorem is completely proved.

3.10. Theorem. *Assume that* $\mathbf{K}\colon I = [0, 1] \times [0, 1] \to L(R_n)$ *satisfies* (3,2), (3,3) *and the matrix* $\mathbf{I} - \mathbf{K}(t, t) - \mathbf{K}(t, t-))$ *is regular for any* $t \in (0, 1]$.

Then there exists a uniquely determined $\boldsymbol{\Gamma}\colon I \to L(R_n)$ *such that the unique solution in* BV_n *to the Volterra-Stieltjes integral equation* (3,1) *with* $\mathbf{f} \in BV_n$ *is given by the relation*

$$\mathbf{x}(t) = \mathbf{f}(t) + \int_0^t \mathrm{d}_s[\boldsymbol{\Gamma}(t, s)]\, \mathbf{f}(s), \qquad t \in [0, 1]. \tag{3,28}$$

The matrix $\boldsymbol{\Gamma}(t, s)$ *satisfies the integral equation*

$$\boldsymbol{\Gamma}(t, s) = \mathbf{K}(t, s) - \mathbf{K}(t, 0) + \int_0^t \mathrm{d}_r[\mathbf{K}(t, r)]\, \boldsymbol{\Gamma}(r, s) \qquad \text{for} \quad 0 \le s \le t \le 1. \tag{3,29}$$

We have $\boldsymbol{\Gamma}(t, s) = \boldsymbol{\Gamma}(t, t)$ *for* $0 \le t < s \le 1$, $\boldsymbol{\Gamma}(t, 0) = \mathbf{0}$, $\operatorname{var}_0^1 \boldsymbol{\Gamma}(0, .) < \infty$ *and* $\mathrm{v}_I(\boldsymbol{\Gamma}) < \infty$.

Proof. By 3.1 the equation (3,1) can be written in the equivalent Fredholm-Stieltjes form

$$\mathbf{x}(t) - \int_0^1 \mathrm{d}_s[\mathbf{K}^\Delta(t, s)]\, \mathbf{x}(s) = \mathbf{f}(t), \qquad t \in [0, 1] \tag{3,30}$$

where $\mathbf{K}^\Delta\colon I \to L(R_n)$ is the corresponding triangular kernel given by (3,4). By 3.9 the homogeneous equation

$$\mathbf{x}(t) - \int_0^1 \mathrm{d}_s[\mathbf{K}^\Delta(t, s)]\, \mathbf{x}(s) = \mathbf{0}, \qquad t \in [0, 1]$$

has only the trivial solution $\mathbf{x} = \mathbf{0}$ in BV_n. Since $\mathbf{K}^\Delta$ satisfies evidently all assumptions of 2.6, we obtain by this theorem the existence of $\boldsymbol{\Gamma}\colon I \to L(R_n)$ such that the solution of (3,30) and consequently also of the equivalent equation (3,1) is given by

$$\mathbf{x}(t) = \mathbf{f}(t) + \int_0^1 \mathrm{d}_t[\boldsymbol{\Gamma}(t, s)]\, \mathbf{f}(s), \qquad t \in [0, 1] \tag{3,31}$$

where $\boldsymbol{\Gamma}(t,s)$ satisfies the matrix integral equation

$$\boldsymbol{\Gamma}(t,s) = \mathbf{K}^{\Delta}(t,s) + \int_0^1 \mathrm{d}_r[\mathbf{K}^{\Delta}(t,r)]\,\boldsymbol{\Gamma}(r,s) \qquad \text{for all} \quad (t,s) \in I\,.$$

Using the definition (3,4) of $\mathbf{K}^{\Delta}(t,s)$ and (3,8) we have

$$\boldsymbol{\Gamma}(t,s) = \mathbf{K}(t,s) - \mathbf{K}(t,0) + \int_0^t \mathrm{d}_r[\mathbf{K}(t,r)]\,\boldsymbol{\Gamma}(r,s) \qquad \text{for} \quad 0 \le s \le t \le 1$$

and (3,29) is satisfied. For $0 \le t < s \le 1$ we have similarly

$$\boldsymbol{\Gamma}(t,s) = \mathbf{K}^{\Delta}(t,t) + \int_0^t \mathrm{d}_r[\mathbf{K}(t,r)]\,\boldsymbol{\Gamma}(r,s)$$

and

$$\boldsymbol{\Gamma}(t,t) = \mathbf{K}^{\Delta}(t,t) + \int_0^t \mathrm{d}_r[\mathbf{K}(t,r)]\,\boldsymbol{\Gamma}(r,t)\,.$$

Hence

$$\boldsymbol{\Gamma}(t,s) - \boldsymbol{\Gamma}(t,t) = \int_0^t \mathrm{d}_r[\mathbf{K}(t,r)]\,(\boldsymbol{\Gamma}(r,s) - \boldsymbol{\Gamma}(r,t))\,,$$

i.e. $\boldsymbol{\Gamma}(t,s) = \boldsymbol{\Gamma}(t,t)$ since Theorem 3.9 yields that the homogeneous equation $\mathbf{x}(t) - \int_0^t \mathrm{d}_s[\mathbf{K}(t,s)]\,\mathbf{x}(s) = \mathbf{0}$ has only the trivial solution $\mathbf{x} = \mathbf{0} \in BV_n$. Similarly we obtain $\boldsymbol{\Gamma}(t,0) = \mathbf{0}$ for all $t \in [0,1]$. The inequalities $\operatorname{var}_0^1 \boldsymbol{\Gamma}(0,.) < \infty$, $\mathrm{v}_I(\boldsymbol{\Gamma}) < \infty$ are immediate consequences of 2.5.

From the equality $\boldsymbol{\Gamma}(t,s) = \boldsymbol{\Gamma}(t,t)$ valid for $t < s$ we get

$$\int_0^1 \mathrm{d}_s[\boldsymbol{\Gamma}(t,s)]\,\mathbf{x}(s) = \int_0^t \mathrm{d}_s[\boldsymbol{\Gamma}(t,s)]\,\mathbf{x}(s)$$

for all $\mathbf{x} \in BV_n$ and hence by (3,31) we obtain (3,28).

3.11. Theorem. *Let* $\mathbf{K}\colon I = [0,1] \times [0,1] \to L(R_n)$ *satisfy* (3,2), (3,3) *and let* $t_0 \in [0,1]$ *be fixed. Then the integral equation*

$$\text{(3,32)} \qquad \mathbf{x}(t) = \int_{t_0}^t \mathrm{d}_s[\mathbf{K}(t,s)]\,\mathbf{x}(s)\,, \qquad t \in [0,1]$$

possesses only the trivial solution $\mathbf{x} = \mathbf{0}$ *in* BV_n *if and only if for any* $t \in (t_0, 1]$ *the matrix* $\mathbf{I} - (\mathbf{K}(t,t) - \mathbf{K}(t,t-))$ *is regular and for any* $t \in [0, t_0)$ *the matrix* $\mathbf{I} + \mathbf{K}(t,t+) - \mathbf{K}(t,t)$ *is regular.*

The proof of this statement can be given by a modification of the proof of 3.9. Since serious technical troubles do not occur we add only a few remarks on this proof. It is evident that $\mathbf{x}(t_0) = \mathbf{0}$ for any solution of (3,32). The proof of the fact that $\mathbf{x}(t) = \mathbf{0}$ for $t \in (t_0, 1]$ if and only if $\mathbf{I} - (\mathbf{K}(t,t) - \mathbf{K}(t,t-))$ is a regular matrix for $t \in (t_0, 1]$, follows exactly the line of the proof of 3.9. For proving "$\mathbf{x}(t) = \mathbf{0}$ for $t \in [0, t_0)$ if and only if $\mathbf{I} + \mathbf{K}(t, t+) - \mathbf{K}(t, t)$ is regular for all $t \in [0, t_0)$", the decomposition $(t \in [0, t_0))$

$$\int_{t_0}^{t} \mathrm{d}_s[\mathbf{K}(t, s)]\, \mathbf{x}(s) = \int_{t_0}^{t} \mathrm{d}_s[\mathbf{M}(t, s)]\, \mathbf{x}(s) - (\mathbf{K}(t, t+) - \mathbf{K}(t, t))\, \mathbf{x}(t)$$

valid for any $\mathbf{x} \in BV_n$ can be used where the integral $\int_{t_0}^{t} \mathrm{d}_s[\mathbf{M}(t, s)]\, \mathbf{x}(s)$ does not depend on the value $\mathbf{x}(t)$. This can be done in the same way as in 3.8 when it is assumed that $\mathbf{M}(t, s) = \mathbf{K}(t, s)$ if $(t, s) \in I$, $t \neq s$, $\mathbf{M}(t, t) = \mathbf{K}(t, t+)$ if $t \in [0, 1)$, $\mathbf{M}(1, 1) = \mathbf{K}(1, 1)$. Using the above decomposition of $\int_{t_0}^{t} \mathrm{d}_s[\mathbf{K}(t, s)]\, \mathbf{x}(s)$ the approach from 3.9 can be used in order to prove the result.

3.12. Corollary. *Let $\mathbf{K}: I \to L(R_n)$ satisfy (3,2), (3,3) and let $t_0 \in [0, 1]$ be fixed. Then the integral equation*

$$\mathbf{x}(t) = \int_{t_0}^{t} \mathrm{d}_s[\mathbf{K}(t, s)]\, \mathbf{x}(s) + \mathbf{f}(t), \qquad t \in [0, 1] \tag{3,33}$$

has a unique solution for every $\mathbf{f} \in BV_n$ if and only if for any $t \in (t_0, 1]$ the matrix $\mathbf{I} - (\mathbf{K}(t, t) - \mathbf{K}(t, t-))$ is regular and for any $t \in [0, t_0)$ the matrix $\mathbf{I} + \mathbf{K}(t, t+) - \mathbf{K}(t, t)$ is regular.

Proof. Let us define a new kernel $\mathbf{K}^{t_0}: I \to L(R_n)$ as follows.

If $t_0 \le t \le 1$, then

$$\begin{aligned} \mathbf{K}^{t_0}(t, s) &= \mathbf{K}(t, s) && \text{for } t_0 \le s \le t, \\ \mathbf{K}^{t_0}(t, s) &= \mathbf{K}(t, t) && \text{for } t < s \le 1, \\ \mathbf{K}^{t_0}(t, s) &= \mathbf{K}(t, t_0) && \text{for } 0 \le s < t_0 \end{aligned}$$

and if $0 \le t < t_0$, then

$$\begin{aligned} \mathbf{K}^{t_0}(t, s) &= -\mathbf{K}(t, s) && \text{for } t \le s \le t_0, \\ \mathbf{K}^{t_0}(t, s) &= -\mathbf{K}(t, t) && \text{for } 0 \le s < t, \\ \mathbf{K}^{t_0}(t, s) &= -\mathbf{K}(t, t_0) && \text{for } t_0 \le s \le 1. \end{aligned}$$

It is a matter of routine to show that $\mathrm{v}_I(\mathbf{K}^{t_0}) < \infty$, $\mathrm{var}_0^1 \mathbf{K}^{t_0}(0, .) < \infty$ and

$$\int_{t_0}^{t} \mathrm{d}_s[\mathbf{K}(t, s)]\, \mathbf{x}(s) = \int_{0}^{1} \mathrm{d}_s[\mathbf{K}^{t_0}(t, s)]\, \mathbf{x}(s),$$

for every $t \in [0, 1]$ and $\mathbf{x} \in BV_n$. Hence the equation (3,33) can be rewritten in the equivalent Fredholm-Stieltjes form

$$\mathbf{x}(t) - \int_{0}^{1} \mathrm{d}_s[\mathbf{K}^{t_0}(t, s)]\, \mathbf{x}(s) = \mathbf{f}(t), \qquad t \in [0, 1].$$

By 3,11 the corresponding homogeneous equation

$$\mathbf{x}(t) - \int_0^1 \mathrm{d}_s[\mathbf{K}^{to}(t, s)]\, \mathbf{x}(s) = \mathbf{0}, \qquad t \in [0, 1]$$

has only the trivial solution if and only if the regularity conditions given in the corollary are satisfied. The corollary follows now immediately from 2.1.

Notes

The Fredholm-Stieltjes integral equation theory is based on the investigations due to Schwabik [2], [5].

The case of Volterra-Stieltjes integral equations was considered by many authors in terms of product integrals, the left and right Cauchy integral or other types of integrals. See e.g. Bitzer [1], Helton [1], Herod [1], Hönig [1], Mac Nerney [2].

III. *Generalized linear differential equations*

1. The generalized linear differential equation and its basic properties

We assume that $\mathbf{A}: [0,1] \to L(R_n)$ is an $n \times n$-matrix valued function such that $\text{var}_0^1 \mathbf{A} < \infty$ and $\mathbf{g} \in BV_n[0,1] = BV_n$.

The generalized linear differential equation will be denoted by the symbol

$$\mathrm{d}\mathbf{x} = \mathrm{d}[\mathbf{A}]\,\mathbf{x} + \mathrm{d}\mathbf{g} \tag{1,1}$$

which is interpreted by the following definition of a solution.

1.1. Definition. Let $[a,b] \subset [0,1]$, $a < b$; a function $\mathbf{x}: [a,b] \to R_n$ is said to be a solution of the *generalized linear differential equation* (1,1) on the interval $[a,b]$ if for any $t, t_0 \in [a,b]$ the equality

$$\mathbf{x}(t) = \mathbf{x}(t_0) + \int_{t_0}^{t} \mathrm{d}[\mathbf{A}(s)]\,\mathbf{x}(s) + \mathbf{g}(t) - \mathbf{g}(t_0) \tag{1,2}$$

is satisfied.

In the original papers of J. Kurzweil (cf. [1], [2]) on generalized differential equations and in other papers in this field the notation

$$\frac{\mathrm{d}\mathbf{x}}{\mathrm{d}\tau} = \mathrm{D}[\mathbf{A}(t)\,\mathbf{x} + \mathbf{g}(t)]$$

was used for the generalized linear differential equation.

It is evident that the generalized linear differential equation can be given on an arbitrary interval $[a,b] \subset R$ instead of $[0,1]$.

If $\mathbf{x}_0 \in R_n$ and $t_0 \in [a,b] \subset [0,1]$ are fixed and $\mathbf{x}: [a,b] \to R_n$ is a solution of (1,1) on $[a,b]$ such that $\mathbf{x}(t_0) = \mathbf{x}_0$, then $\mathbf{x}$ is called the solution of the initial value (Cauchy) problem

$$\mathrm{d}\mathbf{x} = \mathrm{d}[\mathbf{A}]\,\mathbf{x} + \mathrm{d}\mathbf{g}\,, \qquad \mathbf{x}(t_0) = \mathbf{x}_0 \tag{1,3}$$

on $[a,b]$.

1.2. Remark. If $\mathbf{B}: [0,1] \to L(R_n)$ is an $n \times n$-matrix valued function, continuous on $[0,1]$ with respect to the norm of a matrix given in I.1.1 and $\mathbf{h}: [0,1] \to R_n$ is continuous on $[0,1]$, then the initial value problem for the linear ordinary differential equation

$$\mathbf{x}' = \mathbf{B}(t)\,\mathbf{x} + \mathbf{h}(t), \qquad \mathbf{x}(t_0) = \mathbf{x}_0 \tag{1,4}$$

is equivalent to the integral equation

$$\mathbf{x}(t) = \mathbf{x}_0 + \int_{t_0}^{t} \mathbf{B}(s)\,\mathbf{x}(s)\,\mathrm{d}s + \int_{t_0}^{t} \mathbf{h}(s)\,\mathrm{d}s, \qquad t \in [0,1].$$

If we denote $\mathbf{A}(t) = \int_0^t \mathbf{B}(r)\,\mathrm{d}r$, $\mathbf{g}(t) = \int_0^t \mathbf{h}(r)\,\mathrm{d}r$ for $t \in [0,1]$, then this equation can be rewritten into the equivalent Stieltjes form

$$\mathbf{x}(t) = \mathbf{x}_0 + \int_{t_0}^{t} \mathrm{d}[\mathbf{A}(s)]\,\mathbf{x}(s) + \mathbf{g}(t) - \mathbf{g}(t_0), \qquad t \in [0,1].$$

The functions $\mathbf{A}: [0,1] \to L(R_n)$, $\mathbf{g}: [0,1] \to R_n$ are absolutely continuous and therefore also of bounded variation. In this way the initial value problem (1,4) has become the initial value problem of the form (1,3) with $\mathbf{A}, \mathbf{g}$ defined above and both problems are equivalent. Essentially the same reasoning yields the equivalence of the problem (1,4) to an equivalent Stieltjes integral equation when $\mathbf{B}: [0,1] \to L(R_n)$, $\mathbf{h}: [0,1] \to R_n$ are assumed to be Lebesgue integrable and if we look for Carathéodory solutions of (1,4).

1.3. Theorem. *Assume that $\mathbf{A}: [0,1] \to L(R_n)$ is of bounded variation on $[0,1]$, $\mathbf{g} \in BV_n$. Let $\mathbf{x}: [a,b] \to R_n$ be a solution of the generalized linear differential equation (1,1) on the interval $[a,b] \subset [0,1]$. Then $\mathbf{x}$ is of bounded variation on $[a,b]$.*

Proof. By the definition 1.1 of a solution of (1,1) the integral $\int_{t_0}^{t} \mathrm{d}[\mathbf{A}(s)]\,\mathbf{x}(s)$ exists for every $t, t_0 \in [a,b]$. Hence by I.4.12 the limit $\lim_{t \to t_0+} \int_{t_0}^{t} \mathrm{d}[\mathbf{A}(s)]\,\mathbf{x}(s)$ exists for $t_0 \in [a,b)$ and $\lim_{t \to t_0-} \int_{t_0}^{t} \mathrm{d}[\mathbf{A}(s)]\,\mathbf{x}(s)$ exists for $t_0 \in (a,b]$. Hence by (1,2) the solution $\mathbf{x}(t)$ of (1,1) possesses onesided limits at every point $t_0 \in [a,b]$ and for every point $t_0 \in [a,b]$ there exists $\delta > 0$ and a constant M such that $\|\mathbf{x}(t)\| \le M$ for $t \in (t_0 - \delta, t_0 + \delta) \cap [a,b]$. By the Heine-Borel Covering Theorem there exists a finite system of intervals of the type $(t_0 - \delta, t_0 + \delta)$ covering the compact interval $[a,b]$. Hence there exists a constant K such that $\|\mathbf{x}(t)\| \le K$ for every $t \in [a,b]$. If now $a = t_0 < t_1 < \ldots < t_k = b$ is an arbitrary subdivision of $[a,b]$, we have by I.4.27

$$\begin{aligned}\|\mathbf{x}(t_i) - \mathbf{x}(t_{i-1})\| &\le \left\| \int_{t_{i-1}}^{t_i} \mathrm{d}[\mathbf{A}(s)]\,\mathbf{x}(s) \right\| + \|\mathbf{g}(t_i) - \mathbf{g}(t_{i-1})\| \\ &\le K \operatorname{var}_{t_{i-1}}^{t_i} \mathbf{A} + \|\mathbf{g}(t_i) - \mathbf{g}(t_{i-1})\|\end{aligned}$$

for every $i = 1, \ldots, k$. Hence

$$\sum_{i=1}^{k} |\mathbf{x}(t_i) - \mathbf{x}(t_{i-1})| \leq K \operatorname{var}_a^b \mathbf{A} + \operatorname{var}_a^b \mathbf{g}$$

and $\operatorname{var}_a^b \mathbf{x} < \infty$ since the subdivision was arbitrary.

Throughout this chapter we use the notations $\Delta^+ \mathbf{f}(t) = \mathbf{f}(t+) - \mathbf{f}(t)$, $\Delta^- \mathbf{f}(t) = \mathbf{f}(t) - \mathbf{f}(t-)$ for any function possessing the onesided limits $\mathbf{f}(t+) = \lim_{r \to t+} \mathbf{f}(r)$, $\mathbf{f}(t-) = \lim_{r \to t-} \mathbf{f}(r)$. This applies evidently also to matrix valued functions.

Since by definition the initial value problem (1,3) is equivalent to the Volterra-Stieltjes integral equation

$$\mathbf{x}(t) = \mathbf{x}_0 + \int_{t_0}^{t} \mathrm{d}[\mathbf{A}(s)]\, \mathbf{x}(s) + \mathbf{g}(t) - \mathbf{g}(t_0), \qquad t \in [0, 1], \tag{1,5}$$

the following theorem is a direct corollary of II.3.12.

1.4. Theorem. *Assume that* $\mathbf{A}: [0, 1] \to L(R_n)$ *satisfies* $\operatorname{var}_0^1 \mathbf{A} < \infty$. *If* $t_0 \in [0, 1)$, *then the initial value problem* (1,3) *possesses for any* $\mathbf{g} \in BV_n$, $\mathbf{x}_0 \in R_n$ *a unique solution* $\mathbf{x}(t)$ *defined on* $[t_0, 1]$ *if and only if the matrix* $\mathbf{I} - \Delta^- \mathbf{A}(t)$ *is regular for any* $t \in (t_0, 1]$. *If* $t_0 \in (0, 1]$, *then the initial value problem* (1,3) *possesses for any* $\mathbf{g} \in BV_n$, $\mathbf{x}_0 \in R_n$ *a unique solution* $\mathbf{x}(t)$ *defined on* $[0, t_0]$ *if and only if the matrix* $\mathbf{I} + \Delta^+ \mathbf{A}(t)$ *is regular for any* $t \in [0, t_0)$. *If* $t_0 \in [0, 1]$, *then the problem* (1,3) *has for any* $\mathbf{g} \in BV_n$, $\mathbf{x}_0 \in R_n$ *a unique solution* $\mathbf{x}(t)$ *defined on* $[0, 1]$ *if and only if* $\mathbf{I} - \Delta^- \mathbf{A}(t)$ *is regular for any* $t \in (t_0, 1]$ *and* $\mathbf{I} + \Delta^+ \mathbf{A}(t)$ *is regular for any* $t \in [0, t_0)$.

1.5. Remark. Let us mention that by 1.3 the solutions of the problem (1,3) whose existence and uniqueness is stated in Theorem 1.4 are of bounded variation on their intervals of definition. Further, if in the last part of the theorem we have $t_0 = 0$, then the regularity of $\mathbf{I} + \Delta^+ \mathbf{A}(0)$ is not required. Similarly for $t_0 = 1$ and for the regularity of $\mathbf{I} - \Delta^- \mathbf{A}(1)$.

Let us mention also that Theorem 1.4 gives the fundamental existence and unicity result for BV_n-solutions of the initial value problem (1,3).

Let us note that if $\mathbf{A}: [0, 1] \to L(R_n)$ is of bounded variation in $[0, 1]$, then there is a finite set of points t in $[0, 1]$ such that the matrix $\mathbf{I} - \Delta^- \mathbf{A}(t)$ is singular and similarly for the matrix $\mathbf{I} + \Delta^+ \mathbf{A}(t)$. In fact, since $\operatorname{var}_0^1 \mathbf{A} < \infty$ the series $\sum_{t \in (a,b]} \Delta^- \mathbf{A}(t)$ converges. Hence there is a finite set of points $t \in [0, 1]$ such that $|\Delta^- \mathbf{A}(t)| \geq \frac{1}{2}$. For all the remaining points in $[0, 1]$ we have $|\Delta^- \mathbf{A}(t)| < \frac{1}{2}$, and consequently $[\mathbf{I} - \Delta^- \mathbf{A}(t)]^{-1} = \sum_{k=0}^{\infty} (\Delta^- \mathbf{A}(t))^k$ exists since the series on the right-hand side converges at these points. For the matrix $\mathbf{I} + \Delta^+ \mathbf{A}(t)$ this fact can be shown analogously.

1.6. Proposition. *Assume that* $\mathbf{A}: [0,1] \to L(R_n)$, $\operatorname{var}_0^1 \mathbf{A} < \infty$, $\mathbf{g} \in BV_n$. *Let* $\mathbf{x}$ *be a solution of the equation* (1,1) *on some interval* $[a,b] \subset [0,1]$, $a < b$. *Then all the onesided limits* $\mathbf{x}(a+)$, $\mathbf{x}(t+)$, $\mathbf{x}(t-)$, $\mathbf{x}(b-)$, $t \in (a,b)$ *exist and*

$$(1,6) \qquad \begin{aligned} \mathbf{x}(t+) &= [\mathbf{I} + \Delta^+\mathbf{A}(t)]\,\mathbf{x}(t) + \Delta^+\mathbf{g}(t) \qquad \text{for all} \quad t \in [a,b)\,, \\ \mathbf{x}(t-) &= [\mathbf{I} - \Delta^-\mathbf{A}(t)]\,\mathbf{x}(t) - \Delta^-\mathbf{g}(t) \qquad \text{for all} \quad t \in (a,b] \end{aligned}$$

holds.

Proof. Let $t \in [a,b)$. By the definition of the solution $\mathbf{x}: [a,b] \to R_n$ we have

$$\mathbf{x}(t+\delta) = \mathbf{x}(t) + \int_t^{t+\delta} \mathrm{d}[\mathbf{A}(s)]\,\mathbf{x}(s) + \mathbf{g}(t+\delta) - \mathbf{g}(t)$$

for any $\delta > 0$. For $\delta \to 0+$ we obtain by I.4.13 the equality

$$\begin{aligned} \mathbf{x}(t+) &= \mathbf{x}(t) + (\mathbf{A}(t+) - \mathbf{A}(t))\,\mathbf{x}(t) + \mathbf{g}(t+) - \mathbf{g}(t) \\ &= \mathbf{x}(t) + \Delta^+\mathbf{A}(t)\,\mathbf{x}(t) + \Delta^+\mathbf{g}(t) \end{aligned}$$

where the limit on the right-hand side evidently exists. The second equality in (1,6) can be proved similarly.

1.7. Theorem. *Assume that* $\mathbf{A}: [0,1] \to L(R_n)$, $\operatorname{var}_0^1 \mathbf{A} < \infty$, $t_0 \in [0,1]$ *and that* $\mathbf{I} + \Delta^+\mathbf{A}(t)$ *is a regular matrix for all* $t \in [0,t_0)$ *and* $\mathbf{I} - \Delta^-\mathbf{A}(t)$ *is a regular matrix for all* $t \in (t_0, 1]$. *Then there exists a constant* C *such that for any solution* $\mathbf{x}(t)$ *of the initial value problem* (1,3) *with* $\mathbf{g} \in BV_n$ *we have*

$$(1,7) \qquad |\mathbf{x}(t)| \le C(|\mathbf{x}_0| + \operatorname{var}_{t_0}^1 \mathbf{g}) \exp(C \operatorname{var}_{t_0}^t \mathbf{A}) \qquad \text{for} \quad t \in [t_0, 1]$$

and

$$(1,8) \qquad |\mathbf{x}(t)| \le C(|\mathbf{x}_0| + \operatorname{var}_0^{t_0} \mathbf{g}) \exp(C \operatorname{var}_t^{t_0} \mathbf{A}) \qquad \text{for} \quad t \in [0, t_0]\,.$$

Proof. We consider only the case $t < t_0$ and prove (1,8). The proof of (1,7) can be given in an analogous way. Let us set $\mathbf{B}(t) = \mathbf{A}(t+)$ for $t \in [0,t_0)$ and $\mathbf{B}(t_0) = \mathbf{A}(t_0)$. Hence $\mathbf{B}(t) - \mathbf{A}(t) = \Delta^+\mathbf{A}(t)$ for $t \in [0,t_0)$, $\mathbf{B}(t_0) - \mathbf{A}(t_0) = \mathbf{0}$, i.e. $\mathbf{B}(t) - \mathbf{A}(t) = \mathbf{0}$ for all $t \in [0,t_0]$ except for an at most countable set of points in $[0,t_0)$ and evidently $\operatorname{var}_0^{t_0} (\mathbf{B} - \mathbf{A}) < \infty$. Hence for every $\mathbf{x} \in BV_n$ and $t \in [0,t_0)$ we have by I.4.23

$$\int_t^{t_0} \mathrm{d}[\mathbf{B}(s) - \mathbf{A}(s)]\,\mathbf{x}(s) = -\Delta^+\mathbf{A}(t)\,\mathbf{x}(t)$$

and by the definition we obtain

$$\mathbf{x}(t) = \mathbf{x}_0 + \int_t^{t_0} \mathrm{d}[\mathbf{B}(s)]\,\mathbf{x}(s) - \Delta^+\mathbf{A}(t)\,\mathbf{x}(t) + \mathbf{g}(t) - \mathbf{g}(t_0)\,, \qquad t \in [0,t_0)$$

i.e.

$$(1,9) \qquad \mathbf{x}(t) = [\mathbf{I} + \Delta^+\mathbf{A}(t)]^{-1} \left(\mathbf{x}_0 + \mathbf{g}(t) - \mathbf{g}(t_0) + \int_t^{t_0} \mathrm{d}[\mathbf{B}(s)]\,\mathbf{x}(s)\right), \qquad t \in [0,t_0)\,.$$

Let us mention that for all $t \in [0, t_0)$ we have

$$(1,10) \qquad |[\mathbf{I} + \Delta^+ \mathbf{A}(t)]^{-1}| \leq C, \qquad C = \text{const.}$$

This inequality can be proved using the equality $[\mathbf{I} + \Delta^+ \mathbf{A}(t)]^{-1} = \sum_{i=0}^{\infty} (-1)^i (\Delta^+ \mathbf{A}(t))^i$ which holds whenever $|\Delta^+ \mathbf{A}(t)| < 1$. Hence

$$|[\mathbf{I} + \Delta^+ \mathbf{A}(t)]^{-1}| \leq \sum_{i=0}^{\infty} |\Delta^+ \mathbf{A}(t)|^i = \frac{1}{1 - |\Delta^+ \mathbf{A}(t)|} < 2$$

provided $|\Delta^+ \mathbf{A}(t)| < \frac{1}{2}$, i.e. for all $t \in [0, t_0)$ except for a finite set of points in $[0, t_0)$. The estimate (1,10) is in this manner obvious. Using (1,10) we obtain by (1,9) the inequality

$$|\mathbf{x}(t)| \leq C\left(|\mathbf{x}_0| + |\mathbf{g}(t) - \mathbf{g}(t_0)| + \left|\int_{t_0}^{t} \mathrm{d}[\mathbf{B}(s)]\, \mathbf{x}(s)\right|\right)$$

$t \in [0, t_0]$. This inequality together with I.4.27 yields

$$(1,11) \qquad |\mathbf{x}(t)| \leq C\left(|\mathbf{x}_0| + \operatorname{var}_0^{t_0} \mathbf{g} + \int_t^{t_0} |\mathbf{x}(s)|\, \mathrm{d} \operatorname{var}_0^s \mathbf{B}\right)$$

$$= C(|\mathbf{x}_0| + \operatorname{var}_0^{t_0} \mathbf{g}) + C \int_t^{t_0} |\mathbf{x}(s)|\, \mathrm{d}\mathbf{h}(s)$$

where $\mathbf{h}(s) = \operatorname{var}_0^s \mathbf{B}$ is defined on $[0, t_0]$ and is evidently continuous from the right-hand side on $[0, t_0)$ since $\mathbf{B}$ has this property by definition. Using I.4.30 for the inequality (1,11) we obtain

$$|\mathbf{x}(t)| \leq C(|\mathbf{x}_0| + \operatorname{var}_0^{t_0} \mathbf{g}) \exp(C(h(t_0) - h(t)))$$
$$\leq C(|\mathbf{x}_0| + \operatorname{var}_0^{t_0} \mathbf{g}) \exp(C(\operatorname{var}_0^{t_0} \mathbf{B} - \operatorname{var}_0^{t} \mathbf{B}))$$
$$= C(|\mathbf{x}_0| + \operatorname{var}_0^{t_0} \mathbf{g}) \exp(C \operatorname{var}_t^{t_0} \mathbf{B})$$

and this implies (1,8) since $\operatorname{var}_t^{t_0} \mathbf{B} \leq \operatorname{var}_t^{t_0} \mathbf{A}$.

Remark. A slight modification in the proof leads to a refinement of the estimates (1,7), (1,8). It can be proved that

$$|\mathbf{x}(t)| \leq C(|\mathbf{x}_0| + \operatorname{var}_{t_0}^{t} \mathbf{g}) \exp(C \operatorname{var}_{t_0}^{t} \mathbf{A}) \qquad \text{for} \quad t \in [t_0, 1]$$

and

$$|\mathbf{x}(t)| \leq C(|\mathbf{x}_0| + \operatorname{var}_t^{t_0} \mathbf{g}) \exp(C \operatorname{var}_t^{t_0} \mathbf{A}) \qquad \text{for} \quad t \in [0, t_0]$$

holds.

1.8. Corollary. *Let* $\mathbf{A}: [0, 1] \to L(R_n)$ *fulfil the assumptions given in* 1.7 *for some* $t_0 \in [0, 1]$, $\mathbf{g}, \tilde{\mathbf{g}} \in BV_n$, $\mathbf{x}_0, \tilde{\mathbf{x}}_0 \in R_n$. *Then if* $\mathbf{x} \in BV_n$ *is a solution of* (1,3) *and* $\tilde{\mathbf{x}} \in BV_n$ *is a solution of*

$$\mathrm{d}\mathbf{x} = \mathrm{d}[\mathbf{A}]\, \mathbf{x} + \mathrm{d}\tilde{\mathbf{g}}, \qquad \mathbf{x}(t_0) = \tilde{\mathbf{x}}_0,$$

we have

$$(1,12)\quad |\mathbf{x}(t) - \tilde{\mathbf{x}}(t)| \le C(|\mathbf{x}_0 - \tilde{\mathbf{x}}_0| + \operatorname{var}_0^{t_0}(\mathbf{g} - \tilde{\mathbf{g}}))\exp(C \operatorname{var}_t^{t_0} \mathbf{A}) \qquad \text{for} \quad t \in [0, t_0]$$

$$|\mathbf{x}(t) - \tilde{\mathbf{x}}(t)| \le C(|\mathbf{x}_0 - \tilde{\mathbf{x}}_0| + \operatorname{var}_{t_0}^{1}(\mathbf{g} - \tilde{\mathbf{g}}))\exp(C \operatorname{var}_{t_0}^{t} \mathbf{A}) \qquad \text{for} \quad t \in [t_0, 1],$$

where $C \ge 1$ *is a constant. Hence*

$$(1,13)\qquad |\mathbf{x}(t) - \tilde{\mathbf{x}}(t)| \le K(|\mathbf{x}_0 - \tilde{\mathbf{x}}_0| + \operatorname{var}_0^1(\mathbf{g} - \tilde{\mathbf{g}}))$$

for all $t \in [0, 1]$ *where* $K = C\exp(C \operatorname{var}_0^1 \mathbf{A})$.

1.9. Remark. The inequality (1,13) yields evidently $\mathbf{x}(t) = \tilde{\mathbf{x}}(t)$ for all $t \in [0, 1]$ whenever $\mathbf{x}_0 = \tilde{\mathbf{x}}_0$ and $\operatorname{var}_0^1(\mathbf{g} - \tilde{\mathbf{g}}) = 0$. In this way the unicity of solutions of the initial value problem (1,3) is confirmed.

1.10. Theorem. *Assume that* $t_0 \in [0, 1]$ *is fixed. Let* $\mathbf{A}: [0, 1] \to L(R_n)$ *be such that* $\operatorname{var}_0^1 \mathbf{A} < \infty$, $\mathbf{I} - \Delta^- \mathbf{A}(t)$ *is a regular matrix for* $t \in (t_0, 1]$ *and* $\mathbf{I} + \Delta^+ \mathbf{A}(t)$ *is a regular matrix for* $t \in [0, t_0)$. *Then the set of all solutions* $\mathbf{x}: [0, 1] \to R_n$ *of the homogeneous generalized differential equation*

$$(1,14)\qquad \mathrm{d}\mathbf{x} = \mathrm{d}[\mathbf{A}]\,\mathbf{x}$$

with the initial value given at the point $t_0 \in [0, 1]$ *is an n-dimensional subspace in* BV_n.

Proof. The linearity of the set of solutions is evident from the linearity of the integral. Let us set $\mathbf{e}^{(k)} = (0, \ldots, 0, 1, 0, \ldots, 0)^* \in R_n$, $k = 1, \ldots, n$ (the value 1 is in the k-th coordinate of $\mathbf{e}^{(k)} \in R_n$) and let $\boldsymbol{\varphi}^{(k)}: [0, 1] \to R_n$ be the unique solution of (1,14) such that $\boldsymbol{\varphi}^{(k)}(t_0) = \mathbf{e}^{(k)}$, $k = 1, \ldots, n$ (they exist by 1.4). The unicity result from 1.4 yields that $\sum_{k=1}^{n} c_k \boldsymbol{\varphi}^{(k)}(t) = \mathbf{0}$, $c_k \in R$ if and only if $c_k = 0$, $k = 1, \ldots, n$. If $\mathbf{x}: [0, 1] \to R_n$ is an arbitrary solution of (1,14), then clearly

$$\mathbf{x}(t) = \sum_{k=1}^{n} \mathbf{x}_k(t_0)\, \boldsymbol{\varphi}^{(k)}(t)$$

for all $t \in [0, 1]$, i.e. $\mathbf{x}$ is a linear combination of the linearly independent solutions $\boldsymbol{\varphi}^{(k)}$, $k = 1, \ldots, n$ and this is our result.

1.11. Example. We give an example of a generalized linear differential equation which demonstrates the role of the assumptions concerning the regularity of the matrices $\mathbf{I} + \Delta^+ \mathbf{A}(t)$, $\mathbf{I} - \Delta^- \mathbf{A}(t)$ in 1.4. Let us set

$$\mathbf{A}(t) = \begin{pmatrix} 0, & 0 \\ 0, & 0 \end{pmatrix}, \qquad \mathbf{A}(t) = \begin{pmatrix} 0, & 0 \\ 0, & 1 \end{pmatrix}$$

for $0 \le t < \frac{1}{2}$, $\frac{1}{2} \le t \le 1$ respectively; for this 2×2-matrix $\boldsymbol{A}: [0,1] \to L(R_2)$ we have evidently $\Delta^+ \boldsymbol{A}(t) = \boldsymbol{0}$ for all $t \in [0,1)$, $\Delta^- \boldsymbol{A}(t) = \boldsymbol{0}$ for all $t \in (0,1]$, $t \neq \frac{1}{2}$ and

$$\Delta^- \boldsymbol{A}(\tfrac{1}{2}) = \begin{pmatrix} 0, & 0 \\ 0, & 1 \end{pmatrix}.$$

Hence

$$\boldsymbol{I} - \Delta^- \boldsymbol{A}(\tfrac{1}{2}) = \begin{pmatrix} 1, & 0 \\ 0, & 0 \end{pmatrix}$$

is not regular. We consider the initial value problem

$$\text{(1,15)} \qquad \mathrm{d}\boldsymbol{x} = \mathrm{d}[\boldsymbol{A}]\,\boldsymbol{x}, \qquad \boldsymbol{x}(0) = \boldsymbol{x}_0$$

where $\boldsymbol{x}_0 = (c_1, c_2)^* \in R_2$. For a solution $\boldsymbol{x}(t)$ of this problem we have

$$\boldsymbol{x}(t) = \boldsymbol{x}_0 + \int_0^t \mathrm{d}[\boldsymbol{A}(s)]\,\boldsymbol{x}(s) = \boldsymbol{x}_0 = (c_1, c_2)^* \qquad \text{if} \quad t \in [0, \tfrac{1}{2}).$$

Further, by 1.6 we obtain $\boldsymbol{x}(\frac{1}{2}-) = [\boldsymbol{I} - \Delta^- \boldsymbol{A}(\frac{1}{2})]\,\boldsymbol{x}(\frac{1}{2})$, i.e. $(c_1, c_2)^* = [\boldsymbol{I} - \Delta^- \boldsymbol{A}(\frac{1}{2})]\,\boldsymbol{x}(\frac{1}{2}) = (x_1(\frac{1}{2}), 0)^*$. This equality is contradictory for $c_2 \neq 0$. Hence the above problem (1,15) cannot have a solution on $[0, \frac{1}{2}]$ when $\boldsymbol{x}_0 = (c_1, c_2)^* \in R_2$ with $c_2 \neq 0$.

Let us now assume that $\boldsymbol{x}_0 = (c_1, 0)^* \in R_2$. Then we have for $t \ge \frac{1}{2}$

$$\boldsymbol{x}(t) = \boldsymbol{x}_0 + \int_0^t \mathrm{d}[\boldsymbol{A}(s)]\,\boldsymbol{x}(s) = \boldsymbol{x}(\tfrac{1}{2}) + \int_{1/2}^t \mathrm{d}[\boldsymbol{A}(s)]\,\boldsymbol{x}(s) = \boldsymbol{x}(\tfrac{1}{2}).$$

By 1.6 necessarily

$$[\boldsymbol{I} - \Delta^- \boldsymbol{A}(\tfrac{1}{2})]\,\boldsymbol{x}(\tfrac{1}{2}) = \begin{pmatrix} 1, & 0 \\ 0, & 0 \end{pmatrix} \boldsymbol{x}(\tfrac{1}{2}) = \boldsymbol{x}(\tfrac{1}{2}-) = \begin{pmatrix} c_1 \\ 0 \end{pmatrix}.$$

Hence $\boldsymbol{x}(\frac{1}{2}) = (c_1, d)^*$, where $d \in R$ is arbitrary, satisfies this relation. It is easy to show that any vector valued function $\boldsymbol{x}: [0,1] \to R_2$ defined by $\boldsymbol{x}(t) = (c_1, 0)^*$ for $0 \le t < \frac{1}{2}$, $\boldsymbol{x}(t) = (c_1, d)^*$ for $\frac{1}{2} \le t \le 1$, satisfies our equation.

Summarizing these facts we have the following. If $\boldsymbol{x}(0) = (c_1, c_2)^*$ and $c_2 \neq 0$, then a solution of (1,15) does not exist on the whole interval $[0,1]$. If $\boldsymbol{x}(0) = (c_1, 0)^*$, then the equation (1,15) has solutions on the whole interval $[0,1]$ but the uniqueness is violated.

If we consider the initial value problem $\mathrm{d}\boldsymbol{x} = \mathrm{d}[\boldsymbol{A}]\,\boldsymbol{x}$, $\boldsymbol{x}(\frac{1}{2}) = (c_1, c_2)^*$ for the given matrix $\boldsymbol{A}(t)$, then it is easy to show that this problem possesses the unique solution $\boldsymbol{x}(t) = (c_1, 0)^*$ if $t \in [0, \frac{1}{2})$, $\boldsymbol{x}(t) = (c_1, c_2)^*$ if $t \in [\frac{1}{2}, 1]$. Hence the singularity of the matrix $\boldsymbol{I} - \Delta^- \boldsymbol{A}(t)$ for $t = \frac{1}{2}$ is irrelevant for the existence and uniqueness of solutions to the initial value problem mentioned above.

2. Variation of constants formula. The fundamental matrix

In this section we continue the consideration of the initial value problem

$$\mathrm{d}\mathbf{x} = \mathrm{d}[\mathbf{A}]\,\mathbf{x} + \mathrm{d}\mathbf{g}, \qquad \mathbf{x}(t_0) = \mathbf{x}_0 \tag{2,1}$$

with $\mathbf{A}$: $[0,1] \to L(R_n)$, $\operatorname{var}_0^1 \mathbf{A} < \infty$, $\mathbf{g} \in BV_n[0,1] = BV_n$, $t_0 \in [0,1]$, $\mathbf{x}_0 \in R_n$.

2.1. Proposition. *Assume that* $\mathbf{A}$: $[0,1] \to L(R_n)$, $\operatorname{var}_0^1 \mathbf{A} < \infty$, $t_0 \in [0,1]$ *is fixed, the matrix* $\mathbf{I} - \Delta^- \mathbf{A}(t)$ *is regular for all* $t \in (t_0, 1]$ *and the matrix* $\mathbf{I} + \Delta^+ \mathbf{A}(t)$ *is regular for all* $t \in [0, t_0)$.

Then the matrix equation

$$\mathbf{X}(t) = \tilde{\mathbf{X}} + \int_{t_1}^{t} \mathrm{d}[\mathbf{A}(r)]\,\mathbf{X}(r) \tag{2,2}$$

has for every $\tilde{\mathbf{X}} \in L(R_n)$ *a unique solution* $\mathbf{X}(t) \in L(R_n)$ *on* $[t_1, 1]$ *provided* $t_0 \le t_1$ *and on* $[0, t_1]$ *provided* $t_1 \le t_0$.

Proof. Let us denote by $\mathbf{B}_k$ the k-th column of a matrix $\mathbf{B} \in L(R_n)$. For the k-th column of the matrix equation (2,2) we have

$$\mathbf{X}_k(t) = \tilde{\mathbf{X}}_k + \int_{t_1}^{t} \mathrm{d}[\mathbf{A}(r)]\,\mathbf{X}_k(r), \qquad k = 1, \dots, n. \tag{2,3}$$

If $t_0 \le t_1$, then for every $t \in (t_1, 1]$ the matrix $\mathbf{I} - \Delta^- \mathbf{A}(t)$ is regular. Hence by 1.4 the equation (2,3) for $\mathbf{X}_k(t)$ has a unique solution on $[t_1, 1]$ for every $k = 1, \dots, n$ and this implies the existence and unicity of an $n \times n$-matrix $\mathbf{X}(t)$: $[t_1, 1] \to L(R_n)$ satisfying (2,2). The case when $t_1 \le t_0$ can be treated similarly.

2.2. Theorem. *If the assumptions of* 2.1 *are satisfied, then there exists a unique* $n \times n$*-matrix valued function* $\mathbf{U}(t, s)$ *defined for* $t_0 \le s \le t \le 1$ *and* $0 \le t \le s \le t_0$ *such that*

$$\mathbf{U}(t, s) = \mathbf{I} + \int_{s}^{t} \mathrm{d}[\mathbf{A}(r)]\,\mathbf{U}(r, s). \tag{2,4}$$

Proof. If e.g. $t_0 \le s \le 1$ and s is fixed, then the matrix equation

$$\mathbf{X}(t) = \mathbf{I} + \int_{s}^{t} \mathrm{d}[\mathbf{A}(r)]\,\mathbf{X}(r) \tag{2,5}$$

has by 2.1 a uniquely determined solution $\mathbf{X}$: $[s, 1] \to L(R_n)$. If we denote this solution by $\mathbf{U}(t, s)$, then $\mathbf{U}(t, s)$ is uniquely determined for $t_0 \le s \le t \le 1$ and satisfies (2,4).

Similarly if $0 \le s \le t_0$, s being fixed, the matrix equation (2,5) has by 2.1 a unique solution $\mathbf{X}$: $[0, s] \to L(R_n)$ which will be denoted by $\mathbf{U}(t, s)$, and $\mathbf{U}(t, s)$ evidently satisfies (2,4) for $0 \le t \le s \le t_0$.

2.3. Lemma. *Suppose that the assumptions of* 2.1 *are fulfilled. Then there exists a constant* $M > 0$ *such that* $|\mathbf{U}(t, s)| \leq M$ *for all* t, s *such that* $0 \leq t \leq s \leq t_0$ *or* $t_0 \leq s \leq t \leq 1$. *Moreover we have*

$$(2,6) \qquad |\mathbf{U}(t_2, s) - \mathbf{U}(t_1, s)| \leq M \operatorname{var}_{t_1}^{t_2} \mathbf{A}$$

for all $0 \leq t_1 \leq t_2 \leq s$ *if* $s \leq t_0$ *and all* $s \leq t_1 \leq t_2 \leq 1$ *if* $t_0 \leq s$. *Consequently* $\operatorname{var}_0^s \mathbf{U}(., s) \leq M \operatorname{var}_0^s \mathbf{A}$, $\operatorname{var}_s^1 \mathbf{U}(., s) \leq M \operatorname{var}_s^1 \mathbf{A}$ *if* $0 \leq s \leq t_0$, $t_0 \leq s \leq 1$ *respectively.*

Proof. Since $\mathbf{U}(t, s)$ satisfies (2,4) in its domain of definition, the k-th column ($k = 1, \ldots, n$) of $\mathbf{U}(t, s)$ denoted by $\mathbf{U}_k(t, s)$ satisfies the equation

$$\mathbf{U}_k(t, s) = \mathbf{e}^{(k)} + \int_s^t \mathrm{d}[\mathbf{A}(r)]\, \mathbf{U}_k(r, s)$$

for every $t \in [0, s]$ when $s \leq t_0$ ($\mathbf{e}^{(k)}$ means the k-th column of the identity matrix $\mathbf{I} \in L(R_n)$, i.e. $\mathbf{U}_k(t, s)$ is a solution of the problem $\mathrm{d}\mathbf{x} = \mathrm{d}[\mathbf{A}]\,\mathbf{x} + \mathrm{d}\mathbf{g}$, $\mathbf{x}(s) = \mathbf{e}^{(k)}$). Hence by 1.7 we have

$$|\mathbf{U}_k(t, s)| \leq C|\mathbf{e}^{(k)}| \exp(C \operatorname{var}_t^s \mathbf{A}) \leq C \exp(C \operatorname{var}_0^1 \mathbf{A}), \qquad k = 1, \ldots, n$$

for every $0 \leq t \leq s \leq t_0$ where $C \geq 1$ is a constant and evidently also

$$|\mathbf{U}(t, s)| \leq \sum_{k=1}^{n} |\mathbf{U}_k(t, s)| \leq nC \exp(C \operatorname{var}_0^1 \mathbf{A}) = M .$$

If $t_0 \leq s$, then 1.7 yields the same result for $s \leq t \leq 1$ and the boundedness of $\mathbf{U}(t, s)$ is proved.

Assume that $0 \leq t_1 \leq t_2 \leq s \leq t_0$. Then we have by I.4.16

$$|\mathbf{U}(t_2, s) - \mathbf{U}(t_1, s)| = \left| \int_s^{t_2} \mathrm{d}[\mathbf{A}(r)]\, \mathbf{U}(r, s) - \int_s^{t_1} \mathrm{d}[\mathbf{A}(r)]\, \mathbf{U}(r, s) \right|$$

$$= \left| \int_{t_1}^{t_2} \mathrm{d}[\mathbf{A}(r)]\, \mathbf{U}(r, s) \right| \leq M \operatorname{var}_{t_1}^{t_2} \mathbf{A} .$$

A similar inequality holds if $t_0 \leq s \leq t_1 \leq t_2 \leq 1$ and (2,6) is proved.

2.4. Theorem. *Suppose that the assumptions of* 2.1 *are fullfilled and* $t_1 \in [0, 1]$. *Then the unique solution of the homogeneous initial value problem*

$$(2,7) \qquad \mathrm{d}\mathbf{x} = \mathrm{d}[\mathbf{A}]\, \mathbf{x}, \qquad \mathbf{x}(t_1) = \tilde{\mathbf{x}}$$

defined on $[t_1, 1]$ *if* $t_0 \leq t_1$ *and on* $[0, t_1]$ *if* $t_1 \leq t_0$ *is given by the relation*

$$(2,8) \qquad \mathbf{x}(t) = \mathbf{U}(t, t_1)\, \tilde{\mathbf{x}}$$

on the intervals of definition, where $\mathbf{U}$ *is the* $n \times n$*-matrix from* 2.2 *satisfying* (2,4).

Proof. Under the given assumptions the existence and uniqueness of a solution of (2,7) is quaranteed by 1.4. Let us assume that $t_0 \leq t_1$. Since by 2.2 $\mathbf{U}(t, t_1)$ is uniquely defined for $t_1 \leq t \leq 1$, by (2,8) a function $\mathbf{x}\colon [t_1, 1] \to R_n$ is given. By 2.3 we have $\operatorname{var}_{t_1}^1 \mathbf{U}(., t_1) < \infty$ and consequently $\operatorname{var}_{t_1}^1 \mathbf{x} = \operatorname{var}_{t_1}^1 \mathbf{U}(., t_1)\tilde{\mathbf{x}} < \infty$. For $\mathbf{x}\colon [t_1, 1] \to R_n$ given by (2,8) the integral $\int_{t_1}^t \mathrm{d}[\mathbf{A}(s)]\, \mathbf{x}(s)$ evidently exists (see I.4.19) for every $t \in [t_1, 1]$ and by (2,4) we have

$$\int_{t_1}^t \mathrm{d}[\mathbf{A}(s)]\, \mathbf{x}(s) = \int_{t_1}^t \mathrm{d}[\mathbf{A}(s)]\, \mathbf{U}(s, t_1)\, \tilde{\mathbf{x}} = (\mathbf{U}(t, t_1) - \mathbf{I})\, \tilde{\mathbf{x}} = \mathbf{x}(t) - \tilde{\mathbf{x}},$$

i.e. $\mathbf{x}(t) = \mathbf{U}(t, t_1)\, \tilde{\mathbf{x}}$ is a solution of (2,7) on $[t_1, 1]$. The proof of this result for the case $t_1 \leq t_0$ is similar.

2.5. Corollary. *If the assumptions of* 2.1 *are satisfied and* $\mathbf{U}(t, s)$ *is the* $n \times n$*-matrix determined by* (2,4) *for* $t_0 \leq s \leq t \leq 1$ *and* $0 \leq t \leq s \leq t_0$, *then*

(2,9) $$\mathbf{U}(t, s) = \mathbf{U}(t, r)\, \mathbf{U}(r, s)$$

if $t_0 \leq s \leq r \leq t \leq 1$ *or* $0 \leq t \leq r \leq s \leq t_0$ *and*

(2,10) $$\mathbf{U}(t, t) = \mathbf{I}$$

for every $t \in [0, 1]$.

Proof. Let e.g. $0 \leq t \leq r \leq s \leq t_0$, then by (2,4) we obtain

$$\mathbf{U}(t, s) = \mathbf{I} + \int_s^t \mathrm{d}[\mathbf{A}(\varrho)]\, \mathbf{U}(\varrho, s) = \mathbf{I} + \int_s^r \mathrm{d}[\mathbf{A}(\varrho)]\, \mathbf{U}(\varrho, s) + \int_r^t \mathrm{d}[\mathbf{A}(\varrho)]\, \mathbf{U}(\varrho, s)$$

$$= \mathbf{U}(r, s) + \int_r^t \mathrm{d}[\mathbf{A}(\varrho)]\, \mathbf{U}(\varrho, s)$$

for every $0 \leq t \leq r$. Hence $\mathbf{U}(t, s)$ satisfies the matrix equation

$$\mathbf{X}(t) = \mathbf{U}(r, s) + \int_r^t \mathrm{d}[\mathbf{A}(\varrho)]\, \mathbf{X}(\varrho)$$

for $0 \leq t \leq r$ and by 2.4 this solution can be expressed in the form $\mathbf{U}(t, r)\, \mathbf{U}(r, s)$, i.e. (2,9) is satisfied. If $t_0 \leq s \leq r \leq t \leq 1$, then (2,9) can be proved analogously. The relation (2,10) obviously follows from (2,4).

2.6. Lemma. *If the assumptions of* 2.1 *are satisfied, then for* $\mathbf{U}(t, s)$ *given by* 2.2 we have

(2,11) $$|\mathbf{U}(t, s_2) - \mathbf{U}(t, s_1)| \leq M^2 \operatorname{var}_{s_1}^{s_2} \mathbf{A}$$

for any s_1, s_2 *such that* $t_0 \leq s_1 \leq s_2 \leq t \leq 1$ *or* $0 \leq t \leq s_1 \leq s_2 \leq t_0$ *where* M *is the bound of* $\mathbf{U}(t, s)$ (see 2.3). *Hence* $\operatorname{var}_{t_0}^t \mathbf{U}(t, .) \leq M^2 \operatorname{var}_{t_0}^t \mathbf{A}$ *if* $t_0 \leq t$ *and* $\operatorname{var}_t^{t_0} \mathbf{U}(t, .) \leq M^2 \operatorname{var}_t^{t_0} \mathbf{A}$ *if* $t \leq t_0$.

Proof. Let us consider the case when $t_0 \le s_1 \le s_2 \le t$. By (2,4) we have

$$\mathbf{U}(t, s_2) - \mathbf{U}(t, s_1) = \int_{s_2}^{t} \mathrm{d}[\mathbf{A}(r)]\, \mathbf{U}(r, s_2) - \int_{s_1}^{t} \mathrm{d}[\mathbf{A}(r)]\, \mathbf{U}(r, s_1)$$

$$= \int_{s_2}^{t} \mathrm{d}[\mathbf{A}(r)]\, \mathbf{U}(r, s_2) - \int_{s_2}^{t} \mathrm{d}[\mathbf{A}(r)]\, \mathbf{U}(r, s_1) - \int_{s_1}^{s_2} \mathrm{d}[\mathbf{A}(r)]\, \mathbf{U}(r, s_1),$$

i.e. the difference $\mathbf{U}(t, s_2) - \mathbf{U}(t, s_1)$ satisfies the matrix equation

$$\mathbf{X}(t) = - \int_{s_1}^{s_2} \mathrm{d}[\mathbf{A}(r)]\, \mathbf{U}(r, s_1) + \int_{s_2}^{t} \mathrm{d}[\mathbf{A}(r)]\, \mathbf{X}(r)$$

for $s_2 \le t \le 1$. Hence by 2.4 we obtain

$$\mathbf{U}(t, s_2) - \mathbf{U}(t, s_1) = \mathbf{U}(t, s_2)\left(- \int_{s_1}^{s_2} \mathrm{d}[\mathbf{A}(r)]\, \mathbf{U}(r, s_1)\right)$$

and by 2.3 and I.4.16 it is

$$|\mathbf{U}(t, s_2) - \mathbf{U}(t, s_1)| \le M \left| \int_{s_1}^{s_2} \mathrm{d}[\mathbf{A}(r)]\, \mathbf{U}(r, s_1)\right| \le M^2 \operatorname{var}_{s_1}^{s_2} \mathbf{A}.$$

The proof for the case $0 \le t \le s_1 \le s_2 \le t_0$ can be given similarly and (2,11) is valid.

2.7. Lemma. *Suppose that the assumptions of* 2.1 *are satisfied. Let us define*

(2,12)
$$\tilde{\mathbf{U}}(t, s) = \mathbf{U}(t, s) \quad \textit{for} \quad t_0 \le s \le t \le 1,$$
$$\tilde{\mathbf{U}}(t, s) = \mathbf{U}(t, t) = \mathbf{I} \quad \textit{for} \quad t_0 \le t \le s \le 1,$$

and

(2,13)
$$\tilde{\mathbf{U}}(t, s) = \mathbf{U}(t, s) \quad \textit{for} \quad 0 \le t \le s \le t_0,$$
$$\tilde{\mathbf{U}}(t, s) = \mathbf{U}(t, t) = \mathbf{I} \quad \textit{for} \quad 0 \le s \le t \le t_0,$$

where $\mathbf{U}(t, s) \in L(R_n)$ *is given by* 2.2.

Then for the twodimensional variations of $\tilde{\mathbf{U}}$ *on the squares* $[t_0, 1] \times [t_0, 1]$ *and* $[0, t_0] \times [0, t_0]$ *on which* $\tilde{\mathbf{U}}$ *is defined we have* $\mathrm{v}_{[t_0,1]\times[t_0,1]}(\tilde{\mathbf{U}}) < \infty$ *and* $\mathrm{v}_{[0,t_0]\times[0,t_0]}(\tilde{\mathbf{U}}) < \infty$.

Proof. Assume that $t_0 = \alpha_0 < \alpha_1 < \ldots < \alpha_k = 1$ is an arbitrary subdivision of the interval $[t_0, 1]$ and $J_{ij} = [\alpha_{i-1}, \alpha_i] \times [\alpha_{j-1}, \alpha_j]$, $i, j = 1, \ldots, k$ the corresponding net-type subdivision of $[t_0, 1] \times [t_0, 1]$. We consider the sum (see I.6.2, I.6.3)

$$\sum_{i,j=1}^{k} |m_{\tilde{U}}(J_{ij})| = \sum_{i=1}^{k} \left(\sum_{j=1}^{i-1} |m_{\tilde{U}}(J_{ij})| + |m_{\tilde{U}}(J_{ii})| + \sum_{j=i+1}^{k} |m_{\tilde{U}}(J_{ij})| \right)$$

where we use the convention that $\sum_{j=1}^{0} |m_{\tilde{U}}(J_{ij})| = 0$ and $\sum_{j=k+1}^{k} |m_{\tilde{U}}(J_{ij})| = 0$. By (2,12) we have $m_{\tilde{U}}(J_{ij}) = m_U(J_{ij})$ if $j \le i - 1$,

$$m_{\tilde{U}}(J_{ii}) = \tilde{\mathbf{U}}(\alpha_i, \alpha_i) - \tilde{\mathbf{U}}(\alpha_i, \alpha_{i-1}) - \tilde{\mathbf{U}}(\alpha_{i-1}, \alpha_i) + \tilde{\mathbf{U}}(\alpha_{i-1}, \alpha_{i-1})$$
$$= \tilde{\mathbf{U}}(\alpha_i, \alpha_i) - \tilde{\mathbf{U}}(\alpha_i, \alpha_{i-1}) = \mathbf{U}(\alpha_i, \alpha_i) - \mathbf{U}(\alpha_i, \alpha_{i-1})$$

and $m_{\tilde{U}}(J_{ij}) = \mathbf{0}$ if $i + 1 \le j$. Hence

$$\sum_{i,j=1}^{k} |m_{\tilde{U}}(J_{ij})| = \sum_{i=1}^{k} \sum_{j=1}^{i-1} |m_U(J_{ij})| + \sum_{i=1}^{k} |\mathbf{U}(\alpha_i, \alpha_i) - \mathbf{U}(\alpha_i, \alpha_{i-1})| . \tag{2,14}$$

If $j \le i - 1$, then $\alpha_{j-1} < \alpha_j \le \alpha_{i-1} < \alpha_i$ and by 2.5

$$m_U(J_{ij}) = \mathbf{U}(\alpha_i, \alpha_j) - \mathbf{U}(\alpha_i, \alpha_{j-1}) - \mathbf{U}(\alpha_{i-1}, \alpha_j) + \mathbf{U}(\alpha_{i-1}, \alpha_{j-1})$$
$$= \mathbf{U}(\alpha_i, \alpha_{i-1})\, \mathbf{U}(\alpha_{i-1}, \alpha_j) - \mathbf{U}(\alpha_{i-1}, \alpha_j) - \mathbf{U}(\alpha_i, \alpha_{i-1})\, \mathbf{U}(\alpha_{i-1}, \alpha_{j-1}) + \mathbf{U}(\alpha_{i-1}, \alpha_{j-1})$$
$$= [\mathbf{U}(\alpha_i, \alpha_{i-1}) - \mathbf{I}]\, \mathbf{U}(\alpha_{i-1}, \alpha_j) - [\mathbf{U}(\alpha_i, \alpha_{i-1}) - \mathbf{I}]\, \mathbf{U}(\alpha_{i-1}, \alpha_{j-1})$$
$$= [\mathbf{U}(\alpha_i, \alpha_{i-1}) - \mathbf{I}]\, [\mathbf{U}(\alpha_{i-1}, \alpha_j) - \mathbf{U}(\alpha_{i-1}, \alpha_{j-1})]$$
$$= [\mathbf{U}(\alpha_i, \alpha_{i-1}) - \mathbf{U}(\alpha_{i-1}, \alpha_{i-1})]\, [\mathbf{U}(\alpha_{i-1}, \alpha_j) - \mathbf{U}(\alpha_{i-1}, \alpha_{j-1})] .$$

Hence by 2.3 and 2.6 we obtain

$$|m_U(J_{ij})| = |\mathbf{U}(\alpha_i, \alpha_{i-1}) - \mathbf{U}(\alpha_{i-1}, \alpha_{i-1})|\, |\mathbf{U}(\alpha_{i-1}, \alpha_j) - \mathbf{U}(\alpha_{i-1}, \alpha_{j-1})|$$
$$\le M(\operatorname{var}_{\alpha_{i-1}}^{\alpha_i} \mathbf{A})\, M^2 \operatorname{var}_{\alpha_{j-1}}^{\alpha_j} \mathbf{A} = M^3 \operatorname{var}_{\alpha_{i-1}}^{\alpha_i} \mathbf{A} \operatorname{var}_{\alpha_{j-1}}^{\alpha_j} \mathbf{A}$$

and

$$\sum_{i=1}^{k} \sum_{j=1}^{i-1} |m_U(J_{ij})| \le M^3 \sum_{i=1}^{k} \operatorname{var}_{\alpha_{i-1}}^{\alpha_i} \mathbf{A} \sum_{j=1}^{i-1} \operatorname{var}_{\alpha_{j-1}}^{\alpha_j} \mathbf{A} \le M^3 (\operatorname{var}_{t_0}^{1} \mathbf{A})^2 .$$

Further, by (2,11) from 2.6 we have

$$\sum_{i=1}^{k} |\mathbf{U}(\alpha_i, \alpha_i) - \mathbf{U}(\alpha_i, \alpha_{i-1})| \le \sum_{i=1}^{k} M^2 \operatorname{var}_{\alpha_{i-1}}^{\alpha_i} \mathbf{A} = M^2 \operatorname{var}_{t_0}^{1} \mathbf{A} .$$

Hence by (2,14) we have

$$\sum_{i,j=1}^{k} |m_{\tilde{U}}(J_{ij})| \le M^3 (\operatorname{var}_{t_0}^{1} \mathbf{A})^2 + M^2 \operatorname{var}_{t_0}^{1} \mathbf{A}$$

and since the net-type subdivision was chosen arbitrarily, we have by the definition also

$$\mathrm{v}_{[t_0,1]\times[t_0,1]}(\tilde{\mathbf{U}}) \le M^3 (\operatorname{var}_{t_0}^{0} \mathbf{A})^2 + M^2 \operatorname{var}_{t_0}^{1} \mathbf{A} < \infty .$$

The finiteness of $\mathrm{v}_{[0,t_0]\times[0,t_0]}(\tilde{\mathbf{U}})$ can be proved similarly.

2.8. Theorem (variation-of-constants formula). *Let* $\mathbf{A}: [0,1] \to L(R_n)$ *satisfy the assumptions given in* 2.1 *where* $t_0 \in [0,1]$ *is fixed. Then for every* $\mathbf{x}_0 \in R_n$, $\mathbf{g} \in BV_n$ *the unique solution of the initial value problem* (2,1) *can be expressed in the form*

$$\mathbf{x}(t) = \mathbf{U}(t, t_0)\,\mathbf{x}_0 + \mathbf{g}(t) - \mathbf{g}(t_0) - \int_{t_0}^{t} \mathrm{d}_s[\mathbf{U}(t, s)]\,(\mathbf{g}(s) - \mathbf{g}(t_0)) \tag{2,15}$$

where $\mathbf{U}$ *is the uniquely determined matrix satisfying* (2,4) *from* 2.2.

Proof. We verify by computation that $\mathbf{x}: [0,1] \to R_n$ from (2,15) is really a solution of (2,1). Let us assume that $t < t_0$. Then

$$\begin{aligned}\int_{t_0}^{t} \mathrm{d}[\mathbf{A}(r)]\,\mathbf{x}(r) &= \int_{t_0}^{t} \mathrm{d}[\mathbf{A}(r)]\,\mathbf{U}(r, t_0)\,\mathbf{x}_0 + \int_{t_0}^{t} \mathrm{d}[\mathbf{A}(r)]\,(\mathbf{g}(r) - \mathbf{g}(t_0)) \\ &\quad - \int_{t_0}^{t} \mathrm{d}[\mathbf{A}(r)] \int_{t_0}^{r} \mathrm{d}_s[\mathbf{U}(r, s)]\,(\mathbf{g}(s) - \mathbf{g}(t_0)) \\ &= (\mathbf{U}(t, t_0) - \mathbf{I})\,\mathbf{x}_0 + \int_{t_0}^{t} \mathrm{d}[\mathbf{A}(r)]\,(\mathbf{g}(r) - \mathbf{g}(t_0)) - \int_{t_0}^{t} \mathrm{d}[\mathbf{A}(r)] \int_{t_0}^{r} \mathrm{d}_s[\mathbf{U}(r, s)]\,(\mathbf{g}(s) - \mathbf{g}(t_0))\end{aligned} \tag{2,16}$$

since $\mathbf{U}$ satisfies 2.4. Let us now consider the last term from the right-hand side in (2,16). We have

$$\int_{t_0}^{t} \mathrm{d}[\mathbf{A}(r)] \int_{t_0}^{r} \mathrm{d}_s[\mathbf{U}(r, s)]\,(\mathbf{g}(s) - \mathbf{g}(t_0)) = \int_{t}^{t_0} \mathrm{d}[\mathbf{A}(r)] \int_{t}^{t_0} \mathrm{d}_s[\tilde{\mathbf{U}}(r, s)]\,(\mathbf{g}(s) - \mathbf{g}(t_0))$$

where $\tilde{\mathbf{U}}$ is defined in 2.7 and satisfies by 2.7, 2.3 and 2.6 the assumptions of I.6.20 on the square $[t, t_0] \times [t, t_0]$. Hence we interchange by I.6.20 the order of integration and obtain by the definition of $\mathbf{U}$

$$\begin{aligned}\int_{t_0}^{t} \mathrm{d}[\mathbf{A}(r)] \int_{t_0}^{r} \mathrm{d}_s[\mathbf{U}(r, s)]\,(\mathbf{g}(s) - \mathbf{g}(t_0)) &= \int_{t}^{t_0} \mathrm{d}_s\left[\int_{t}^{t_0} \mathrm{d}[\mathbf{A}(r)]\,\tilde{\mathbf{U}}(r, s)\right](\mathbf{g}(s) - \mathbf{g}(t_0)) \\ &= \int_{t}^{t_0} \mathrm{d}_s\left[\int_{t}^{s} \mathrm{d}[\mathbf{A}(r)]\,\tilde{\mathbf{U}}(r, s) + \int_{s}^{t_0} \mathrm{d}[\mathbf{A}(r)]\,\tilde{\mathbf{U}}(r, s)\right](\mathbf{g}(s) - \mathbf{g}(t_0)) \\ &= \int_{t_0}^{t} \mathrm{d}_s\left[\int_{s}^{t} \mathrm{d}[\mathbf{A}(r)]\,\mathbf{U}(r, s) + \int_{t_0}^{s} \mathrm{d}[\mathbf{A}(r)]\right](\mathbf{g}(s) - \mathbf{g}(t_0)) \\ &= \int_{t_0}^{t} \mathrm{d}_s[\mathbf{U}(t, s) - \mathbf{I} + \mathbf{A}(s) - \mathbf{A}(t_0)]\,(\mathbf{g}(s) - \mathbf{g}(t_0)) \\ &= \int_{t_0}^{t} \mathrm{d}_s[\mathbf{U}(t, s)]\,(\mathbf{g}(s) - \mathbf{g}(t_0)) + \int_{t_0}^{t} \mathrm{d}[\mathbf{A}(s)]\,(\mathbf{g}(s) - \mathbf{g}(t_0)).\end{aligned}$$

Using this expression we obtain by (2,16)

$$\int_{t_0}^{t} \mathrm{d}[\mathbf{A}(r)]\,\mathbf{x}(r) = \mathbf{U}(t,t_0)\,\mathbf{x}_0 - \mathbf{x}_0 + \int_{t_0}^{t} \mathrm{d}[\mathbf{A}(r)]\,(\mathbf{g}(r) - \mathbf{g}(t_0))$$

$$- \int_{t_0}^{t} \mathrm{d}_s[\mathbf{U}(t,s)]\,(\mathbf{g}(s) - \mathbf{g}(t_0)) - \int_{t_0}^{t} \mathrm{d}[\mathbf{A}(s)]\,(\mathbf{g}(s) - \mathbf{g}(t_0))$$

$$= \mathbf{U}(t,t_0)\,\mathbf{x}_0 + \mathbf{g}(t) - \mathbf{g}(t_0) - \int_{t_0}^{t} \mathrm{d}_s[\mathbf{U}(t,s)]\,(\mathbf{g}(s) - \mathbf{g}(t_0)) - (\mathbf{g}(t) - \mathbf{g}(t_0)) - \mathbf{x}_0$$

$$= \mathbf{x}(t) - \mathbf{x}_0 - (\mathbf{g}(t) - \mathbf{g}(t_0)).$$

Hence $\mathbf{x}(t)$ is a solution of (2,1) for $t \le t_0$. For the case $t_0 \le t$ the proof can be given analogously. Using 1.4 the solutions of (2,1) are uniquely determined and this completes the proof.

2.9. Remark. Let us mention that the operator $\mathbf{x} \in BV_n \to \int_{t_0}^{t} \mathrm{d}[\mathbf{A}(s)]\,\mathbf{x}(s)$ appearing in the definition of the generalized linear differential equation (2,1) can be written in the Fredholm-Stieltjes form $\int_0^1 \mathrm{d}_s\,[\mathbf{K}(t,s)]\,\mathbf{x}(s)$ where $\mathbf{K}\colon [0,1] \times [0,1] \to L(R_n)$ is defined as follows: if $t_0 \le t \le 1$, then

$$\begin{aligned}
\mathbf{K}(t,s) &= \mathbf{A}(t_0) &&\text{for } 0 \le s \le t_0,\\
\mathbf{K}(t,s) &= \mathbf{A}(s) &&\text{for } t_0 \le s \le t,\\
\mathbf{K}(t,s) &= \mathbf{A}(t) &&\text{for } t \le s \le 1,
\end{aligned}$$

and if $0 \le t \le t_0$, then

$$\begin{aligned}
\mathbf{K}(t,s) &= -\mathbf{A}(t) &&\text{for } 0 \le s \le t,\\
\mathbf{K}(t,s) &= -\mathbf{A}(s) &&\text{for } t \le s \le t_0,\\
\mathbf{K}(t,s) &= -\mathbf{A}(t_0) &&\text{for } t_0 \le s \le 1.
\end{aligned}$$

If this fact is used and II.2.5 is taken into account, then the solution of the equation (2,1) can be given by the resolvent formula (II.2.16) in the form

$$\mathbf{x}(t) = \mathbf{x}_0 + \mathbf{g}(t) - \mathbf{g}(t_0) + \int_0^1 \mathrm{d}_s[\boldsymbol{\Gamma}(t,s)]\,(\mathbf{x}_0 + \mathbf{g}(t) - \mathbf{g}(t_0)), \tag{2,17}$$

for $t \in [0,1]$ since (2,1) has a solution uniquely defined for every $\mathbf{x}_0 \in R_n$, $\mathbf{g} \in BV_n$. The resolvent kernel $\boldsymbol{\Gamma}\colon [0,1] \times [0,1] \to L(R_n)$ satisfies

$$\boldsymbol{\Gamma}(t,s) = \mathbf{K}(t,s) + \int_0^1 \mathrm{d}_r[\mathbf{K}(t,r)]\,\boldsymbol{\Gamma}(r,t).$$

If we set $\mathbf{U}(t,s) = \mathbf{I} + \boldsymbol{\Gamma}(t,s) - \boldsymbol{\Gamma}(t,t)$, then the variation-of-constants formula (2,15) can be derived from (2,17).

In the following we consider the initial value problem (2,1) with the assumptions on $\mathbf{A}: [0,1] \to L(R_n)$ strengthened.

2.10. Theorem. *Assume that the matrix* $\mathbf{A}: [0,1] \to L(R_n)$, $\operatorname{var}_0^1 \mathbf{A} < \infty$ *is such that* $\mathbf{I} - \Delta^- \mathbf{A}(t)$ *is regular for all* $t \in (0,1]$ *and* $\mathbf{I} + \Delta^+ \mathbf{A}(t)$ *is regular for all* $t \in [0,1)$.

Then there exists a unique $n \times n$*-matrix valued function* $\mathbf{U}: [0,1] \times [0,1] \to L(R_n)$ *such that*

$$\mathbf{U}(t,s) = \mathbf{I} + \int_s^t \mathrm{d}[\mathbf{A}(r)]\,\mathbf{U}(r,s) \tag{2,18}$$

for all $t, s \in [0,1]$.

The matrix $\mathbf{U}(t,s)$ *determined by* (2,18) *has the following properties.*

(i) $\mathbf{U}(t,t) = \mathbf{I}$ *for all* $t \in [0,1]$.

(ii) *There exists a constant* $M > 0$ *such that* $|\mathbf{U}(t,s)| \le M$ *for all* $t, s \in [0,1]$, $\operatorname{var}_0^1 \mathbf{U}(t,.) \le M$, $\operatorname{var}_0^1 \mathbf{U}(.,s) \le M$ *for all* $t, s \in [0,1]$.

(iii) *For any* $r, s, t \in [0,1]$ *the relation*

$$\mathbf{U}(t,s) = \mathbf{U}(t,r)\,\mathbf{U}(r,s) \tag{2,19}$$

holds.

(iv) $\mathbf{U}(t+,s) = [\mathbf{I} + \Delta^+ \mathbf{A}(t)]\,\mathbf{U}(t,s)$ *for* $t \in [0,1)$, $s \in [0,1]$,

$\mathbf{U}(t-,s) = [\mathbf{I} - \Delta^- \mathbf{A}(t)]\,\mathbf{U}(t,s)$ *for* $t \in (0,1]$, $s \in [0,1]$,

$\mathbf{U}(t,s+) = \mathbf{U}(t,s)\,[\mathbf{I} + \Delta^+ \mathbf{A}(s)]^{-1}$ *for* $t \in [0,1]$, $s \in [0,1)$,

$\mathbf{U}(t,s-) = \mathbf{U}(t,s)\,[\mathbf{I} - \Delta^- \mathbf{A}(s)]^{-1}$ *for* $t \in [0,1]$, $s \in (0,1]$.

(v) *The matrix* $\mathbf{U}(t,s)$ *is regular for any* $t, s \in [0,1]$.

(vi) *The matrices* $\mathbf{U}(t,s)$ *and* $\mathbf{U}(s,t)$ *are mutually reciprocal, i.e.* $[\mathbf{U}(t,s)]^{-1} = \mathbf{U}(s,t)$ *for every* $t, s \in [0,1]$.

(vii) *The twodimensional variation of* $\mathbf{U}$ *is finite on* $[0,1] \times [0,1]$, *i.e.* $v_{[0,1]\times[0,1]}(\mathbf{U}) < \infty$.

Proof. By 2.1 for every fixed $s \in [0,1]$ the matrix equation

$$\mathbf{X}(t) = \tilde{\mathbf{X}} + \int_s^t \mathrm{d}[\mathbf{A}(r)]\,\mathbf{X}(r), \qquad \tilde{\mathbf{X}} \in L(R_n)$$

has a unique solution $\mathbf{X}: [0,1] \to L(R_n)$, which is defined on the whole interval $[0,1]$. Hence the existence of $\mathbf{U}(t,s)$ satisfying (2,18) is quaranteed.

(i) is obvious from (2,18). (ii) follows immediately from 2.3 and 2.6. For (iii) we have

$$\mathbf{U}(t,s) = \mathbf{I} + \int_s^t \mathrm{d}[\mathbf{A}(\varrho)]\,\mathbf{U}(\varrho,s) = \mathbf{I} + \int_s^r \mathrm{d}[\mathbf{A}(\varrho)]\,\mathbf{U}(\varrho,s) + \int_r^t \mathrm{d}[\mathbf{A}(\varrho)]\,\mathbf{U}(\varrho,s)$$

$$= \mathbf{U}(r,s) + \int_r^t \mathrm{d}[\mathbf{A}(\varrho)]\,\mathbf{U}(\varrho,s),$$

i.e. $\mathbf{U}(t,s)$ satisfies the matrix equation

$$\mathbf{X}(t) = \mathbf{U}(r,s) + \int_r^t \mathrm{d}[\mathbf{A}(r)]\,\mathbf{X}(r).$$

Hence by 2.4 we obtain $\mathbf{U}(t,s) = \mathbf{U}(t,r)\,\mathbf{U}(r,s)$ for every $r, s, t \in [0,1]$, and (2,19) is satisfied.

The first two relations in (iv) are simple consequences of 1.6. To prove the third relation in (iv) let us mention that for any $t \in [0,1]$, $s \in [0,1)$ and sufficiently small $\delta > 0$ we have by definition

$$\mathbf{U}(t, s+\delta) - \mathbf{U}(t,s) = \int_{s+\delta}^t \mathrm{d}[\mathbf{A}(r)]\,\mathbf{U}(r, s+\delta) - \int_s^t \mathrm{d}[\mathbf{A}(r)]\,\mathbf{U}(r,s)$$

$$= \int_{s+\delta}^t \mathrm{d}[\mathbf{A}(r)]\,(\mathbf{U}(r, s+\delta) - \mathbf{U}(r,s)) - \int_s^{s+\delta} \mathrm{d}[\mathbf{A}(r)]\,\mathbf{U}(r,s),$$

i.e. the difference $\mathbf{U}(t, s+\delta) - \mathbf{U}(t,s)$ satisfies the matrix equation

$$\mathbf{X}(t) = -\int_s^{s+\delta} \mathrm{d}[\mathbf{A}(r)]\,\mathbf{U}(r,s) + \int_{s+\delta}^t \mathrm{d}[\mathbf{A}(r)]\,\mathbf{X}(r)$$

and consequently by 2.4 it is

$$\mathbf{U}(t, s+\delta) - \mathbf{U}(t,s) = \mathbf{U}(t, s+\delta)\left(-\int_s^{s+\delta} \mathrm{d}[\mathbf{A}(r)]\,\mathbf{U}(r,s)\right).$$

For $\delta \to 0+$ this equality yields

$$\mathbf{U}(t, s+) - \mathbf{U}(t,s) = -\mathbf{U}(t, s+)\,\Delta^+\mathbf{A}(s)\,\mathbf{U}(s,s) = -\mathbf{U}(t, s+)\,\Delta^+\mathbf{A}(s).$$

Hence $\mathbf{U}(t,s) = \mathbf{U}(t,s+)\,[\mathbf{I} + \Delta^+\mathbf{A}(s)]$ for any $t \in [0,1]$, $s \in [0,1)$ and the assumption of the regularity of the matrix $\mathbf{I} + \Delta^+\mathbf{A}(s)$ gives the existence of the inverse $[\mathbf{I} + \Delta^+\mathbf{A}(s)]^{-1}$ and also the third equality from (iv). The fourth equality in (iv) can be proved analogously.

By (iii) we have $\mathbf{U}(t,s)\,\mathbf{U}(s,t) = \mathbf{I}$ and $\mathbf{U}(s,t)\,\mathbf{U}(t,s) = \mathbf{I}$ for every $t, s \in [0,1]$. Hence $\mathbf{U}(t,s) = \mathbf{U}(s,t)^{-1}$ and $\mathbf{U}(s,t) = \mathbf{U}(t,s)^{-1}$ and (vi) is proved. From (vi) the statement (v) follows immediately. (In this connection we note that a direct proof of (v) can be given without using (iii), see Schwabik [1].)

Finally by (iii) we have $\mathbf{U}(t,s) = \mathbf{U}(t,0)\,\mathbf{U}(0,s)$ for every $(t,s) \in [0,1] \times [0,1]$. By (ii) it is $\operatorname{var}_0^1 \mathbf{U}(.,0) < \infty$ and $\operatorname{var}_0^1 \mathbf{U}(0,.) < \infty$. Hence by I.6.4 we have $v_{[0,1]\times[0,1]}(\mathbf{U}) < \infty$ and (vii) is also proved.

2.11. Corollary. *If* $\mathbf{A}\colon [0,1] \to L(R_n)$, $\operatorname{var}_0^1 \mathbf{A} < \infty$, *satisfies the assumptions given in* 2.10, *then*

(2,20) $$\mathbf{U}(t,s) = \mathbf{X}(t)\,\mathbf{X}^{-1}(s) \quad \textit{for every} \quad s, t \in [0,1]$$

where $\mathbf{X}: [0,1] \to L(R_n)$ *satisfies the matrix equation*

$$\mathbf{X}(t) = \mathbf{I} + \int_0^t \mathrm{d}[\mathbf{A}(r)]\,\mathbf{X}(r), \qquad t \in [0,1]. \tag{2,21}$$

Proof. Since the matrix equation (2,21) has a unique solution, it is easy to compare it with (2,18) and state that $\mathbf{X}(t) = \mathbf{U}(t,0)$. By (iii) from 2.10 we have $\mathbf{U}(t,s) = \mathbf{U}(t,0)\,\mathbf{U}(0,s)$ and by (vi) from 2.10 it follows $\mathbf{U}(0,s) = [\mathbf{U}(s,0)]^{-1} = \mathbf{X}^{-1}(s)$. Hence (2,20) hold.

2.12. Remark. If the matrix $\mathbf{A}: [0,1] \to L(R_n)$ satisfies the assumptions of 2.10, then evidently the assumptions of 1.4, 2.1–2.8 are satisfied for every $t_0 \in [0,1]$. Hence by 1.4 the initial value problem (2,1) has for every $t_0 \in [0,1]$, $\mathbf{x}_0 \in R_n$, $\mathbf{g} \in BV_n$ a unique solution $\mathbf{x}: [0,1] \to R_n$ defined on the whole interval $[0,1]$.

The variation-of-constants formula 2.8 leads to the following.

2.13. Theorem (variation-of constants formula). *Let us assume that* $\mathbf{A}: [0,1] \to L(R_n)$ *satisfies the conditions given in* 2.10. *Then for any* $t_0 \in [0,1]$, $\mathbf{x}_0 \in R_n$, $\mathbf{g} \in BV_n$ *the solution of the nonhomogeneous initial value problem* (2,1) *is given by the expression*

$$\mathbf{x}(t) = \mathbf{U}(t,t_0)\,\mathbf{x}_0 + \mathbf{g}(t) - \mathbf{g}(t_0) - \int_{t_0}^t \mathrm{d}_s[\mathbf{U}(t,s)]\,(\mathbf{g}(s) - \mathbf{g}(t_0)), \qquad t \in [0,1]$$

where $\mathbf{U}(t,s): [0,1] \times [0,1] \to L(R_n)$ *is the matrix whose existence was stated in* 2.10.

The proof follows immediately from 2.8.

2.14. Corollary. *If* $\mathbf{A}: [0,1] \to L(R_n)$ *satisfies the assumptions from* 2.10, *then the above variation-of-constants formula can be written in the form*

$$\mathbf{x}(t) = \mathbf{g}(t) - \mathbf{g}(t_0) + \mathbf{X}(t)\left\{\mathbf{X}^{-1}(t_0)\,\mathbf{x}_0 - \int_{t_0}^t \mathrm{d}_s[\mathbf{X}^{-1}(s)]\,(\mathbf{g}(s) - \mathbf{g}(t_0))\right\} \tag{2,22}$$

for $t \in [0,1]$ *where* $\mathbf{X}: [0,1] \to L(R_n)$ *is the uniquely determined solution of the matrix equation* (2,21).

The proof follows immediately from 2.13 and from the product decomposition (2,20) given in 2.11.

2.15. Proposition. *If* $\mathbf{A}: [0,1] \to L(R_n)$ *satisfies the assumptions given in* 2.10 *and* $\mathbf{X}: [0,1] \to L(R_n)$ *is the unique solution of the matrix equation* (2,21), *then*

$$\mathbf{X}^{-1}(s) = \mathbf{I} + \mathbf{A}(0) - \mathbf{X}^{-1}(s)\,\mathbf{A}(s) + \int_0^s \mathrm{d}[\mathbf{X}^{-1}(r)]\,\mathbf{A}(r) \tag{2,23}$$

for every $s \in [0,1]$.

Proof. For $\mathbf{X}$: $[0,1] \to L(R_n)$ we have by (2,21)

$$\mathbf{X}(s) - \mathbf{I} = \int_0^s \mathrm{d}[\mathbf{A}(r)]\,\mathbf{X}(r) = \int_0^s \mathrm{d}[\mathbf{A}(r)]\,(\mathbf{X}(r) - \mathbf{I}) + \mathbf{A}(s) - \mathbf{A}(0)$$

for every $s \in [0,1]$. Using the variation-of-constants formula (2,22) in the matrix form we get

$$\mathbf{X}(s) - \mathbf{I} = \mathbf{A}(s) - \mathbf{A}(0) - \mathbf{X}(s) \int_0^s \mathrm{d}[\mathbf{X}^{-1}(r)]\,(\mathbf{A}(r) - \mathbf{A}(0))$$

$$= \mathbf{A}(s) - \mathbf{A}(0) - \mathbf{X}(s) \int_0^s \mathrm{d}[\mathbf{X}^{-1}(r)]\,\mathbf{A}(r) + \mathbf{X}(s)\,[\mathbf{X}^{-1}(s) - \mathbf{X}^{-1}(0)]\,\mathbf{A}(0)$$

$$= \mathbf{A}(s) - \mathbf{X}(s)\,\mathbf{A}(0) - \mathbf{X}(s) \int_0^s \mathrm{d}[\mathbf{X}^{-1}(r)]\,\mathbf{A}(r).$$

Multiplying this relation from the left by $\mathbf{X}^{-1}(s)$ we obtain for every $s \in [0,1]$

$$\mathbf{I} - \mathbf{X}^{-1}(s) = -\mathbf{A}(0) + \mathbf{X}^{-1}(s)\,\mathbf{A}(s) - \int_0^s \mathrm{d}[\mathbf{X}^{-1}(r)]\,\mathbf{A}(r)$$

and (2,23) is satisfied.

2.16. Definition. The matrix $\mathbf{U}(t,s)$: $[0,1] \times [0,1] \to L(R_n)$ given by 2.10 is called the *fundamental matrix* (or *transition matrix*) for the homogeneous generalized linear differential equation $\mathrm{d}\mathbf{x} = \mathrm{d}[\mathbf{A}]\,\mathbf{x}$.

2.17. Remark. If $\mathbf{B}$: $[0,1] \to L(R_n)$ is an $n \times n$-matrix, continuous on $[0,1]$ and $\mathbf{x} = \mathbf{B}(t)\,\mathbf{x}$ is the corresponding ordinary linear differential system, then in the theory of ordinary differential equations the transition matrix $\boldsymbol{\Phi}(t, t_0)$ is defined as a solution of the matrix differential equation

$$\mathbf{X}' = \mathbf{B}(t)\,\mathbf{X}$$

satisfying the condition $\mathbf{X}(t_0) = \mathbf{I} \in L(R_n)$. Hence for $\boldsymbol{\Phi}$ we have

$$\boldsymbol{\Phi}(t, t_0) = \mathbf{I} + \int_{t_0}^t \mathbf{B}(\tau)\,\boldsymbol{\Phi}(\tau, t_0)\,\mathrm{d}\tau\,,$$

i.e. $\boldsymbol{\Phi}$ satisfies the generalized matrix differential equation

$$\boldsymbol{\Phi}(t, t_0) = \mathbf{I} + \int_{t_0}^t \mathrm{d}[\mathbf{A}(\tau)]\,\boldsymbol{\Phi}(\tau, t_0)$$

where $\mathbf{A}(t) = \int_0^t \mathbf{B}(\tau)\,\mathrm{d}\tau$ (see also 1.3). The variation-of-constant formula for the generalized linear differential equation

$$\mathrm{d}\mathbf{x} = \mathrm{d}[\mathbf{A}]\,\mathbf{x} + \mathrm{d}\mathbf{g}\,, \qquad \mathbf{x}(t_0) = \mathbf{x}_0$$

where $\mathbf{g}(t) = \int_{t_0}^{t} \mathbf{h}(s)\,\mathrm{d}s$, which corresponds by 1.3 to the ordinary linear system

$$\mathbf{x}' = \mathbf{B}(t)\,\mathbf{x} + \mathbf{h}(t)\,, \qquad \mathbf{x}(t_0) = \mathbf{x}_0$$

has the form

$$\mathbf{x}(t) = \boldsymbol{\Phi}(t, t_0)\,\mathbf{x}_0 + \mathbf{g}(t) - \mathbf{g}(t_0) - \int_{t_0}^{t} \mathrm{d}_s[\boldsymbol{\Phi}(t, s)]\,(\mathbf{g}(s) - \mathbf{g}(t_0))$$

$$= \boldsymbol{\Phi}(t, t_0)\,\mathbf{x}_0 + \int_{t_0}^{t} \mathbf{h}(s)\,\mathrm{d}s + \int_{t_0}^{t} \boldsymbol{\Phi}(t, s)\,\mathrm{d}\left(\int_{t_0}^{s} \mathbf{h}(\sigma)\,\mathrm{d}\sigma\right) - \boldsymbol{\Phi}(t, t) \int_{t_0}^{t} \mathbf{h}(s)\,\mathrm{d}s$$

$$= \boldsymbol{\Phi}(t, t_0)\,\mathbf{x}_0 + \int_{t_0}^{t} \boldsymbol{\Phi}(t, s)\,\mathbf{h}(s)\,\mathrm{d}s\,.$$

This is the usual form of the variation-of-constants formula for ordinary linear differential equations.

2.18. Definition. The $n \times n$-matrix $\mathbf{U}(t, s)$ defined for $t, s \in [0, 1]$ is called *harmonic* if $\operatorname{var}_0^1 \mathbf{U}(t, .) < \infty$ for every $t \in [0, 1]$, $\operatorname{var}_0^1 \mathbf{U}(., s) < \infty$ for every $s \in [0, 1]$.

(2,19) $\quad \mathbf{U}(t, s) = \mathbf{U}(t, r)\,\mathbf{U}(r, s)$ for any three points $r, s, t \in [0, 1]$,

(2,24) $\quad \mathbf{U}(t, t) = \mathbf{I}$ for any $t \in [0, 1]$.

For the concept of harmonic matrices see e.g. Hildebrandt [2], Mac Nerney [1], Wall [1].

As was shown in 2.10 for $\mathbf{A}\colon [0, 1] \to L(R_n)$, $\operatorname{var}_0^1 \mathbf{A} < \infty$ with the matrices $\mathbf{I} - \Delta^- \mathbf{A}(t)$, $\mathbf{I} + \Delta^+ \mathbf{A}(t)$ regular for $t \in (0, 1]$, $t \in [0, 1)$ respectively, the corresponding fundamental matrix $\mathbf{U}(t, s)$ is harmonic (see (i), (ii) and (iii) in 2.10). In other words, to any $n \times n$-matrix valued function $\mathbf{A}\colon [0, 1] \to L(R_n)$ with the above mentioned properties through the relation

$$\mathbf{U}(t, s) = \mathbf{I} + \int_{s}^{t} \mathrm{d}[\mathbf{A}(r)]\,\mathbf{U}(r, s)\,, \qquad t, s \in [0, 1]$$

a uniquely determined harmonic matrix $\mathbf{U}(t, s)$ corresponds. In the opposite direction the following holds.

2.19. Theorem. *If the $n \times n$-matrix $\mathbf{U}(t, s)\colon [0, 1] \times [0, 1] \to L(R_n)$ is harmonic, then there exists $\mathbf{A}\colon [0, 1] \to L(R_n)$ such that $\operatorname{var}_0^1 \mathbf{A} < \infty$, the matrices $\mathbf{I} - \Delta^- \mathbf{A}(t)$, $\mathbf{I} + \Delta^+ \mathbf{A}(t)$ are regular for all $t \in (0, 1]$, $t \in [0, 1)$, respectively and $\mathbf{U}$ satisfies the relation*

$$\mathbf{U}(t, s) = \mathbf{I} + \int_{s}^{t} \mathrm{d}[\mathbf{A}(r)]\,\mathbf{U}(r, s)\,, \qquad t, s \in [0, 1]\,, \tag{2,25}$$

i.e. $\mathbf{U}(t, s)$ is the fundamental matrix for the homogeneous generalized linear differential equation with the matrix $\mathbf{A}$ (see 2.16).

Proof. Let us set

$$\mathbf{A}(t,\tau) = \int_0^t \mathrm{d}_r[\mathbf{U}(r,\tau)]\,\mathbf{U}(\tau,r)$$

for $t, \tau \in [0,1]$. This integral exists for every t, τ by I.4.19. For every $t, \tau \in [0,1]$ we have by (2,19) and (2,24)

$$\mathbf{A}(t,\tau) = \int_0^t \mathrm{d}_r[\mathbf{U}(r,\tau)\,\mathbf{U}(\tau,0)]\,\mathbf{U}(0,\tau)\,\mathbf{U}(\tau,r) = \int_0^t \mathrm{d}_r[\mathbf{U}(r,0)]\,\mathbf{U}(0,r) = \mathbf{A}(t,0).$$

Hence the matrix $\mathbf{A}(t,\tau)$ is independent of τ and we denote $\mathbf{A}(t) = \mathbf{A}(t,\tau) = \mathbf{A}(t,0)$ for $t \in [0,1]$. Evidently $\mathrm{var}_0^1 \mathbf{A} < \infty$ by I.4.27. Further we have by the definition of $\mathbf{A}$, by the substitution theorem I.4.25 and by (2,19), (2,24)

$$\int_s^t \mathrm{d}[\mathbf{A}(r)]\,\mathbf{U}(r,s) = \int_s^t \mathrm{d}_r\left[\int_0^r \mathrm{d}_\varrho[\mathbf{U}(\varrho,0)]\,\mathbf{U}(0,\varrho)\right]\mathbf{U}(r,s)$$

$$= \int_s^t \mathrm{d}_r[\mathbf{U}(r,0)]\,\mathbf{U}(0,r)\,\mathbf{U}(r,s) = \int_s^t \mathrm{d}_r[\mathbf{U}(r,0)]\,\mathbf{U}(0,s)$$

$$= (\mathbf{U}(t,0) - \mathbf{U}(s,0))\,\mathbf{U}(0,s) = \mathbf{U}(t,s) - \mathbf{I},$$

i.e. $\mathbf{U}(t,s)$ satisfies (2,25) for every $t, s \in [0,1]$. Finally we show that $\mathbf{A}\colon [0,1] \to L(R_n)$ satisfies the regularity conditions for $\mathbf{I} - \Delta^- \mathbf{A}(t)$, $\mathbf{I} + \Delta^+ \mathbf{A}(t)$. By definition we have for $t \in (0,1]$

$$\Delta^- \mathbf{A}(t) = \mathbf{A}(t) - \lim_{\delta \to 0+} \mathbf{A}(t-\delta)$$

$$= \int_0^t \mathrm{d}_r[\mathbf{U}(r,0)]\,\mathbf{U}(0,r) - \lim_{\delta \to 0+} \int_0^{t-\delta} \mathrm{d}_r[\mathbf{U}(r,0)]\,\mathbf{U}(0,r)$$

$$= \lim_{\delta \to 0+} \int_{t-\delta}^t \mathrm{d}_r[\mathbf{U}(r,0)]\,\mathbf{U}(0,r) = \lim_{\delta \to 0+} (\mathbf{U}(t,0) - \mathbf{U}(t-\delta,0))\,\mathbf{U}(0,t)$$

$$= \mathbf{U}(t,0)\,\mathbf{U}(0,t) - \lim_{\delta \to 0+} \mathbf{U}(t-\delta,0)\,\mathbf{U}(0,t) = \mathbf{I} - \lim_{\delta \to 0+} \mathbf{U}(t-\delta,t),$$

where I.4.13 was used. Hence

$$\mathbf{I} - \Delta^- \mathbf{A}(t) = \lim_{\delta \to 0+} \mathbf{U}(t-\delta,t) = \mathbf{U}(t-,t) \tag{2,26}$$

for every $t \in (0,1]$. Since $\mathbf{U}$ is assumed to be harmonic, we have $\mathbf{U}(t-\delta,t)\,\mathbf{U}(t,t-\delta) = \mathbf{I}$ for any sufficiently small $\delta > 0$. $\mathbf{U}(t,s)$ is of bounded variation in each variable, the limits $\lim_{\delta \to 0+} \mathbf{U}(t-\delta,t) = \mathbf{U}(t-,t)$ and $\lim_{\delta \to 0+} \mathbf{U}(t,t-\delta) = \mathbf{U}(t,t-)$ exist. Hence

$$\mathbf{U}(t-,t)\,\mathbf{U}(t,t-) = \lim_{\delta \to 0+} \mathbf{U}(t-\delta,t)\,\mathbf{U}(t,t-\delta) = \mathbf{I}$$

and the matrix $\mathbf{U}(t-,t)$ is evidently regular since it has an inverse $[\mathbf{U}(t-,t)]^{-1} = \mathbf{U}(t,t-)$. This yields by (2,26) the regularity of $\mathbf{I} - \Delta^- \mathbf{A}(t)$ for every $t \in (0,1]$. The regularity of $\mathbf{I} + \Delta^+ \mathbf{A}(t)$ for every $t \in [0,1)$ can be proved analogously.

3. Generalized linear differential equations on the whole real axis

In this section let us assume that $\mathbf{A}: R \to L(R_n)$ is an $n \times n$-matrix defined on the whole real axis R and is of locally bounded variation in R, i.e. $\operatorname{var}_a^b \mathbf{A} < \infty$ for every compact interval $[a, b] \subset R$. We consider the generalized linear differential equation

$$\text{(3,1)} \qquad \mathrm{d}\mathbf{x} = \mathrm{d}[\mathbf{A}]\,\mathbf{x} + \mathrm{d}\mathbf{g}$$

where $\mathbf{g}: R \to R_n$ is of locally bounded variation in R.

The basic existence and uniqueness result follows from 1.4.

3.1. Theorem. *Assume that $\mathbf{A}: R \to L(R_n)$ is of locally bounded variation in R and $\mathbf{I} - \Delta^- \mathbf{A}(t)$, $\mathbf{I} + \Delta^+ \mathbf{A}(t)$ are regular matrices for all $t \in R$. Then for any $t_0 \in R$, $\mathbf{x}_0 \in R_n$ and $\mathbf{g}: R \to R_n$ of locally bounded variation in R there is a unique solution $\mathbf{x}: R \to R_n$ of the equation (3,1) with $\mathbf{x}(t_0) = \mathbf{x}_0$ and this solution is of locally bounded variation in R.*

Proof. This theorem follows immediately from 1.4 and 1.7 since evidently the assumptions of 1.4 are satisfied on every compact interval $[a, b] \subset R$.

In this way our preceding arguments on generalized linear differential equations are applicable to the case of equations on the whole real axis R. Especially the fundamental matrix $\mathbf{U}(t, s)$ determined uniquely by the equation

$$\mathbf{U}(t, s) = \mathbf{I} + \int_s^t \mathrm{d}[\mathbf{A}(r)]\,\mathbf{U}(r, s)$$

is defined for all $t, s \in R$, has the properties (i), (iii), (iv), (v), (vi) from 2.10 and is of locally bounded variation in R in each variable separately (see (ii) in 2.10). Moreover, the twodimensional variation of $\mathbf{U}$ on every compact interval $I = [a, b] \times [c, d] \subset R_2$ is finite.

Now we prove a result which is analogous to the Floquet theory for linear systems of ordinary differential equations.

3.2. Theorem. *Assume that $\mathbf{A}: R \to L(R_n)$ is of locally bounded variation in R such that $\mathbf{I} - \Delta^- \mathbf{A}(t)$, $\mathbf{I} + \Delta^+ \mathbf{A}(t)$ are regular matrices for every $t \in R$. Moreover let*

$$\mathbf{A}(t + \omega) - \mathbf{A}(t) = \mathbf{C} \qquad \text{for every} \quad t \in R$$

where $\omega > 0$ and $\mathbf{C} \in L(R_n)$ is a constant $n \times n$-matrix. If $\mathbf{X}: R \to L(R_n)$ is the solution of the matrix equation

$$\mathbf{X}(t) = \mathbf{I} + \int_0^t \mathrm{d}[\mathbf{A}(r)]\,\mathbf{X}(r), \qquad t \in R$$

(i.e. $\mathbf{X}(t) = \mathbf{U}(t, 0)$) then there exists a regular $n \times n$-matrix $\mathbf{P}: R \to L(R_n)$, which is

periodic with the period ω $(\mathbf{P}(t+\omega)=\mathbf{P}(t))$ and a constant $n \times n$-matrix $\mathbf{Q} \in L(R_n)$ such that

$$\mathbf{X}(t)=\mathbf{P}(t)\,\mathrm{e}^{t\mathbf{Q}}$$

is satisfied for every $t \in R$.

Proof. By definition we have

$$\mathbf{X}(t+\omega)=\mathbf{I}+\int_0^{t+\omega}\mathrm{d}[\mathbf{A}(r)]\,\mathbf{X}(r)=\mathbf{X}(\omega)+\int_\omega^{t+\omega}\mathrm{d}[\mathbf{A}(r)]\,\mathbf{X}(r)$$

$$=\mathbf{X}(\omega)+\int_0^t\mathrm{d}[\mathbf{A}(r+\omega)]\,\mathbf{X}(r+\omega)=\mathbf{X}(\omega)+\int_0^t\mathrm{d}[\mathbf{A}(r)+\mathbf{C}]\,\mathbf{X}(r+\omega)$$

$$=\mathbf{X}(\omega)+\int_0^t\mathrm{d}[\mathbf{A}(r)]\,\mathbf{X}(r+\omega)$$

for every $t \in R$. Using the variation of constants formula 2.14 in the matrix form we get

$$\mathbf{X}(t+\omega)=\mathbf{X}(t)\,\mathbf{X}(\omega) \qquad \text{for every} \quad t \in R\,.$$

By (v) from 2.10 the matrix $\mathbf{X}(\omega)=\mathbf{U}(\omega,0)$ is regular. Using the standard argument we conclude that there is a constant real $n \times n$-matrix $\mathbf{Q} \in L(R_n)$ ($\mathbf{Q}$ is not unique) such that $\mathbf{X}(\omega)=\mathrm{e}^{\omega\mathbf{Q}}$ (see e.g. Coddington, Levinson [1], III.1.), i.e.

$$\mathbf{X}(t+\omega)=\mathbf{X}(t)\,\mathrm{e}^{\omega\mathbf{Q}}.$$

Let us define $\mathbf{P}(t)=\mathbf{X}(t)\,\mathrm{e}^{-t\mathbf{Q}}$ for every $t \in R$. We have

$$\mathbf{P}(t+\omega)=\mathbf{X}(t+\omega)\,\mathrm{e}^{-(t+\omega)\mathbf{Q}}=\mathbf{X}(t)\,\mathrm{e}^{\omega\mathbf{Q}}\,\mathrm{e}^{-\omega\mathbf{Q}}\,\mathrm{e}^{-t\mathbf{Q}}=\mathbf{X}(t)\,\mathrm{e}^{-t\mathbf{Q}}=\mathbf{P}(t)$$

for all $t \in R$, i.e. $\mathbf{P}$ is periodic with the period ω. The regularity of $\mathbf{P}(t)$ is obvious by the regularity of $\mathbf{X}(t)$ and $\mathrm{e}^{-t\mathbf{Q}}$. Hence $\mathbf{X}(t)=\mathbf{P}(t)\,\mathrm{e}^{t\mathbf{Q}}$ and the result is proved.

Remark. This theorem is a basis for more detailed considerations concerning the linear system (3,1) with $\mathbf{A}\colon R \to L(R_n)$ satisfying the "periodicity" condition $\mathbf{A}(t+\omega)-\mathbf{A}(t)=\text{const}$. Some special results are contained in Hnilica [1].

4. Formally adjoint equation

Let $\mathbf{B}\colon [0,1] \to L(R_n)$, $\operatorname{var}_0^1 \mathbf{B} < \infty$ and $\mathbf{g} \in BV_n$. Let us consider the generalized linear differential equation for a row n-vector valued function $\mathbf{y}^*$

$$(4{,}1) \qquad \mathrm{d}\mathbf{y}^* = -\mathbf{y}^*\,\mathrm{d}[\mathbf{B}] + \mathrm{d}\mathbf{g}^* \quad \text{on } [0,1]\,,$$

which is equivalent to the integral equation

$$\mathbf{y}^*(s)=\mathbf{y}^*(s_0)-\int_{s_0}^s\mathbf{y}^*(t)\,\mathrm{d}[\mathbf{B}(t)]+\mathbf{g}^*(s)-\mathbf{g}^*(s_0)\,, \qquad s, s_0 \in [0,1]\,.$$

Obviously, $\mathbf{y}^*\colon [0,1] \to R_n$ is a solution to (4,1) on $[a,b] \subset [0,1]$ if and only if $\mathbf{y}$ verifies the equation

$$\mathbf{y}(s) = \mathbf{y}(s_0) - \int_{s_0}^{s} \mathrm{d}[\mathbf{B}^*(t)]\, \mathbf{y}(t) + \mathbf{g}(s) - \mathbf{g}(s_0) \tag{4,2}$$

for every $s, s_0 \in [a,b]$. Thus taking into account that $\mathbf{I} - \Delta^-(-\mathbf{B}^*)(s) = [\mathbf{I} + \Delta^-\mathbf{B}(s)]^*$ on $(0,1]$, $\mathbf{I} + \Delta^+(-\mathbf{B}^*)(s) = [\mathbf{I} - \Delta^+\mathbf{B}(s)]^*$ on $[0,1)$ we may easily obtain the basic results for the equation (4,1) as consequences of the corresponding theorems from the foregoing sections.

Given $\mathbf{y}_0^* \in R_n^*$, the equation (4,1) possesses a unique solution $\mathbf{y}^*$ on $[0,1]$ such that $\mathbf{y}^*(1) = \mathbf{y}_0^*$ or $\mathbf{y}^*(0) = \mathbf{y}_0^*$ if and only if

$$\det[\mathbf{I} - \Delta^+\mathbf{B}(s)] \neq 0 \qquad \text{on } [0,1) \tag{4,3}$$

or

$$\det[\mathbf{I} + \Delta^-\mathbf{B}(s)] \neq 0 \qquad \text{on } (0,1], \tag{4,4}$$

respectively (cf. 1.4).

If (4,3) holds, then by 2.2 there exists a unique $n \times n$-matrix valued function $\mathbf{W}(t,s)$ defined for $t, s \in [0,1]$ such that $s \geq t$ and fulfilling for all such t, s the relation

$$\mathbf{W}(t,s) = \mathbf{I} - \int_s^t \mathrm{d}[\mathbf{B}^*(r)]\, \mathbf{W}(r,s).$$

Furthermore, given $t, s \in [0,1]$, $\operatorname{var}_0^s \mathbf{W}(.,s) + \operatorname{var}_t^1 \mathbf{W}(t,.) < \infty$, $\mathbf{W}(t+,s) = [\mathbf{I} - \Delta^+\mathbf{B}(t)]^*\, \mathbf{W}(t,s)$ if $t < s$ and $\mathbf{W}(t-,s) = [\mathbf{I} + \Delta^-\mathbf{B}(t)]^*\, \mathbf{W}(t,s)$ if $t \leq s$ (cf. 2.10). It follows that the function $\mathbf{V}(t,s) = \mathbf{W}^*(s,t)$ for $t \geq s$ is a unique $n \times n$-matrix valued function which fulfils for $t, s \in [0,1]$, $t \geq s$ the relation

$$\mathbf{V}(t,s) = \mathbf{I} + \int_s^t \mathbf{V}(t,r)\, \mathrm{d}[\mathbf{B}(r)]. \tag{4,5}$$

Moreover, given $t, s \in [0,1]$

$$\operatorname{var}_0^t \mathbf{V}(t,.) + \operatorname{var}_s^1 \mathbf{V}(.,s) < \infty$$

and

$$\mathbf{V}(t,s+) = \mathbf{V}(t,s)\,[\mathbf{I} - \Delta^+\mathbf{B}(s)] \qquad \text{if } t > s, \tag{4,6}$$

$$\mathbf{V}(t,s-) = \mathbf{V}(t,s)\,[\mathbf{I} + \Delta^-\mathbf{B}(s)] \qquad \text{if } t \geq s. \tag{4,7}$$

If $\mathbf{y}_0^* \in R_n^*$ is given, the unique solution $\mathbf{y}^*$ of (4,1) on $[0,1]$ with $\mathbf{y}^*(1) = \mathbf{y}_0^*$ is given on $[0,1]$ by

$$\mathbf{y}^*(s) = \mathbf{y}_0^*\, \mathbf{V}(1,s) + \mathbf{g}^*(s) - \mathbf{g}^*(1) + \int_s^1 (\mathbf{g}^*(t) - \mathbf{g}^*(1))\, \mathrm{d}_t[\mathbf{V}(t,s)] \tag{4,8}$$

(cf. 2.8).

If (4,4) holds, then the fundamental matrix $\mathbf{V}(t,s)$ for (4,1) is defined and fulfils (4,5) for $t \le s$, (4,6) holds for $t \le s$ and (4,7) holds for $t < s$. Furthermore, $\operatorname{var}_0^s \mathbf{V}(.,s) + \operatorname{var}_t^1 \mathbf{V}(t,.) < \infty$ for all $t, s \in [0,1]$ and given $\mathbf{y}_0^* \in R_n^*$, the unique solution $\mathbf{y}^*$ of (4,1) on $[0,1]$ with $\mathbf{y}^*(0) = \mathbf{y}_0^*$ is given on $[0,1]$ by

$$(4,9) \qquad \mathbf{y}^*(s) = \mathbf{y}_0^* \mathbf{V}(0,s) + \mathbf{g}^*(s) - \mathbf{g}^*(0) - \int_0^s (\mathbf{g}^*(t) - \mathbf{g}^*(0))\, \mathrm{d}_t[\mathbf{V}(t,s)]\,.$$

If both (4,3) and (4,4) hold, then there exists $M < \infty$ such that given $t, s \in [0,1]$

$$|\mathbf{V}(t,s)| + \operatorname{var}_0^1 \mathbf{V}(t,.) + \operatorname{var}_0^1 \mathbf{V}(.,s) + \mathrm{v}_{[0,1]\times[0,1]}(\mathbf{V}) \le M < \infty\,.$$

Moreover, in this case, given $t, s, r \in [0,1]$,

$$(4,10) \qquad \mathbf{V}(t,r)\,\mathbf{V}(r,s) = \mathbf{V}(t,s) \quad \text{and} \quad \mathbf{V}(t,t) = \mathbf{I}$$

(cf. 2.10).

The equation (4,1) is said to be *formally adjoint* to (1,1) if

$$(4,11) \qquad \mathbf{B}(t+) - \mathbf{A}(t+) = \mathbf{B}(t-) - \mathbf{A}(t-) = \mathbf{B}(0) - \mathbf{A}(0) \quad \text{on } [0,1]\,.$$

(According to the convention introduced in I.3 we have

$$\mathbf{B}(0-) - \mathbf{A}(0-) = \mathbf{B}(0) - \mathbf{A}(0) = \mathbf{B}(1+) - \mathbf{A}(1+) = \mathbf{B}(1) - \mathbf{A}(1)\,.)$$

The condition (4,11) ensures that

$$(4,12) \qquad \int_0^1 \mathbf{y}^*(t)\, \mathrm{d}[\mathbf{B}(t) - \mathbf{A}(t)]\, \mathbf{x}(t) = 0 \qquad \text{for all} \quad \mathbf{x}, \mathbf{y} \in BV_n$$

(cf. I.4.23). (4,11) holds e.g. if $\mathbf{B}(t) \equiv \mathbf{A}(t)$ on $[0,1]$ or

$$(4,13) \qquad \begin{aligned} &\mathbf{B}(t) = \mathbf{A}_*(t) = \mathbf{A}(t-) + \Delta^+ \mathbf{A}(t) \qquad \text{on } (0,1)\,,\\ &\mathbf{B}(0) = \mathbf{A}_*(0) = \mathbf{A}(0)\,, \qquad \mathbf{B}(1) = \mathbf{A}_*(1) = \mathbf{A}(1)\,. \end{aligned}$$

Without any loss of generality we may assume that $\mathbf{A}(0) = \mathbf{B}(0)$.

4.1. Theorem. *Let the* $n \times n$*-matrix valued functions* $\mathbf{A}$, $\mathbf{B}$ *be of bounded variation on* $[0,1]$ *and such that* (4,11) *with* $\mathbf{A}(0) = \mathbf{B}(0)$ *holds.*

(i) *If*

$$(4,14) \qquad \det(\mathbf{I} - \Delta^- \mathbf{A}(t)) \det(\mathbf{I} - \Delta^+ \mathbf{B}(t)) \det(\mathbf{I} + \Delta^+ \mathbf{A}(t)) \neq 0 \qquad \text{on } [0,1]$$

or

$$(4,15) \qquad \det(\mathbf{I} - \Delta^- \mathbf{A}(t)) \det(\mathbf{I} - \Delta^+ \mathbf{B}(t)) \det(\mathbf{I} + \Delta^- \mathbf{B}(t)) \neq 0 \qquad \text{on } [0,1]\,,$$

then the fundamental matrices $\mathbf{U}(t,s)$ *to* (1,1) *and* $\mathbf{V}(t,s)$ *to* (4,1) *fulfil the relation*

$$\begin{aligned}(4{,}16)\quad \mathbf{V}(t,s) = {} & \mathbf{U}(t,s) + \mathbf{V}(t,s)\left[\mathbf{A}(s) - \mathbf{B}(s)\right] - \left[\mathbf{A}(t) - \mathbf{B}(t)\right]\mathbf{U}(t,s) \\ & + \mathbf{V}(t,s)\,\Delta^+\mathbf{B}(s)\,\Delta^+\mathbf{A}(s) - \Delta^-\mathbf{B}(t)\,\Delta^-\mathbf{A}(t)\,\mathbf{U}(t,s) \\ & + \sum_{s<\tau<t} \mathbf{V}(t,\tau)\left[\Delta^+\mathbf{B}(\tau)\,\Delta^+\mathbf{A}(\tau) - \Delta^-\mathbf{B}(\tau)\,\Delta^-\mathbf{A}(\tau)\right]\mathbf{U}(\tau,s) \quad \textit{if } t > s, \\ \mathbf{V}(t,t) = {} & \mathbf{U}(t,t) = \mathbf{I}.\end{aligned}$$

(ii) *If*

$$(4{,}17)\quad \det(\mathbf{I} + \Delta^+\mathbf{A}(t))\det(\mathbf{I} + \Delta^-\mathbf{B}(t))\det(\mathbf{I} - \Delta^+\mathbf{B}(t)) \neq 0 \qquad \textit{on } [0,1]$$

or

$$(4{,}18)\quad \det(\mathbf{I} + \Delta^+\mathbf{A}(t))\det(\mathbf{I} + \Delta^-\mathbf{B}(t))\det(\mathbf{I} - \Delta^-\mathbf{A}(t)) \neq 0 \qquad \textit{on } [0,1],$$

then

$$\begin{aligned}(4{,}19)\quad \mathbf{V}(t,s) = {} & \mathbf{U}(t,s) + \mathbf{V}(t,s)\left[\mathbf{A}(s) - \mathbf{B}(s)\right] - \left[\mathbf{A}(t) - \mathbf{B}(t)\right]\mathbf{U}(t,s) \\ & + \mathbf{V}(t,s)\,\Delta^-\mathbf{B}(s)\,\Delta^-\mathbf{A}(s) - \Delta^+\mathbf{B}(t)\,\Delta^+\mathbf{A}(t)\,\mathbf{U}(t,s) \\ & + \sum_{t<\tau<s} \mathbf{V}(t,\tau)\left[\Delta^-\mathbf{B}(\tau)\,\Delta^-\mathbf{A}(\tau) - \Delta^+\mathbf{B}(\tau)\,\Delta^+\mathbf{A}(\tau)\right]\mathbf{U}(\tau,s) \quad \textit{if } t < s, \\ \mathbf{V}(t,t) = {} & \mathbf{U}(t,t) = \mathbf{I}.\end{aligned}$$

(*In* (4,14)–(4,19) $\Delta^-\mathbf{A}(0) = \Delta^-\mathbf{B}(0) = \mathbf{0}$ *and* $\Delta^+\mathbf{A}(1) = \Delta^+\mathbf{B}(1) = \mathbf{0}$.)

Proof. Let e.g. (4,14) hold. Then $\mathbf{U}(t,s)$ is defined for all $t,s \in [0,1]$ and $\mathbf{V}(t,s)$ is defined for $t \geq s$. Let $t,s \in [0,1]$, $t > s$ be given and let us consider the expression

$$\mathbf{W} = \int_s^t \mathrm{d}_\tau[\mathbf{V}(t,\tau)]\,\mathbf{U}(\tau,t) + \int_s^t \mathbf{V}(t,\tau)\,\mathrm{d}_\tau[\mathbf{U}(\tau,t)].$$

Inserting into $\mathbf{W}$ from (2,4) and (4,5) and making use of the subsitution theorem I.4.25 we easily obtain

$$\mathbf{W} = \int_s^t \mathbf{V}(t,\tau)\,\mathrm{d}[\mathbf{A}(\tau) - \mathbf{B}(\tau)]\,\mathbf{U}(\tau,t)$$

and according to (4,11) and I.4.23

$$\begin{aligned}\mathbf{W} &= \mathbf{V}(t,s)\left[\Delta^+\mathbf{A}(s) - \Delta^+\mathbf{B}(s)\right]\mathbf{U}(s,t) + \left[\Delta^-\mathbf{A}(t) - \Delta^-\mathbf{B}(t)\right] \\ &= -\mathbf{V}(t,s)\left[\mathbf{A}(s) - \mathbf{B}(s)\right]\mathbf{U}(s,t) + \left[\mathbf{A}(t) - \mathbf{B}(t)\right]\end{aligned}$$

because the components of $\mathbf{A}(t) - \mathbf{B}(t)$ are evidently break functions on $[0,1]$. On the other hand, the integration-by-parts theorem I.4.33 yields

$$\begin{aligned}\mathbf{W} = {} & \mathbf{I} - \mathbf{V}(t,s)\,\mathbf{U}(s,t) - \Delta_2^+\mathbf{V}(t,s)\,\Delta_1^+\mathbf{U}(s,t) + \Delta_2^-\mathbf{V}(t,t)\,\Delta_1^-\mathbf{U}(t,t) \\ & + \sum_{s<\tau<t}\left[\Delta_2^-\mathbf{V}(t,\tau)\,\Delta_1^-\mathbf{U}(\tau,t) - \Delta_2^+\mathbf{V}(t,\tau)\,\Delta_1^+\mathbf{U}(\tau,t)\right],\end{aligned}$$

where $\Delta_1^+ \boldsymbol{Z}(t,s) = \boldsymbol{Z}(t+,s) - \boldsymbol{Z}(t,s)$, $\Delta_2^+ \boldsymbol{Z}(t,s) = \boldsymbol{Z}(t,s+) - \boldsymbol{Z}(t,s)$, $\Delta_1^- \boldsymbol{Z}(t,s) = \boldsymbol{Z}(t,s) - \boldsymbol{Z}(t-,s)$ and $\Delta_2^- \boldsymbol{Z}(t,s) = \boldsymbol{Z}(t,s) - \boldsymbol{Z}(t,s-)$ for $\boldsymbol{Z} = \boldsymbol{U}$ and $\boldsymbol{Z} = \boldsymbol{V}$. Taking into account the relations (4,6), (4,7), (4,10) and 2.10 we obtain immediately (4,16).

The remaining cases can be treated similarly. If (4,17) or (4,18) holds, then instead of the expression $\boldsymbol{W}$ we should handle the expression

$$\int_s^t \mathrm{d}_\tau[\boldsymbol{V}(s,\tau)]\,\boldsymbol{U}(\tau,s) + \int_s^t \boldsymbol{V}(s,\tau)\,\mathrm{d}_\tau[\boldsymbol{U}(\tau,s)]\,.$$

4.2. Theorem (Lagrange identity). *Let* $\boldsymbol{A}\colon [0,1] \to L(R_n)$ *and* $\boldsymbol{B}\colon [0,1] \to L(R_n)$ *be of bounded variation on* $[0,1]$ *and let* (4,11) *hold. Then for any* $\boldsymbol{x} \in BV_n$ *left-continuous on* $(0,1]$ *and right-continuous at* 0 *and any* $\boldsymbol{y} \in BV_n$ *right-continuous on* $[0,1)$ *and left-continuous at* 1

$$\int_0^1 \boldsymbol{y}^*(t)\,\mathrm{d}\Big[\boldsymbol{x}(t) - \int_0^t \mathrm{d}[\boldsymbol{A}(s)]\,\boldsymbol{x}(s)\Big] + \int_0^1 \mathrm{d}\Big[\boldsymbol{y}^*(s) - \int_s^1 \boldsymbol{y}^*(t)\,\mathrm{d}[\boldsymbol{B}(t)]\Big]\,\boldsymbol{x}(s) = \boldsymbol{y}^*(1)\,\boldsymbol{x}(1) - \boldsymbol{y}^*(0)\,\boldsymbol{x}(0)\,. \tag{4,20}$$

Proof. Applying the substitution theorem I.4.25 the left-hand side of (4,20) reduces to

$$\int_0^1 \boldsymbol{y}^*(t)\,\mathrm{d}[\boldsymbol{x}(t)] + \int_0^1 \mathrm{d}[\boldsymbol{y}^*(t)]\,\boldsymbol{x}(t) + \int_0^1 \boldsymbol{y}^*(t)\,\mathrm{d}[\boldsymbol{B}(t) - \boldsymbol{A}(t)]\,\boldsymbol{x}(t)\,.$$

The integration-by-parts formula I.4.33 yields

$$\int_0^1 \boldsymbol{y}^*(t)\,\mathrm{d}[\boldsymbol{x}(t)] + \int_0^1 \mathrm{d}[\boldsymbol{y}^*(t)]\,\boldsymbol{x}(t) = \boldsymbol{y}^*(1)\,\boldsymbol{x}(1) - \boldsymbol{y}^*(0)\,\boldsymbol{x}(0)$$

whence by (4,11) and (4,12) our assertion follows.

4.3. Remark. The relations (4,16) and (4,19) are considerably simplified if

$$\Delta^+\boldsymbol{B}(t)\,\Delta^+\boldsymbol{A}(t) = \Delta^-\boldsymbol{B}(t)\,\Delta^-\boldsymbol{A}(t) \qquad \text{on } [0,1]\,. \tag{4,21}$$

This together with (4,11) and $\boldsymbol{A}(0) = \boldsymbol{B}(0)$ is true e.g. if

(i) $\boldsymbol{B} = \boldsymbol{A}$ and $(\Delta^+\boldsymbol{A}(t))^2 = (\Delta^-\boldsymbol{A}(t))^2$ on $[0,1]$, or
(ii) $\boldsymbol{B} = \boldsymbol{A}_*$ (cf. (4,13)), $(\Delta^+\boldsymbol{A}(0))^2 = (\Delta^-\boldsymbol{A}(1))^2 = \boldsymbol{0}$ and $\Delta^+\boldsymbol{A}(t)\,\Delta^-\boldsymbol{A}(t) = \Delta^-\boldsymbol{A}(t)\,\Delta^+\boldsymbol{A}(t)$ on $(0,1)$.

5. Two-point boundary value problem

Let $\boldsymbol{M}$ and $\boldsymbol{N}$ be $m \times n$-matrices and $\boldsymbol{r} \in R_m$. The problem of determining a solution $\boldsymbol{x}\colon [0,1] \to R_n$ to

$$\mathrm{d}\boldsymbol{x} = \mathrm{d}[\boldsymbol{A}]\,\boldsymbol{x} + \mathrm{d}\boldsymbol{f} \tag{5,1}$$

on $[0,1]$, which fulfils in addition the relation

$$\mathbf{M}\,\mathbf{x}(0)+\mathbf{N}\,\mathbf{x}(1)=\mathbf{r}, \tag{5,2}$$

is called the *two-point boundary value problem.*

5.1. Assumptions. *Throughout the section,* $\mathbf{A},\mathbf{B}$ *are* $n\times n$*-matrix valued functions of bounded variation on* $[0,1]$. *Moreover we suppose that* (4,11) *with* $\mathbf{A}(0)=\mathbf{B}(0)$, (4,21) *and at least one of the conditions* (4,14), (4,15), (4,17), (4,18) *are satisfied. (In particular, the assumptions of* 4.1 *are fulfilled.)* $\mathbf{M}$ *and* $\mathbf{N}$ *are* $m\times n$*-matrices,* $\mathbf{f}\in BV_n$ *and* $\mathbf{r}\in R_m$, $m\geq 1$.

Making use of the variation-of-constants formula (2,15) we may reduce the boundary value problem (5,1), (5,2) to a linear nonhomogeneous algebraic equation.

5.2. Lemma. *If* (4,14) *or* (4,15) *holds, then* $\mathbf{x}\colon[0,1]\to R_n$ *is a solution of the problem* (5,1), (5,2) *if and only if*

$$\mathbf{x}(t)=\mathbf{U}(t,0)\,\mathbf{c}+\mathbf{f}(t)-\mathbf{f}(0)-\int_0^t \mathrm{d}_s[\mathbf{U}(t,s)]\,(\mathbf{f}(s)-\mathbf{f}(0)) \quad \text{on } [0,1], \tag{5,3}$$

where $\mathbf{c}\in R_n$ *is a solution to the algebraic equation*

$$[\mathbf{M}+\mathbf{N}\,\mathbf{V}(1,0)]\,\mathbf{c}=\mathbf{r}+\mathbf{N}\left\{\mathbf{V}(1,0)\,\mathbf{f}(0)-\mathbf{f}(1)+\int_0^1 \mathrm{d}_s[\mathbf{V}(1,s)]\,\mathbf{f}(s)\right\}.$$

If (4,17) *or* (4,18) *holds, then* $\mathbf{x}\colon[0,1]\to R_n$ *is a solution to* (5,1), (5,2) *if and only if*

$$\mathbf{x}(t)=\mathbf{U}(t,1)\,\mathbf{c}+\mathbf{f}(t)-\mathbf{f}(1)+\int_t^1 \mathrm{d}_s[\mathbf{U}(t,s)]\,(\mathbf{f}(s)-\mathbf{f}(1)) \quad \text{on } [0,1],$$

where

$$[\mathbf{M}\,\mathbf{V}(0,1)+\mathbf{N}]\,\mathbf{c}=\mathbf{r}+\mathbf{M}\left\{-\mathbf{f}(0)+\mathbf{V}(0,1)\,\mathbf{f}(1)-\int_0^1 \mathrm{d}_s[\mathbf{V}(0,s)\,\mathbf{f}(s)]\right\}.$$

Proof. Let (4,14) or (4,15) hold. Then by 2,15 $\mathbf{x}\colon[0,1]\to R_n$ is a solution of the given problem if and only if it is given by (5,3), where $\mathbf{c}\in R_n$ fulfils the equation

$$[\mathbf{M}+\mathbf{N}\,\mathbf{U}(1,0)]\,\mathbf{c}=\mathbf{r}+\mathbf{N}\left\{\mathbf{U}(1,0)\,\mathbf{f}(0)-\mathbf{f}(1)+\int_0^1 \mathrm{d}_s[\mathbf{U}(1,s)]\,\mathbf{f}(s)\right\}.$$

By (4,16) and (4,21)

$$\mathbf{V}(1,s)=\mathbf{U}(1,s)+\mathbf{V}(1,s)\,(\mathbf{A}(s)-\mathbf{B}(s))+\mathbf{V}(1,s)\,\Delta^+\mathbf{B}(s)\,\Delta^+\mathbf{A}(s) \tag{5,4}$$

and thus

$$\mathbf{V}(1,s+)-\mathbf{U}(1,s+)=\mathbf{V}(1,s-)-\mathbf{U}(1,s-)$$

for any $s \in [0,1]$. (In particular $\mathbf{V}(1,0) = \mathbf{U}(1,0)$, $\mathbf{V}(1,1) = \mathbf{U}(1,1)$.) This implies by I.4.23

$$\int_0^1 \mathrm{d}_s[\mathbf{U}(1,s)]\,\mathbf{v}(s) = \int_0^1 \mathrm{d}_s[\mathbf{V}(1,s)]\,\mathbf{v}(s) \qquad \text{for any} \quad \mathbf{v} \in BV_n$$

wherefrom our assertion follows.

The cases (4,17) and (4,18) could be treated analogously. ($\mathbf{V}(0,s) = \mathbf{U}(0,s) + \mathbf{V}(0,s)(\mathbf{A}(s) - \mathbf{B}(s)) + \mathbf{V}(0,s)\,\Delta^-\mathbf{B}(s)\,\Delta^-\mathbf{A}(s)$ on $[0,1]$.)

5.3. Remark. Consequently, in the cases (4,14) or (4,15) the problem (5,1), (5,2) has a solution if and only if

$$\lambda^*[\mathbf{M} + \mathbf{N}\,\mathbf{V}(1,0)] = \mathbf{0} \tag{5,5}$$

implies

$$\lambda^*\mathbf{N}\,\mathbf{V}(1,1)\,\mathbf{f}(1) - \lambda^*\mathbf{N}\,\mathbf{V}(1,0)\,\mathbf{f}(0) - \int_0^1 \mathrm{d}_s[\lambda^*\mathbf{N}\,\mathbf{V}(1,s)]\,\mathbf{f}(s) = \lambda^*\mathbf{r}\,. \tag{5,6}$$

Let us denote $\mathbf{y}_\lambda^*(s) = \lambda^*\mathbf{N}\,\mathbf{V}(1,s)$ for $s \in [0,1]$ and $\lambda \in R_m$. Then (5,6) becomes

$$\mathbf{y}_\lambda^*(1)\,\mathbf{f}(1) - \mathbf{y}_\lambda^*(0)\,\mathbf{f}(0) - \int_0^1 \mathrm{d}[\mathbf{y}_\lambda^*(s)]\,\mathbf{f}(s) = \lambda^*\mathbf{r}\,.$$

By (4,8) for any $\lambda^* \in R_m^*$ and $s, s_0 \in [0,1]$

$$\mathbf{y}_\lambda^*(s) = \mathbf{y}_\lambda^*(s_0) + \int_s^{s_0} \mathbf{y}_\lambda^*(t)\,\mathrm{d}[\mathbf{B}(t)]\,.$$

Moreover, if $\lambda^* \in R_m^*$ verifies (5,5), then $\mathbf{y}_\lambda^*(0) = \lambda^*\mathbf{N}\,\mathbf{V}(1,0) = -\lambda^*\mathbf{M}$ and $\mathbf{y}_\lambda^*(1) = \lambda^*\mathbf{N}$. Analogously, if (4,17) or (4,18) holds, the problem (5,1), (5,2) possesses a solution if and only if $\lambda^*[\mathbf{M}\,\mathbf{V}(0,1) + \mathbf{N}] = \mathbf{0}$ implies

$$\mathbf{y}_\lambda^*(1)\,\mathbf{f}(1) - \mathbf{y}_\lambda^*(0)\,\mathbf{f}(0) - \int_0^1 \mathrm{d}[\mathbf{y}_\lambda^*(s)]\,\mathbf{f}(s) = \lambda^*\mathbf{r}\,,$$

where $\mathbf{y}_\lambda^*(s) = -\lambda^*\mathbf{M}\,\mathbf{V}(0,s)$ on $[0,1]$.

5.4. Lemma. *Let $\mathbf{g} \in BV_n$ and $\mathbf{p}, \mathbf{q} \in R_n$. If (4,14) or (4,15) holds, then $\mathbf{y}^*\colon [0,1] \to R_n^*$ is a solution to the generalized differential equation*

$$\mathrm{d}\mathbf{y}^* = -\mathbf{y}^*\,\mathrm{d}[\mathbf{B}] + \mathrm{d}\mathbf{g}^* \qquad \text{on } [0,1] \tag{5,7}$$

and together with $\lambda^* \in R_m^*$ verifies the relations

$$\mathbf{y}^*(0) + \lambda^*\mathbf{M} = \mathbf{p}^*\,, \qquad \mathbf{y}^*(1) - \lambda^*\mathbf{N} = \mathbf{q}^* \tag{5,8}$$

if and only if

$$\mathbf{y}^*(s) = (\lambda^*\mathbf{N} + \mathbf{q}^*)\,\mathbf{V}(1,s) + \mathbf{g}^*(s) - \mathbf{g}^*(1) + \int_s^1 (\mathbf{g}^*(t) - \mathbf{g}^*(1))\,\mathrm{d}_t[\mathbf{V}(t,s)] \tag{5,9}$$

on $[0,1]$ *and*

$$\lambda^*[\mathbf{M} + \mathbf{N}\,\mathbf{U}(1,0)]$$

$$= \mathbf{p}^* - \mathbf{q}^*\,\mathbf{U}(1,0) - \mathbf{g}^*(0) - \mathbf{g}^*(1)\,\mathbf{U}(1,0) + \int_0^1 \mathbf{g}^*(t)\,\mathrm{d}_t[\mathbf{U}(t,0)]\,.$$

(*By* (4,16) $\mathbf{V}(t,0) - \mathbf{U}(t,0) = (\mathbf{A}(t) - \mathbf{B}(t))\,\mathbf{U}(t,0) + \Delta^-\mathbf{B}(t)\,\Delta^-\mathbf{A}(t)\,\mathbf{U}(t,0)$.)

If (4,17) *or* (4,18) *holds, then* $\mathbf{y}^*\colon [0,1] \to R_n^*$ *and* $\lambda^* \in R_m^*$ *verify the system* (5,7), (5,8) *if and only if*

(5,10) $$\mathbf{y}^*(s) = (\mathbf{p}^* - \lambda^*\mathbf{M})\,\mathbf{V}(0,s) + \mathbf{g}^*(s) - \mathbf{g}^*(0) - \int_0^s (\mathbf{g}^*(t) - \mathbf{g}^*(0))\,\mathrm{d}_t[\mathbf{V}(t,s)]$$

on $[0,1]$

and

$$\lambda^*[\mathbf{M}\,\mathbf{U}(0,1) + \mathbf{N}]$$

$$= \mathbf{p}^*\,\mathbf{U}(0,1) - \mathbf{q}^* + \mathbf{g}^*(1) - \mathbf{g}^*(0)\,\mathbf{U}(0,1) - \int_0^1 \mathbf{g}^*(t)\,\mathrm{d}_t[\mathbf{U}(t,1)]\,.$$

($\mathbf{V}(t,1) - \mathbf{U}(t,1) = (\mathbf{A}(t) - \mathbf{B}(t))\,\mathbf{U}(t,1) + \mathbf{V}(t,1)\,\Delta^+\mathbf{B}(t)\,\Delta^+\mathbf{A}(t)$ *by* (4,19).)

Proof. In virtue of our assumption (4,21) the fundamental matrices $\mathbf{U}(t,s)$ and $\mathbf{V}(t,s)$ fulfil the relation (5,4). Inserting (4,8) or (4,9) into (5,8) we complete the proof.

5.5. Theorem. *Under the assumptions* 5.1 *the given problem* (5,1), (5,2) *possesses a solution if and only if*

(5,11) $$\mathbf{y}^*(1)\,\mathbf{f}(1) - \mathbf{y}^*(0)\,\mathbf{f}(0) - \int_0^1 \mathrm{d}[\mathbf{y}^*(t)]\,\mathbf{f}(t) = \lambda^*\mathbf{r}$$

for any solution $(\mathbf{y}^*, \lambda^*)$ *of the homogeneous system*

(5,12) $$\mathrm{d}\mathbf{y}^* = -\mathbf{y}^*\,\mathrm{d}[\mathbf{B}] \quad \text{on } [0,1]\,,$$

(5,13) $$\mathbf{y}^*(0) + \lambda^*\mathbf{M} = \mathbf{0}\,, \qquad \mathbf{y}^*(1) - \lambda^*\mathbf{N} = \mathbf{0}\,.$$

Proof follows immediately from 5.2 (cf. also 5.3).

5.6. Theorem. *Let* $\mathbf{A}$, $\mathbf{B}$, $\mathbf{M}$, $\mathbf{N}$ *fulfil* 5.1. *Then given* $\mathbf{g} \in BV_n$ *and* $\mathbf{p}, \mathbf{q} \in R_n$ *the system* (5,7), (5,8) *possesses a solution if and only if*

$$\mathbf{g}^*(1)\,\mathbf{x}(1) - \mathbf{g}^*(0)\,\mathbf{x}(0) - \int_0^1 \mathbf{g}^*(s)\,\mathrm{d}[\mathbf{x}(s)] = \mathbf{q}^*\,\mathbf{x}(1) - \mathbf{p}^*\,\mathbf{x}(0)$$

for any solution $\mathbf{x}$ *of the homogeneous equation*

(5,14) $$\mathrm{d}\mathbf{x} = \mathrm{d}[\mathbf{A}]\,\mathbf{x} \quad \text{on } [0,1]$$

which fulfils also

(5,15) $$\mathbf{M}\,\mathbf{x}(0) + \mathbf{N}\,\mathbf{x}(1) = \mathbf{0}\,.$$

Proof. If (4,14) or (4,15) holds, then by 5.4 the system (5,7), (5,8) possesses a solution if and only if

$$[\mathbf{M} + \mathbf{N}\,\mathbf{U}(1,0)]\,\mathbf{c} = \mathbf{0} \tag{5,16}$$

implies

$$\mathbf{q}^*\,\mathbf{x}_c(1) - \mathbf{p}^*\,\mathbf{x}_c(0) = \mathbf{g}^*(1)\,\mathbf{x}_c(1) - \mathbf{g}^*(0)\,\mathbf{x}_c(0) - \int_0^1 \mathbf{g}^*(s)\,\mathrm{d}[\mathbf{x}_c(s)]\,,$$

where $\mathbf{x}_c(t) = \mathbf{U}(t,0)\,\mathbf{c}$ for $t \in [0,1]$ and $\mathbf{c} \in R_n$. By 5.2 $\mathbf{x}$: $[0,1] \to R_n$ is a solution to (5,14), (5,15) if and only if $\mathbf{x}(t) = \mathbf{U}(t,0)\,\mathbf{c}$ on $[0,1]$ where $\mathbf{c} \in R_n$ verifies (5,16). Now, our assertion follows readily.

5.7. Definition. The system (5,12), (5,13) of equations for $\mathbf{y}^*$: $[0,1] \to R_n^*$ and $\boldsymbol{\lambda}^* \in R_m^*$ is called the *adjoint boundary value problem* to the problem (5,1), (5,2) (or (5,14), (5,15)).

5.8. Definition. The homogeneous problem (5,14), (5,15) (or (5,12), (5,13)) has *exactly k linearly independent solutions* if it has at least k linearly independent solutions on $[0,1]$, while any set of its solutions which contains at least $k+1$ elements is linearly dependent on $[0,1]$.

Another interesting question is the *index of the boundary value problem*, i.e. the relationship between the number of linearly independent solutions to the homogeneous problem (5,14), (5,15) and its adjoint.

5.9. Remark. Without any loss of generality we may assume rank $[\mathbf{M}, \mathbf{N}] = m$. In fact, if rank $[\mathbf{M}, \mathbf{N}] = m_1 < m$, then there exists a regular $m \times n$-matrix $\boldsymbol{\Theta}$ such that

$$\boldsymbol{\Theta}[\mathbf{M}, \mathbf{N}] = \begin{bmatrix} \mathbf{M}_1, & \mathbf{N}_1 \\ \mathbf{0}, & \mathbf{0} \end{bmatrix},$$

where $\mathbf{M}_1, \mathbf{N}_1 \in L(R_n, R_{m_1})$ are such that rank $[\mathbf{M}_1, \mathbf{N}_1] = m_1$. Let $\mathbf{r} \in R_m$, $\boldsymbol{\Theta}\mathbf{r} = \begin{pmatrix} \mathbf{r}_1 \\ \mathbf{r}_2 \end{pmatrix}$, $\mathbf{r}_1 \in R_{m_1}$ and $\mathbf{r}_2 \in R_{m-m_1}$. Then either $\mathbf{r}_2 \neq 0$ and the equation for $\mathbf{d} \in R_{2n}$

$$[\mathbf{M}, \mathbf{N}]\,\mathbf{d} = \mathbf{r} \tag{5,17}$$

possesses no solution or $\mathbf{r}_2 = \mathbf{0}$ and (5,17) is equivalent to $[\mathbf{M}_1, \mathbf{N}_1]\,\mathbf{d} = \mathbf{r}_1$.

5.10. Theorem. *Let* $\mathbf{A}, \mathbf{B}, \mathbf{M}, \mathbf{N}$ *fulfil* 5.1 *and* rank $[\mathbf{M}, \mathbf{N}] = m$. *Then both the homogeneous problem* (5,14), (5,15) *and its adjoint* (5,12), (5,13) *possesses at most a finite number of linearly independent solutions on* $[0,1]$. *Let* (5,14), (5,15) *possess exactly k linearly independent solutions on* $[0,1]$ *and let* (5,12), (5,13) *possess exactly* k^* *linearly independent solutions on* $[0,1]$. *Then* $k^* - k = m - n$.

Proof. Let us assume e.g. (4,14). By 5.2 the system (5,14), (5,15) possesses exactly $k = n - \operatorname{rank}[\mathbf{M} + \mathbf{N}\,\mathbf{U}(1,0)]$ linearly independent solutions on $[0,1]$. (If $\mathbf{c}_j \in R_n$ are linearly independent solutions to (5,16), then since $\mathbf{U}(0,0) = \mathbf{I}$, the functions $\mathbf{x}_j(t) = \mathbf{U}(t,0)\,\mathbf{c}_j$ are linearly independent solutions on $[0,1]$ of the system (5,14), (5,15).)

On the other hand, the equation (5,5) has exactly $m - \operatorname{rank}[\mathbf{M} + \mathbf{N}\,\mathbf{U}(1,0)] = h$ linearly independent solutions. Let Λ denote an arbitrary $h \times n$-matrix whose rows $\boldsymbol{\lambda}_1^*, \boldsymbol{\lambda}_2^*, \ldots, \boldsymbol{\lambda}_h^*$ are linearly independent solutions of (5,5). Let us assume that the functions $\mathbf{y}_j^*(s) = \boldsymbol{\lambda}_j^*\mathbf{N}\,\mathbf{V}(1,s)$ are linearly dependent on $[0,1]$, i.e. there is $\boldsymbol{\alpha} \in R_h$, $\boldsymbol{\alpha} \neq \mathbf{0}$ such that $\boldsymbol{\alpha}^*\Lambda\mathbf{N}\,\mathbf{V}(1,s) \equiv \mathbf{0}$ on $[0,1]$. In particular, $\mathbf{0} = \boldsymbol{\alpha}^*\Lambda\mathbf{N}\,\mathbf{V}(1,1) = \boldsymbol{\alpha}^*\Lambda\mathbf{N}$ and $\mathbf{0} = \boldsymbol{\alpha}^*\Lambda\mathbf{N}\,\mathbf{V}(1,0) = -\boldsymbol{\alpha}^*\Lambda\mathbf{M}$. Since (5,17), $\boldsymbol{\alpha}^*\Lambda = \mathbf{0}$ and by the definition of Λ it is $\boldsymbol{\alpha} = \mathbf{0}$. This being a contradiction, $k^* = m - \operatorname{rank}[\mathbf{M} + \mathbf{N}\,\mathbf{U}(1,0)]$ and $k^* - k = m - n$.

5.11. Definition. Given $m \times n$-matrices $\mathbf{M}$, $\mathbf{N}$ with $\operatorname{rank}[\mathbf{M}, \mathbf{N}] = m$, any $(2n - m) \times n$-matrices $\mathbf{M}^c$, $\mathbf{N}^c$ such that

$$\det\begin{bmatrix} \mathbf{M}, & \mathbf{N} \\ \mathbf{M}^c, & \mathbf{N}^c \end{bmatrix} \neq 0 \tag{5,18}$$

are called the *complementary matrices to* $[\mathbf{M}, \mathbf{N}]$.

5.12. Proposition. *Let* $\mathbf{M}, \mathbf{N} \in L(R_n, R_m)$, $\operatorname{rank}[\mathbf{M}, \mathbf{N}] = m$ *and let* $\mathbf{M}^c, \mathbf{N}^c \in L(R_n, R_{2n-m})$ *be arbitrary matrices complementary to* $[\mathbf{M}, \mathbf{N}]$. *Then there exist uniquely determined matrices* $\mathbf{P}, \mathbf{Q} \in L(R_{2n-m}, R_n)$ *and* $\mathbf{P}^c, \mathbf{Q}^c \in L(R_m, R_n)$ *such that*

$$\det\begin{bmatrix} \mathbf{P}^c, & \mathbf{Q} \\ \mathbf{Q}^c, & \mathbf{Q} \end{bmatrix} \neq 0 \tag{5,19}$$

and $\mathbf{y}_1^*\mathbf{x}_1 - \mathbf{y}_0^*\mathbf{x}_0 = (\mathbf{y}_0^*\mathbf{P}^c + \mathbf{y}_1^*\mathbf{Q}^c)(\mathbf{M}\mathbf{x}_0 + \mathbf{N}\mathbf{x}_1) + (\mathbf{y}_0^*\mathbf{P} + \mathbf{y}_1^*\mathbf{Q})(\mathbf{M}^c\mathbf{x}_0 + \mathbf{N}^c\mathbf{x}_1)$ *for all* $\mathbf{x}_0, \mathbf{x}_1, \mathbf{y}_0, \mathbf{y}_1 \in R_n$.

Proof. Let $\mathbf{P}, \mathbf{Q} \in L(R_{2n-m}, R_n)$ and $\mathbf{P}^c, \mathbf{Q}^c \in L(R_m, R_n)$ be such that

$$\begin{bmatrix} \mathbf{M}, & \mathbf{N} \\ \mathbf{M}^c, & \mathbf{N}^c \end{bmatrix}^{-1} = \begin{bmatrix} -\mathbf{P}^c, & -\mathbf{P} \\ \mathbf{Q}^c, & \mathbf{Q} \end{bmatrix}. \tag{5,20}$$

Then

$$-\mathbf{P}^c\mathbf{M} - \mathbf{P}\mathbf{M}^c = \mathbf{I}_n, \qquad -\mathbf{P}^c\mathbf{N} - \mathbf{P}\mathbf{N}^c = \mathbf{0}, \tag{5,21}$$
$$\mathbf{Q}^c\mathbf{M} + \mathbf{Q}\mathbf{M}^c = \mathbf{0}, \qquad \mathbf{Q}^c\mathbf{N} + \mathbf{Q}\mathbf{N}^c = \mathbf{I}_n$$

and

$$\begin{bmatrix} \mathbf{P}^c, & \mathbf{P} \\ \mathbf{Q}^c, & \mathbf{Q} \end{bmatrix}\begin{bmatrix} \mathbf{M}, & \mathbf{N} \\ \mathbf{M}^c, & \mathbf{N}^c \end{bmatrix} = \begin{bmatrix} -\mathbf{I}_n, & \mathbf{0} \\ \mathbf{0}, & \mathbf{I}_n \end{bmatrix}. \tag{5,22}$$

Thus, given $\mathbf{x}_0, \mathbf{x}_1, \mathbf{y}_0, \mathbf{y}_1 \in R_n$,

$$\mathbf{y}_1^*\mathbf{x}_1 - \mathbf{y}_0^*\mathbf{x}_0 = (\mathbf{y}_0^*, \mathbf{y}_1^*)\begin{bmatrix} -\mathbf{I}_n, & \mathbf{0} \\ \mathbf{0}, & \mathbf{I}_n \end{bmatrix}\begin{pmatrix} \mathbf{x}_0 \\ \mathbf{x}_1 \end{pmatrix}$$
$$= (\mathbf{y}_0^*, \mathbf{y}_1^*)\begin{bmatrix} \mathbf{P}^c, & \mathbf{P} \\ \mathbf{Q}^c, & \mathbf{Q} \end{bmatrix}\begin{bmatrix} \mathbf{M}, & \mathbf{N} \\ \mathbf{M}^c, & \mathbf{N}^c \end{bmatrix}\begin{pmatrix} \mathbf{x}_0 \\ \mathbf{x}_1 \end{pmatrix}$$
$$= (\mathbf{y}_0^*\mathbf{P}^c + \mathbf{y}_1^*\mathbf{Q}^c)(\mathbf{M}\mathbf{x}_0 + \mathbf{N}\mathbf{x}_1) + (\mathbf{y}_0^*\mathbf{P} + \mathbf{y}_1^*\mathbf{Q})(\mathbf{M}^c\mathbf{x}_0 + \mathbf{N}^c\mathbf{x}_1).$$

5.13. Remark. It follows from (5,20) that according to 5.12 the matrices $\mathbf{P}, \mathbf{Q} \in L(R_{2n-m}, R_n)$ and $\mathbf{P}^c, \mathbf{Q}^c \in L(R_m, R_n)$ associated to $\mathbf{M}, \mathbf{N}, \mathbf{M}^c, \mathbf{N}^c$ fulfil besides (5,21), (5,22) also

$$\begin{bmatrix} -\mathbf{M}, & \mathbf{N} \\ -\mathbf{M}^c, & \mathbf{N}^c \end{bmatrix}\begin{bmatrix} \mathbf{P}^c, & \mathbf{P} \\ \mathbf{Q}^c, & \mathbf{Q} \end{bmatrix} = \mathbf{I}_{2n},$$

i.e.

(5,23) $$-\mathbf{M}\mathbf{P}^c + \mathbf{N}\mathbf{Q}^c = \mathbf{I}_m, \qquad -\mathbf{M}\mathbf{P} + \mathbf{N}\mathbf{Q} = \mathbf{0},$$

(5,24) $$-\mathbf{M}^c\mathbf{P}^c + \mathbf{N}^c\mathbf{Q}^c = \mathbf{0}, \qquad -\mathbf{M}^c\mathbf{P} + \mathbf{N}^c\mathbf{Q} = \mathbf{I}_{2n-m}.$$

The following assertion is evident.

5.14. Proposition. *Let* $\mathbf{M}, \mathbf{N} \in L(R_n, R_m)$, rank $[\mathbf{M}, \mathbf{N}] = m$ *and let* $\mathbf{P}, \mathbf{Q} \in L(R_{2n-m}, R_n)$ *and* $\mathbf{P}^c, \mathbf{Q}^c \in L(R_m, R_n)$ *be such that* (5,19) *and* (5,23) *hold. Then* $\mathbf{P}_1, \mathbf{Q}_1 \in L(R_{2n-m}, R_n)$ *and* $\mathbf{P}_1^c, \mathbf{Q}_1^c \in L(R_m, R_n)$ *fulfil also* (5,19) *and* (5,23) *if and only if there exist a regular matrix* $\mathbf{E} \in L(R_{2n-m})$ *and* $\mathbf{F} \in L(R_m, R_{2n-m})$ *such that*

(5,25) $$\mathbf{P}_1 = \mathbf{P}\mathbf{E}, \qquad \mathbf{Q}_1 = \mathbf{Q}\mathbf{E}$$

and

(5,26) $$\mathbf{P}_1^c = \mathbf{P}^c + \mathbf{P}\mathbf{F}, \qquad \mathbf{Q}_1^c = \mathbf{Q}^c + \mathbf{Q}\mathbf{F}.$$

5.15. Definition. Let $\mathbf{M}, \mathbf{N} \in L(R_n, R_m)$ and let $\mathbf{P}, \mathbf{Q} \in L(R_{2n-m}, R_n)$ and $\mathbf{P}^c, \mathbf{Q}^c \in L(R_m, R_n)$ be such that (5,19) and (5,23) hold. Then the matrices $\mathbf{P}, \mathbf{Q}$ are called *adjoint matrices associated to* $[\mathbf{M}, \mathbf{N}]$ and the matrices $\mathbf{P}^c, \mathbf{Q}^c$ are called *complementary adjoint matrices associated to* $[\mathbf{M}, \mathbf{N}]$.

5.16. Remark. If $\mathbf{M}, \mathbf{N} \in L(R_m, R_n)$, rank $[\mathbf{M}, \mathbf{N}] = m$ and if $\mathbf{P}, \mathbf{Q} \in L(R_{2n-m}, R_n)$ are arbitrary adjoint matrices associated to $\mathbf{M}, \mathbf{N}$, then

(5,27) $$\text{rank}\begin{bmatrix} \mathbf{P} \\ \mathbf{Q} \end{bmatrix} = 2n - m$$

and the rows of the $m \times 2n$-matrix $[-\mathbf{M}, \mathbf{N}]$ form a basis in the space of all solutions $\mathbf{d}^* \in R_{2n}^*$ to the equation

(5,28) $$\mathbf{d}^*\begin{bmatrix} \mathbf{P} \\ \mathbf{Q} \end{bmatrix} = \mathbf{0}.$$

5.17. Remark. Let $\mathbf{M}, \mathbf{N} \in L(R_n, R_m)$ and $\operatorname{rank}[\mathbf{M}, \mathbf{N}] = m$. Let $\mathbf{P}, \mathbf{Q}$ and $\mathbf{P}^c, \mathbf{Q}^c$ be respectively adjoint and complementary adjoint matrices to $[\mathbf{M}, \mathbf{N}]$. If $\mathbf{y}^*\colon [0,1] \to R_n^*$ and $\lambda^* \in R_m^*$ fulfil (5,13), then

$$\mathbf{y}^*(0)\,\mathbf{P} + \mathbf{y}^*(1)\,\mathbf{Q} = \mathbf{0} \tag{5,29}$$

and

$$\mathbf{y}^*(0)\,\mathbf{P}^c + \mathbf{y}^*(1)\,\mathbf{Q}^c = \lambda^*. \tag{5,30}$$

On the other hand, if $\mathbf{y}^*\colon [0,1] \to R_n^*$ fulfils (5,29), then there exists $\lambda^* \in R_m^*$ such that (5,13) and consequently also (5,30) hold (cf. 5.16).

5.18. Corollary. *Let the assumptions 5,1 be fulfilled. Then the boundary value problem (5,1), (5,2) has a solution if and only if*

$$\mathbf{y}^*(1)\,\mathbf{f}(1) - \mathbf{y}^*(0)\,\mathbf{f}(0) - \int_0^1 \mathrm{d}[\mathbf{y}^*(s)]\,\mathbf{f}(s) = [\mathbf{y}^*(0)\,\mathbf{P}^c + \mathbf{y}^*(1)\,\mathbf{Q}^c]\,\mathbf{r} \tag{5,31}$$

for any solution $\mathbf{y}^*\colon [0,1] \to R_n^*$ *of the system* (5,12), (5,29) *where* $\mathbf{P}, \mathbf{Q}$ *and* $\mathbf{P}^c, \mathbf{Q}^c$ *are respectively adjoint and complementary adjoint matrices associated to* $[\mathbf{M}, \mathbf{N}]$.

Proof follows immediately from 5.5 and 5.17.

5.19. Remark. If $\mathbf{P}_1, \mathbf{Q}_1$ and $\mathbf{P}_1^c, \mathbf{Q}_1^c$ are also adjoint and complementary adjoint matrices associated to $[\mathbf{M}, \mathbf{N}]$, then by 5.14 there exist a regular matrix $\mathbf{E} \in L(R_{2n-m})$ and $\mathbf{F} \in L(R_m, R_{2n-m})$ such that for all $\mathbf{y}_0^*, \mathbf{y}_1^* \in R_n^*$ we have $\mathbf{y}_0^*\mathbf{P}_1 + \mathbf{y}_1^*\mathbf{Q}_1 = [\mathbf{y}_0^*\mathbf{P} + \mathbf{y}_1^*\mathbf{Q}]\,\mathbf{E}$ and $\mathbf{y}_0^*\mathbf{P}_1^c + \mathbf{y}_1^*\mathbf{Q}_1^c = \mathbf{y}_0^*\mathbf{P}^c + \mathbf{y}_1^*\mathbf{Q}^c + [\mathbf{y}_0^*\mathbf{P} + \mathbf{y}_1^*\mathbf{Q}]\,\mathbf{F}$. Thus $\mathbf{y}_0^*\mathbf{P} + \mathbf{y}_1^*\mathbf{Q} = \mathbf{0}$ and $\mathbf{y}_0^*\mathbf{P}^c + \mathbf{y}_1^*\mathbf{Q}^c = \lambda^*$ if and only if also $\mathbf{y}_0^*\mathbf{P}_1 + \mathbf{y}_1^*\mathbf{Q}_1 = \mathbf{0}$ and $\mathbf{y}_0^*\mathbf{P}_1^c + \mathbf{y}_1^*\mathbf{Q}_1^c = \lambda^*$. This means that neither the boundary condition (5,29) nor the condition (5,31) depend on the choice of the adjoint and complementary adjoint matrices associated to $[\mathbf{M}, \mathbf{N}]$.

5.20. Remark. The matrix valued functions $\mathbf{A}\colon [0,1] \to L(R_n)$ and $\mathbf{B}\colon [0,1] \to L(R_n)$ of bounded variation on $[0,1]$ fulfil 5.1 e.g. if

(i) $\mathbf{A}$ is left-continuous on $(0,1]$ and right-continuous at 0, $\det[\mathbf{I} + \Delta^+\mathbf{A}(t)] \neq 0$ on $[0,1]$ and $\mathbf{B} = \mathbf{A}_*$ (cf. (4,13)), or

(ii) $(\Delta^+\mathbf{A}(0))^2 = (\Delta^-\mathbf{A}(1))^2 = \mathbf{0}$, $(\Delta^+\mathbf{A}(t))^2 = (\Delta^-\mathbf{A}(t))^2$ on $(0,1)$, $\det[\mathbf{I} - (\Delta^+\mathbf{A}(t)]^2 \neq 0$ on $[0,1]$ and $\mathbf{B} = \mathbf{A}$, or

(iii) $\Delta^+\mathbf{A}(t) = \Delta^-\mathbf{A}(t)$ on $[0,1]$, $(\Delta^+\mathbf{A}(t))^2 = \mathbf{0}$ on $[0,1]$ and $\mathbf{B} = \mathbf{A}$.

(In the case (iii)

$$[\mathbf{I} + \Delta^+\mathbf{A}(t)]\,[\mathbf{I} - \Delta^-\mathbf{A}(t)] = \mathbf{I} - (\Delta^+\mathbf{A}(t))^2 = \mathbf{I}.)$$

We shall see later that the problems of the type (5,1), (5,2) cover also problems with a more general side condition (cf. V.7.19).

Notes

The theory of generalized differential equations was initiated by J. Kurzweil [1], [2], [4]. It is based on the generalization of the concept of the Perron integral; special results needed in the linear case are given in I.4. Differential equations with discontinuous solutions are considered e.g. in Stallard [2], Ligęza [2].

The paper by Hildebrandt [2] is devoted to linear differentio-Stieltjes integral equations. These equations are essentially generalized linear differential equations in our setting where the Young integral is used for the definition of a solution. Some results for the equations of this type can be found in Atkinson [1], Hönig [1], Schwabik [1], [4], Schwabik, Tvrdý [1], Mac Nerney [1], Wall [1].

Boundary value problems for generalized differential equations were for the first time mentioned in Atkinson [1] (Chapter XI). They appeared also in Halanay, Moro [1] as adjoints to boundary value problems with Stieltjes integral side conditions. A systematic study of such problems was initiated in Vejvoda, Tvrdý [1] and Tvrdý [1], [2]. Further related references are Krall [6], [8], Ligęza [1] and Zimmerberg [1], [2].

IV. *Linear boundary value problems for ordinary differential equations*

1. Preliminaries

This chapter is concerned with boundary value problems for linear nonhomogeneous vector ordinary differential equations

$$\mathbf{x}' - \mathbf{A}(t)\,\mathbf{x} = \mathbf{f}(t) \tag{1,1}$$

and the corresponding homogeneous equation

$$\mathbf{x}' - \mathbf{A}(t)\,\mathbf{x} = \mathbf{0}\,. \tag{1,2}$$

The differential equations (1,1) and (1,2) are considered in the sense of Carathéodory.

In the theory of ordinary differential equations the locution "*boundary value problems*" (BVP) refers to finding solutions to an ordinary differential equation which, in addition, satisfy some (additional) side conditions. In general, such conditions may require that the sought solution should belong to a prescribed set of functions. Very often this set is given as a set of solutions of a certain, generally nonlinear operator equation. In this chapter we restrict ourselves to the case of linear Stieltjes-integral side conditions of the form

$$\mathbf{S}\mathbf{x} = \mathbf{M}\,\mathbf{x}(0) + \mathbf{N}\,\mathbf{x}(1) + \int_0^1 \mathrm{d}[\mathbf{K}(t)]\,\mathbf{x}(t) = \mathbf{r} \tag{1,3}$$

or

$$\mathbf{M}\,\mathbf{x}(0) + \mathbf{N}\,\mathbf{x}(1) + \int_0^1 \mathrm{d}[\mathbf{K}(t)]\,\mathbf{x}(t) = \mathbf{0}\,. \tag{1,4}$$

Throughout the chapter the following hypotheses are kept to.

1.1. Assumptions. $\mathbf{A}$: $[0,1] \to L(R_n)$ *and* $\mathbf{f}$: $[0,1] \to R_n$ *are L-integrable on* $[0,1]$ $(\mathbf{f} \in L_n^1)$; $\mathbf{M}$ *and* $\mathbf{N} \in L(R_n, R_m)$, $\mathbf{r} \in R_m$, $m \geq 1$ *and* $\mathbf{K}$: $[0,1] \to L(R_n, R_m)$ *is of bounded variation on* $[0,1]$.

1.2. Definition. A function $\mathbf{x}: [0,1] \to R_n$ is a solution to the equation (1,1) on $[0,1]$ if it is absolutely continuous on $[0,1]$ ($\mathbf{x} \in AC_n$) and verifies $\mathbf{x}'(t) - \mathbf{A}(t)\,\mathbf{x}(t) = \mathbf{f}(t)$ a.e. on $[0,1]$.

1.3. Remark. Consequently $\mathbf{x}: [0,1] \to R_n$ is a solution to (1,1) on $[0,1]$ if and only if for any $t, t_0 \in [0,1]$

$$\mathbf{x}(t) = \mathbf{x}(t_0) + \int_{t_0}^{t} \mathbf{A}(s)\,\mathbf{x}(s)\,\mathrm{d}s + \int_{t_0}^{t} \mathbf{f}(s)\,\mathrm{d}s\,,$$

i.e. (1,1) is a special case of the linear generalized differential equation

$$\text{(1,5)} \qquad \mathrm{d}\mathbf{x} = \mathrm{d}[\mathbf{B}]\,\mathbf{x} + \mathrm{d}\mathbf{g} \qquad \left(\mathbf{B}(t) = \int_0^t \mathbf{A}(s)\,\mathrm{d}s\,, \quad \mathbf{g}(t) = \int_0^t \mathbf{f}(s)\,\mathrm{d}s\right).$$

1.4. Definition. A function $\mathbf{x}: [0,1] \to R_n$ is a solution to the *nonhomogeneous boundary value problem* (*BVP*) (1,1), (1,3) (verifies the system (1,1), (1,3)) if it is a solution of (1,1) on $[0,1]$ and satisfies (1,3). The problem of finding a solution $\mathbf{x}: [0,1] \to R_n$ of the homogeneous equation (1,2) on $[0,1]$ which fulfils also (1,4) is called the *homogeneous BVP* (1,2), (1,4).

1.5. Notation. Throughout the chapter $\mathbf{U}: [0,1] \times [0,1] \to L(R_n)$ is the fundamental matrix for the equation (1,5) defined by III.2.2 and $\mathbf{X}(t) = \mathbf{U}(t,0)$.

Let us recall that $\det \mathbf{X}(t) \neq 0$ on $[0,1]$, $\mathbf{U}(t,s) = \mathbf{X}(t)\,\mathbf{X}^{-1}(s)$ on $[0,1] \times [0,1]$,

$$\text{(1,6)} \qquad \mathbf{X}(t)\,\mathbf{X}^{-1}(s) = \mathbf{I} + \int_s^t \mathbf{A}(\tau)\,\mathbf{X}(\tau)\,\mathbf{X}^{-1}(s)\,\mathrm{d}\tau \qquad \text{for all} \quad t, s \in [0,1]$$

and

$$\text{(1,7)} \qquad \mathbf{X}(t)\,\mathbf{X}^{-1}(s) = \mathbf{I} + \int_s^t \mathbf{X}(t)\,\mathbf{X}^{-1}(\tau)\,\mathbf{A}(\tau)\,\mathrm{d}\tau \qquad \text{for all} \quad t, s \in [0,1]\,.$$

Both $\mathbf{X}(t)$ and $\mathbf{X}^{-1}(s)$ are absolutely continuous on $[0,1]$. The variation-of-constants formula reduces to

$$\text{(1,8)} \qquad \mathbf{x}(t) = \mathbf{U}(t,t_0)\,\mathbf{x}(t_0) + \int_{t_0}^{t} \mathbf{U}(t,s)\,\mathbf{f}(s)\,\mathrm{d}s \qquad \text{for all} \quad t, t_0 \in [0,1]\,.$$

1.6. Remark. Since $\mathbf{A}(t)$ is supposed to be L-integrable on $[0,1]$, for any $\mathbf{x} \in AC_n$ the function $\mathbf{x}'(t) - \mathbf{A}(t)\,\mathbf{x}(t)$ is defined a.e. on $[0,1]$ and is L-integrable on $[0,1]$. Hence the operator

$$\text{(1,9)} \qquad \mathbf{L}: \mathbf{x} \in AC_n \to \mathbf{L}\mathbf{x}\,, \qquad (\mathbf{L}\mathbf{x})(t) = \mathbf{x}'(t) - \mathbf{A}(t)\,\mathbf{x}(t) \qquad \text{a.e. on } [0,1]$$

maps AC_n into L_n^1. Obviously it is linear and

$$\|\mathbf{L}\mathbf{x}\|_{L^1} = \int_0^1 |\mathbf{x}'(t) - \mathbf{A}(t)\,\mathbf{x}(t)|\,\mathrm{d}t \le \int_0^1 |\mathbf{x}'(t)|\,\mathrm{d}t + \left(\int_0^1 |\mathbf{A}(t)|\,\mathrm{d}t\right) \sup_{t\in[0,1]} |\mathbf{x}(t)|$$

$$\le \left(1 + \int_0^1 |\mathbf{A}(t)|\,\mathrm{d}t\right) \|\mathbf{x}\|_{AC}$$

for any $\mathbf{x} \in AC_n$. Moreover, for a given $\mathbf{x} \in C_n$

$$\left|\mathbf{M}\,\mathbf{x}(0) + \mathbf{N}\,\mathbf{x}(1) + \int_0^1 \mathrm{d}[\mathbf{K}(t)]\,\mathbf{x}(t)\right| \le (|\mathbf{M}| + |\mathbf{N}| + \mathrm{var}_0^1\,\mathbf{K})\,\|\mathbf{x}\|_C$$

and the operator

$$\mathbf{S}\colon \mathbf{x} \in C_n \to \mathbf{M}\,\mathbf{x}(0) + \mathbf{N}\,\mathbf{x}(1) + \int_0^1 \mathrm{d}[\mathbf{K}(t)]\,\mathbf{x}(t) \in R_m \tag{1,10}$$

is linear and bounded. Consequently, under the assumptions 1.1

$$\mathscr{L}\colon \mathbf{x} \in AC_n \to \begin{bmatrix} \mathbf{Lx} \\ \mathbf{Sx} \end{bmatrix} \in L_n^1 \times R_m$$

is linear and bounded. The given BVP (1,1), (1,3) may be now rewritten as the linear operator equation

$$\mathscr{L}\mathbf{x} = \begin{pmatrix} \mathbf{f} \\ \mathbf{r} \end{pmatrix}.$$

1.7. Proposition. *Given $\mathbf{c} \in R_n$ and $\mathbf{f} \in L_n^1$, the unique solution $\mathbf{x}$ to (1,1) on $[0, 1]$ such that $\mathbf{x}(0) = \mathbf{c}$ can be expressed in the form*

$$\mathbf{x}(t) = (\mathbf{Uc})(t) + (\mathbf{Vf})(t) \qquad \text{on } [0, 1],$$

where

$$\mathbf{U}\colon \mathbf{c} \in R_n \to \mathbf{X}(t)\,\mathbf{c} \in AC_n \tag{1,11}$$

and

$$\mathbf{V}\colon \mathbf{f} \in L_n^1 \to \mathbf{X}(t) \int_0^t \mathbf{X}^{-1}(s)\,\mathbf{f}(s)\,\mathrm{d}s \in AC_n \tag{1,12}$$

are linear and bounded operators.

Proof. The linearity is obvious. Let $\mathbf{c} \in R_n$ and $\mathbf{f} \in L_n^1$. Then

$$\|\mathbf{Uc}\|_{AC} \le \left(1 + \int_0^1 |\mathbf{X}'(t)|\,\mathrm{d}t\right)|\mathbf{c}| = \varkappa_1 |\mathbf{c}|, \qquad \varkappa_1 < \infty$$

and

$$\|\mathbf{Vf}\|_{AC} = \int_0^1 \left|\mathbf{X}'(t)\left(\int_0^t \mathbf{X}^{-1}(s)\,\mathbf{f}(s)\,\mathrm{d}s\right) + \mathbf{f}(t)\right| \mathrm{d}t$$

$$\le \left[\left(\int_0^1 |\mathbf{X}'(t)|\,\mathrm{d}t\right)\left(\sup_{s\in[0,1]} |\mathbf{X}^{-1}(s)|\right) + 1\right] \|\mathbf{f}\|_{L^1} = \varkappa_2 \|\mathbf{f}\|_{L^1}, \qquad \varkappa_2 < \infty .$$

1.8. Remark. By the Riesz Representation Theorem an arbitrary linear bounded mapping $\mathbf{S}\colon C_n \to R_m$ may be expressed in the form (1,10), where $\mathbf{M} = \mathbf{N} = \mathbf{0}$.

If $\mathbf{K}(t)$ is the sum of a series of the simple jump functions of bounded variation on $[0, 1]$ with the jumps $\Delta\mathbf{K}_j$ at $t = \tau_j \in [0, 1]$ $(j = 1, 2, \ldots)$, then (1,3) reduces to the *infinite point condition* (cf. I.4.23)

$$\mathbf{M}\,\mathbf{x}(0) + \sum_{j=1}^{\infty} \Delta\mathbf{K}_j\,\mathbf{x}(\tau_j) + \mathbf{N}\,\mathbf{x}(1) = \mathbf{r} \qquad \Big(\sum_{j=1}^{\infty} |\Delta\mathbf{K}_j| < \infty\Big).$$

In particular, if $\mathbf{K}(t)$ is a finite-step function on $[0, 1]$ $(\mathbf{K}(t) = \mathbf{K}_j$ for $\tau_{j-1} \le t < \tau_j$ $(j = 1, 2, \ldots, p-1)$, $\mathbf{K}(t) = \mathbf{K}_p$ for $\tau_{p-1} \le t \le 1$ where $0 = \tau_0 < \tau_1 < \ldots < \tau_p = 1)$, then (1,3) reduces to the *multipoint condition*

$$\mathbf{M}\,\mathbf{x}(0) + \sum_{j=1}^{p-1} \Delta\mathbf{K}_j\,\mathbf{x}(\tau_j) + \mathbf{N}\,\mathbf{x}(1) = \mathbf{r} \qquad (\Delta\mathbf{K}_j = \mathbf{K}_{j+1} - \mathbf{K}_j)$$

or even to the *two-point boundary conditions* (if $\Delta\mathbf{K}_j = \mathbf{0}$, $j = 1, 2, \ldots, p-1$).

The problem of determining a function $\mathbf{x}$: $[0, 1] \to R_n$ absolutely continuous on each subinterval (τ_j, τ_{j+1}) $(j = 0, 1, \ldots, p-1,\ 0 = \tau_0 < \tau_1 < \ldots < \tau_p = 1)$ and such that $\mathbf{x}'(t) - \mathbf{A}(t)\,\mathbf{x}(t) = \mathbf{f}(t)$ a.e. on $[0, 1]$ and

$$\mathbf{M}\,\mathbf{x}(0) + \sum_{j=0}^{p-1} [\mathbf{M}_j\,\mathbf{x}(\tau_j+) + \mathbf{N}_j\,\mathbf{x}(\tau_{j+1}-)] + \mathbf{N}\,\mathbf{x}(1) = \mathbf{r}$$

is called the *interface problem* and is to be dealt with separately.

1.9. Remark. If we put $\mathbf{K}_0(t) = \mathbf{K}(t+) - \mathbf{K}(1-)$ for $t \in [0, 1)$ and $\mathbf{K}_0(1) = \mathbf{0}$, then $\mathbf{K} - \mathbf{K}_0$ is of bounded variation on $[0, 1]$, $\Delta^+\mathbf{K}_0(t) = \mathbf{0}$ on $[0, 1)$, $\mathbf{K}_0(1-) = \mathbf{K}_0(1) = \mathbf{0}$, $\mathbf{K}(t+) - \mathbf{K}_0(t+) = \mathbf{K}(t-) - \mathbf{K}_0(t-) = \mathbf{0}$ on $[0, 1]$, $\mathbf{K}(1) - \mathbf{K}_0(1) = \mathbf{K}(1)$, $\mathbf{K}(0) - \mathbf{K}_0(0) = -\Delta^+\mathbf{K}(0) - \mathbf{K}(1-)$ and hence for any $\mathbf{x} \in C_n$ (cf. I.4.23 and I.5.5)

$$\mathbf{M}\,\mathbf{x}(0) + \mathbf{N}\,\mathbf{x}(1) + \int_0^1 \mathrm{d}[\mathbf{K}(t)]\,\mathbf{x}(t) = \mathbf{M}_0\,\mathbf{x}(0) + \mathbf{N}_0\,\mathbf{x}(1) + \int_0^1 \mathrm{d}[\mathbf{K}_0(t)]\,\mathbf{x}(t)$$

$(\mathbf{M}_0 = \mathbf{M} - \Delta^+\mathbf{K}(0) - \mathbf{K}(1-), \qquad \mathbf{N}_0 = \mathbf{N} + \mathbf{K}(1).)$

Thus, without any loss of generality we may add the following hypotheses to 1.1.

1.10. Assumptions. $\mathbf{K}(t)$ is right-continuous on $[0, 1)$, *left-continuous at* 1 *and* $\mathbf{K}(1) = \mathbf{0}$.

1.11. Definition. The side condition (1,3) $(\mathbf{S}\mathbf{x} = \mathbf{r})$ is *linearly dependent* if there exists $\mathbf{q} \in R_m$, $\mathbf{q} \neq \mathbf{0}$ such that $\mathbf{q}^*(\mathbf{S}\mathbf{x}) = 0$ for all $\mathbf{x} \in AC_n$. It is *linearly independent* if it is not linearly dependent.

1.12. Proposition. *Let* $\mathbf{M}, \mathbf{N}$ *and* $\mathbf{K}(t)$ *fulfil the hypotheses* 1.1 *and* 1.10. *Then the side condition* (1,3) *is linearly dependent if and only if there is* $\mathbf{q} \in R_m$, $\mathbf{q} \neq \mathbf{0}$ *such that*

$$\mathbf{q}^*\mathbf{M} = \mathbf{q}^*\mathbf{N} \equiv \mathbf{q}^*\,\mathbf{K}(t) \equiv \mathbf{0} \qquad \text{on } [0, 1].$$

Proof. Let $\boldsymbol{q} \neq \boldsymbol{0}$ and let

$$(1,13) \quad \boldsymbol{q}^*[\boldsymbol{M}\,\boldsymbol{x}(0) + \boldsymbol{N}\,\boldsymbol{x}(1) + \int_0^1 \mathrm{d}[\boldsymbol{K}(t)]\,\boldsymbol{x}(t)] = 0 \qquad \text{for each} \quad \boldsymbol{x} \in AC_n .$$

Then for every $\boldsymbol{x} \in AC_n$ with $\boldsymbol{x}(0) = \boldsymbol{x}(1) = \boldsymbol{0}$ we have

$$\int_0^1 \mathrm{d}[\boldsymbol{q}^*\,\boldsymbol{K}(t)]\,\boldsymbol{x}(t) = 0 .$$

By I.5.17 this implies $\boldsymbol{q}^*\,\boldsymbol{K}(t) = \boldsymbol{0}$ on $[0,1]$ and hence (1,13) reduces to

$$\boldsymbol{q}^*[\boldsymbol{M}\boldsymbol{c} + \boldsymbol{N}\boldsymbol{d}] = \boldsymbol{0} \qquad \text{for all} \quad \boldsymbol{c}, \boldsymbol{d} \in R_n .$$

Choosing $\boldsymbol{c} = \boldsymbol{0}$ and $\boldsymbol{d} \in R_n$ arbitrary or $\boldsymbol{d} = \boldsymbol{0}$ and $\boldsymbol{c} \in R_n$ arbitrary, we obtain $\boldsymbol{q}^*\boldsymbol{N} = \boldsymbol{0}$ or $\boldsymbol{q}^*\boldsymbol{M} = \boldsymbol{0}$, respectively.

1.13. Definition. The side condition (1,3) is said to be *nonzero* if the corresponding operator $\boldsymbol{S}$ given by (1,10) is nonzero. Given $\boldsymbol{r} \in R_m$, the side condition (1,3) is *reasonable* if $\boldsymbol{q}^*\boldsymbol{r} = 0$ for any $\boldsymbol{q} \in R_m$ such that $\boldsymbol{q}^*(\boldsymbol{S}\boldsymbol{x}) = 0$ for all $\boldsymbol{x} \in AC_n$. (Obviously, given $\begin{pmatrix}\boldsymbol{f}\\ \boldsymbol{r}\end{pmatrix} \in L_n^1 \times R_m$, BVP (1,1), (1,3) may be solvable only if the side condition (1,3) is reasonable.)

Given $\boldsymbol{x} \in AC_n$, let $S_j\boldsymbol{x}$ $(j = 1, 2, \dots, m)$ denote the components of the vector $\boldsymbol{S}\boldsymbol{x} \in R_m$. Then S_j: $\boldsymbol{x} \in AC_n \to S_j\boldsymbol{x} \in R$ are linear bounded functionals on AC_n and the side condition (1,3) may be rewritten as the system of equations $S_j\boldsymbol{x} = r_j$ $(j = 1, 2, \dots, m)$, where r_j are components of the vector $\boldsymbol{r}$. The side condition (1,3) is linearly dependent if and only if the functionals $S_j \in AC_n^*$ $(j = 1, 2, \dots, m)$ are linearly dependent. Since the linear subspace of AC_n spanned on $\{S_1, S_2, \dots, S_m\}$ is finite dimensional, the following assertion is obvious.

1.14. Proposition. *If the side condition (1,3) is nonzero and reasonable, then there exist a natural number l, matrices $\boldsymbol{M}_0, \boldsymbol{N}_0 \in L(R_n, R_l)$, $\boldsymbol{r}_0 \in R_l$ and a function $\boldsymbol{K}_0$: $[0,1] \to L(R_n, R_l)$ of bounded variation on $[0,1]$ such that the condition*

$$\boldsymbol{S}_0\boldsymbol{x} = \boldsymbol{M}_0\,\boldsymbol{x}(0) + \boldsymbol{N}_0\,\boldsymbol{x}(1) + \int_0^1 \mathrm{d}[\boldsymbol{K}_0(t)]\,\boldsymbol{x}(t) = \boldsymbol{r}_0$$

is linearly independent, while $\boldsymbol{S}\boldsymbol{x} = \boldsymbol{r}$ for $\boldsymbol{x} \in AC_n$ if and only if $\boldsymbol{S}_0\,\boldsymbol{x} = \boldsymbol{r}_0$.

Henceforth let us assume that the side condition (1,3) is reasonable, linearly independent and fulfils the hypotheses 1.1 and 1.10. Let us denote by p the dimension of the linear subspace spanned on the rows of $\boldsymbol{K}(t)$. If $0 < p < m$, then there exists a regular $m \times m$-matrix $\boldsymbol{\Sigma}_1$ such that

$$\boldsymbol{\Sigma}_1\,\boldsymbol{K}(t) \equiv \begin{bmatrix}\boldsymbol{0}\\ \tilde{\boldsymbol{K}}(t)\end{bmatrix} \qquad \text{on } [0,1],$$

where the rows of $\tilde{\mathbf{K}}: [0,1] \to L(R_n, R_p)$ are linearly independent on $[0,1]$. Let us denote $m_0 = m - p$ and let the matrices $\mathbf{M}_0, \mathbf{N}_0 \in L(R_n, R_{m_0})$ and $\tilde{\mathbf{M}}, \tilde{\mathbf{N}} \in L(R_n, R_p)$ be such that

$$\Sigma_1[\mathbf{M}, \mathbf{N}, \mathbf{K}(t)] = \begin{bmatrix} \mathbf{M}_0, & \mathbf{N}_0, & \mathbf{0} \\ \tilde{\mathbf{M}}, & \tilde{\mathbf{N}}, & \tilde{\mathbf{K}}(t) \end{bmatrix}.$$

If there were $\boldsymbol{\alpha}^*[\mathbf{M}_0, \mathbf{N}_0] = \mathbf{0}$, then $\boldsymbol{\beta}^*\Sigma_1[\mathbf{M}, \mathbf{N}, \mathbf{K}(t)] \equiv \mathbf{0}$ or according to 1.12 $\boldsymbol{\beta}^*\Sigma_1 = \mathbf{0}$ should hold for $\boldsymbol{\beta}^* = (\boldsymbol{\alpha}^*, \mathbf{0}) \in R_m^*$. As Σ_1 is regular, $\boldsymbol{\beta}^*\Sigma_1 = \mathbf{0}$ implies $\boldsymbol{\beta}^* = \mathbf{0}$ and hence also $\boldsymbol{\alpha}^* = \mathbf{0}$. This means that the $m_0 \times 2n$-matrix $[\mathbf{M}_0, \mathbf{N}_0]$ has a full rank (rank $[\mathbf{M}_0, \mathbf{N}_0] = m_0$). If rank $[\mathbf{M}, \mathbf{N}] = m_0 + m_1$, i.e.

$$\operatorname{rank} \begin{bmatrix} \mathbf{M}_0, & \mathbf{N}_0 \\ \tilde{\mathbf{M}}, & \tilde{\mathbf{N}} \end{bmatrix} = m_0 + m_1 \qquad (0 \le m_1 \le p),$$

then there exists a regular $p \times p$-matrix $\boldsymbol{\Sigma}_2$ such that

$$\begin{bmatrix} \mathbf{I}_{m_0}, & \mathbf{0} \\ \mathbf{0}, & \Sigma_2 \end{bmatrix} \begin{bmatrix} \mathbf{M}_0, & \mathbf{N}_0 \\ \tilde{\mathbf{M}}, & \tilde{\mathbf{N}} \end{bmatrix} = \begin{bmatrix} \mathbf{M}_0, & \mathbf{N}_0 \\ \mathbf{M}_1, & \mathbf{N}_1 \\ \mathbf{0}, & \mathbf{0} \end{bmatrix},$$

where $\mathbf{M}_1, \mathbf{N}_1 \in L(R_n, R_{m_1})$ are such that

(1,15)
$$\operatorname{rank} \begin{bmatrix} \mathbf{M}_0, & \mathbf{N}_0 \\ \mathbf{M}_1, & \mathbf{N}_1 \end{bmatrix} = m_0 + m_1 .$$

Denoting

$$\boldsymbol{\Theta} = \begin{bmatrix} \mathbf{I}_{m_0}, & \mathbf{0} \\ \mathbf{0}, & \Sigma_2 \end{bmatrix} \Sigma_1 ,$$

we obtain

(1,16)
$$\boldsymbol{\Theta}[\mathbf{M}, \mathbf{N}, \mathbf{K}(t)] \equiv \begin{bmatrix} \mathbf{M}_0, & \mathbf{N}_0, & \mathbf{0} \\ \mathbf{M}_1, & \mathbf{N}_1, & \mathbf{K}_1(t) \\ \mathbf{0}, & \mathbf{0}, & \mathbf{K}_2(t) \end{bmatrix},$$

where $\mathbf{K}_1: [0,1] \to L(R_n, R_{m_1})$, $\mathbf{K}_2: [0,1] \to L(R_n, R_{m_2})$ $(m_1 + m_2 = p)$ are given by

$$\begin{bmatrix} \mathbf{K}_1(t) \\ \mathbf{K}_2(t) \end{bmatrix} \equiv \mathbf{K}^0(t) \equiv \Sigma_2 \tilde{\mathbf{K}}(t) \qquad \text{on } [0,1].$$

As $\boldsymbol{\Sigma}_2$ is regular, the rows of the $p \times n$-matrix $\mathbf{K}^0(t)$ are linearly independent on $[0,1]$. Finally, let us notice that the $m \times m$-matrix $\boldsymbol{\Theta}$ is regular. To summarize:

1.15. Theorem. *Any linearly independent and reasonable Stieltjes-integral side condition* (1,3) *fulfilling* 1.1 *and* 1.10 *is equivalent to the system*

(1,17)
$$\begin{aligned} \mathbf{M}_0\, \mathbf{x}(0) + \mathbf{N}_1\, \mathbf{x}(1) \qquad\qquad &= \mathbf{r}_0 , \\ \mathbf{M}_1\, \mathbf{x}(0) + \mathbf{N}_1\, \mathbf{x}(1) + \int_0^1 \mathrm{d}[\mathbf{K}_1(t)]\, \mathbf{x}(t) &= \mathbf{r}_1 , \\ \int_0^1 \mathrm{d}[\mathbf{K}_2(t)]\, \mathbf{x}(t) &= \mathbf{r}_2 , \end{aligned}$$

where $\mathbf{r}_0 \in R_{m_0}$, $\mathbf{r}_1 \in R_{m_1}$, $\mathbf{r}_2 \in R_{m_2}$ *and the* $m_0 \times n$*-matrices* $\mathbf{M}_0, \mathbf{N}_0$, *the* $m_1 \times n$*-matrices* $\mathbf{M}_1, \mathbf{N}_1, \mathbf{K}_1(t)$ *and the* $m_2 \times n$*-matrix* $\mathbf{K}_2(t)$ *are such that* (1,15) *holds and the rows of the* $(m_1 + m_2) \times n$*-matrix* $[\mathbf{K}_1^*(t), \mathbf{K}_2^*(t)]^*$ *are linearly independent on* $[0, 1]$. *There exists a regular* $m \times m$*-matrix* $\boldsymbol{\Theta}$ *such that* (1,16) *and* $\boldsymbol{\Theta}\mathbf{r} = (\mathbf{r}_0^*, \mathbf{r}_1^*, \mathbf{r}_2^*)^*$ *hold.*

1.16. Definition. The system (1,17) associated to (1,3) by 1.15 is said to be the *canonical form* of (1,3).

1.17. Remark. By I.5.16 the general form of the linear bounded operator $\mathbf{S}$: $AC_n \to R_m$ is

$$\text{(1,18)} \qquad \mathbf{S}: \mathbf{x} \in AC_n \to \mathbf{M}\,\mathbf{x}(0) + \int_0^1 \mathbf{K}(t)\,\mathbf{x}'(t)\,\mathrm{d}t\,,$$

where $\mathbf{M} \in L(R_n, R_m)$ and $\mathbf{K}$: $[0, 1] \to L(R_n, R_m)$ is measurable and essentially bounded on $[0, 1]$. If $\mathbf{K}$ is of bounded variation on $[0, 1]$, then by integrating by parts we may easily reduce $\mathbf{S}$ to the form (1,10).

Most of the results given in this chapter may be extended to BVP with the side operator $\mathbf{S}$ of the form (1,18). Some of them are formulated and proved in the following chapter for more general BVP which include integro-differential equations, the rest is left to the reader.

2. Duality theory

Let us consider BVP (1,1), (1,3), i.e. the system

$$\text{(1,1)} \quad \mathbf{x}' - \mathbf{A}(t)\,\mathbf{x} = \mathbf{f}(t)\,, \qquad \text{(1,3)} \quad \mathbf{M}\,\mathbf{x}(0) + \mathbf{N}\,\mathbf{x}(1) + \int_0^1 \mathrm{d}[\mathbf{K}(t)]\,\mathbf{x}(t) = \mathbf{r}\,,$$

where $\mathbf{A}$: $[0, 1] \to L(R_n)$, $\mathbf{M}$ and $\mathbf{N} \in L(R_n, R_m)$ and $\mathbf{K}$: $[0, 1] \to L(R_n, R_m)$ fulfil 1.1 and 1.10. Moreover, we suppose that (1,3) is nonzero and reasonable (see 1.13).

Let $\mathbf{f} \in L_n^1$ and $\mathbf{r} \in R_m$. By the variation-of-constants formula 1.7 a function $\mathbf{x}$: $[0, 1] \to R_n$ is a solution to BVP (1,1), (1,3) if and only

$$\mathbf{x} = \mathbf{U}\mathbf{c} + \mathbf{V}\mathbf{f}$$

and

$$\text{(2,1)} \qquad (\mathbf{S}\mathbf{U})\,\mathbf{c} = \mathbf{r} - (\mathbf{S}\mathbf{V})\,\mathbf{f}\,,$$

where $\mathbf{U}$: $R_n \to AC_n$ and $\mathbf{V}$: $L_n^1 \to AC_n$ are the linear bounded operators respectively given by (1,11) and (1,12),

$$\mathbf{S}\mathbf{U} = \mathbf{M}\,\mathbf{X}(0) + \mathbf{N}\,\mathbf{X}(1) + \int_0^1 \mathrm{d}[\mathbf{K}(t)]\,\mathbf{X}(t)$$

and

$$\text{(2,2)} \quad (\mathbf{S}\mathbf{V})\,\mathbf{f} = \mathbf{N}\,\mathbf{X}(1) \int_0^1 \mathbf{X}^{-1}(t)\,\mathbf{f}(t)\,\mathrm{d}t + \int_0^1 \mathrm{d}[\mathbf{K}(t)]\,\mathbf{X}(t) \int_0^t \mathbf{X}^{-1}(s)\,\mathbf{f}(s)\,\mathrm{d}s\,.$$

This yields immediately the following necessary and sufficient condition for the existence of a solution to BVP (1,1), (1,3).

2.1. Theorem. *BVP* (1,1), (1,3) *has a solution if and only if*

$$\lambda^*(\mathbf{SU}) = \mathbf{0} \tag{2,3}$$

implies

$$\lambda^*(\mathbf{SV})\,\mathbf{f} = \lambda^*\mathbf{r}\,. \tag{2,4}$$

2.2. Remark. Applying the Dirichlet formula I.4.32 to (2,2) we obtain for any $\mathbf{f} \in L_n^1$

$$(\mathbf{SV})\,\mathbf{f} = \int_0^1 \mathbf{F}(t)\,\mathbf{f}(t)\,\mathrm{d}t\,, \tag{2,5}$$

where

$$\mathbf{F}(t) = \left(\mathbf{N}\,\mathbf{X}(1) + \int_t^1 \mathrm{d}[\mathbf{K}(s)]\,\mathbf{X}(s)\right)\mathbf{X}^{-1}(t) \quad \text{on } [0,1]\,. \tag{2,6}$$

Hence the condition (2,4) may be rewritten as

$$\int_0^1 \lambda^*\,\mathbf{F}(t)\,\mathbf{f}(t)\,\mathrm{d}t = \lambda^*\mathbf{r}\,.$$

By (III. 4,8) the n-vector valued function $\mathbf{y}^*(t) = \lambda^*\,\mathbf{F}(t)$ is for any $\lambda^* \in R_m^*$ a unique solution of the initial value problem

$$\mathrm{d}\mathbf{y}^* = -\mathbf{y}^*\,\mathrm{d}[\mathbf{B}] - \mathrm{d}(\lambda^*\mathbf{K}) \quad \text{on } [0,1] \quad \left(\mathbf{B}(t) = \int_0^t \mathbf{A}(s)\,\mathrm{d}s\right), \qquad \mathbf{y}^*(1) = \lambda^*\mathbf{N}\,. \tag{2,7}$$

(In fact, if $\mathbf{h} \in BV_n$ is right-continuous on $[0,1)$ and left-continuous at 1, then integrating by parts (cf. I.4.33) we reduce the variation-of-constants formula for the initial value problem

$$\mathrm{d}\mathbf{y}^* = -\mathbf{y}^*\,\mathrm{d}[\mathbf{B}] - \mathrm{d}\mathbf{h}^*\,, \qquad \mathbf{y}^*(1) = \mathbf{y}_1^*$$

to the form

$$\mathbf{y}^*(t) = \left(\mathbf{y}_1^*\,\mathbf{X}(1) + \int_t^1 \mathrm{d}[\mathbf{h}^*(s)]\,\mathbf{X}(s)\right)\mathbf{X}^{-1}(t) \quad \text{on } [0,1]\,.) \tag{2,7a}$$

Furthermore, if $\lambda^*(\mathbf{SU}) = \mathbf{0}$, then

$$\mathbf{y}^*(0) = \lambda^*\left(\mathbf{N}\,\mathbf{X}(1) + \int_0^1 \mathrm{d}[\mathbf{K}(t)]\,\mathbf{X}(t)\right) = -\lambda^*\mathbf{M}\,.$$

On the other hand, it follows from the variation-of-constants formula that if $\mathbf{y}^*\colon [0,1] \to R_n^*$ and $\lambda^* \in R_m^*$ solve (2,7) on $[0,1]$ and

$$\mathbf{y}^*(0) + \lambda^*\mathbf{M} = \mathbf{0}\,, \qquad \mathbf{y}^*(1) - \lambda^*\mathbf{N} = \mathbf{0}\,, \tag{2,8}$$

then

$$\text{(2,9)} \qquad \mathbf{y}^*(t) = \boldsymbol{\lambda}^* \mathbf{F}(t) \quad \text{on } [0,1] \text{ and } \boldsymbol{\lambda}^*(\mathbf{SU}) = -\mathbf{y}^*(0) + \mathbf{y}^*(0) = \mathbf{0}.$$

This completes the proof of the following theorem.

2.3. Theorem. *BVP (1,1), (1,3) has a solution if and only if*

$$\text{(2,10)} \qquad \int_0^1 \mathbf{y}^*(t)\, \mathbf{f}(t)\, \mathrm{d}t = \boldsymbol{\lambda}^* \mathbf{r}$$

for any solution $(\mathbf{y}^*, \boldsymbol{\lambda}^*)$ *of the system (2,7), (2,8).*

2.4. Definition. The system (2,7), (2,8) of equations for $\mathbf{y}^*: [0,1] \to R_n^*$ and $\boldsymbol{\lambda}^* \in R_m^*$ is called the *adjoint boundary value problem to BVP* (1,1), (1,3) (or (1,2), (1,4)).

The following assertion provides the necessary and sufficient condition for the existence of a solution to the nonhomogeneous BVP corresponding to BVP (2,7), (2,8).

2.5. Theorem. *Let* $\mathbf{p}, \mathbf{q} \in R_n$ *and let* $\mathbf{g} \in BV_n$ *be right-continuous on* $[0,1)$ *and left-continuous at* 1. *Then the system*

$$\text{(2,11)} \qquad \mathrm{d}\mathbf{y}^* = \mathbf{y}^* \mathrm{d}[-\mathbf{B}] - \mathrm{d}(\boldsymbol{\lambda}^* \mathbf{K}) + \mathrm{d}\mathbf{g}^* \quad \text{on } [0,1],$$

$$\text{(2,12)} \qquad \mathbf{y}^*(0) + \boldsymbol{\lambda}^* \mathbf{M} = \mathbf{p}^*, \qquad \mathbf{y}^*(1) - \boldsymbol{\lambda}^* \mathbf{N} = \mathbf{q}^*$$

has a solution if and only if

$$\int_0^1 \mathrm{d}[\mathbf{g}^*(t)]\, \mathbf{x}(t) = \mathbf{q}^* \mathbf{x}(1) - \mathbf{p}^* \mathbf{x}(0)$$

for any solution $\mathbf{x}$ *of the homogeneous BVP (1,2), (1,4).*

Proof. Inserting (2,7a), where $\mathbf{h}^*(t) = \boldsymbol{\lambda}^* \mathbf{K}(t) - \mathbf{g}^*(t)$ into (2,12) we easily obtain that $\mathbf{y}^*: [0,1] \to R_n^*$ and $\boldsymbol{\lambda}^* \in R_m^*$ verify (2,11), (2,12) if and only if

$$\text{(2,13)} \qquad \mathbf{y}^*(t) = \boldsymbol{\lambda}^* \mathbf{F}(t) + \mathbf{q}^* \mathbf{X}(1) \mathbf{X}^{-1}(t) - \int_t^1 \mathrm{d}[\mathbf{g}^*(s)]\, \mathbf{X}(s) \mathbf{X}^{-1}(t) \quad \text{on } [0,1]$$

($\mathbf{F}(t)$ given by (2,6)) and

$$\boldsymbol{\lambda}^*(\mathbf{SU}) = \mathbf{p}^* \mathbf{X}(0) - \mathbf{q}^* \mathbf{X}(1) + \int_0^1 \mathrm{d}[\mathbf{g}^*(t)]\, \mathbf{X}(t).$$

Since all the solutions of BVP (1,2), (1,4) are of the form $\mathbf{X}(t)\, \mathbf{c}$ where $(\mathbf{SU})\, \mathbf{c} = \mathbf{0}$, the theorem follows immediately.

2.6. Remark. Let us notice that under our assumptions all the solutions $\mathbf{y}^*: [0,1] \to R_n$ of (2,11) on $[0,1]$ are of bounded variation on $[0,1]$, right-continuous on $[0,1]$ and left-continuous at 1.

2.7. Theorem. *The homogeneous problems* (1,2), (1,4) *and* (2,7), (2,8) *possess exactly* $k = n - \operatorname{rank}(\mathbf{SU})$ *and* $k^* = m - \operatorname{rank}(\mathbf{SU})$ *linearly independent solutions, respectively.*

Proof. The homogeneous algebraic equation

$$(\mathbf{SU})\,\mathbf{c} = \mathbf{0} \tag{2,14}$$

has exactly $k = n - \operatorname{rank}(\mathbf{SU})$ linearly independent solutions. Let $\mathbf{C}_0$ be an arbitrary $n \times k$-matrix whose columns form a basis in the space of all solutions to (2,14). ($(\mathbf{SU})\,\mathbf{C}_0 = \mathbf{0}$ and $\operatorname{rank}(\mathbf{C}_0) = k$.) This obviously implies that the columns of the $n \times k$-matrix valued function

$$\mathbf{X}_0(t) = \mathbf{X}(t)\,\mathbf{C}_0 \qquad \text{on } [0,1]$$

form a basis in the space of all solutions of BVP (1,2), (1,4).

The latter assertion follows from the fact that $\mathbf{y}^*: [0,1] \to R_n^*$ and $\boldsymbol{\lambda}^* \in R_m^*$ verify the system (2,7), (2,8) if and only if $\mathbf{y}^*(t) = \boldsymbol{\lambda}^*\,\mathbf{F}(t)$ on $[0,1]$ and (2,3) holds (cf. 2.3 and its proof). In fact, since (2,3) has exactly $k^* = m - \operatorname{rank}(\mathbf{SU})$ linearly independent solutions, BVP (2,7), (2,8) has also exactly k^* linearly independent solutions on $[0,1]$. In particular, given an arbitrary $\Lambda_0 \in L(R_n, R_{k^*})$ whose rows form a basis in the space of all solutions to (2,3), the rows of $(\Lambda_0\,\mathbf{F}(t), \Lambda_0)$ form a basis in the space of all solutions to BVP (2,7), (2,8).

2.8. Remark. From the proof of 2.7 it follows that all the solutions to BVP (1,2), (1,4) or BVP (2,7), (2,8) are of the form

$$\mathbf{x}(t) = \mathbf{X}_0(t)\,\mathbf{d}\,, \quad \mathbf{d} \in R_k \qquad \text{or} \qquad (\mathbf{y}^*(t), \boldsymbol{\lambda}^*) = \boldsymbol{\delta}^*(\Lambda_0\,\mathbf{F}(t), \Lambda_0)\,, \quad \boldsymbol{\delta}^* \in R_{k^*}^*\,,$$

respectively. Furthermore, by the definition of $\mathbf{X}_0(t)$, Λ_0, $\mathbf{F}(t)$

$$\operatorname{rank}(\mathbf{X}_0(t)) \equiv k \quad \text{and} \quad \operatorname{rank}(\Lambda_0\,\mathbf{F}(t), \Lambda_0) \equiv k^* \quad \text{on } [0,1]\,.$$

2.9. Remark. The number $k^* - k = m - n$ is called the *index of BVP* (1,2), (1,4).

2.10. Remark. If we added one zero row to the matrices $\mathbf{M}$, $\mathbf{N}$, $\mathbf{K}(t)$ in (1,4), we should obtain the equivalent problem. Let us assume that it has exactly k linearly independent solutions. Then by 2.8 its adjoint should have exactly both $k + m - n$ and $k + (m + 1) - n$ linearly independent solutions. This seems to be confusing. But we must take into account that while in the former case the adjoint problem has solutions $(\mathbf{y}^*, \boldsymbol{\lambda}^*)$ with $\boldsymbol{\lambda}^* \in R_m^*$, in the latter case the adjoint has solutions $(\mathbf{y}^*, \boldsymbol{\mu}^*)$, where $\boldsymbol{\mu}^*$ is an $(m+1)$-vector, with an arbitrary last component. Nevertheless it can be seen that it is reasonable to remove from (1,4) all the linearly dependent rows and to consider the given BVP with linearly independent side conditions.

2.11. Remark. Given $\boldsymbol{x} \in AC_n$, $\boldsymbol{y}^* \in L_n^\infty$ and $\boldsymbol{\lambda}^* \in R_m^*$, we have by I.4.33

$$(2{,}15)\quad \int_0^1 \boldsymbol{y}^*(t)\,[\boldsymbol{x}'(t) - \boldsymbol{A}(t)\,\boldsymbol{x}(t)]\,\mathrm{d}t - \boldsymbol{\lambda}^*\left[\boldsymbol{M}\,\boldsymbol{x}(0) + \boldsymbol{N}\,\boldsymbol{x}(1) + \int_0^1 \mathrm{d}[\boldsymbol{K}(t)]\,\boldsymbol{x}(t)\right]$$
$$= \int_0^1 \left[\boldsymbol{y}^*(t) + \int_t^1 \boldsymbol{y}^*(s)\,\boldsymbol{A}(s)\,\mathrm{d}s + \boldsymbol{\lambda}^*(\boldsymbol{K}(t) - \boldsymbol{N})\right]\boldsymbol{x}'(t)\,\mathrm{d}t$$
$$+ \left[\int_0^1 \boldsymbol{y}^*(s)\,\boldsymbol{A}(s)\,\mathrm{d}s - \boldsymbol{\lambda}^*(\boldsymbol{M} + \boldsymbol{N} - \boldsymbol{K}(0))\right]\boldsymbol{x}(0)\,.$$

In particular, applying again I.4.33 to the right-hand side of (2,15), we obtain that

$$(2{,}16)\quad \int_0^1 \boldsymbol{y}^*(t)\,[\boldsymbol{x}'(t) - \boldsymbol{A}(t)\,\boldsymbol{x}(t)]\,\mathrm{d}t - \boldsymbol{\lambda}^*\left[\boldsymbol{M}\,\boldsymbol{x}(0) + \boldsymbol{N}\,\boldsymbol{x}(1) + \int_0^1 \mathrm{d}[\boldsymbol{K}(t)]\,\boldsymbol{x}(t)\right]$$
$$= \int_0^1 \mathrm{d}\left[-\boldsymbol{y}^*(t) + \boldsymbol{y}^*(1) - \int_t^1 \boldsymbol{y}^*(s)\,\boldsymbol{A}(s)\,\mathrm{d}s - \boldsymbol{\lambda}^*\,\boldsymbol{K}(t)\right]\boldsymbol{x}(t)$$
$$- [\boldsymbol{y}^*(0) + \boldsymbol{\lambda}^*\boldsymbol{M}]\,\boldsymbol{x}(0) + [\boldsymbol{y}^*(1) - \boldsymbol{\lambda}^*\boldsymbol{N}]\,\boldsymbol{x}(1) \qquad \text{for all } \boldsymbol{x} \in AC_n,\quad \boldsymbol{y} \in BV_n,\ \boldsymbol{\lambda}^* \in R_m^*\,.$$

The formulas (2,15) and (2,16) will be called the *Green formulas.*

The adjoint BVP (2,7), (2,8) is a system of equations for an n-vector valued function $\boldsymbol{y}^*(t)$ of bounded variation on $[0, 1]$ and an m-vector parameter. Our wish is now to disclose the relationship between $\boldsymbol{y}^*$ and $\boldsymbol{\lambda}^*$ if $(\boldsymbol{y}^*, \boldsymbol{\lambda}^*)$ solves BVP (2,7), (2,8). To this end it appears to be convenient to consider BVP (1,1), (1,3) with the side condition in its canonical form (see 1.16)

$$(2{,}17)\qquad \begin{aligned} \boldsymbol{M}_0\,\boldsymbol{x}(0) + \boldsymbol{N}_0\,\boldsymbol{x}(1) &= \boldsymbol{r}_0\,,\\ \boldsymbol{M}_1\,\boldsymbol{x}(0) + \boldsymbol{N}_1\,\boldsymbol{x}(1) + \int_0^1 \mathrm{d}[\boldsymbol{K}_1(t)]\,\boldsymbol{x}(t) &= \boldsymbol{r}_1\,,\\ \int_0^1 \mathrm{d}[\boldsymbol{K}_2(t)]\,\boldsymbol{x}(t) &= \boldsymbol{r}_2\,. \end{aligned}$$

In this case the adjoint BVP (2,7), (2,8) reduces to the system of equations for $\boldsymbol{y} \in BV_n$, $\boldsymbol{\varkappa}_0^* \in R_{m_0}^*$, $\boldsymbol{\varkappa}_1^* \in R_{m_1}^*$ and $\boldsymbol{\varkappa}^* \in R_{m_2}^*$

$$(2{,}18)\qquad \mathrm{d}\boldsymbol{y}^* = \boldsymbol{y}^*\,\mathrm{d}[-\boldsymbol{B}] - \mathrm{d}(\boldsymbol{\varkappa}_1^*\boldsymbol{K}_1 + \boldsymbol{\varkappa}^*\boldsymbol{K}_2) \qquad \text{on } [0, 1]\,,$$

$$(2{,}19)\qquad \boldsymbol{y}^*(0) + \boldsymbol{\varkappa}_0^*\boldsymbol{M}_0 + \boldsymbol{\varkappa}_1^*\boldsymbol{M}_1 = \boldsymbol{0}\,, \qquad \boldsymbol{y}^*(1) - \boldsymbol{\varkappa}_0^*\boldsymbol{N}_0 - \boldsymbol{\varkappa}_1^*\boldsymbol{N}_1 = \boldsymbol{0}\,.$$

2.12. Remark. Let $\boldsymbol{\Theta}$ be a regular $m \times m$-matrix such that

$$\boldsymbol{\Theta}[\boldsymbol{M}, \boldsymbol{N}, \boldsymbol{K}(t)] \equiv \begin{bmatrix} \boldsymbol{M}_0, & \boldsymbol{N}_0, & \boldsymbol{0} \\ \boldsymbol{M}_1, & \boldsymbol{N}_1, & \boldsymbol{K}_1(t) \\ \boldsymbol{0}, & \boldsymbol{0}, & \boldsymbol{K}_2(t) \end{bmatrix} \qquad \text{on } [0, 1]\,.$$

Given $\boldsymbol{\lambda}^* \in R_m^*$, let $\boldsymbol{\varkappa}_0^* \in R_{m_0}^*$, $\boldsymbol{\varkappa}_1^* \in R_{m_1}^*$ and $\boldsymbol{\varkappa}^* \in R_{m_2}^*$ be such that $\boldsymbol{\lambda}^* = (\boldsymbol{\varkappa}_0^*, \boldsymbol{\varkappa}_1^*, \boldsymbol{\varkappa}^*)\,\boldsymbol{\Theta}$.

Then

$$\boldsymbol{\lambda}^*\mathbf{M} = \boldsymbol{\varkappa}_0^*\mathbf{M}_0 + \boldsymbol{\varkappa}_1^*\mathbf{M}_1, \qquad \boldsymbol{\lambda}^*\mathbf{N} = \boldsymbol{\varkappa}_0^*\mathbf{N}_0 + \boldsymbol{\varkappa}_1^*\mathbf{N}_1$$

and

$$\boldsymbol{\lambda}^*\,\mathbf{K}(t) \equiv \boldsymbol{\varkappa}_1^*\,\mathbf{K}_1(t) + \boldsymbol{\varkappa}^*\,\mathbf{K}_2(t) \qquad \text{on } [0,1].$$

It follows that $\mathbf{y}^*\colon [0,1] \to R_n^*$ and $\boldsymbol{\lambda}^* \in R_m^*$ satisfy (2,7), (2,8) if and only if $(\mathbf{y}^*, \boldsymbol{\varkappa}_0^*, \boldsymbol{\varkappa}_1^*, \boldsymbol{\varkappa}^*)$, where $\boldsymbol{\varkappa}_0^* \in R_{m_0}^*$, $\boldsymbol{\varkappa}_0^* \in R_{m_1}^*$ and $\boldsymbol{\varkappa}^* \in R_{m_2}^*$ are such that $\boldsymbol{\lambda}^* = (\boldsymbol{\varkappa}_0^*, \boldsymbol{\varkappa}_1^*, \boldsymbol{\varkappa}^*)\,\boldsymbol{\Theta}$, satisfy (2,18), (2,19).

2.13. Notation. In the following $\mathbf{C}$ and $\mathbf{D}$ denote the $l \times n$-matrices such that $\mathbf{C}^* = [\mathbf{M}_0^*, \mathbf{M}_1^*]$ and $\mathbf{D}^* = [\mathbf{N}_0^*, \mathbf{N}_1^*]$ $(l = m_0 + m_1)$, $\mathbf{M}^c$ and $\mathbf{N}^c$ being arbitrary $(2n - l) \times n$-matrices complementary to $[\mathbf{C}, \mathbf{D}]$ (cf. III.5.11). Let $\mathbf{P}, \mathbf{Q} \in L(R_{2n-l}, R_n)$ and $\mathbf{P}^c, \mathbf{Q}^c \in L(R_l, R_n)$ be associated to $\mathbf{C}, \mathbf{D}, \mathbf{M}^c, \mathbf{N}^c$ by III.5.12.

Furthermore, let $\mathbf{P}_0^c, \mathbf{Q}_0^c \in L(R_{m_0}, R_n)$ and $\mathbf{P}_1^c, \mathbf{Q}_1^c \in L(R_{m_1}, R_n)$ be such that $\mathbf{P}^c = [\mathbf{P}_0^c, \mathbf{P}_1^c]$ and $\mathbf{Q}^c = [\mathbf{Q}_0^c, \mathbf{Q}_1^c]$. (By 1.16 rank $[\mathbf{C}, \mathbf{D}] = l$.)

Let us recall that according to III.5.12

$$(2{,}20) \qquad \begin{bmatrix} -\mathbf{M}_0, & \mathbf{N}_0 \\ -\mathbf{M}_1, & \mathbf{N}_1 \\ -\mathbf{M}^c, & \mathbf{N}^c \end{bmatrix} \begin{bmatrix} \mathbf{P}_0^c, & \mathbf{P}_1^c, & \mathbf{P} \\ \mathbf{Q}_0^c, & \mathbf{Q}_1^c, & \mathbf{Q} \end{bmatrix} = \begin{bmatrix} \mathbf{I}_{m_0}, & \mathbf{0}, & \mathbf{0} \\ \mathbf{0}, & \mathbf{I}_{m_1}, & \mathbf{0} \\ \mathbf{0}, & \mathbf{0}, & \mathbf{I}_{2n-l} \end{bmatrix}$$

and

$$(2{,}21) \qquad \begin{bmatrix} \mathbf{P}_0^c, & \mathbf{P}_1^c, & \mathbf{P} \\ \mathbf{Q}_0^c, & \mathbf{Q}_1^c, & \mathbf{Q} \end{bmatrix} \begin{bmatrix} \mathbf{M}_0, & \mathbf{N}_0 \\ \mathbf{M}_1, & \mathbf{N}_1 \\ \mathbf{M}^c, & \mathbf{N}^c \end{bmatrix} = \begin{bmatrix} -\mathbf{I}_n, & \mathbf{0} \\ \mathbf{0}, & \mathbf{I}_n \end{bmatrix}.$$

Analogously as in III.5.17 it is easy to show on the basis of (2,20) that (2,19) holds if and only if

$$(2{,}22) \qquad \begin{aligned} \mathbf{y}^*(0)\,\mathbf{P} \;+ \mathbf{y}^*(1)\,\mathbf{Q} &= \mathbf{0}, \\ \mathbf{y}^*(0)\,\mathbf{P}_0^c + \mathbf{y}^*(1)\,\mathbf{Q}_0^c &= \boldsymbol{\varkappa}_0^*, \\ \mathbf{y}^*(0)\,\mathbf{P}_1^c + \mathbf{y}^*(1)\,\mathbf{Q}_1^c &= \boldsymbol{\varkappa}_1^*. \end{aligned}$$

This implies that BVP (2,18), (2,19) is equivalent to the problem of determining $\mathbf{y}^*\colon [0,1] \to R_n^*$ and $\boldsymbol{\varkappa}^* \in R_{m_2}^*$ such that $\mathbf{y}^*$ is a solution to

$$(2{,}23) \qquad \mathrm{d}\mathbf{y}^* = \mathbf{y}^*\,\mathrm{d}[-\mathbf{B}] - \mathrm{d}[(\mathbf{y}^*(0)\,\mathbf{P}_1^c + \mathbf{y}^*(1)\,\mathbf{Q}_1^c)\,\mathbf{K}_1 + \boldsymbol{\varkappa}^*\mathbf{K}_2] \qquad \text{on } [0,1]$$

and

$$(2{,}24) \qquad \mathbf{y}^*(0)\,\mathbf{P} + \mathbf{y}^*(1)\,\mathbf{Q} = \mathbf{0}.$$

In particular, if $(\mathbf{y}^*, \boldsymbol{\varkappa}^*)$ is a solution to (2,23), (2,24) and $\boldsymbol{\varkappa}_0^* \in R_{m_0}^*$ and $\boldsymbol{\varkappa}_1^* \in R_{m_1}^*$ are given by (2,22), then $(\mathbf{y}^*, \boldsymbol{\varkappa}_0^*, \boldsymbol{\varkappa}_1^*, \boldsymbol{\varkappa}^*)$ is a solution to (2,18), (2,19). On the other hand, if $(\mathbf{y}^*, \boldsymbol{\varkappa}_0^*, \boldsymbol{\varkappa}_1^*, \boldsymbol{\varkappa}^*)$ is a solution to (2,18), (2,19), then $(\mathbf{y}^*, \boldsymbol{\varkappa}^*)$ solves (2,23), (2,24) and $\boldsymbol{\varkappa}_0^* \in R_{m_0}^*$ and $\boldsymbol{\varkappa}_1^* \in R_{m_1}^*$ are given by (2,22).

2.14. Corollary. *BVP* (1,1), (2,17) *has a solution if and only if*

$$\int_0^1 \mathbf{y}^*(t)\,\mathbf{f}(t)\,\mathrm{d}t = (\mathbf{y}^*(0)\,\mathbf{P}_0^c + \mathbf{y}^*(1)\,\mathbf{Q}_0^c)\,\mathbf{r}_0 + (\mathbf{y}^*(0)\,\mathbf{P}_1^c + \mathbf{y}^*(1)\,\mathbf{Q}_1^c)\,\mathbf{r}_1 + \boldsymbol{\varkappa}^*\mathbf{r}_2$$

for any solution $(\mathbf{y}^*, \boldsymbol{\varkappa}^*)$ *of BVP* (2,23), (2,24).

Proof follows immediately from 2.3 and (2,22).

2.15. Remark. It is easy to see that BVP (2,23), (2,24) also possesses exactly $k^* = m - \operatorname{rank}(\mathbf{SU})$ linearly independent solutions.

2.16. Remark. The $m_2 \times m_2$-matrix

$$\mathbf{T} = \int_0^1 \mathbf{K}_2(t)\,\mathbf{K}_2^*(t)\,\mathrm{d}t$$

is regular. In fact, if there were $\mathbf{d}^*\mathbf{T} = \mathbf{0}$ for some $\mathbf{d}^* \in R_{m_2}^*$, then we should have also $\mathbf{d}^*\mathbf{T}\mathbf{d} = 0$, i.e.

$$0 = \int_0^1 \mathbf{h}^*(t)\,\mathbf{h}(t)\,\mathrm{d}t = \sum_{j=1}^{n} \int_0^1 (h_j(t))^2\,\mathrm{d}t\,,$$

where $\mathbf{h}^*(t) = (h_1(t), h_2(t), \ldots, h_n(t)) = \mathbf{d}^*\,\mathbf{K}_2(t)$ is of bounded variation on $[0, 1]$. This may happen if and only if $\mathbf{h}^*(t) = \mathbf{d}^*\,\mathbf{K}_2(t) = \mathbf{0}$ a.e. on $[0, 1]$. Since by the assumption $\mathbf{K}_2$ is right-continuous on $[0, 1)$ and left-continuous at 1, we have even $\mathbf{d}^*\,\mathbf{K}_2(t) \equiv \mathbf{0}$ on $[0, 1]$ and in virtue of the linear independence on $[0, 1]$ of the rows in $\mathbf{K}_2(t)$, it is $\mathbf{d}^* = \mathbf{0}$.

Let us put

$$\mathbf{L}_2(t) = -\int_0^t \mathbf{K}_2^*(s)\,\mathbf{T}^{-1}\,\mathrm{d}s \quad \text{on } [0, 1]\,.$$

For $\mathbf{K}_2(1) = \mathbf{0}$ and $\mathbf{L}_2(0) = \mathbf{0}$, the integration-by-parts formula I.4.33 yields

$$(2,25) \qquad \int_0^1 \mathrm{d}[\mathbf{K}_2(t)]\,\mathbf{L}_2(t) = \left(\int_0^1 \mathbf{K}_2(t)\,\mathbf{K}_2^*(t)\,\mathrm{d}t\right)\mathbf{T}^{-1} = \mathbf{T}\mathbf{T}^{-1} = \mathbf{I}_{m_2}\,.$$

This enables us to express also the parameter $\boldsymbol{\varkappa}^*$ in (2,23), (2,24) in terms of $\mathbf{y}^*$. Let $(\mathbf{y}^*, \boldsymbol{\varkappa}^*)$ verify (2,23) on $[0, 1]$, then by (2,25)

$$\boldsymbol{\Phi}_2\mathbf{y} = \int_0^1 \mathrm{d}\left[\mathbf{y}^*(t) - \int_t^1 \mathbf{y}^*(s)\,\mathbf{A}(s)\,\mathrm{d}s - (\mathbf{y}^*(0)\,\mathbf{P}_1^c + \mathbf{y}^*(1)\,\mathbf{Q}_1^c)\,\mathbf{K}_1(t)\right]\mathbf{L}_2(t)$$

$$= \int_0^1 \mathrm{d}[\boldsymbol{\varkappa}^*\,\mathbf{K}_2(t)]\,\mathbf{L}_2(t) = \boldsymbol{\varkappa}^*\,.$$

The operator $\mathbf{y} \in BV_n \to \boldsymbol{\Phi}_2 \mathbf{y} \in R^*_{m_2}$ is linear and bounded. In fact, given $\mathbf{y} \in BV_n$,

$$|\boldsymbol{\Phi}_2 \mathbf{y}| \le [\mathrm{var}_0^1 \mathbf{y}^* + (\sup_{t\in[0,1]} |\mathbf{y}^*(t)|)] \left(\int_0^1 |\mathbf{A}(s)| \, \mathrm{d}s + (|\mathbf{P}_1^c| + |\mathbf{Q}_1^c|) \, \mathrm{var}_0^1 \mathbf{K}_1 \right) (\sup_{t\in[0,1]} |\mathbf{L}_2(t)|)$$

$$\le \left[1 + \int_0^1 |\mathbf{A}(s)| \, \mathrm{d}s + (|\mathbf{P}_1^c| + |\mathbf{Q}_1^c|) \, \mathrm{var}_0^1 \mathbf{K}_1 \right] \sup_{t\in[0,1]} |\mathbf{L}_2(t)| \, \|\mathbf{y}^*\|_{BV} .$$

The adjoint BVP (2,23), (2,24) to BVP (1,1), (2,17) may be thus written in the form

$$\mathrm{d}\mathbf{y}^* = \mathbf{y}^* \, \mathrm{d}[-\mathbf{B}] - \mathrm{d}[(\boldsymbol{\Phi}_1 \mathbf{y}) \, \mathbf{K}_1 - (\boldsymbol{\Phi}_2 \mathbf{y}) \, \mathbf{K}_2] \quad \text{on } [0,1],$$
$$\mathbf{y}^*(0) \, \mathbf{P} + \mathbf{y}^*(1) \, \mathbf{Q} = \mathbf{0},$$

where $\boldsymbol{\Phi}_j$: $BV_n \to R^*_{m_j}$ $(j = 1, 2)$ are known linear bounded operators ($\boldsymbol{\Phi}_1 \mathbf{y} = \mathbf{y}^*(0) \, \mathbf{P}_1^c + \mathbf{y}^*(1) \, \mathbf{Q}_1^c$).

3. Generalized Green's functions

Let us continue the investigation of BVP (1,1), (1,3). In addition to 1.1 we assume throughout the paragraph that 1.10 holds ($\mathbf{K}$ is right-continuous on $[0,1)$ and left-continuous at 1 and $\mathbf{K}(1) = 0$).

Let $\mathscr{L}$ denote the linear bounded operator

$$\text{(3,1)} \qquad \mathscr{L}: \mathbf{x} \in AC_n \to \begin{bmatrix} \mathbf{x}'(t) - \mathbf{A}(t) \, \mathbf{x}(t) \\ \mathbf{M} \, \mathbf{x}(0) + \mathbf{N} \, \mathbf{x}(1) + \int_0^1 \mathrm{d}[\mathbf{K}(t)] \, \mathbf{x}(t) \end{bmatrix} \in L_n^1 \times R_m$$

(cf. 1.6). It may be shown from 2.3 that its range $R(\mathscr{L})$ is closed in $L_n^1 \times R_m$ and consequently $R(\mathscr{L})$ equipped with the norm of $L_n^1 \times R_m$ becomes a Banach space. We shall show this fact directly, without making use of Theorem 2.3. The symbols $\mathbf{U}$, $\mathbf{V}$ are again defined by (1,11) and (1,12).

3.1. Theorem. *The range $R(\mathscr{L})$ of the operator (3,1) is closed in $L_n^1 \times R_m$.*

Proof. A couple $\begin{pmatrix} \mathbf{f} \\ \mathbf{r} \end{pmatrix} \in L_n^1 \times R_m$ belongs to $R(\mathscr{L})$ if and only if (2,1) has a solution $\mathbf{c} \in R_n$, i.e. if and only if $\mathbf{r} - (\mathbf{SV}) \, \mathbf{f} \in R(\mathbf{SU})$. $R(\mathbf{SU})$ being finite dimensional, it is closed. Since

$$\mathbf{W}: \begin{pmatrix} \mathbf{f} \\ \mathbf{r} \end{pmatrix} \in L_n^1 \times R_m \to \mathbf{r} - (\mathbf{SV}) \, \mathbf{f} \in R_m$$

is a continuous operator, the set $\mathbf{W}_{-1}(R(\mathbf{SU})) = R(\mathscr{L})$ of all $\begin{pmatrix} \mathbf{f} \\ \mathbf{r} \end{pmatrix} \in L_n^1 \times R_m$ such that $\mathbf{W} \begin{pmatrix} \mathbf{f} \\ \mathbf{r} \end{pmatrix} \in R(\mathbf{SU})$ is also closed.

The concept of the generalized inverse matrix introduced in the Section I.2 and in particular theorems I.2.6 and I.2.7 enables us to give the necessary and sufficient condition for the existence of a solution to BVP (1,1), (1,3) in the following form.

3.2. Theorem. *BVP* (1,1), (1,3) *possesses a solution if and only if*

$$[\mathbf{I}_m - (\mathbf{SU})(\mathbf{SU})^{\#}][\mathbf{r} - (\mathbf{SV})\mathbf{f}] = \mathbf{0},$$

where $(\mathbf{SU})^{\#}$ *is the generalized inverse matrix to* $(\mathbf{SU})$. *If this condition is satisfied, then any solution* $\mathbf{x}$ *of BVP* (1,1), (1,3) *is of the form*

(3,2)
$$\mathbf{x}(t) = \mathbf{X}(t)[\mathbf{I}_n - (\mathbf{SU})^{\#}(\mathbf{SU})]\mathbf{d} + \mathbf{X}(t)(\mathbf{SU})^{\#}[\mathbf{r} - (\mathbf{SV})\mathbf{f}] + (\mathbf{Vf})(t) \quad \text{on } [0,1],$$

where $\mathbf{d} \in R_n$ *may be arbitrary.*

Proof follows by I.2.6 and I.2.7 from the equivalence between BVP (1,1), (1,3) and the equation (2,1) $((\mathbf{SU})\mathbf{c} = \mathbf{r} - (\mathbf{SV})\mathbf{f})$.

3.3. Remark. By 2.3 the homogeneous BVP (1,2), (1,4) has only the trivial solution if and only if $\operatorname{rank}(\mathbf{SU}) = n$. Consequently, BVP (1,1), (1,3) is uniquely solvable for any $\begin{pmatrix}\mathbf{f}\\ \mathbf{r}\end{pmatrix} \in R(\mathscr{L})$ if and only if $\operatorname{rank}(\mathbf{SU}) = n$.

On the other hand, BVP (1,1), (1,3) has a solution for any $\mathbf{f} \in L_n^1$ and $\mathbf{r} \in R_m$ if and only if $(\mathbf{SU})\mathbf{c} = \mathbf{q}$ is solvable for any $\mathbf{q} \in R_m$. ((2,1) has to be solvable for any $\mathbf{r} \in R_m$ and $\mathbf{f}(t) \equiv \mathbf{0}$ on $[0,1]$.) This holds if and only if (2,3) has only the trivial solution, i.e. if and only if $\operatorname{rank}(\mathbf{SU}) = m$.

In particular, BVP (1,1), (1,3) has a unique solution for any $\mathbf{f} \in L_n^1$ and $\mathbf{r} \in R_m$ if and only if $m = n$ and $\det(\mathbf{SU}) \neq 0$.

3.4. Theorem. *Let BVP* (1,1), (1,3) *have a solution. Then all its solutions are of the form*

(3,3)
$$\mathbf{x}(t) = \mathbf{x}_0(t) + \mathbf{H}_0(t)\mathbf{r} + \int_0^1 \mathbf{G}_0(t,s)\mathbf{f}(s)\,\mathrm{d}s \quad \text{on } [0,1],$$

where $\mathbf{x}_0(t) = \mathbf{X}(t)[\mathbf{I} - (\mathbf{SU})^{\#}(\mathbf{SU})]\mathbf{d}$ $(\mathbf{d} \in R_n)$ *is an arbitrary solution to the homogeneous BVP* (1,2), (1,4),

(3,4)
$$\mathbf{H}_0(t) = \mathbf{X}(t)(\mathbf{SU})^{\#} \quad \text{for } t \in [0,1],$$

$$\mathbf{G}_0(t,s) = \mathbf{X}(t)\Delta(t,s)\mathbf{X}^{-1}(s) - \mathbf{X}(t)(\mathbf{SU})^{\#}\mathbf{F}(s) \quad \text{for } t, s \in [0,1],$$

$$\Delta(t,s) = \mathbf{0} \quad \text{for } t < s, \qquad \Delta(t,s) = \mathbf{I}_n \quad \text{for } t \geq s$$

and

(3,5)
$$\mathbf{F}(s) = \left[\mathbf{N}\mathbf{X}(1) + \int_s^1 \mathrm{d}[\mathbf{K}(\tau)]\,\mathbf{X}(\tau)\right]\mathbf{X}^{-1}(s) \quad \text{for } s \in [0,1].$$

Proof follows immediately from 3.2 and (2,5), (2,6).

3.5. Remark. Let us notice that the representations (3,2) or (3,3) of the solutions to BVP (1,1), (1,3) are true even if the generalized inverse matrix $(\boldsymbol{SU})^{\#}$ to $(\boldsymbol{SU})$ is replaced by an arbitrary $n \times m$-matrix $\boldsymbol{B}$ such that $(\boldsymbol{SU})\,\boldsymbol{B}(\boldsymbol{SU}) = (\boldsymbol{SU})$ (see I.2.11).

3.6. Lemma. *The $n \times m$-matrix valued function $\boldsymbol{H}(t) = \boldsymbol{H}_0(t)$ and the $n \times n$-matrix valued function $\boldsymbol{G}(t, s) = \boldsymbol{G}_0(t, s)$ defined by (3,4) possess the following properties*

(i) *$\boldsymbol{H}(t)$ is absolutely continuous on $[0, 1]$,*

(ii) *$\boldsymbol{G}(t, s)$ is measurable in (t, s) on $[0, 1] \times [0, 1]$, $\operatorname{var}_0^1 \boldsymbol{G}(., s) < \infty$ for a.e. $s \in [0, 1]$ and $\boldsymbol{G}(t, .)$ is for any $t \in [0, 1]$ measurable and essentially bounded on $[0, 1]$,*

(iii) *$\gamma(s) = |\boldsymbol{G}(0, s)| + \operatorname{var}_0^1 \boldsymbol{G}(., s)$ is measurable and essentially bounded on $[0, 1]$,*

(iv) *$\boldsymbol{G}(t, s) = \boldsymbol{G}_a(t, s) - \boldsymbol{G}_b(t, s)$ on $[0, 1] \times [0, 1]$, where for any $s \in [0, 1]$ $\boldsymbol{G}_a(., s)$ is absolutely continuous on $[0, 1]$ and $\boldsymbol{G}_b(., s)$ is a simple jump function with the jump $\boldsymbol{I}_n$ at $t = s$.*

Proof. The assertions (i) and (ii) are obvious. Furthermore, $\boldsymbol{F}$ is of bounded variation on $[0, 1]$ and for any $s \in [0, 1]$

$$\begin{aligned}\gamma(s) &\le |\boldsymbol{X}^{-1}(s)| + |(\boldsymbol{SU})^{\#}|\,|\boldsymbol{F}(s)| + (\operatorname{var}_0^1 \boldsymbol{X})\,(|\boldsymbol{X}^{-1}(s)| + |(\boldsymbol{SU})^{\#}|\,|\boldsymbol{F}(s)|)\\ &\le (1 + \operatorname{var}_0^1 \boldsymbol{X}) \sup_{s \in [0,1]} (|\boldsymbol{X}^{-1}(s)| + |(\boldsymbol{SU})^{\#}|\,|\boldsymbol{F}(s)|) = \varkappa < \infty .\end{aligned}$$

The last assertion is proved by putting $\boldsymbol{G}_a(t, s) = \boldsymbol{G}_0(t, s)$ if $t \ge s$, $\boldsymbol{G}_a(t, s) = \boldsymbol{G}_0(t, s) + \boldsymbol{I}_n$ if $t < s$ and $\boldsymbol{G}_b(t, s) = \boldsymbol{0}$ if $t \ge s$, $\boldsymbol{G}_b(t, s) = \boldsymbol{I}_n$ if $t < s$.

3.7. Remark. Let us notice that actually we have proved that $\gamma(s)$ is bounded on $[0, 1]$ and hence also $\boldsymbol{G}_0(t, s)$ is bounded on $[0, 1] \times [0, 1]$ ($|\boldsymbol{G}_0(t, s)| \le \gamma(s) \le \varkappa < \infty$ on $[0, 1] \times [0, 1]$). Moreover, by (3,4) $\operatorname{var}_0^1 \boldsymbol{G}_0(t, .) + \operatorname{var}_0^1 \boldsymbol{G}_0(., s) < \infty$ for all $t, s \in [0, 1]$.

3.8. Lemma. *Let $\boldsymbol{H}\colon [0, 1] \to L(R_m, R_n)$ and $\boldsymbol{G}\colon [0, 1] \times [0, 1] \to L(R_n)$ fulfil (i)–(iii) from 3.6. Then for any couple $\begin{pmatrix}\boldsymbol{f}\\ \boldsymbol{r}\end{pmatrix} \in L_n^1 \times R_m$ the n-vector valued function*

$$\boldsymbol{h}(t) = \boldsymbol{H}(t)\,\boldsymbol{r} + \int_0^1 \boldsymbol{G}(t, s)\,\boldsymbol{f}(s)\,\mathrm{d}s$$

is of bounded variation on $[0, 1]$ and the linear operator

$$\mathcal{M}\colon \begin{pmatrix}\boldsymbol{f}\\ \boldsymbol{r}\end{pmatrix} \in L_n^1 \times R_m \to \boldsymbol{h} \in BV_n$$

is bounded.

Proof. Given $\boldsymbol{f}\in L^1_n$, $\boldsymbol{r}\in R_m$ and a subdivision $\{0=t_0<t_1<\ldots<t_p=1\}$ of $[0,1]$,

$$\sum_{j=1}^{p}\left|\int_0^1 \boldsymbol{G}(t_j,s)\,\boldsymbol{f}(s)\,\mathrm{d}s-\int_0^1 \boldsymbol{G}(t_{j-1},s)\,\boldsymbol{f}(s)\,\mathrm{d}s\right|+\left|\int_0^1 \boldsymbol{G}(0,s)\,\boldsymbol{f}(s)\,\mathrm{d}s\right|$$
$$\leq \int_0^1\left(\sum_{j=1}^{p}|\boldsymbol{G}(t_j,s)-\boldsymbol{G}(t_{j-1},s)|+|\boldsymbol{G}(0,s)|\right)|\boldsymbol{f}(s)|\,\mathrm{d}s$$
$$\leq \int_0^1 \gamma(s)\,|\boldsymbol{f}(s)|\,\mathrm{d}s\leq(\sup_{s\in[0,1]}|\gamma(s)|)\,\|\boldsymbol{f}\|_{L^1}\leq \varkappa\|\boldsymbol{f}\|_{L^1}\,.$$

Hence also

$$\operatorname{var}_0^1\left(\int_0^1 \boldsymbol{G}(.,s)\,\boldsymbol{f}(s)\,\mathrm{d}s\right)+\left|\int_0^1 \boldsymbol{G}(0,s)\,\boldsymbol{f}(s)\,\mathrm{d}s\right|\leq \varkappa\|\boldsymbol{f}\|_{L^1}$$

and

$$\left\|\mathscr{M}\begin{pmatrix}\boldsymbol{f}\\ \boldsymbol{r}\end{pmatrix}\right\|_{BV}\leq c(\|\boldsymbol{f}\|_{L^1}+|\boldsymbol{r}|)=c\left\|\begin{pmatrix}\boldsymbol{f}\\ \boldsymbol{r}\end{pmatrix}\right\|_{L^1\times R}$$

where

$$c=\varkappa+|\boldsymbol{H}(0)|+\int_0^1|\boldsymbol{H}'(t)|\,\mathrm{d}t<\infty\,.$$

3.9. Remark. Let the operator $\mathscr{L}^{\oplus}$ be defined by

$$\mathscr{L}^{\oplus}\colon \begin{pmatrix}\boldsymbol{f}\\ \boldsymbol{r}\end{pmatrix}\in L^1_n\times R_m\to \boldsymbol{H}_0(t)\,\boldsymbol{r}+\int_0^1 \boldsymbol{G}_0(t,s)\,\boldsymbol{f}(s)\,\mathrm{d}s\in AC_n\,, \tag{3,6}$$

where the matrix valued functions $\boldsymbol{G}_0(t,s)$ and $\boldsymbol{H}_0(t)$ are given by (3,4) ($R(\mathscr{L}^{\oplus})\subset AC_n$ due to 3.4). According to (3,2) and (3,4)

$$\mathscr{L}^{\oplus}\begin{pmatrix}\boldsymbol{f}\\ \boldsymbol{r}\end{pmatrix}=\boldsymbol{U}(\boldsymbol{SU})^{\#}\,(\boldsymbol{r}-\boldsymbol{SVf})+\boldsymbol{Vf}=\boldsymbol{U}(\boldsymbol{SU})^{\#}\,\boldsymbol{r}+\boldsymbol{Vf}-\boldsymbol{U}(\boldsymbol{SU})^{\#}\,(\boldsymbol{SV})\,\boldsymbol{f}$$

for any $\boldsymbol{f}\in L^1_n$ and $\boldsymbol{r}\in R_m$. Consequently $\mathscr{L}^{\oplus}$ is linear and bounded (cf. also 3.6 and 3.8). Moreover, for any $\boldsymbol{f}\in L^1_n$ and $\boldsymbol{r}\in R_m$ such that BVP (1,1), (1,3) has a solution $\left(\begin{pmatrix}\boldsymbol{f}\\ \boldsymbol{r}\end{pmatrix}\in R(\mathscr{L})\right)$ $\mathscr{L}\mathscr{L}^{\oplus}\begin{pmatrix}\boldsymbol{f}\\ \boldsymbol{r}\end{pmatrix}=\begin{pmatrix}\boldsymbol{f}\\ \boldsymbol{r}\end{pmatrix}$ and hence $\mathscr{L}\mathscr{L}^{\oplus}\mathscr{L}\boldsymbol{x}=\mathscr{L}\boldsymbol{x}$ for any $\boldsymbol{x}\in AC_n$, i.e. $\mathscr{L}\mathscr{L}^{\oplus}\mathscr{L}=\mathscr{L}$ ($\mathscr{L}^{\oplus}$ is a generalized inverse operator to $\mathscr{L}$).

In particular, if $m=n$ and rank $(\boldsymbol{SU})=n$, then by 3.3 $\mathscr{L}^{\oplus}$ becomes a bounded inverse operator to $\mathscr{L}$. In this case the functions $\boldsymbol{G}_0(t,s)$, $\boldsymbol{H}_0(t)$ are called the Green couple of BVP (1,1), (1,3) (or (1,2), (1,4)), while the function $\boldsymbol{G}_0(t,s)$ is the Green function of BVP (1,1), (1,3).

3.10. Definition. A couple $\boldsymbol{G}(t,s)$, $\boldsymbol{H}(t)$ of matrix valued functions fulfilling (i)–(iii) of 3.6 is called the *generalized Green couple* if for all $\boldsymbol{f}\in L^1_n$, $\boldsymbol{r}\in R_m$ such that BVP

(1,1), (1,3) has a solution, the function

$$\mathbf{x}(t) = \mathbf{H}(t)\,\mathbf{r} + \int_0^1 \mathbf{G}(t,s)\,\mathbf{f}(s)\,\mathrm{d}s$$

is also a solution to BVP (1,1), (1,3).

3.11. Remark. By 3.4 and 3.6 the couple $\mathbf{G}_0(t,s)$, $\mathbf{H}_0(t)$ given by (3,4) is a generalized Green couple of BVP (1,1), (1,3).

3.12. Theorem. *A linear bounded operator $\mathscr{L}^+\colon R(\mathscr{L}) \to AC_n$ fulfils $\mathscr{L}\mathscr{L}^+\mathscr{L} = \mathscr{L}$ if and only if there exists a generalized Green couple $\mathbf{G}(t,s)$, $\mathbf{H}(t)$ such that $\mathscr{L}^+$ is given by*

$$\mathscr{L}^+\colon \begin{pmatrix}\mathbf{f}\\ \mathbf{r}\end{pmatrix} \in R(\mathscr{L}) \to \mathbf{H}(t)\,\mathbf{r} + \int_0^1 \mathbf{G}(t,s)\,\mathbf{f}(s)\,\mathrm{d}s \in AC_n\,. \tag{3,7}$$

Proof. Let $\mathscr{L}\mathscr{L}^+\mathscr{L} = \mathscr{L}$ and let $\mathscr{L}^{\oplus}$ be given by (3,6). According to 3.9 $\mathscr{L}(\mathscr{L}^+ - \mathscr{L}^{\oplus})\begin{pmatrix}\mathbf{f}\\ \mathbf{r}\end{pmatrix} = \mathbf{0}$, i.e. $(\mathscr{L}^+ - \mathscr{L}^{\oplus})\begin{pmatrix}\mathbf{f}\\ \mathbf{r}\end{pmatrix} \in N(\mathscr{L})$ for each $\mathbf{f} \in L^1_n$ and $\mathbf{r} \in R_m$. In particular, $\mathscr{L}^+ = \mathscr{L}^{\oplus}$ on $R(\mathscr{L})$ if $N(\mathscr{L}) = \{\mathbf{0}\}$. If $k = \dim N(\mathscr{L}) = n - \operatorname{rank}(\mathbf{SU})$ > 0, let $\mathbf{X}_0(t)$ be defined as in the proof of 2.7. Then $\operatorname{rank}(\mathbf{X}_0(t)) = k$ on $[0,1]$ (cf. 2.8) and given $\begin{pmatrix}\mathbf{f}\\ \mathbf{r}\end{pmatrix} \in R(\mathscr{L})$, there exists $\mathbf{d} \in R_k$ such that

$$(\mathscr{L}^+ - \mathscr{L}^{\oplus})\begin{pmatrix}\mathbf{f}\\ \mathbf{r}\end{pmatrix}(t) \equiv \mathbf{X}_0(t)\,\mathbf{d} \qquad \text{on } [0,1]\,. \tag{3,8}$$

By I.2.6, I.2.7 and I.2.15 this is possible if and only if

$$\mathbf{d} \equiv \mathbf{X}_0^{\#}(t)\,(\mathscr{L}^+ - \mathscr{L}^{\oplus})\begin{pmatrix}\mathbf{f}\\ \mathbf{r}\end{pmatrix}(t) \qquad \text{on } [0,1]\,.$$

By the definition $\mathbf{X}_0(t) = \mathbf{X}(t)\,\mathbf{C}_0$ on $[0,1]$, where $\mathbf{C}_0 \in L(R_k, R_n)$ has a full rank ($\operatorname{rank}(\mathbf{C}_0) = k$). According to 2.16 $\mathbf{X}_0^{\#}(t) = \mathbf{C}_0^{\#}\,\mathbf{X}^{-1}(t)$. It follows immediately that the mapping

$$\boldsymbol{\Phi}\colon \begin{pmatrix}\mathbf{f}\\ \mathbf{r}\end{pmatrix} \in R(\mathscr{L}) \to \mathbf{d} = \mathbf{X}_0^{\#}(t)\,(\mathscr{L}^+ - \mathscr{L}^{\oplus})\begin{pmatrix}\mathbf{f}\\ \mathbf{r}\end{pmatrix}(t) \in R_k$$

is a linear bounded vector valued functional on $R(\mathscr{L})$. Let $\boldsymbol{\Psi}$ be its arbitrary extension on the whole space $L^1_n \times R_m$. ($\boldsymbol{\Psi}$ is defined and bounded on $L^1_n \times R_m$ and $\boldsymbol{\Psi} = \boldsymbol{\Phi}$ on $R(\mathscr{L})$.) Then there exist a function $\boldsymbol{\Theta}_1\colon [0,1] \to L(R_n, R_k)$ essentially bounded and measurable on $[0,1]$ and $\boldsymbol{\Theta}_2 \in L(R_m, R_k)$ such that

$$\boldsymbol{\Psi}\begin{pmatrix}\mathbf{f}\\ \mathbf{r}\end{pmatrix} = \int_0^1 \boldsymbol{\Theta}_1(s)\,\mathbf{f}(s)\,\mathrm{d}s + \boldsymbol{\Theta}_2\mathbf{r} \qquad \text{for all } \begin{pmatrix}\mathbf{f}\\ \mathbf{r}\end{pmatrix} \in L^1_n \times R_m\,.$$

Together with (3,6) and (3,8) this yields (3,7), where

$$(3,9)\quad \mathbf{G}(t,s) = \mathbf{G}_0(t,s) + \mathbf{X}_0(t)\,\boldsymbol{\Theta}_1(s) \qquad \text{for all} \quad t\in[0,1] \quad \text{and a.e.} \ s\in[0,1],$$
$$\mathbf{H}(t) = \mathbf{H}_0(t) + \mathbf{X}_0(t)\,\boldsymbol{\Theta}_2 \qquad \text{on } [0,1]$$

obviously fulfil the conditions (i)–(iii) of 3.6.

The proof will be completed by taking into account the obvious fact that if $\mathbf{G}(t, s)$, $\mathbf{H}(t)$ is a generalized Green couple, then the operator (3,7) fulfils $\mathscr{L}\mathscr{L}^{+}\mathscr{L} = \mathscr{L}$.

3.13. Proposition. *A couple* $(\mathbf{z}^*, \boldsymbol{\lambda}^*) \in L_n^\infty \times R_m^*$ *fulfils*

$$(3,10)\qquad \int_0^1 \mathbf{z}^*(t)\,\mathbf{f}(t)\,\mathrm{d}t = \boldsymbol{\lambda}^*\mathbf{r} \qquad \textit{for all} \quad \begin{pmatrix}\mathbf{f}\\ \mathbf{r}\end{pmatrix} \in R(\mathscr{L})$$

if and only if there exists $\mathbf{y}^*\colon [0,1] \to R_n^*$ *such that* $\mathbf{y}^*(t) = \mathbf{z}^*(t)$ *a.e. on* $[0,1]$ *and* $(\mathbf{y}^*, \boldsymbol{\lambda}^*)$ *is a solution of BVP* (2,7), (2,8).

Proof. Let $\mathbf{z}^* \in L_n^\infty$ and $\boldsymbol{\lambda}^* \in R_m^*$. Then by the Green formula (2,15), (3,10) holds if and only if for any $\mathbf{x} \in AC_n$

$$(3,11)\qquad \int_0^1 \left[\mathbf{z}^*(t) + \int_t^1 \mathbf{z}^*(s)\,\mathbf{A}(s)\,\mathrm{d}s + \boldsymbol{\lambda}^*(\mathbf{K}(t) - \mathbf{N})\right] \mathbf{x}'(t)\,\mathrm{d}t$$
$$+ \left[\int_0^1 \mathbf{z}^*(s)\,\mathbf{A}(s)\,\mathrm{d}s - \boldsymbol{\lambda}^*(\mathbf{M} + \mathbf{N} - \mathbf{K}(0))\right] \mathbf{x}(0) = 0\,.$$

In particular, if $\mathbf{x}(t) \equiv \mathbf{x}(0)$ on $[0,1]$, (3,11) means that

$$\left[\int_0^1 \mathbf{z}^*(s)\,\mathbf{A}(s)\,\mathrm{d}s - \boldsymbol{\lambda}^*(\mathbf{M} + \mathbf{N} - \mathbf{K}(0))\right] \mathbf{c} = 0$$

for each $\mathbf{c} \in R_n$, i.e.

$$(3,12)\qquad \int_0^1 \mathbf{z}^*(s)\,\mathbf{A}(s)\,\mathrm{d}s = \boldsymbol{\lambda}^*(\mathbf{M} + \mathbf{N} - \mathbf{K}(0))\,.$$

Consequently (3,11) holds for each $\mathbf{x} \in AC_n$ if and only if

$$\int_0^1 \left[\mathbf{z}^*(t) + \int_t^1 \mathbf{z}^*(s)\,\mathbf{A}(s)\,\mathrm{d}s + \boldsymbol{\lambda}^*(\mathbf{K}(t) - \mathbf{N})\right] \mathbf{v}(t)\,\mathrm{d}t = 0 \qquad \text{for any} \quad \mathbf{v} \in L_n^1$$

or $\mathbf{z}^*(t) = \mathbf{u}^*(t)$ a.e. on $[0,1]$, where

$$(3,13)\qquad \mathbf{u}^*(t) = -\int_t^1 \mathbf{z}^*(s)\,\mathbf{A}(s)\,\mathrm{d}s - \boldsymbol{\lambda}^*(\mathbf{K}(t) - \mathbf{N}) \qquad \text{on } [0,1]\,.$$

Let us put $\mathbf{y}^*(t) \equiv \mathbf{u}^*(t)$ on $(0,1)$, $\mathbf{y}^*(0) = \mathbf{u}^*(0+)$ and $\mathbf{y}^*(1) = \mathbf{u}^*(1-)$. Then owing to (3,13) and (3,12)

$$\mathbf{y}^*(1) = \boldsymbol{\lambda}^*\mathbf{N} \quad \text{and} \quad \mathbf{y}^*(0) = -\boldsymbol{\lambda}^*\mathbf{M}$$

and $(\mathbf{z}^*, \boldsymbol{\lambda}^*)$ fulfils (3,10) if and only if $\mathbf{z}^*(t) = \mathbf{y}^*(t)$ a.e. on $[0, 1]$. The proof will be completed by taking into account that

$$\int_t^1 \mathbf{y}^*(s)\, \mathbf{A}(s)\, ds = \int_t^1 \mathbf{z}^*(s)\, \mathbf{A}(s)\, ds \qquad \text{for any} \quad t \in [0, 1]$$

and hence

$$\mathbf{y}^*(t) = \mathbf{y}^*(1) + \int_t^1 \mathbf{y}^*(s)\, \mathbf{A}(s)\, ds + \boldsymbol{\lambda}^*\, \mathbf{K}(t) \qquad \text{on } [0, 1].$$

The latter implication follows from 2.3.

The set of all generalized Green couples is characterized in the following theorem.

If $k = n - \text{rank}\,(\mathbf{SU}) > 0$, then $\mathbf{C}_0$ is an arbitrary $n \times k$-matrix whose columns form a basis in the space of all solutions to $(\mathbf{SU})\,\mathbf{c} = \mathbf{0}$ and $\mathbf{X}_0(t) = \mathbf{X}(t)\,\mathbf{C}_0$ on $[0, 1]$.

If $k^* = m - \text{rank}\,(\mathbf{SU}) > 0$, then Λ_0 is an arbitrary $k^* \times n$-matrix whose rows form a basis in the space of all solutions to $\boldsymbol{\lambda}^*(\mathbf{SU}) = \mathbf{0}$ and $\mathbf{Y}_0(t) = \Lambda_0\, \mathbf{F}(t)$ on $[0, 1]$, where $\mathbf{F}(t)$ is given by (3,5).

3.14. Theorem. *A couple* $\mathbf{G}\colon [0, 1] \times [0, 1] \to L(R_n)$, $\mathbf{H}\colon [0, 1] \to L(R_m, R_n)$ *is a generalized Green couple to BVP* (1,1), (1,3) *if and only if there exist a function* $\Theta_1\colon [0, 1] \to L(R_n, R_k)$ *essentially bounded and measurable on* $[0, 1]$, *a function* $\Sigma\colon [0, 1] \to L(R_{k^*}, R_n)$ *of bounded variation on* $[0, 1]$ *and* $\Theta_2 \in L(R_m, R_k)$ *such that*

$$\mathbf{G}(t, s) = \mathbf{G}_0(t, s) + \mathbf{X}_0(t)\, \Theta_1(s) + \Sigma(t)\, \mathbf{Y}_0(s) \tag{3,14}$$

$$\textit{for all} \quad t \in [0, 1] \quad \textit{and a.e.} \quad s \in [0, 1],$$

$$\mathbf{H}(t) = \mathbf{H}_0(t) + \mathbf{X}_0(t)\, \Theta_2 - \Sigma(t)\, \Lambda_0 \qquad \textit{on } [0, 1],$$

where $\mathbf{G}_0(t, s)$ *and* $\mathbf{H}_0(t)$ *are given by* (3,4), *the terms* $\mathbf{X}_0(t)\, \Theta_1(s)$ *and* $\mathbf{X}_0(t)\, \Theta_2$ *vanish if* $k = 0$ *and the terms* $\Sigma(t)\, \mathbf{Y}_0(s)$ *and* $\Sigma(t)\, \Lambda_0$ *vanish if* $k^* = 0$.

Proof. Let us assume that $k > 0$ and $k^* > 0$.

(a) Let $\mathbf{G}(t, s)$, $\mathbf{H}(t)$ be a generalized Green couple of BVP (1,1), (1,3). Then by 3.12 and its proof there exist $\Theta_1\colon [0, 1] \to L(R_n, R_k)$ essentially bounded on $[0, 1]$ and $\Theta_2 \in L(R_m, R_k)$ such that for all $\begin{pmatrix} \mathbf{f} \\ \mathbf{r} \end{pmatrix} \in R(\mathscr{L})$

$$\mathbf{H}(t)\, \mathbf{r} + \int_0^1 \mathbf{G}(t, s)\, \mathbf{f}(s)\, ds$$

$$= [\mathbf{H}_0(t) + \mathbf{X}_0(t)\, \Theta_2]\, \mathbf{r} + \int_0^1 [\mathbf{G}_0(t, s) + \mathbf{X}_0(t)\, \Theta_1(s)]\, \mathbf{f}(s)\, ds \qquad \text{on } [0, 1].$$

By 3.13 and 2.8 this holds if and only if there exists $\Sigma\colon [0, 1] \to L(R_{k^*}, R_n)$ such that (3,14) holds. According to 2.8 and I.2.15 $[\mathbf{Y}_0(s), \Lambda_0]\, [\mathbf{Y}_0(s), \Lambda_0]^{\#} = \mathbf{I}_{k^*}$ for any $s \in [0, 1]$. The functions $\mathbf{P}(t, s) = \mathbf{G}(t, s) - \mathbf{G}_0(t, s) - \mathbf{X}_0(t)\, \Theta_1(s)$ and $\mathbf{Q}(t)$

$= -\mathbf{H}(t) + \mathbf{H}_0(t) + \mathbf{X}_0(t)\,\boldsymbol{\Theta}_2$ as functions of t for a.e. $s \in [0,1]$ are of bounded variation on $[0,1]$. Therefore the function $\boldsymbol{\Sigma}(t) = [\mathbf{P}(t,s), \mathbf{Q}(s)]\,[\mathbf{Y}_0(s), \boldsymbol{\Lambda}_0]^{\#}$ for all $t \in [0,1]$ and a.e. $s \in [0,1]$ (cf. I.2.6 and I.2.7) has also a bounded variation on $[0,1]$.

(b) Let $\boldsymbol{\Theta}_1\colon [0,1] \to L(R_n, R_k)$ be essentially bounded on $[0,1]$, $\boldsymbol{\Theta}_2 \in L(R_m, R_k)$ and let $\boldsymbol{\Sigma}\colon [0,1] \to L(R_{k*}, R_n)$ be of bounded variation on $[0,1]$. Then the functions $\mathbf{G}(t,s)$, $\mathbf{H}(t)$ given by (3,14) are sure to fulfil (i)–(iii) from 3.6 and since by 2.3

$$\int_0^1 \mathbf{Y}_0(t)\,\mathbf{f}(t)\,\mathrm{d}t = \boldsymbol{\Lambda}_0\mathbf{r} \qquad \text{for all} \quad \begin{pmatrix}\mathbf{f}\\ \mathbf{r}\end{pmatrix} \in R(\mathscr{L}),$$

it is easy to verify that $\mathbf{G}(t,s)$, $\mathbf{H}(t)$ is a generalized Green couple.

The modification of the proof if $k = 0$ and/or $k^* = 0$ is obvious.

3.15. Theorem. *Let $\mathbf{G}_0(t,s)$ and $\mathbf{H}_0(t)$ be given by* (3,4). *Then $\mathbf{G}(t,s) = \mathbf{G}_0(t,s)$ and $\mathbf{H}(t) = \mathbf{H}_0(t)$ fulfil for any $s \in (0,1)$ the relations*

$$\text{(3,15)} \qquad \mathbf{G}(t,s) - \mathbf{G}(0,s) - \int_0^t \mathbf{A}(\tau)\,\mathbf{G}(\tau,s)\,\mathrm{d}\tau = \Delta(t,s) \qquad \textit{for all} \quad t \in [0,1],$$

$$\text{(3,16)} \qquad \mathbf{M}\,\mathbf{G}(0,s) + \mathbf{N}\,\mathbf{G}(1,s) + \int_0^1 \mathrm{d}[\mathbf{K}(\tau)]\,\mathbf{G}(\tau,s) = [\mathbf{I} - (\mathbf{SU})(\mathbf{SU})^{\#}]\,\mathbf{F}(s)$$

and

$$\text{(3,17)} \qquad \mathbf{H}(t) - \mathbf{H}(0) - \int_0^t \mathbf{A}(\tau)\,\mathbf{H}(\tau)\,\mathrm{d}\tau = \mathbf{0} \qquad \textit{on} \quad [0,1],$$

$$\text{(3,18)} \qquad \mathbf{M}\,\mathbf{H}(0) + \mathbf{N}\,\mathbf{H}(1) + \int_0^1 \mathrm{d}[\mathbf{K}(\tau)]\,\mathbf{H}(\tau) = (\mathbf{SU})(\mathbf{SU})^{\#}.$$

Proof follows easily by inserting (3,4) into (3,15)–(3,18) and making use of (1,6) and

$$\int_0^1 \mathrm{d}[\mathbf{K}(\tau)]\,\mathbf{X}(\tau)\,\Delta(\tau,s)\,\mathbf{X}^{-1}(s) = \int_s^1 \mathrm{d}[\mathbf{K}(\tau)]\,\mathbf{X}(\tau)\,\mathbf{X}^{-1}(s) + \Delta^+\mathbf{K}(s)$$

(cf. also III.2.13).

3.16. Remark. Let us notice that $\mathbf{F}(1) = \mathbf{F}(1-) = \mathbf{N}$ and by (1,7) and the Dirichlet formula I.4.32

$$\text{(3,19)} \qquad \int_s^1 \mathbf{F}(\sigma)\,\mathbf{A}(\sigma)\,\mathrm{d}\sigma = \mathbf{F}(s) - \mathbf{F}(1) + \mathbf{K}(s) \qquad \text{on} \quad [0,1].$$

3.17. Theorem. *The functions* $\mathbf{G}(t,s) = \mathbf{G}_0(t,s)$, $\mathbf{H}(t) = \mathbf{H}_0(t)$ *given by* (3,4) *fulfil for any* $t \in (0,1)$ *the relations*

$$\mathbf{G}(t,s) - \mathbf{G}(t,1) - \int_s^1 \mathbf{G}(t,\sigma)\,\mathbf{A}(\sigma)\,\mathrm{d}\sigma - \mathbf{H}(t)\,\mathbf{K}(s) = \Delta(t,s) \quad \text{for any} \quad s \in [0,1], \tag{3,20}$$

$$\mathbf{G}(t,0) - \mathbf{H}(t)\,\mathbf{M} = \mathbf{X}(t)\left[\mathbf{I} - (\mathbf{SU})^{\#}(\mathbf{SU})\right], \qquad \mathbf{G}(t,1) + \mathbf{H}(t)\,\mathbf{N} = \mathbf{0}. \tag{3,21}$$

Proof. Given $t \in (0,1)$ and $s \in [0,1]$,

$$\mathbf{G}_0(t,s) = -\mathbf{H}_0(t)\,\mathbf{N}\,\mathbf{X}(1)\,\mathbf{X}^{-1}(s)$$
$$-\mathbf{H}_0(t)\int_s^1 \mathrm{d}[\mathbf{K}(\tau)]\,\mathbf{X}(\tau)\,\mathbf{X}^{-1}(s) - \int_s^1 \mathrm{d}_\sigma[\Delta(t,\sigma)]\,\mathbf{X}(\sigma)\,\mathbf{X}^{-1}(s).$$

Our assertion follows readily taking into account the variation-of-constants formula for the initial value problem $\mathrm{d}\mathbf{y}^* = -\mathbf{y}^*\,\mathrm{d}[\mathbf{B}] - \mathrm{d}\mathbf{h}^*$, $\mathbf{y}^*(1) = \mathbf{y}_1^*$ (cf. also the proof of 2.3).

On the other hand, we have

3.18. Theorem. *Let* $\mathbf{G}\colon [0,1] \times [0,1] \to L(R_n)$ *and* $\mathbf{H}\colon [0,1] \to L(R_m, R_n)$ *fulfil for any* $s \in (0,1)$ *the relations* (3,15)–(3,18) *and let* $\gamma(s) = |\mathbf{G}(0,s)| + \operatorname{var}_0^1 \mathbf{G}(.,s) \le \gamma_0 < \infty$ *on* $[0,1]$, $\mathbf{G}$ *being measurable* $[0,1] \times [0,1]$. *Then* $\mathbf{G}(t,s)$, $\mathbf{H}(t)$ *is a generalized Green couple for BVP* (1,1), (1,3).

Proof. Let $\begin{pmatrix}\mathbf{f}\\ \mathbf{r}\end{pmatrix} \in R(\mathscr{L})$ and

$$\mathbf{x}(t) = \mathbf{H}(t)\,\mathbf{r} + \int_0^1 \mathbf{G}(t,s)\,\mathbf{f}(s)\,\mathrm{d}s \quad \text{on } [0,1].$$

$$\int_0^1 \left(\int_0^1 |\mathbf{A}(\tau)\,\mathbf{G}(\tau,s)\,\mathbf{f}(s)|\,\mathrm{d}s\right)\mathrm{d}\tau \le \int_0^1 |\mathbf{A}(\tau)| \left(\int_0^1 \gamma_0 |\mathbf{f}(s)|\,\mathrm{d}s\right)\mathrm{d}\tau$$
$$\le \left(\int_0^1 |\mathbf{A}(\tau)|\,\mathrm{d}\tau\right)\gamma_0 \|\mathbf{f}\|_{L^1} < \infty,$$

the Tonelli-Hobson theorem I.4.36 yields

$$\int_0^t \mathbf{A}(\tau)\left(\int_0^1 \mathbf{G}(\tau,s)\,\mathbf{f}(s)\,\mathrm{d}s\right)\mathrm{d}\tau = \int_0^1 \left(\int_0^t \mathbf{A}(\tau)\,\mathbf{G}(\tau,s)\,\mathrm{d}\tau\right)\mathbf{f}(s)\,\mathrm{d}s$$

for any $t \in [0,1]$. Consequently in virtue of (3,15) and (3,17)

$$\mathbf{x}(t) - \mathbf{x}(0) - \int_0^t \mathbf{A}(\tau)\,\mathbf{x}(\tau)\,\mathrm{d}\tau = \int_0^1 \Delta(t,s)\,\mathbf{f}(s)\,\mathrm{d}s = \int_0^t \mathbf{f}(\tau)\,\mathrm{d}\tau.$$

Finally, taking into account that $\begin{pmatrix} \mathbf{f} \\ \mathbf{r} \end{pmatrix} \in R(\mathscr{L})$ if and only if (cf. 3.1)

$$[\mathbf{I} - (\mathbf{SU})(\mathbf{SU})^{\#}]\left[\mathbf{r} - \int_0^1 \mathbf{F}(s)\,\mathbf{f}(s)\,\mathrm{d}s\right] = \mathbf{0}$$

and applying I.4.38 it is not difficult to check that $\mathbf{x}(t)$ verifies also the side condition (1,3).

3.19. Remark. By the variation-of-constants formula III.2.13 for generalized differential equations $\mathbf{G}$: $[0,1] \times [0,1] \to L(R_n)$ fulfils (3,15) and (3,16) for all $s \in (0,1)$ if and only if there exists $\mathbf{C}$: $[0,1] \to L(R_n)$ such that

$$\mathbf{G}(t,s) = \mathbf{X}(t)\,\Delta(t,s)\,\mathbf{X}^{-1}(s) + \mathbf{X}(t)\,\mathbf{C}(s) \qquad \text{for all} \quad t \in [0,1] \quad \text{and} \quad s \in (0,1)$$

and

$$(\mathbf{SU})\,\mathbf{C}(s) = -(\mathbf{SU})(\mathbf{SU})^{\#}\,\mathbf{F}(s) \qquad \text{on } (0,1).$$

Hence according to I.2.6 $\mathbf{G}(t,s)$ fulfils (3,15) and (3,16) if and only if

(3,22) $$\mathbf{G}(t,s) = \mathbf{G}_0(t,s) + \mathbf{X}_0(t)\,\mathbf{D}(s),$$

where $\mathbf{X}_0(t)$ has the same meaning as in 3.14 and vanishes if $k = n - \operatorname{rank}(\mathbf{SU}) = 0$ and $\mathbf{D}(s)$ is an arbitrary $k \times n$-matrix valued function defined on $(0,1)$.

Analogously $\mathbf{H}$: $[0,1] \to L(R_m, R_n)$ fulfils (3,17), (3,18) if and only if

(3,23) $$\mathbf{H}(t) = \mathbf{H}_0(t) + \mathbf{X}_0(t)\,\mathbf{\Gamma} \qquad \text{on } [0,1],$$

where $\mathbf{\Gamma} \in L(R_m, R_k)$ may be arbitrary.

Since $\operatorname{rank}(\mathbf{X}_0(t)) \equiv k$ on $[0,1]$ (cf. 2.8), we have by I.2.6.

(3,24) $$\mathbf{D}(s) = \mathbf{X}_0^{\#}(t)\,(\mathbf{G}(t,s) - \mathbf{G}_0(t,s)) \qquad \text{for all} \quad s \in (0,1) \quad \text{and} \quad t \in [0,1],$$
$$\mathbf{\Gamma} = \mathbf{X}_0^{\#}(t)\,(\mathbf{H}(t) - \mathbf{H}_0(t)) \qquad \text{for all} \quad t \in [0,1].$$

Now, let $\mathbf{G}(t,s)$, $\mathbf{H}(t)$ satisfy also (3,20), (3,21) for any $t \in (0,1)$. Then $\operatorname{var}_0^1 \mathbf{G}(t,.) < \infty$ for any $t \in [0,1]$. Moreover, by (3,23) and (3,24)

(3,25) $$\mathbf{D}(0+) = \mathbf{X}_0^{\#}(t)\,(\mathbf{G}(t,0+) - \mathbf{G}_0(t,0+)) = \mathbf{X}_0^{\#}(t)\,(\mathbf{H}(t) - \mathbf{H}_0(t))\,\mathbf{M} = \mathbf{\Gamma M}$$

and

(3,26) $$\mathbf{D}(1-) = -\mathbf{\Gamma N}.$$

Putting $\mathbf{D}(0) = \mathbf{D}(0+)$, $\mathbf{D}(1) = \mathbf{D}(1-)$, $\mathbf{D}$ will be of bounded variation on $[0,1]$. By (3,20)

$$\mathbf{D}(s) - \mathbf{D}(1) - \int_s^1 \mathbf{D}(\tau)\,\mathbf{A}(\tau)\,\mathrm{d}\tau - \mathbf{\Gamma K}(s) = \mathbf{0} \qquad \text{on } [0,1].$$

This together with (3,25), (3,26) may hold if and only if there is $\mathbf{W} \in L(R_{k*}, R_k)$ such that $\mathbf{D}(s) \equiv \mathbf{W Y}_0(s)$ on $[0,1]$ and $\mathbf{\Gamma} = \mathbf{W}\Lambda_0$ (cf. 2.8). To summarize:

$$\mathbf{G}: [0,1] \times [0,1] \to L(R_n) \quad \text{and} \quad \mathbf{H}: [0,1] \to L(R_m, R_n)$$

satisfy (3,15)–(3,18) for any $s \in (0,1)$ and (3,20), (3,21) for any $t \in (0,1)$ if and only if

$$\mathbf{G}(t,s) = \mathbf{G}_0(t,s) + \mathbf{X}_0(t)\,\mathbf{W}\mathbf{Y}_0(s) \quad \text{on } [0,1]\times[0,1],$$
$$\mathbf{H}(t) = \mathbf{H}_0(t) - \mathbf{X}_0(t)\,\mathbf{W}\mathbf{\Lambda}_0 \quad \text{on } [0,1],$$

where $\mathbf{W} \in L(R_{k^*}, R_k)$ may be arbitrary.

3.20. Remark. By the definition (3,4) of $\mathbf{G}_0(t,s)$ and $\mathbf{H}_0(t)$ and 3.17

$$(3{,}27) \qquad \begin{aligned} &\mathbf{G}_0(t,0+) - \mathbf{H}_0(t)\,\mathbf{M} = \mathbf{X}(t)\,[\mathbf{I} - (\mathbf{SU})^{\#}\,(\mathbf{SU})] \quad \text{if } t > 0,\\ &\mathbf{G}_0(0,0+) - \mathbf{H}_0(0)\,\mathbf{M} = -(\mathbf{SU})^{\#}\,(\mathbf{SU}),\\ &\mathbf{G}_0(t,1-) + \mathbf{H}_0(t)\,\mathbf{N} = \mathbf{0} \quad \text{if } t < 1,\\ &\mathbf{G}_0(1,1-) + \mathbf{H}_0(1)\,\mathbf{N} = \mathbf{I}. \end{aligned}$$

In particular, for any $\mathbf{g} \in BV_n$ right-continuous on $[0,1)$ and left-continuous at 1

$$(3{,}28) \qquad \int_0^1 \mathrm{d}[\mathbf{g}^*(\tau)]\,(\mathbf{G}_0(\tau,0+) - \mathbf{H}_0(\tau)\,\mathbf{M}) = \int_0^1 \mathrm{d}[\mathbf{g}^*(\tau)]\,\mathbf{X}(\tau)\,[\mathbf{I} - (\mathbf{SU})^{\#}\,(\mathbf{SU})]$$

and

$$(3{,}29) \qquad \int_0^1 \mathrm{d}[\mathbf{g}^*(\tau)]\,(\mathbf{G}_0(\tau,1-) + \mathbf{H}_0(\tau)\,\mathbf{N}) = \mathbf{0}.$$

We shall conclude this section by proving that the couple $\mathbf{G}_0(t,s)$, $\mathbf{H}_0(t)$ has also the meaning of a generalized Green couple for the adjoint nonhomogeneous BVP (2,11), (2,12).

3.21. Theorem. *Let $\mathbf{g} \in BV_n$ be right-continuous on $[0,1)$ and left-continuous at 1 and let $\mathbf{p}, \mathbf{q} \in R_n$. Then, if BVP (2,11), (2,12) has a solution, the couple $(\mathbf{y}^*, \boldsymbol{\lambda}^*)$ given by*

$$(3{,}30) \qquad \begin{aligned} &\mathbf{y}^*(s) = \mathbf{q}^*\,\mathbf{G}_0(1,s) - \mathbf{p}^*\,\mathbf{G}_0(0,s) - \int_0^1 \mathrm{d}[\mathbf{g}^*(\tau)]\,\mathbf{G}_0(\tau,s) \quad \textit{on } (0,1),\\ &\mathbf{y}^*(0) = \mathbf{y}^*(0+), \qquad \mathbf{y}^*(1) = \mathbf{y}^*(1-),\\ &\boldsymbol{\lambda}^* = -\mathbf{q}^*\,\mathbf{H}_0(1) + \mathbf{p}^*\,\mathbf{H}_0(0) + \int_0^1 \mathrm{d}[\mathbf{g}^*(\tau)]\,\mathbf{H}_0(\tau) \end{aligned}$$

is also its solution.

Proof. (a) By (3,4) $\mathbf{G}_0(1,s) = \mathbf{X}(1)\,(\mathbf{X}^{-1}(s) - (\mathbf{SU})^{\#}\,\mathbf{F}(s))$ on $[0,1]$ and owing to (1,7) and (3,19)

$$\int_s^1 \mathbf{G}_0(1,\sigma)\,\mathbf{A}(\sigma)\,\mathrm{d}\sigma$$
$$= \mathbf{X}(1)\,\mathbf{X}^{-1}(s) - \mathbf{X}(1)\,(\mathbf{SU})^{\#}\,\mathbf{F}(s) - \mathbf{I} + \mathbf{X}(1)\,(\mathbf{SU})^{\#}\,\mathbf{F}(1) - \mathbf{X}(1)\,(\mathbf{SU})^{\#}\,\mathbf{K}(s)$$

or

$$(3{,}31) \qquad \mathbf{G}_0(1,s) = \mathbf{G}_0(1,1) + \int_s^1 \mathbf{G}_0(1,\sigma)\,\mathbf{A}(\sigma)\,\mathrm{d}\sigma + \mathbf{H}_0(1)\,\mathbf{K}(s) \quad \text{on } [0,1].$$

Let us notice that $\mathbf{G}_0(1,1-) = \mathbf{G}_0(1,1) = \mathbf{I} - \mathbf{X}(1)(\mathbf{SU})^+ \mathbf{F}(1)$ and $\mathbf{G}_0(1,0+) = \mathbf{G}_0(1,0) = \mathbf{X}(1) - \mathbf{X}(1)(\mathbf{SU})^\# \mathbf{F}(0)$, $\mathbf{F}(1) = \mathbf{N}$, $\mathbf{F}(0) = (\mathbf{SU}) - \mathbf{M}$. Furthermore,

$$\mathbf{G}_0(0,0) = \mathbf{I} - (\mathbf{SU})^\# \mathbf{F}(0) = \mathbf{I} - (\mathbf{SU})^\# (\mathbf{SU}) + (\mathbf{SU})^\# \mathbf{M},$$
$$\mathbf{G}_0(0,s) = -(\mathbf{SU})^\# \mathbf{F}(s)$$

if $s > 0$. In particular, $\mathbf{G}_0(0,1-) = \mathbf{G}_0(0,1) = -(\mathbf{SU})^\# \mathbf{F}(1) = -(\mathbf{SU})^\# \mathbf{N}$. Hence, making use of (3,19)

$$\text{(3,32)} \quad \int_s^1 \mathbf{G}_0(0,\sigma)\,\mathbf{A}(\sigma)\,d\sigma = \mathbf{G}_0(0,s) - \mathbf{G}_0(0,1) - \mathbf{H}_0(0)\,\mathbf{K}(s) \quad \text{on } [0,1].$$

Now, if $\mathbf{y}^*\colon [0,1] \to R_n^*$ and $\boldsymbol{\lambda}^* \in R_m^*$ are given by (3,30), then by (3,15), (3,20), (3,31), (3,32) and I.4.32

$$\mathbf{y}^*(s) - \mathbf{y}^*(1) - \int_s^1 \mathbf{y}^*(\sigma)\,\mathbf{A}(\sigma)\,d\sigma + \boldsymbol{\lambda}^*\,\mathbf{K}(s)$$
$$= -\int_0^1 d[\mathbf{g}^*(\tau)]\left(\mathbf{G}_0(\tau,s) - \mathbf{G}_0(\tau,1) - \int_s^1 \mathbf{G}_0(\tau,\sigma)\,\mathbf{A}(\sigma)\,d\sigma - \mathbf{H}_0(\tau)\,\mathbf{K}(s)\right)$$
$$= \mathbf{g}^*(s) - \mathbf{g}^*(1) \quad \text{on } [0,1].$$

(Here we have also made use of the assumption $\mathbf{g}^*(1-) = \mathbf{g}^*(1)$, $\mathbf{g}^*(0+) = \mathbf{g}^*(0)$ and of the fact that $\mathbf{G}_0(0,s+) = \mathbf{G}_0(0,s)$ if $s > 0$ and $\mathbf{G}_0(1,s-) = \mathbf{G}_0(1,s)$ if $s < 1$.)

(b) By (3,21), (3,27) and (3,29)

$$\mathbf{y}^*(1) - \boldsymbol{\lambda}^*\mathbf{N} = \mathbf{q}^*[\mathbf{G}_0(1,1-) + \mathbf{H}_0(1)\,\mathbf{N}] - \mathbf{p}^*[\mathbf{G}_0(0,1-) + \mathbf{H}_0(0)\,\mathbf{N}]$$
$$- \int_0^1 d[\mathbf{g}^*(\tau)]\,(\mathbf{G}_0(\tau,1-) + \mathbf{H}_0(\tau)\,\mathbf{N}) = \mathbf{q}^*.$$

(c) Finally, by (3,21), (3,27) and (3,28)

$$\mathbf{y}^*(0) + \boldsymbol{\lambda}^*\mathbf{M} = \mathbf{q}^*[\mathbf{G}_0(1,0+) - \mathbf{H}_0(1)\,\mathbf{M}] - \mathbf{p}^*[\mathbf{G}_0(0,0+) - \mathbf{H}_0(0)\,\mathbf{M}]$$
$$- \int_0^1 d[\mathbf{g}^*(\tau)]\,(\mathbf{G}_0(\tau,0+) - \mathbf{H}_0(\tau)\,\mathbf{M})$$
$$= \mathbf{p}^* + \left[\mathbf{q}^*\,\mathbf{X}(1) - \mathbf{p}^* - \int_0^1 d[\mathbf{g}^*(\tau)]\,\mathbf{X}(\tau)\right][\mathbf{I} - (\mathbf{SU})^\# (\mathbf{SU})].$$

Since $\mathbf{x}_0(t) = \mathbf{X}(t)\,[\mathbf{I} - (\mathbf{SU})^\# (\mathbf{SU})]$ is a solution to the homogeneous BVP (1,2), (1,4), the last expression reduces to $\mathbf{p}^*$ (cf. 2.5).

Notes

Canonical form of Stieltjes integral conditions (IV.3.15) is due to Zimmerberg [2]. Section IV.2 is based on Vejvoda, Tvrdý [1] and Tvrdý, Vejvoda [1]. In IV.2.16 the idea of Pagni [1] is utilized. For writing IV.3, the paper Brown [1] was stimulating.

Bryan [1], Cole [2], Halanay, Moro [1], Krall [1] – [4] and Tucker [1] are related references to IV.2, while Reid [2], [3], Chitwood [1], Zubov [1], [2] and Bradley [1] concern IV.3. For a historical survey of the subject and a more complete bibliography the reader is referred e.g. to Whyburn [2], Conti [2], Reid [1] and Krall [9]. More detail concerning some special questions (as e.g. two-point problems, second order and n-th order equations, selfadjointness, expansion theorems) as well as examples may be found in the monographs Coddington, Levinson [1], Reid [1], Najmark [1] and Cole [1].

The interface problems were treated in Conti [3], Krall [2], [3], Parhimovič [3], Stallard [1] and Zettl [1]. Boundary problems in the L^p-setting were dealt with in Krall [1] – [8], Brown [1], [3], Brown, Krall [1], [3]. Expansion theorems for problems with a multipoint or Stieltjes integral side conditions are to be found in Krall [5], Brown, Green, Krall [1] and Coddington, Dijksma [1]. For applications to controllability, minimization problems and splines see Brown [2], Brown, Krall [2], Halanay [1] and Marchiò [1].

V. *Integro-differential operators*

1. Fredholm-Stieltjes integro-differential operator

The most part of this chapter is devoted to the *Fredholm-Stieltjes integro-differential operators* of the form

$$\mathbf{x} \to \mathbf{x}'(t) - \int_0^1 \mathrm{d}_s[\mathbf{P}(t,s)]\,\mathbf{x}(s).$$

The kernel $\mathbf{P}(t,s)$ *is assumed to be an* $n \times n$*-matrix valued function defined for a.e.* $t \in [0,1]$ *and any* $s \in [0,1]$ *and such that* $\mathbf{P}(.,s)$ *is measurable on* $[0,1]$ *for any* $s \in [0,1]$,

$$(1,1) \qquad \varrho(t) = |\mathbf{P}(t,0)| + \operatorname{var}_0^1 \mathbf{P}(t,.) = \|\mathbf{P}(t,.)\|_{BV} < \infty \qquad \text{a.e. on } [0,1]$$

and

$$(1,2) \qquad \|\varrho\|_{L^p} = \left(\int_0^1 (\varrho(t))^p \,\mathrm{d}t\right)^{1/p} < \infty,$$

where $1 \le p < \infty$.

Such kernels will be called $L^p[BV]$*-kernels.*

1.1. Remark. For $L^r \subset L^p$ if $p \le r$, any $L^r[BV]$-kernel is also an $L^p[BV]$-kernel for each p, $1 \le p \le r$. Furthermore

$$|\mathbf{P}(t,s)| \le |\mathbf{P}(t,0)| + |\mathbf{P}(t,s) - \mathbf{P}(t,0)| \le \varrho(t)$$

for all $s \in [0,1]$ and a.e. $t \in [0,1]$. Hence by (1,2)

$$\int_0^1 |\mathbf{P}(t,s)|^p \,\mathrm{d}t < \infty \qquad \text{for any} \quad s \in [0,1].$$

1.2. Proposition. *If* $\mathbf{P}(t,s)$ *is an* $L^p[BV]$*-kernel, then the function*

$$\mathbf{Px}\colon t \in [0,1] \to \int_0^1 \mathrm{d}_s[\mathbf{P}(t,s)]\,\mathbf{x}(s) \in R_n$$

belongs to L_n^p *for any* $\mathbf{x} \in BV_n$ *and the operator*

$$\mathbf{P}\colon \mathbf{x} \in BV_n \to \int_0^1 \mathrm{d}_s[\mathbf{P}(t,s)]\,\mathbf{x}(s) \in L_n^p \tag{1,3}$$

is linear and bounded.

Proof. By I.4.27 and I.4.37 $\mathbf{Px} \in L_n^p$ and

$$|(\mathbf{Px})(t)| \leq \varrho(t)\,(\sup_{s\in[0,1]} |\mathbf{x}(s)|) \qquad \text{a.e. on } [0,1] \tag{1,4}$$

for any $\mathbf{x} \in BV_n$. Since $\varrho \in L^p$ and $\sup_{s\in[0,1]} |\mathbf{x}(s)| \leq \|\mathbf{x}\|_{BV}$, our assertion follows immediately.

1.3. Remark. Since (1,4) holds also for any $\mathbf{x} \in C_n$, the mapping $\mathbf{x} \to \mathbf{Px}$ is bounded as an operator $C_n \to L_n^p$, as well. Let us notice, furthermore, that if $\mathbf{x}_k, \mathbf{x} \in C_n$ $(k = 1, 2, \ldots)$ and $\lim_{k\to\infty} \|\mathbf{x}_k - \mathbf{x}\|_C = 0$, then in virtue of (1,4) $\lim_{k\to\infty} (\mathbf{Px}_k)(t) = (\mathbf{Px})(t)$ a.e. on $[0,1]$. In other words, $\mathbf{P}$ maps sequences converging uniformly on $[0,1]$ onto seuqences converging a.e. on $[0,1]$. It was shown in Kantorovič, Pinsker, Vulich [1] that

$$\mathbf{x} \in C_n \to \int_0^1 \mathrm{d}_s[\mathbf{P}(t,s)]\,\mathbf{x}(s) \in L_n^1,$$

with the $L^1[BV]$-kernel $\mathbf{P}(t,s)$, is a general form of operators $C_n \to L_n^1$ possessing this property.

1.4. Proposition. *If* $\mathbf{P}(t,s)$ *is an* $L^p[BV]$*-kernel, then the operator* $\mathbf{P}\colon BV_n \to L_n^p$ *given by* (1,3) *is compact.*

Proof. Let $\mathbf{x}_k \in BV_n$ and $\|\mathbf{x}\|_{BV} \leq 1$ for each $k = 1, 2, \ldots$. By the Helly Choice Theorem the sequence $\{\mathbf{x}_k\}_{k=1}^{\infty}$ contains a subsequence $\{\mathbf{x}_{k_l}\}_{l=1}^{\infty}$ such that

$$\lim_{l\to\infty} \mathbf{x}_{k_l}(t) = \mathbf{x}(t) \qquad \text{on } [0,1]$$

for some $\mathbf{x} \in BV_n$. For $t, s \in [0,1]$ let us denote

$$p(t,s) = \mathrm{var}_0^s\, \mathbf{P}(t,.)$$

and

$$z_l(t) = \int_0^1 \mathrm{d}_s[p(t,s)]\,|\mathbf{x}_{k_l}(s) - \mathbf{x}(s)|\,.$$

Given $l = 1, 2, \ldots$ and $s \in [0,1]$,

$$|\mathbf{x}_{k_l}(s) - \mathbf{x}(s)| \leq \|\mathbf{x}_{k_l} - \mathbf{x}\|_{BY} \leq \|\mathbf{x}_{k_l}\|_{BV} + \|\mathbf{x}\|_{BV} \leq 1 + \|\mathbf{x}\|_{BV} < \infty$$

and hence by I.4.27

$$|z_l(t)| \leq (\mathrm{var}_0^1\, \mathbf{P}(t,.))\,(1 + \|\mathbf{x}\|_{BV}) \leq (1 + \|\mathbf{x}\|_{BV})\,\varrho(t) \qquad \text{a.e. on } [0,1].$$

Moreover, according to I.4.24

$$\lim_{l\to\infty} z_l(t) = 0 \qquad \text{a.e. on } [0,1].$$

By the assumption $\varrho \in L^p$ and hence applying the classical Lebesgue Convergence Theorem we obtain

$$\lim_{l\to\infty} \int_0^1 |z_l(t)|^p \, dt = 0. \tag{1,5}$$

Since for any $l = 1, 2, \ldots$

$$\int_0^1 \left| \int_0^1 d_s[\mathbf{P}(t,s)]\,(\mathbf{x}_{k_l}(s) - \mathbf{x}(s)) \right|^p dt \le \int_0^1 |z_l(t)|^p \, dt,$$

(1,5) implies

$$\lim_{l\to\infty} \|\mathbf{P}\mathbf{x}_{k_l} - \mathbf{P}\mathbf{x}\|_{L^p} = 0$$

and this completes the proof.

1.5. Notation. Throughout the chapter $\mathbf{P}$ denotes the operator defined by (1,3) or its restriction on W_n^p $(1 \le p < \infty)$, where W_n^p stands for the Sobolev space defined in I.5.10. Furthermore,

$$\mathbf{D}\colon \mathbf{x} \in W_n^p \to \mathbf{x}' \in L_n^p \tag{1,6}$$

and

$$\mathbf{L} = \mathbf{D} - \mathbf{P}\colon \mathbf{x} \in W_n^p \to \mathbf{x}' - \mathbf{P}\mathbf{x} \in L_n^p \tag{1,7}$$

for any $p \in R$, $p \ge 1$.

1.6. Remark. Clearly, $\mathbf{D}$ is linear and bounded for any $p \in R$, $p \ge 1$. Hence if $\mathbf{P}(t,s)$ is an $L^p[BV]$-kernel, then $\mathbf{L}$ is also linear and bounded. We shall show that it has a closed range and hence by I.3.14 it is normally solvable.

1.7. Proposition. *Let $\mathbf{P}\colon [0,1] \times [0,1] \to L(R_n)$ be an $L^p[BV]$-kernel $(1 \le p < \infty)$. Then the operator $\mathbf{L}\colon W_n^p \to L_n^p$ given by (1,7) has a closed range in L_n^p.*

Proof. Let $\mathbf{f} \in L_n^p$. Then $\mathbf{f} \in R(\mathbf{L})$ if and only if there exists $\mathbf{x} \in W_n^p$ such that

$$\mathbf{x}(t) - \mathbf{x}(0) - \int_0^t \left(\int_0^1 d_s[\mathbf{P}(\tau,s)]\,\mathbf{x}(s) \right) d\tau = \int_0^t \mathbf{f}(\tau)\, d\tau. \tag{1,8}$$

Hence denoting

$$\boldsymbol{\Psi}\colon \mathbf{h} \in L_n^p \to \int_0^t \mathbf{h}(\tau)\, d\tau \in W_n^p, \tag{1,9}$$

$$\boldsymbol{\Pi}\colon \mathbf{x} \in W_n^p \to \mathbf{z}(t) \equiv \mathbf{x}(0) \in W_n^p,$$

we have $\mathbf{f} \in R(\mathbf{L})$ if and only if $\boldsymbol{\Psi}\mathbf{f} \in R(\mathbf{I} - (\boldsymbol{\Pi} + \boldsymbol{\Psi}\mathbf{P}))$, where $\mathbf{I}$ stands for the identity operator on W_n^p.

The operators $\boldsymbol{\Pi}$ and $\boldsymbol{\Psi}$ are evidently linear and bounded. As $R(\boldsymbol{\Pi})$ is finite dimensional, $\boldsymbol{\Pi}$ is compact (cf. I.3.21). Since, given $\mathbf{x} \in W_n^p$, $\|\mathbf{x}\|_{BV} \leq \|\mathbf{x}\|_{W^p}$, it follows from 1.4 that also $\mathbf{P}$: $W_n^p \to L_n^p$ is compact. Hence the operator $\boldsymbol{\Theta} = \boldsymbol{\Pi} + \boldsymbol{\Psi}\mathbf{P}$: $W_n^p \to W_n^p$ is linear bounded and compact. Consequently $R(\mathbf{I} - \boldsymbol{\Theta})$ is closed (cf. I.3.20). Since $\boldsymbol{\Psi}(R(\mathbf{L})) = R(\mathbf{I} - \boldsymbol{\Theta})$, $R(\mathbf{L})$ is closed.

1.8. Proposition. *If* $\mathbf{P}(t, s)$ *is an* $L^p[BV]$*-kernel, then*

$$n \leq \dim N(\mathbf{L}) < \infty\,,$$

while $\dim N(\mathbf{L}) = n$ *if and only if* $R(\mathbf{L}) = L_n^p$.

Proof. By the proof of 1.7 the equation $\mathbf{L}\mathbf{x} = \mathbf{f}$ is equivalent to the equation

$$\mathbf{x} - \boldsymbol{\Theta}\mathbf{x} = \boldsymbol{\Psi}\mathbf{f}\,,$$

where $\boldsymbol{\Theta} = \boldsymbol{\Pi} + \boldsymbol{\Psi}\mathbf{P}$: $W_n^p \to W_n^p$ is defined by (1,9). Since $\boldsymbol{\Theta}$ is compact, by I.3.20 we have $\dim N(\mathbf{L}) = \dim N(\mathbf{I} - \boldsymbol{\Theta}) < \infty$ and

$$\dim N(\mathbf{L}) = \operatorname{codim} R(\mathbf{I} - \boldsymbol{\Theta}) = \dim W_n^p / R(\mathbf{I} - \boldsymbol{\Theta})\,. \tag{1,10}$$

It follows from the definition of $\boldsymbol{\Theta}$ that

$$R(\mathbf{I} - \boldsymbol{\Theta}) \subset \{\mathbf{g} \in W_n^p;\ \mathbf{g}(0) = \mathbf{0}\} = V_n^p\,.$$

Consequently

$$\dim W_n^p / R(\mathbf{I} - \boldsymbol{\Theta}) \geq \dim W_n^p / V_n^p\,.$$

If $\{\mathbf{e}_1, \mathbf{e}_2, \ldots, \mathbf{e}_n\}$ is a basis in R_n and $\boldsymbol{\xi}_j(t) = \mathbf{e}_j$ on $[0, 1]$ $(j = 1, 2, \ldots, n)$, then the system of equivalence classes $\boldsymbol{\xi}_j + V_n^p$ $(j = 1, 2, \ldots, n)$ forms a basis in W_n^p / V_n^p. Hence

$$\dim W_n^p / V_n^p = n$$

and by (1,10) $\dim N(\mathbf{L}) = n$ if and only if

$$\dim W_n^p / V_n^p = \dim W_n^p / R(\mathbf{I} - \boldsymbol{\Theta})\,.$$

Since $R(\mathbf{I} - \boldsymbol{\Theta}) = V_n^p$ if and only if $R(\mathbf{L}) = L_n^p$, the proof will be completed by means of the following assertion.

1.9. Lemma. *Given a Banach space* X *and its closed linear subspaces* M, N *such that* $M \subset N \subset X$, $\dim X/M = \dim X/N < \infty$ *holds if and only if* $M = N$.

Proof. Let $\dim X/M = \dim X/N = k < \infty$ and let $\mathbf{x} \in N \setminus M$. Let $\Xi_j = \xi_j + N$ $(j = 1, 2, \ldots, k)$ be a basis in X/N and let

$$\alpha\mathbf{x} + \sum_{j=1}^{k} \lambda_j \xi_j \in M \subset N$$

for some real numbers α, λ_j $(j = 1, 2, \dots, k)$. Since $\alpha\mathbf{x} \in N$, this may happen only if $\lambda_1\xi_1 + \lambda_2\xi_2 + \dots + \lambda_k\xi_k \in N$, i.e. $\lambda_1 = \lambda_2 = \dots = \lambda_k = 0$. Thus $\alpha\mathbf{x} \in M$ and for $\mathbf{x} \notin M$, $\alpha = 0$. This means that the classes $\{\mathbf{x} + M,\ \xi_j + M;\ j = 1, 2, \dots, k\}$ are linearly independent in X/M and $\dim X/M \geq k + 1 > \dim X/N$. This being contradictory to the assumption, we have $M = N$.

1.10. Remark. By 1.8 there exists an $n \times k$-matrix valued function $\mathbf{X}$ $(k = \dim N(\mathbf{L}))$ such that $\mathbf{x} \in W_n^p$ is a solution to $\mathbf{Lx} = \mathbf{0}$ if and only if $\mathbf{x}_0(t) \equiv \mathbf{X}(t)\,\mathbf{c}$ on $[0, 1]$ for some $\mathbf{c} \in R_k$. Unfortunately, even if $k = n$, it need not be $\det(\mathbf{X}(t)) \neq 0$ on $[0, 1]$. For example, the equation

$$\mathbf{x}'(t) - 4\int_0^1 \mathbf{x}(\tau)\,\mathrm{d}\tau = \mathbf{f}(t) \quad \text{a.e. on } [0, 1] \tag{1,11}$$

possesses for any $\mathbf{f} \in L_n^1$ and $\mathbf{c} \in R_n$ the unique solution

$$\mathbf{x}(t) = \mathbf{I}(1 - 4t)\,\mathbf{c} + 4t\int_0^1 \left(\int_0^s \mathbf{f}(\tau)\,\mathrm{d}\tau\right)\mathrm{d}s + \int_0^t \mathbf{f}(\tau)\,\mathrm{d}\tau \qquad \text{on } [0, 1]$$

such that $\mathbf{x}(0) = \mathbf{c}$. In particular, $\mathbf{x} \in AC_n$ is a solution of the corresponding homogeneous equation if and only if $\mathbf{x}(t) = \mathbf{I}(1 - 4t)\,\mathbf{c}$ for some $\mathbf{c} \in R_n$ and $\mathbf{X}(t) = \mathbf{I}(1 - 4t)$ is the fundamental matrix solution for (1,11). Let us notice that $\mathbf{X}(\frac{1}{4}) = \mathbf{0}$.

1.11. Remark. Putting $\mathbf{R}(t, s) = \mathbf{P}(t, s+) - \mathbf{P}(t, 1)$ for $s \in (0, 1)$, $\mathbf{R}(t, 0) = \mathbf{P}(t, 0) - \mathbf{P}(t, 1)$ and $\mathbf{R}(t, 1) = \mathbf{0}$, we would obtain

$$\begin{aligned} \mathbf{R}(t, s+) &= \mathbf{P}(t, s+) - \mathbf{P}(t, 1) \quad &&\text{if } s \in [0, 1),\\ \mathbf{R}(t, s-) &= \mathbf{P}(t, s-) - \mathbf{P}(t, 1) \quad &&\text{if } s \in (0, 1] \end{aligned}$$

and hence according to I.5.5

$$\int_0^1 \mathrm{d}_s[\mathbf{P}(t, s)]\,\mathbf{x}(s) = \int_0^1 \mathrm{d}_s[\mathbf{R}(t, s)]\,\mathbf{x}(s) \qquad \text{for each } \mathbf{x} \in AC_n.$$

Given a subdivision $\sigma = \{0 = s_0 < s_1 < \dots < s_m = 1\}$ of $[0, 1]$ and $\delta > 0$ such that $0 = s_0 < s_0 + \delta < s_1 < s_1 + \delta < \dots < s_{m-1} < s_{m-1} + \delta < s_m = 1$, we have

$$\begin{aligned} V_\delta(t) = {} & |\mathbf{P}(t, s_0 + \delta) - \mathbf{P}(t, 0)| + \sum_{j=1}^{m-1} |\mathbf{P}(t, s_j + \delta) - \mathbf{P}(t, s_{j-1} + \delta)| \\ & + |\mathbf{P}(t, 1) - \mathbf{P}(t, s_{m-1} + \delta)| \leq \varrho(t) \qquad \text{a.e. on } [0, 1]. \end{aligned}$$

Consequently

$$\sum_{j=1}^{m} |\mathbf{R}(t, s_j) - \mathbf{R}(t, s_{j-1})| = \lim_{\delta \to 0+} V_\delta(t) \leq \varrho(t)$$

and $\operatorname{var}_0^1 \mathbf{R}(t, .) \leq \varrho(t)$ a.e. on $[0, 1]$. Since $|\mathbf{R}(t, 0)| \leq 2\,\varrho(t)$ a.e. on $[0, 1]$ (cf. 1.1), it follows that $\mathbf{R}\colon [0, 1] \times [0, 1] \to L(R_n)$ is also an $L^p[BV]$-kernel.

This means that without any loos of generality we may assume that $\mathbf{P}(t,.)$ is right-continuous on $(0,1)$ and $\mathbf{P}(t,1)=\mathbf{0}$ for almost all $t\in[0,1]$.

1.12. Remark. Let

$$\mathbf{P}(t,s)=\begin{cases}-\mathbf{A}(t)-\mathbf{C}(t)-\mathbf{D}(t) & \text{if} \quad s=0\,,\\ -\mathbf{A}(t)-\mathbf{D}(t) & \text{if} \quad 0<s<t\,,\\ -\mathbf{D}(t) & \text{if} \quad t\le s<1\,,\\ \mathbf{0} & \text{if} \quad s=1\,,\end{cases}$$

where $\mathbf{A}, \mathbf{C}, \mathbf{D}$ are $n\times n$-matrix valued functions whose columns are elements of L^p_n. Then

$$\operatorname{var}_0^1 \mathbf{P}(t,.)=|\mathbf{A}(t)|+|\mathbf{C}(t)|+|\mathbf{D}(t)| \qquad \text{a.e. on } [0,1]$$

and hence $\mathbf{P}(t,s)$ is an $L^p[BV]$-kernel. Furthermore, given $\mathbf{x}\in AC_n$,

$$\int_0^1 \mathrm{d}_s[\mathbf{P}(t,s)]\,\mathbf{x}(s)=\mathbf{A}(t)\,\mathbf{x}(t)+\mathbf{C}(t)\,\mathbf{x}(0)+\mathbf{D}(t)\,\mathbf{x}(1) \qquad \text{a.e. on } [0,1]$$

and the integro-differential operator $\mathbf{L}=\mathbf{D}-\mathbf{P}$ reduces to the *differential-boundary operator*

$$\mathbf{x}\in W^p_n \to \mathbf{x}'(t)-\mathbf{A}(t)\,\mathbf{x}(t)-\mathbf{C}(t)\,\mathbf{x}(0)-\mathbf{D}(t)\,\mathbf{x}(1)\in L^p_n\,.$$

2. Duality theory

Our wish is now to establish the duality theory for BVP

$$\mathbf{x}'(t)-\int_0^1 \mathrm{d}_s[\mathbf{P}(t,s)]\,\mathbf{x}(s)=\mathbf{f}(t) \qquad \text{a.e. on } [0,1]\,, \tag{2,1}$$

$$\mathbf{S}\mathbf{x}\equiv\mathbf{M}\,\mathbf{x}(0)+\int_0^1 \mathbf{K}(t)\,\mathbf{x}'(t)\,\mathrm{d}t=\mathbf{r}\,. \tag{2,2}$$

In particular, we shall show the normal solvability and evaluate the index of this boundary value problem under the following assumptions.

2.1. Assumptions. *$\mathbf{P}$: $[0,1]\times[0,1]\to L(R_n)$ is an L^p-$[BV]$-kernel, $1\le p<\infty$, $\mathbf{f}\in L^p_n$, $\mathbf{M}\in L(R_n,R_m)$, $\mathbf{K}$: $[0,1]\to L(R_n,R_m)$, $\|\mathbf{K}\|_{L^q}<\infty$, $q=p/(p-1)$ if $p>1$, $q=\infty$ if $p=1$ and $\mathbf{r}\in R_m$.*

2.2. Definition. A function $\mathbf{x}$: $[0,1]\to R_n$ is said to be a *solution of BVP* (2,1), (2,2) if $\mathbf{x}\in AC_n$ and (2,1), (2,2) hold for a.e. $t\in[0,1]$.

2.3. Remark. According to 1.13 we may assume that for a.e. $t\in[0,1]$ $\mathbf{P}(t,.)$ is right-continuous on $(0,1)$ and $\mathbf{P}(t,1)=\mathbf{0}$. Furthermore, let us mention, that if

$\mathbf{P}(t, s)$ is an $L^p[BV]$-kernel and $\mathbf{f} \in L_n^p$, then obviously $\mathbf{x}' \in L_n^p$ for any solution $\mathbf{x} \in AC_n$ of the integro-differential equation (2,1). Thus given a solution $\mathbf{x}$ of BVP (2,1), (2,2), $\mathbf{x} \in W_n^p$.

2.4. Notations. *The operators* $\mathbf{D} \in B(W_n^p, L_n^p)$ *and* $\mathbf{P} \in K(W_n^p, L_n^p)$ *are defined by* (1,3) *and* (1,6),

$$\mathbf{S}: \mathbf{x} \in W_n^p \to \mathbf{M}\,\mathbf{x}(0) + \int_0^1 \mathbf{K}(t)\,\mathbf{x}'(t)\,\mathrm{d}t \in R_m$$

and

$$\mathscr{L}: \mathbf{x} \in W_n^p \to \begin{bmatrix} \mathbf{Dx} - \mathbf{Px} \\ \mathbf{Sx} \end{bmatrix} \in L_n^p \times R_m. \tag{2,3}$$

Making use of 2.4, we may reformulate BVP (2,1), (2,2) as the operator equation

$$\mathscr{L}\mathbf{x} = \begin{pmatrix} \mathbf{f} \\ \mathbf{r} \end{pmatrix}. \tag{2,4}$$

It appears to be convenient to handle instead of (2,4) the operator equation for $\xi = \begin{pmatrix} \mathbf{x} \\ \mathbf{d} \end{pmatrix} \in W_n^p \times R_m$

$$\xi - \mathbf{T}\xi = \varphi, \tag{2,5}$$

where

$$\boldsymbol{\Psi}: \mathbf{u} \in L_n^p \to \int_0^1 \mathbf{u}(\tau)\,\mathrm{d}\tau \subset W_n^p, \qquad \boldsymbol{\Phi}: \mathbf{x} \in W_n^p \to \mathbf{v}(t) \equiv \mathbf{x}(0) \in W_n^p, \tag{2,6}$$

$$\mathbf{T}: \begin{pmatrix} \mathbf{x} \\ \mathbf{d} \end{pmatrix} \in W_n^p \times R_m \to \begin{bmatrix} \boldsymbol{\Phi}\mathbf{x} + \boldsymbol{\Psi}\mathbf{Px} \\ \mathbf{d} - \mathbf{Sx} \end{bmatrix} \in W_n^p \times R_m \quad \text{and} \quad \varphi = \begin{pmatrix} \boldsymbol{\Psi}\mathbf{f} \\ \mathbf{r} \end{pmatrix} \in W_n^p \times R_m.$$

Clearly, $\mathbf{x} \in W_n^p$ is a solution to BVP (2,1), (2,2) if and only if for an arbitrary $\mathbf{d} \in R_m$ the couple $\xi = \begin{pmatrix} \mathbf{x} \\ \mathbf{d} \end{pmatrix}$ is a solution of (2,5). In particular,

$$\dim N(\mathbf{I} - \mathbf{T}) = \dim N(\mathscr{L}) + m. \tag{2,7}$$

Furthermore, $\begin{pmatrix} \mathbf{f} \\ \mathbf{r} \end{pmatrix} \in L_n^p \times R_m$ belongs to $R(\mathscr{L})$ if and only if $\begin{pmatrix} \boldsymbol{\Psi}\mathbf{f} \\ \mathbf{r} \end{pmatrix} \in R(\mathbf{I} - \mathbf{T})$. As according to 1.4 and I.3.21 the linear operator $\mathbf{T}$ given by (2,6) is compact and the linear operator $\mathbf{W}: \begin{pmatrix} \mathbf{f} \\ \mathbf{r} \end{pmatrix} \in L_n^p \times R_m \to \begin{pmatrix} \boldsymbol{\Psi}\mathbf{f} \\ \mathbf{r} \end{pmatrix} \in W_n^p \times R_m$ is obviously bounded, we have

2.5. Proposition. *Under the assumptions* 2.1 *the operator* $\mathscr{L}$ *given by* (2,3) *has a closed range in* $L_n^p \times R_m$.

Since by I.5.13 the dual space $(W_n^p)^*$ to W_n^p is isometrically isomorphic with $L_n^q \times R_n^*$ and $(L_n^p \times R_m)^*$ is isometrically isomorphic with $L_n^q \times R_m^*$ (cf. I.3.9 and

I.3.10), the adjoint operator to $\mathscr{L}$ may be represented analytically by the linear bounded operator

$$\mathscr{L}^*: (\mathbf{y}^*, \boldsymbol{\lambda}^*) \in L_n^q \times R_m^* \to (\mathbf{L}_1^*(\mathbf{y}^*, \boldsymbol{\lambda}^*), \mathbf{L}_2^*(\mathbf{y}^*, \boldsymbol{\lambda}^*)) \in L_n^q \times R_n^* \tag{2,8}$$

which is defined by the relation

$$\int_0^1 \mathbf{y}^*(t)\,[\mathbf{D}\mathbf{x} - \mathbf{P}\mathbf{x}]\,(t)\,\mathrm{d}t + \boldsymbol{\lambda}^*[\mathbf{S}\mathbf{x}] = \int_0^1 \mathbf{L}_1^*(\mathbf{y}^*, \boldsymbol{\lambda}^*)\,(t)\,\mathbf{x}'(t)\,\mathrm{d}t + \mathbf{L}_2^*(\mathbf{y}^*, \boldsymbol{\lambda}^*)\,\mathbf{x}(0) \tag{2,9}$$

$$\text{for all} \quad \mathbf{x} \in W_n^p, \quad \mathbf{y}^* \in L_n^q \quad \text{and} \quad \boldsymbol{\lambda}^* \in R_m^*.$$

Analogously, the operator

$$\begin{aligned} \mathbf{T}^*: (\mathbf{y}^*, \boldsymbol{\varkappa}^*, \boldsymbol{\lambda}^*) &\in L_n^q \times R_n^* \times R_m^* \\ &\to (\mathbf{T}_1^*(\mathbf{y}^*, \boldsymbol{\varkappa}^*, \boldsymbol{\lambda}^*), \mathbf{T}_2^*(\mathbf{y}^*, \boldsymbol{\varkappa}^*, \boldsymbol{\lambda}^*), \mathbf{T}_3^*(\mathbf{y}^*, \boldsymbol{\varkappa}^*, \boldsymbol{\lambda}^*)) \in L_n^q \times R_n^* \times R_m^* \end{aligned} \tag{2,10}$$

defined by

$$\begin{aligned} &\int_0^1 \mathbf{y}^*(t)\,(\mathbf{P}\mathbf{x})\,(t)\,\mathrm{d}t + \boldsymbol{\varkappa}^*\,\mathbf{x}(0) + \boldsymbol{\lambda}^*(\mathbf{d} - \mathbf{S}\mathbf{x}) \\ &= \int_0^1 \mathbf{T}_1^*(\mathbf{y}^*, \boldsymbol{\varkappa}^*, \boldsymbol{\lambda}^*)\,\mathbf{x}'(t)\,\mathrm{d}t + \mathbf{T}_2^*(\mathbf{y}^*, \boldsymbol{\varkappa}^*, \boldsymbol{\lambda}^*)\,\mathbf{x}(0) + \mathbf{T}_3^*(\mathbf{y}^*, \boldsymbol{\varkappa}^*, \boldsymbol{\lambda}^*)\,\mathbf{d} \end{aligned} \tag{2,11}$$

$$\text{for all} \quad \mathbf{x} \in W_n^p, \quad \mathbf{d} \in R_m, \quad \mathbf{y}^* \in L_n^q, \quad \boldsymbol{\varkappa}^* \in R_n^*, \quad \boldsymbol{\lambda}^* \in R_m^*$$

represents analytically the adjoint operator to the operator $\mathbf{T}$.

2.6. Theorem. *If 2,1 holds and $\mathbf{P}(t, 1) = \mathbf{0}$ a.e. on $[0, 1]$, then the operator $\mathscr{L}^*: L_n^q \times R_m^* \to L_n^q \times R_n^*$ given by (2,8) verifies (2,9) if and only if*

$$\mathbf{L}_1^*(\mathbf{y}^*, \boldsymbol{\lambda}^*)\,(t) = \mathbf{y}^*(t) + \int_0^1 \mathbf{y}^*(s)\,\mathbf{P}(s, t)\,\mathrm{d}s + \boldsymbol{\lambda}^*\,\mathbf{K}(t) \qquad \textit{a.e. on } [0, 1], \tag{2,12}$$

$$\mathbf{L}_2^*(\mathbf{y}^*, \boldsymbol{\lambda}^*) = \boldsymbol{\lambda}^*\mathbf{M} + \int_0^1 \mathbf{y}^*(s)\,\mathbf{P}(s, 0)\,\mathrm{d}s. \tag{2,13}$$

Proof. Let $\mathbf{x} \in W_n^p$, $\mathbf{y}^* \in L_n^q$ and $\boldsymbol{\lambda}^* \in R_m^*$. By I.4.38

$$\int_0^1 \mathbf{y}^*(t)\,(\mathbf{P}\mathbf{x})\,(t)\,\mathrm{d}t = \int_0^1 \mathrm{d}_t\left[\int_0^1 \mathbf{y}^*(s)\,\mathbf{P}(s, t)\,\mathrm{d}s\right]\mathbf{x}(t).$$

Furthermore, integrating by parts (I.4.33) and taking into account the assumption $\mathbf{P}(t, 1) = \mathbf{0}$ a.e. on $[0, 1]$, we obtain

$$\int_0^1 \mathbf{y}^*(t)\,(\mathbf{P}\mathbf{x})\,(t)\,\mathrm{d}t = -\left(\int_0^1 \mathbf{y}^*(s)\,\mathbf{P}(s, 0)\,\mathrm{d}s\right)\mathbf{x}(0) - \int_0^1 \left(\int_0^1 \mathbf{y}^*(s)\,\mathbf{P}(s, t)\,\mathrm{d}s\right)\mathbf{x}'(t)\,\mathrm{d}t.$$

Hence

$$\int_0^1 \mathbf{y}^*(t)\,[\mathbf{D}\mathbf{x} - \mathbf{P}\mathbf{x}]\,(t)\,dt + \boldsymbol{\lambda}^*(\mathbf{S}\mathbf{x})$$

$$= \left[\boldsymbol{\lambda}^*\mathbf{M} + \int_0^1 \mathbf{y}^*(s)\,\mathbf{P}(s,0)\,ds\right]\mathbf{x}(0) + \int_0^1 \left[\mathbf{y}^*(t) + \int_0^1 \mathbf{y}^*(s)\,\mathbf{P}(s,t)\,ds + \boldsymbol{\lambda}^*\,\mathbf{K}(t)\right]\mathbf{x}'(t)\,dt$$

for all $\mathbf{x} \in W_n^p$, $\mathbf{y}^* \in L_n^q$ and $\boldsymbol{\lambda}^* \in R_m^*$.

In virtue of (2,9) this yields that

$$\int_0^1 \left[\mathbf{L}_1^*(\mathbf{y}^*, \boldsymbol{\lambda}^*)(t) - \mathbf{y}^*(t) - \int_0^1 \mathbf{y}^*(s)\,\mathbf{P}(s,t)\,ds - \boldsymbol{\lambda}^*\,\mathbf{K}(t)\right]\mathbf{x}'(t)\,dt$$

$$+ \left[\mathbf{L}_2^*(\mathbf{y}^*, \boldsymbol{\lambda}^*) - \boldsymbol{\lambda}^*\mathbf{M} - \int_0^1 \mathbf{y}^*(s)\,\mathbf{P}(s,0)\,ds\right]\mathbf{x}(0) = 0$$

holds for all $\mathbf{x} \in W_n^p$, $\mathbf{y}^* \in L_n^q$ and $\boldsymbol{\lambda}^* \in R_m^*$.

The proof will be completed by making use of I.5.15.

Similarly

2.7. Proposition. *If* 2.1 *holds and* $\mathbf{P}(t,1) = \mathbf{0}$ *a.e. on* $[0,1]$, *then the operator* $\mathbf{T}^*\colon L_n^q \times R_n^* \times R_m^* \to L_n^q \times R_n^* \times R_m^*$ *given by* (2,10) *verifies* (2,11) *if and only if*

$$\mathbf{T}_1^*(\mathbf{y}^*, \boldsymbol{\varkappa}^*, \boldsymbol{\lambda}^*)(t) = -\mathbf{L}_1^*(\mathbf{y}^*, \boldsymbol{\lambda}^*)(t) + \mathbf{y}^*(t) \qquad \text{a.e. on } [0,1],$$

$$\mathbf{T}_2^*(\mathbf{y}^*, \boldsymbol{\varkappa}^*, \boldsymbol{\lambda}^*) = -\mathbf{L}_2^*(\mathbf{y}^*, \boldsymbol{\lambda}^*) + \boldsymbol{\varkappa}^*, \qquad \mathbf{T}_3^*(\mathbf{y}^*, \boldsymbol{\varkappa}^*, \boldsymbol{\lambda}^*) = \boldsymbol{\lambda}^*$$

for all $\mathbf{y}^* \in L_n^q$, $\boldsymbol{\varkappa}^* \in R_n^*$ *and* $\boldsymbol{\lambda}^* \in R_m^*$.

2.8. Corollary. $\dim N(\mathscr{L}^*) = \dim N(\mathbf{I} - \mathbf{T}^*) - n < \infty$.

Proof follows readily from 2.6, 2.7 and I.3.20.

2.9. Theorem. *If* 2.1 *holds and* $\mathbf{P}(t,1) = \mathbf{0}$ *a.e. on* $[0,1]$, *then*

$$\operatorname{ind}(\mathscr{L}) = \dim N(\mathscr{L}^*) - \dim N(\mathscr{L}) = m - n\,.$$

Proof. By 2.5 and I.3.15 $\operatorname{codim} R(\mathscr{L}) = \dim N(\mathscr{L}^*)$. Hence by (2,7) and 2.8 and I.3.20

$$\operatorname{ind}(\mathscr{L}) = \dim N(\mathbf{I} - \mathbf{T}^*) - n - \dim N(\mathbf{I} - \mathbf{T}) + m = m - n\,.$$

2.10. Remark. The relation (2,9), where $\mathbf{L}_1^*(\mathbf{y}^*, \boldsymbol{\lambda}^*)$ and $\mathbf{L}_2^*(\mathbf{y}^*, \boldsymbol{\lambda}^*)$ are given by (2,12) and (2,13) is the *Green formula* for BVP (2,1), (2,2).

2.11. Remark. Let $\mathbf{A}, \mathbf{C}, \mathbf{D}\colon [0,1] \to L(R_n)$ be Lebesgue integrable on $[0,1]$, let $\mathbf{P}\colon [0,1] \times [0,1] \to L(R_n)$ be an $L^1[BV]$-kernel and let $\mathbf{K}\colon [0,1] \to L(R_n, R_m)$ be of bounded variation on $[0,1]$ and $\mathbf{M}, \mathbf{N} \in L(R_n, R_m)$. Let us consider the problem of determining $\mathbf{x} \in AC_n$ which verifies the system

$$\mathbf{x}'(t) - \mathbf{A}(t)\,\mathbf{x}(t) - [\mathbf{C}(t)\,\mathbf{x}(0) + \mathbf{D}(t)\,\mathbf{x}(1)] - \int_0^1 \mathrm{d}_s[\mathbf{P}(t,s)]\,\mathbf{x}(s) = \mathbf{f}(t) \quad \text{a.e. on } [0,1] \tag{2,14}$$

and

$$\mathbf{M}\,\mathbf{x}(0) + \mathbf{N}\,\mathbf{x}(1) + \int_0^1 \mathrm{d}[\mathbf{K}(t)]\,\mathbf{x}(t) = \mathbf{r}\,, \tag{2,15}$$

where $\mathbf{f} \in L_n^1$ and $\mathbf{r} \in R_m$. Again we may assume that $\mathbf{P}(t,\,.)$ is for almost all $t \in [0,1]$ right-continuous on $(0,1)$. Moreover, if we put

$$\mathbf{P}_0(t,s) = \begin{cases} \mathbf{P}(t,0+) - \mathbf{P}(t,1-) & \text{if} \quad s = 0\,, \\ \mathbf{P}(t,s) \quad - \mathbf{P}(t,1-) & \text{if} \quad 0 < s < 1\,, \\ \mathbf{0} & \text{if} \quad s = 1 \end{cases}$$

and $\mathbf{C}_0(t) = \mathbf{C}(t) - [\mathbf{P}(t,0+) - \mathbf{P}(t,0)]$, $\mathbf{D}_0(t) = \mathbf{D}(t) - [\mathbf{P}(t,1) - \mathbf{P}(t,1-)]$, for any $\mathbf{x} \in AC_n$ we should obtain

$$\mathbf{C}(t)\,\mathbf{x}(0) + \mathbf{D}(t)\,\mathbf{x}(1) + \int_0^1 \mathrm{d}_s[\mathbf{P}(t,s)]\,\mathbf{x}(s)$$
$$= \mathbf{C}_0(t)\,\mathbf{x}(0) + \mathbf{D}_0(t)\,\mathbf{x}(1) + \int_0^1 \mathrm{d}_s[\mathbf{P}_0(t,s)]\,\mathbf{x}(s)\,.$$

Hence, without any loss of generality we may assume that for almost all $t \in [0,1]$ $\mathbf{P}(t,\,.)$ is right-continuous on $[0,1)$, left-continuous at 1 and $\mathbf{P}(t,1) = \mathbf{0}$. Analogously, $\mathbf{K}$ may be assumed right-continuous on $[0,1)$, left-continuous at 1 and $\mathbf{K}(1) = \mathbf{0}$.

According to 1.12 we may rewrite the equation (2,14) in the form

$$\mathbf{x}'(t) - \int_0^1 \mathrm{d}_s[\mathbf{R}(t,s)]\,\mathbf{x}(s) = \mathbf{f}(t) \qquad \text{a.e. on } [0,1]\,, \tag{2,16}$$

where

$$\mathbf{R}(t,s) = \mathbf{P}(t,s) + \begin{cases} -\mathbf{A}(t) - \mathbf{C}(t) - \mathbf{D}(t) & \text{if} \quad s = 0\,, \\ -\mathbf{A}(t) - \mathbf{D}(t) & \text{if} \quad 0 < s < t\,, \\ -\mathbf{D}(t) & \text{if} \quad t \le s < 1\,, \\ \mathbf{0} & \text{if} \quad s = 1 \end{cases}$$

is again an $L^1[BV]$-kernel. Furthermore, applying the integration-by-parts formula and taking into account that $\mathbf{K}(1) = \mathbf{0}$ and

$$\mathbf{x}(1) = \mathbf{x}(0) + \int_0^1 \mathbf{x}'(\tau)\,\mathrm{d}\tau \qquad \text{for any} \quad \mathbf{x} \in AC_n\,,$$

we transfer the side condition (2,15) into

$$\mathbf{H}\,\mathbf{x}(0) + \int_0^1 \mathbf{F}(t)\,\mathbf{x}'(t)\,\mathrm{d}t = \mathbf{r}\,, \tag{2,17}$$

where

$$\mathbf{H} = \mathbf{M} + \mathbf{N} - \mathbf{K}(0)\,, \qquad \mathbf{F}(t) = \mathbf{N} - \mathbf{K}(t)\,.$$

The system (2,16), (2,17) may be written as the operator equation

$$\mathcal{R}\mathbf{x} = \begin{pmatrix} \mathbf{f} \\ \mathbf{r} \end{pmatrix}$$

with $\mathcal{R}$: $AC_n \to L_n^1 \times R_m$ defined in an obvious way. Now, proceeding analogously as in the close of the proof of IV.3.13 we may deduce from 2.6 that $(\mathbf{y}^*, \boldsymbol{\lambda}^*) \in N(\mathcal{R}^*)$ if and only if there exists $\mathbf{z} \in BV_n$ such that $\mathbf{z}(t) = \mathbf{y}(t)$ a.e. on $[0, 1]$, $\mathbf{z}(0+) = \mathbf{z}(0)$, $\mathbf{z}(1-) = \mathbf{z}(1)$ and

$$\mathbf{z}^*(t) + \int_0^1 \mathbf{z}^*(s)\,\mathbf{R}(s, t)\,\mathrm{d}s + \boldsymbol{\lambda}^*\,\mathbf{F}(t) = \mathbf{0} \qquad \text{on } (0, 1)\,, \tag{2,18}$$

$$\boldsymbol{\lambda}^*\mathbf{H} + \int_0^1 \mathbf{z}^*(s)\,\mathbf{R}(s, 0)\,\mathrm{d}s = \mathbf{0}\,. \tag{2,19}$$

As $\mathbf{F}(1-) = \mathbf{F}(1) = \mathbf{N}$ and $\mathbf{R}(t, 1-) = -\mathbf{D}(t)$ for almost all $t \in [0, 1]$, we have by (2,18)

$$\mathbf{z}^*(1) = \int_0^1 \mathbf{z}^*(s)\,\mathbf{D}(s)\,\mathrm{d}s - \boldsymbol{\lambda}^*\mathbf{N}\,. \tag{2,20}$$

Since $\mathbf{F}(0+) = \mathbf{F}(0) = \mathbf{N} - \mathbf{K}(0)$ and $\mathbf{R}(t, 0+) = \mathbf{P}(t, 0) - \mathbf{A}(t) - \mathbf{D}(t)$ for almost all $t \in [0, 1]$, the relations (2,18) and (2,19) imply

$$\begin{aligned}
\mathbf{z}^*(0) &= -\int_0^1 \mathbf{z}^*(s)\,\mathbf{P}(s, 0)\,\mathrm{d}s + \int_0^1 \mathbf{z}^*(s)\,\mathbf{D}(s)\,\mathrm{d}s + \int_0^1 \mathbf{z}^*(s)\,\mathbf{A}(s)\,\mathrm{d}s - \boldsymbol{\lambda}^*\mathbf{N} + \boldsymbol{\lambda}^*\,\mathbf{K}(0) \\
&= -\left[\boldsymbol{\lambda}^*\mathbf{H} + \int_0^1 \mathbf{z}^*(s)\,\mathbf{R}(s, 0)\,\mathrm{d}s\right] - \int_0^1 \mathbf{z}^*(s)\,\mathbf{C}(s)\,\mathrm{d}s + \boldsymbol{\lambda}^*\mathbf{M} \\
&= -\int_0^1 \mathbf{z}^*(s)\,\mathbf{C}(s)\,\mathrm{d}s + \boldsymbol{\lambda}^*\mathbf{M}\,.
\end{aligned}$$

By the definition of $\mathbf{R}$ and $\mathbf{F}$ we have for any $\mathbf{z} \in BV_n$ and $\boldsymbol{\lambda} \in R_m$ fulfilling (2,20)

$$\begin{aligned}
&\int_0^1 \mathbf{z}^*(s)\,\mathbf{R}(s, t)\,\mathrm{d}s + \boldsymbol{\lambda}^*\,\mathbf{F}(t) \\
&= \left(\boldsymbol{\lambda}^*\mathbf{N} - \int_0^1 \mathbf{z}^*(s)\,\mathbf{D}(s)\,\mathrm{d}s\right) - \int_t^1 \mathbf{z}^*(s)\,\mathbf{A}(s)\,\mathrm{d}s + \int_0^1 \mathbf{z}^*(s)\,\mathbf{P}(s, t)\,\mathrm{d}s - \boldsymbol{\lambda}^*\,\mathbf{K}(t) \\
&= -\mathbf{z}^*(1) - \int_t^1 \mathbf{z}^*(s)\,\mathbf{A}(s)\,\mathrm{d}s + \int_0^1 \mathbf{z}^*(s)\,\mathbf{P}(s, t)\,\mathrm{d}s - \boldsymbol{\lambda}^*\,\mathbf{K}(t) \qquad \text{on } [0, 1]\,.
\end{aligned}$$

Thus, the adjoint problem to BVP (2,14), (2,15) is equivalent to the problem of determining $\mathbf{z} \in BV_n$ and $\boldsymbol{\lambda}^* \in R_m^*$ such that

$$(2,21)\quad \mathbf{z}^*(t) = \mathbf{z}^*(1) + \int_t^1 \mathbf{z}^*(s)\,\mathbf{A}(s)\,\mathrm{d}s - \int_0^1 \mathbf{z}^*(s)\,\mathbf{P}(s,t)\,\mathrm{d}s - \boldsymbol{\lambda}^*\,\mathbf{K}(t) \qquad \text{on } [0,1]$$

$$(2,22)\qquad \mathbf{z}^*(0) + \boldsymbol{\lambda}^*\mathbf{M} + \int_0^1 \mathbf{z}^*(s)\,\mathbf{C}(s)\,\mathrm{d}s = \mathbf{0}\,,$$

$$\mathbf{z}^*(1) - \boldsymbol{\lambda}^*\mathbf{N} - \int_0^1 \mathbf{z}^*(s)\,\mathbf{D}(s)\,\mathrm{d}s = \mathbf{0}\,.$$

2.12. Theorem. *Let us assume* 2.1 *and* $\mathbf{P}(t,1) = \mathbf{0}$ *a.e. on* $[0,1]$. *Then for given* $\mathbf{f} \in L_n^p$ *and* $\mathbf{r} \in R_m$ *BVP* (2,1), (2,2) *possesses a solution if and only if*

$$\int_0^1 \mathbf{y}^*(t)\,\mathbf{f}(t)\,\mathrm{d}t + \boldsymbol{\lambda}^*\mathbf{r} = 0$$

for any couple $(\mathbf{y}^*, \boldsymbol{\lambda}^*) \in L_n^q \times R_m^*$ *which verifies the adjoint system*

$$(2,23)\qquad \mathbf{y}^*(t) + \int_0^1 \mathbf{y}^*(s)\,\mathbf{P}(s,t)\,\mathrm{d}s + \boldsymbol{\lambda}^*\,\mathbf{K}(t) = \mathbf{0} \qquad \text{a.e. on } [0,1]\,,$$

$$(2,24)\qquad \boldsymbol{\lambda}^*\mathbf{M} + \int_0^1 \mathbf{y}^*(s)\,\mathbf{P}(s,0)\,\mathrm{d}s = \mathbf{0}\,.$$

Proof follows from 2.5, 2.6 and I.3.14 (cf. I.3.23).

2.13. Theorem. *Let us assume* 2.1 *and* $\mathbf{P}(t,1) = \mathbf{0}$ *a.e. on* $[0,1]$. *Then for given* $\mathbf{g}^* \in L_n^q$ *and* $\mathbf{q}^* \in R_n^*$, *the system*

$$\mathbf{y}^*(t) + \int_0^1 \mathbf{y}^*(s)\,\mathbf{P}(s,t)\,\mathrm{d}s + \boldsymbol{\lambda}^*\,\mathbf{K}(t) = \mathbf{g}^*(t) \qquad \text{a.e. on } [0,1]\,,$$

$$\boldsymbol{\lambda}^*\mathbf{M} + \int_0^1 \mathbf{y}^*(s)\,\mathbf{P}(s,0)\,\mathrm{d}s = \mathbf{q}^*$$

possesses a solution $(\mathbf{y}^*, \boldsymbol{\lambda}^*) \in L_n^q \times R_m^*$ *if and only if*

$$\int_0^1 \mathbf{g}^*(t)\,\mathbf{x}'(t)\,\mathrm{d}t + \mathbf{q}^*\,\mathbf{x}(0) = 0$$

holds for any solution $\mathbf{x} \in W_n^p$ *of the homogeneous problem* $\mathscr{L}\mathbf{x} = \mathbf{0}$.

Proof follows again from 2,5, 2.6 and I.3.14.

2.14. Remark. Let us notice that the side condition (2,2) is linearly dependent if there exists $\mathbf{q} \in R_m$ such that $\mathbf{q}^*\mathbf{M} = \mathbf{q}^*\,\mathbf{K}(t) = \mathbf{0}$ a.e. on $[0,1]$ $(\mathbf{q}^*(\mathbf{S}\mathbf{x}) = 0$ for all

$\mathbf{x} \in W_n^p$ implies that

$$\mathbf{x} \in W_n^p \to \mathbf{q}^*(\mathbf{S}\mathbf{x}) = (\mathbf{q}^*\mathbf{M})\,\mathbf{x}(0) + \int_0^1 (\mathbf{q}^*\,\mathbf{K}(t))\,\mathbf{x}'(t)\,\mathrm{d}t \in R$$

is the zero functional on W_n^p).

Analogously as in the case of Stieltjes-integral side conditions (cf. IV.1.14, where no use of the special form of side conditions was made), we can also show that to any nonzero linear operator $\mathbf{S}_0$: $W_n^p \to R_k$ and $\mathbf{r}_0 \in R_k$ such that $\mathbf{q}^*(\mathbf{S}_0\mathbf{x}) = \mathbf{0}$ for any $\mathbf{x} \in AC_n$ implies $\mathbf{q}^*\mathbf{r}_0 = \mathbf{0}$, there exist $m \leq k$, $\mathbf{S}$: $W_n^p \to R_m$ and $\mathbf{r} \in R_m$ such that the condition $\mathbf{S}\mathbf{x} = \mathbf{r}$ is linearly independent and equivalent to $\mathbf{S}_0\mathbf{x} = \mathbf{r}_0$.

2.15. Remark. It follows from the proof of IV.1.15 that if (2,2) is reasonable and linearly independent, then there exists a regular $m \times m$-matrix $\boldsymbol{\Theta}$ such that

$$\boldsymbol{\Theta}[\mathbf{M}, \mathbf{K}(t)] = \begin{bmatrix} \mathbf{M}_0, & \mathbf{0} \\ \mathbf{M}_1, & \mathbf{K}_1(t) \\ \mathbf{0}, & \mathbf{K}_2(t) \end{bmatrix} \qquad \text{a.e. on } [0, 1],$$

where $\mathbf{M}_0 \in L(R_n, R_{m_0})$, $\mathbf{M}_1$ and $\mathbf{K}_1(t) \in L(R_n, R_{m_1})$ and $\mathbf{K}_2(t) \in L(R_n, R_{m_2})$ are such that $m_0 + m_1 + m_2 = m$, $\operatorname{rank} \begin{bmatrix} \mathbf{M}_0 \\ \mathbf{M}_1 \end{bmatrix} = m_0 + m_1$ and the rows of $\begin{bmatrix} \mathbf{K}_1(t) \\ \mathbf{K}_2(t) \end{bmatrix}$ are linearly independent in L_n^q, i.e.

$$\mathbf{q}^* \begin{bmatrix} \mathbf{K}_1(t) \\ \mathbf{K}_2(t) \end{bmatrix} = \mathbf{0} \qquad \text{a.e. on } [0, 1]$$

implies $\mathbf{q}^* = \mathbf{0}$. The system

$$\begin{aligned} \mathbf{M}_0\,\mathbf{x}(0) &= \mathbf{r}_0\,, \\ \mathbf{M}_1\,\mathbf{x}(0) + \int_0^1 \mathbf{K}_1(t)\,\mathbf{x}'(t)\,\mathrm{d}t &= \mathbf{r}_1 \qquad \left(\begin{pmatrix} \mathbf{r}_0 \\ \mathbf{r}_1 \\ \mathbf{r}_2 \end{pmatrix} = \boldsymbol{\Theta}\mathbf{r} \right), \\ \int_0^1 \mathbf{K}_2(t)\,\mathbf{x}'(t)\,\mathrm{d}t &= \mathbf{r}_2 \end{aligned}$$

is the canonical form of the side condition (2,2).

2.16. Remark. Another possible functional analytic way of attacking BVP (2,1), (2,2) with $\mathbf{r} \in R_m$ fixed consists in considering the linear operator $\mathscr{L}_r$ defined on $D(\mathscr{L}_r) = \{\mathbf{x} \in W_n^p;\ \mathbf{S}\mathbf{x} = \mathbf{r}\} \subset W_n^p$ by

$$\mathscr{L}_r\colon\ \mathbf{x} \in D(\mathscr{L}_r) \to \mathbf{D}\mathbf{x} - \mathbf{P}\mathbf{x} \in L_n^p\,.$$

BVP (2,1), (2,2) may be rewritten as the operator equation

$$\mathscr{L}_r\mathbf{x} = \mathbf{f}\,.$$

As $R(\mathscr{L}_r)$ is the set of all $\mathbf{f} \in L_n^p$ for which $\begin{pmatrix}\mathbf{f}\\ \mathbf{r}\end{pmatrix} \in R(\mathscr{L})$ and $R(\mathscr{L})$ is closed by 2.5, $R(\mathscr{L}_r)$ is also closed. By 2.12 $R(\mathscr{L}_r)$ is the set of all $\mathbf{f} \in L_n^p$ which fulfil the relation

$$\int_0^1 \mathbf{y}^*(t)\,\mathbf{f}(t)\,\mathrm{d}t + \boldsymbol{\lambda}^*\mathbf{r} = 0$$

for all couples $(\mathbf{y}^*, \boldsymbol{\lambda}^*) \in N(\mathscr{L}^*) \subset L_n^q \times R_m^*$. In particular, if N_0^* denotes the set of all $\mathbf{y}^* \in L_n^q$ for which there exists $\boldsymbol{\lambda}^* \in R_m^*$ such that $(\mathbf{y}^*, \boldsymbol{\lambda}^*) \in N(\mathscr{L}^*)$, then

$$R(\mathscr{L}_0) = {}^{\perp}(N_0^*)$$

(the set of all $\mathbf{f} \in L_n^p$ for which $\langle \mathbf{f}, \mathbf{y}^* \rangle_L = 0$ for any $\mathbf{y}^* \in N_0^*$).

2.17. Proposition. *$R(\mathscr{L}_0)^{\perp} = N_0^*$, where $R(\mathscr{L}_0)^{\perp}$ denotes the set of all $\mathbf{y}^* \in L_n^q$ such that*

$$\int_0^1 \mathbf{y}^*(t)\,\mathbf{f}(t)\,\mathrm{d}t = 0 \qquad \textit{for any} \quad \mathbf{f} \in R(\mathscr{L}_0)$$

and N_0^* is the set of all $\mathbf{y}^* \in L_n^q$ for which there exists $\boldsymbol{\lambda}^* \in R_m^*$ such that $(\mathbf{y}^*, \boldsymbol{\lambda}^*) \in N(\mathscr{L}^*)$ (i.e. (2,23), (2,24) hold).

Proof. Let $\mathbf{y}^* \in L_n^q$. Then $\mathbf{y}^* \in R(\mathscr{L}_0)^{\perp}$ if and only if

$$0 = \int_0^1 \mathbf{y}^*(t)\,[\mathbf{D}\mathbf{x} - \mathbf{P}\mathbf{x}]\,(t)\,\mathrm{d}t$$

$$= \int_0^1 \left[\mathbf{y}^*(t) + \int_0^1 \mathbf{y}^*(s)\,\mathbf{P}(s, t)\,\mathrm{d}s\right] \mathbf{x}'(t)\,\mathrm{d}t + \left[\int_0^1 \mathbf{y}^*(s)\,\mathbf{P}(s, 0)\,\mathrm{d}s\right] \mathbf{x}(0)$$

holds for every $\mathbf{x} \in D(\mathscr{L}_0) = N(\mathbf{S})$.

This is true if and only if $(\mathbf{u}^*, \mathbf{v}^*) \in N(\mathbf{S})^{\perp}$, where

$$\mathbf{u}^*(t) = \mathbf{y}^*(t) + \int_0^1 \mathbf{y}^*(s)\,\mathbf{P}(s, t)\,\mathrm{d}s \qquad \text{on } [0, 1], \tag{2,25}$$

$$\mathbf{v}^* = \int_0^1 \mathbf{y}^*(s)\,\mathbf{P}(s, 0)\,\mathrm{d}s\,.$$

Since $R(\mathbf{S})$ is a linear subspace in R_m, it is certainly closed and thus according to I.3.14 $N(\mathbf{S})^{\perp} = R(\mathbf{S}^*)$, where

$$\mathbf{S}^*\colon \boldsymbol{\lambda}^* \in R_m^* \to (\mathbf{S}_1^*\boldsymbol{\lambda}^*, \mathbf{S}_2^*\boldsymbol{\lambda}^*) \in L_n^q \times R_n^*$$

is the adjoint of $\mathbf{S}$ defined by the relation

$$\boldsymbol{\lambda}^*(\mathbf{S}\mathbf{x}) = \int_0^1 (\mathbf{S}_1^*\boldsymbol{\lambda}^*)\,(t)\,\mathbf{x}'(t)\,\mathrm{d}t + (\mathbf{S}_2^*\boldsymbol{\lambda}^*)\,\mathbf{x}(0) \qquad \text{for all} \quad \mathbf{x} \in W_n^p \quad \text{and} \quad \boldsymbol{\lambda}^* \in R_m^*\,.$$

Obviously, $(\mathbf{S}_1^*\boldsymbol{\lambda}^*)(t) = \boldsymbol{\lambda}^* \mathbf{K}(t)$ a.e. on $[0,1]$ and $\mathbf{S}_2^*\boldsymbol{\lambda}^* = \boldsymbol{\lambda}^*\mathbf{M}$. This means that $(\mathbf{u}^*, \mathbf{v}^*) \in N(\mathbf{S})^\perp$ if and only if there exists $\boldsymbol{\lambda}^* \in R_m^*$ such that

$$\mathbf{u}^*(t) = \boldsymbol{\lambda}^* \mathbf{K}(t) \quad \text{a.e. on } [0,1], \qquad \mathbf{v}^* = \boldsymbol{\lambda}^*\mathbf{M},$$

wherefrom $R(\mathscr{L}_0)^\perp = N_0^*$ follows immediately by (2,25).

2.18. Remark. Since by 2.8 $\dim N_0^* < \infty$, Proposition 2.17 is a consequence of the following general assertion due to J. Dieudonné (cf. Goldberg II.3.6).

If Y is a linear normed space, $N \subset Y^*$, $\dim N < \infty$, then $({}^\perp N)^\perp = N$.

3. Green's function

Let us continue the investigation of the operator

$$\mathscr{L}\colon \mathbf{x} \in W_n^p \to \begin{bmatrix} \mathbf{Dx} - \mathbf{Px} \\ \mathbf{Sx} \end{bmatrix} \in L_n^p \times R_m,$$

given by (2,15). (cf. also (1,6), (1,3) and (2,2).) We assume again that 2.1 holds. Moreover, we assume that $\mathbf{P}(t,1) = \mathbf{0}$ a.e. on $[0,1]$ (cf. 1.15 and 2.2).

Of particular interest is the case when the operator equation

$$\mathscr{L}\mathbf{x} = \begin{pmatrix} \mathbf{f} \\ \mathbf{r} \end{pmatrix} \tag{3,1}$$

(or BVP (2,1), (2,2)) has a unique solution for any $\mathbf{f} \in L_n^p$ and $\mathbf{r} \in R_m$.

3.1. Notation. Throughout the section $l = \dim N(\mathbf{D} - \mathbf{P})$, $\mathbf{X}(t)$ is an arbitrary $n \times k$-matrix valued function whose columns form a basis in $N(\mathbf{D} - \mathbf{P})$ and $(\mathbf{SX})$ is the $m \times l$-matrix

$$(\mathbf{SX}) = \mathbf{M}\,\mathbf{X}(0) + \int_0^1 \mathbf{K}(t)\,\mathbf{X}'(t)\,\mathrm{d}t. \tag{3,2}$$

(According to 1.8 $n \le l < \infty$.)

3.2. Lemma. $\dim N(\mathscr{L}) = l - \operatorname{rank}(\mathbf{SX})$.

Proof. By the definition of $\mathbf{X}(t)$ we have $\mathbf{x} \in N(\mathscr{L})$ if and only if $\mathbf{x}(t) \equiv \mathbf{X}(t)\,\mathbf{c}$ on $[0,1]$, where $\mathbf{c} \in R_l$ is such that

$$(\mathbf{SX})\,\mathbf{c} = \mathbf{0}. \tag{3,3}$$

Obviously, the functions $\mathbf{X}(t)\,\mathbf{c}_j$ with $\mathbf{c}_j \in R_l$ $(j = 1, 2, \ldots, \nu)$ are linearly dependent in W_n^p if and only if the vectors $\mathbf{c}_j$ $(j = 1, 2, \ldots, \nu)$ are linearly dependent. The assertion of the lemma follows immediately.

3.3. Remark. Since rank $(\mathbf{SX}) \le m$ and $l \ge n$, 3.2 implies

$$\dim N(\mathscr{L}) \ge n - m\,.$$

3.4. Lemma. $R(\mathscr{L}) = L_n^p \times R_m$ *if and only if* $\dim N(\mathscr{L}) = n - m$.

Proof. Since by 2.5 $R(\mathscr{L})$ is closed in $L_n^p \times R_m$, $R(\mathscr{L}) = L_n^p \times R_m$ if and only if

$$0 = \operatorname{codim} R(\mathscr{L}) = \dim ((L_n^p \times R_m)/R(\mathscr{L})) = \dim N(\mathscr{L}^*) \tag{3,4}$$

(cf. I.3.11). According to 2.9

$$\dim N(\mathscr{L}^*) = \dim N(\mathscr{L}) + m - n$$

wherefrom by (3,4) the assertion of the lemma follows.

3.5. Corollary. *BVP* (2,1), (2,2) *possesses a unique solution for any* $\mathbf{f} \in L_n^p$ *and* $\mathbf{r} \in R_m$ *if and only if*

$$m = n \quad \textit{and} \quad \dim N(\mathscr{L}) = 0\,. \tag{3,5}$$

Proof follows from 3.4 taking into account that (3,1) has a unique solution for any $\begin{pmatrix}\mathbf{f}\\ \mathbf{r}\end{pmatrix} \in R(\mathscr{L})$ if and only if $\dim N(\mathscr{L}) = 0$.

Analogously as in the case of ordinary differential equations we want to represent solutions to (3,1) in the form

$$\mathbf{x}(t) = \int_0^1 \mathbf{G}(t,s)\,\mathbf{f}(s)\,\mathrm{d}s + \mathbf{H}(t)\,\mathbf{r} \qquad \text{on } [0,1]\,. \tag{3,6}$$

3.6. Definition. A couple of functions $\mathbf{G}\colon [0,1] \times [0,1] \to L(R_n)$ and $\mathbf{H}\colon [0,1] \to L(R_n)$ is said to be a *Green couple of BVP* (2,1), (2,2) if for any $t \in [0,1]$ the rows of $\mathbf{G}(t,.)$ are elements of L_n^q and the function (3,6) is for any $\mathbf{f} \in L_n^p$ and $\mathbf{r} \in R_n$ the unique solution of BVP (2,1), (2,2).

Clearly, (3,6) verifies (3,1) for any $\mathbf{f} \in L_n^p$ and $\mathbf{r} \in R_n$ if and only if

$$\begin{aligned}\mathbf{x}(t) = {} & \int_0^1 \mathbf{G}(t,s)\left[\mathbf{x}'(s) - \int_0^1 \mathrm{d}_\sigma[\mathbf{P}(s,\sigma)]\,\mathbf{x}(\sigma)\right]\mathrm{d}s \\ & + \mathbf{H}(t)\left[\mathbf{M}\,\mathbf{x}(0) + \int_0^1 \mathbf{K}(s)\,\mathbf{x}'(s)\,\mathrm{d}s\right] \qquad \text{on } [0,1]\end{aligned} \tag{3,7}$$

holds for any $\mathbf{x} \in W_n^p$. If for any $t \in [0,1]$ the rows of $\mathbf{G}(t,.)$ are elements of L_n^q, then by I.4.33 and I.4.38

$$\begin{aligned}\int_0^1 \mathbf{G}(t,\sigma)\left(\int_0^1 \mathrm{d}_s[\mathbf{P}(\sigma,s)]\,\mathbf{x}(s)\,\mathrm{d}\sigma\right) &= \int_0^1 \mathrm{d}_s\left[\int_0^1 \mathbf{G}(t,\sigma)\,\mathbf{P}(\sigma,s)\,\mathrm{d}\sigma\right]\mathbf{x}(s) \\ &= -\left(\int_0^1 \mathbf{G}(t,\sigma)\,\mathbf{P}(\sigma,0)\,\mathrm{d}\sigma\right)\mathbf{x}(0) - \int_0^1 \left(\int_0^1 \mathbf{G}(t,\sigma)\,\mathbf{P}(\sigma,s)\,\mathrm{d}\sigma\right)\mathbf{x}'(s)\,\mathrm{d}s\end{aligned}$$

for any $t \in [0,1]$ and any $\mathbf{x} \in W_n^p$. (We assume $\mathbf{P}(t,1) = \mathbf{0}$.) Consequently the right-hand side of (3,7) becomes

$$\int_0^1 \left[\mathbf{G}(t,s) + \int_0^1 \mathbf{G}(t,\sigma)\,\mathbf{P}(\sigma,s)\,\mathrm{d}\sigma + \mathbf{H}(t)\,\mathbf{K}(s) \right] \mathbf{x}'(s)\,\mathrm{d}s$$
$$+ \left[\mathbf{H}(t)\,\mathbf{M} + \int_0^1 \mathbf{G}(t,\sigma)\,\mathbf{P}(\sigma,0)\,\mathrm{d}\sigma \right] \mathbf{x}(0).$$

Thus, since for any $\mathbf{x} \in W_n^p$

$$\mathbf{x}(t) = \mathbf{x}(0) + \int_0^t \mathbf{x}'(\tau)\,\mathrm{d}\tau = \mathbf{x}(0) + \int_0^1 \Delta(t,s)\,\mathbf{x}'(s)\,\mathrm{d}s \qquad \text{on } [0,1],$$

where

$$\Delta(t,s) = \begin{cases} \mathbf{0} & \text{if } t < s, \\ \mathbf{I} & \text{if } t \geq s, \end{cases} \tag{3,8}$$

the relation (3,7) may be rewritten as follows

$$\int_0^1 \left[\mathbf{G}(t,s) + \int_0^1 \mathbf{G}(t,\sigma)\,\mathbf{P}(\sigma,s)\,\mathrm{d}\sigma + \mathbf{H}(t)\,\mathbf{K}(s) - \Delta(t,s) \right] \mathbf{x}'(s)\,\mathrm{d}s \tag{3,9}$$
$$+ \left[\mathbf{H}(t)\,\mathbf{M} + \int_0^1 \mathbf{G}(t,\sigma)\,\mathbf{P}(\sigma,0)\,\mathrm{d}\sigma - \mathbf{I} \right] \mathbf{x}(0) = \mathbf{0} \qquad \text{for any} \quad \mathbf{x} \in W_n^p.$$

Applying I.5.15 we complete the proof of the following

3.7. Proposition. *Let us assume* 2.1 *and* $\mathbf{P}(t,1) = \mathbf{0}$ *a.e. on* $[0,1]$. *Let* $\mathbf{G}$: $[0,1] \times [0,1] \to L(R_n)$ *and* $\mathbf{H}$: $[0,1] \to L(R_n)$ *and let* $\mathbf{G}(t,.)$ *be* L^q*-intergrable on* $[0,1]$ *for any* $t \in [0,1]$. *Then* $\mathbf{G}(t,s)$, $\mathbf{H}(t)$ *is a Green couple of BVP* (2,1), (2,2) *if and only if* (3,5) *holds and for any* $t \in [0,1]$

$$\mathbf{G}(t,s) + \int_0^1 \mathbf{G}(t,\sigma)\,\mathbf{P}(\sigma,s)\,\mathrm{d}\sigma + \mathbf{H}(t)\,\mathbf{K}(s) = \Delta(t,s) \qquad \textit{for a.e.} \quad s \in [0,1], \tag{3,10}$$
$$\mathbf{H}(t)\,\mathbf{M} + \int_0^1 \mathbf{G}(t,\sigma)\,\mathbf{P}(\sigma,0)\,\mathrm{d}\sigma = \mathbf{I},$$

where $\Delta(t,s)$ *is given by* (3,8).

Moreover, we have

3.8. Proposition. *Let the assumptions of* 3.7 *be satisfied. If* $m = n$ *and for any* $t \in [0,1]$ $\mathbf{G}(t,s)$ *and* $\mathbf{H}(t)$ *satisfy the system* (3,10), *then* $\mathbf{G}(t,s)$, $\mathbf{H}(t)$ *is a Green couple of BVP* (2,1), (2,2).

Proof. Since (3,10) implies that (3,9) and consequently also (3,7) hold for any $\mathbf{x} \in W_n^p$, it is easy to see that then (3,6) is a solution to BVP (2,1), (2,2) for any couple

$\begin{pmatrix} \boldsymbol{f} \\ \boldsymbol{r} \end{pmatrix} \in R(\mathscr{L})$. Furthermore, if $\boldsymbol{x}_1, \boldsymbol{x}_2 \in W_n^p$ and $\mathscr{L}\boldsymbol{x}_1 = \mathscr{L}\boldsymbol{x}_2 = \begin{pmatrix} \boldsymbol{f} \\ \boldsymbol{r} \end{pmatrix}$, then inserting $\boldsymbol{x} = \boldsymbol{x}_1$ and $\boldsymbol{x} = \boldsymbol{x}_2$ into (3,7) we obtain

$$\boldsymbol{x}_1(t) \equiv \int_0^1 \boldsymbol{G}(t, s)\, \boldsymbol{f}(s)\, \mathrm{d}s + \boldsymbol{H}(t)\, \boldsymbol{r} \equiv \boldsymbol{x}_2(t) \qquad \text{on } [0, 1],$$

i.e. $\dim N(\mathscr{L}) = 0$. If $m = n$, then by 2.9 $\operatorname{codim} R(\mathscr{L}) = \dim N(\mathscr{L}) = 0$. Thus $R(\mathscr{L}) = L_n^p \times R_m$ (cf. 1.9) and this completes the proof.

Let $\mathscr{L}^*$: $L_n^q \times R_m^* \to L_n^q \times R_n^*$ denote again the analytical representation of the adjoint operator to $\mathscr{L}$ given by 2.6.

3.9. Lemma. *If* (3,5) *holds, then* $\dim N(\mathscr{L}^*) = 0$ *and* $R(\mathscr{L}^*) = L_n^q \times R_m^*$.

Proof. By 2.9 (3,5) implies $0 = \dim N(\mathscr{L}) = \operatorname{codim} R(\mathscr{L}^*) = \dim N(\mathscr{L}^*)$ and the proof will be completed by means of 1.9.

Lemma 3.9 together with the Bounded Inverse Theorem I.3.4 yields

3.10. Proposition. *The operator* $\mathscr{L}^*$: $L_n^q \times R_n^* \to L_n^q \times R_n^*$ *defined by* 2.9 *possesses a bounded inverse.*

3.11. Theorem. *Let us assume* 2.1 *with* $\boldsymbol{P}(t, 1) = \boldsymbol{0}$ *a.e. on* $[0, 1]$ *and* (3,5). *Then there exist functions* $\boldsymbol{G}$: $[0, 1] \times [0, 1] \to L(R_n)$ *and* $\boldsymbol{H}$: $[0, 1] \to L(R_n)$ *which verify the system* (3,10) *for any* $t \in [0, 1]$. *Moreover,*

(i) *given* $t \in [0, 1]$, $\|\boldsymbol{G}(t, .)\|_{L^q} < \infty$ $(q = p/(p-1)$ *if* $p > 1$, $q = \infty$ *if* $p = 1)$,

(ii) *there exists* $\beta \in R$ *such that*

$$\|\boldsymbol{G}(t, .)\|_{L^q} + |\boldsymbol{H}(t)| \le \beta < \infty \qquad \textit{for any} \quad t \in [0, 1],$$

(iii) *if* $\tilde{\boldsymbol{G}}$: $[0, 1] \times [0, 1] \to L(R_n)$ *and* $\tilde{\boldsymbol{H}}$: $[0, 1] \to L(R_n)$ *also fulfil* (3,10) *for any* $t \in [0, 1]$, (i) *and* (ii), *then* $\tilde{\boldsymbol{G}}(t, s) = \boldsymbol{G}(t, s)$ *and* $\tilde{\boldsymbol{H}}(t) = \boldsymbol{H}(t)$ *for all* $t \in [0, 1]$ *and for a.e.* $s \in [0, 1]$.

Proof. Let $\boldsymbol{\delta}_j^*(t, s)$ and $\mathbf{e}_j^*$ $(j = 1, 2, \ldots, n)$ be the rows of $\Delta(t, s)$ and $\boldsymbol{I}$, respectively. By 3.10 any equation from the system

$$\mathscr{L}^*(\boldsymbol{g}^*, \boldsymbol{h}^*) = (\boldsymbol{\delta}_j^*(t, .), \mathbf{e}_j^*), \qquad t \in [0, 1], \quad j = 1, 2, \ldots, n \tag{3,11}$$

has a unique solution $(\boldsymbol{g}_j^*(t, .), \boldsymbol{h}_j^*(t))$ in $L_n^q \times R_n^*$ and

$$\|\boldsymbol{g}_j^*(t, .)\|_{L^q} + |\boldsymbol{h}_j^*(t)| \le \varkappa(\|\boldsymbol{\delta}_j^*(t, .)\|_{L^q} + |\mathbf{e}_j^*|) \quad \text{for any } t \in [0, 1] \text{ and } j = 1, 2, \ldots, n, \tag{3,12}$$

where $\varkappa = \|(\mathscr{L}^*)^{-1}\| < \infty$. Let us put

$$\mathbf{G}(t,s) = [\mathbf{g}_1(t,s), \mathbf{g}_2(t,s), \dots, \mathbf{g}_n(t,s)]^* \qquad \text{on } [0,1] \times [0,1],$$
$$\mathbf{H}(t) = [\mathbf{h}_1(t), \mathbf{h}_2(t), \dots, \mathbf{h}_n(t)]^* \qquad \text{on } [0,1].$$

Then, given $t \in [0,1]$, the couple $(\mathbf{G}(t,s), \mathbf{H}(t))$ verifies (3,10). By (3,12)

$$\|\mathbf{G}(t,.)\|_{L^q} + |\mathbf{H}(t)| \le n\varkappa < \infty \qquad \text{for any} \quad t \in [0,1]$$

whence (ii) follows. The assertion (iii) is a consequence of the uniqueness of solutions to the equations (3,11).

3.12. Corollary. *Under the assumptions of* 3.11 *the given operator* $\mathscr{L}$ *possesses a bounded inverse*

$$\mathscr{L}^{-1}: \begin{pmatrix} \mathbf{f} \\ \mathbf{r} \end{pmatrix} \in L_n^p \times R_m \to \int_0^1 \mathbf{G}(t,s)\,\mathbf{f}(s)\,\mathrm{d}s + \mathbf{H}(t)\,\mathbf{r} \in W_n^p .$$

3.13. Theorem. *Let us assume* 2.1 *with* $\mathbf{P}(t,1) = \mathbf{0}$ *and* (3,5). *Then the couple* $\mathbf{G}(t,s)$, $\mathbf{H}(t)$ *given by* 3.11 *is a Green couple of BVP* (3,1). *If* $\tilde{\mathbf{G}}(t,s)$, $\tilde{\mathbf{H}}(t)$ *is also a Green couple to* (3,1), *then* $\tilde{\mathbf{G}}(t,s) = \mathbf{G}(t,s)$ *and* $\tilde{\mathbf{H}}(t) = \mathbf{H}(t)$ *for all* $t \in [0,1]$ *and almost all* $s \in [0,1]$.

Proof follows from 3.7 and 3.11.

3.14. Remark. Let $\mathbf{r} \in R_n$. According to the definition 3.1 of $\mathbf{X}$, $\mathbf{x} \in W_n^p$ is a solution to

$$\mathbf{Dx} - \mathbf{Px} = \mathbf{0}, \qquad \mathbf{Sx} = \mathbf{r} \tag{3,13}$$

if and only if $\mathbf{x}(t) = \mathbf{X}(t)\,\mathbf{c}$ on $[0,1]$, where $\mathbf{c} \in R_l$ fulfils $(\mathbf{SX})\,\mathbf{c} = \mathbf{r}$. In particular, if we assume (3,5), then by 1.8 $l = n$ and by 3.2 $\det(\mathbf{SX}) \neq 0$, i.e. $\mathbf{x} \in W_n^p$ verifies (3,13) if and only if $\mathbf{x}(t) = \tilde{\mathbf{H}}(t)\,\mathbf{r}$ on $[0,1]$, where

$$\tilde{\mathbf{H}}(t) = \mathbf{X}(t)\,(\mathbf{SX})^{-1} \qquad \text{on } [0,1].$$

On the other hand, if $\mathbf{G}(t,s)$, $\mathbf{H}(t)$ is the Green couple of BVP (2,1), (2,2), then $\mathbf{x}(t) = \mathbf{H}(t)\,\mathbf{r}$ on $[0,1]$ is for any $\mathbf{r} \in R_m$ the unique solution of (3,13) on W_n^p. Hence $(\mathbf{H}(t) - \tilde{\mathbf{H}}(t))\,\mathbf{r} = \mathbf{0}$ on $[0,1]$ for any $\mathbf{r} \in R_n$ or

$$\mathbf{H}(t) = \mathbf{X}(t)\,(\mathbf{SX})^{-1} \qquad \text{on } [0,1].$$

Let us notice that the columns of $\mathbf{X}$ being elements of W_n^p, the columns of $\mathbf{H}(t)$ are also elements of W_n^p.

4. Generalized Green's couples

If $\mathbf{P}: [0,1] \times [0,1] \to L(R_n)$ is an $L^2[BV]$-kernel, then obviously

$$\int_0^1 |\mathbf{P}(\tau,s)|^2\,\mathrm{d}\tau + \int_0^1 |\mathbf{P}(t,\sigma)|^2\,\mathrm{d}\sigma < \infty$$

for almost all $t, s \in [0,1]$ (cf. 1.1). Moreover, according to the assumptions (1,1) and (1,2) (where $p = 2$)

$$\int_0^1 \left(\int_0^1 |\mathbf{P}(t,s)|^2 \, ds \right) dt \le \int_0^1 \varrho^2(t) \, dt < \infty .$$

By the Tonelli-Hobson Theorem I.4.36 this implies that if an $L^2[BV]$-kernel $\mathbf{P}(t,s)$ is measurable in (t,s) on $[0,1] \times [0,1]$, then

$$\|\!|\mathbf{P}|\!\| = \iint_{[0,1]\times[0,1]} |\mathbf{P}(t,s)|^2 \, dt \, ds = \int_0^1 \left(\int_0^1 |\mathbf{P}(t,s)|^2 \, ds \right) dt < \infty . \tag{4,1}$$

4.1. L^2**-kernels.** The function $\mathbf{P}: [0,1] \times [0,1] \to L(R_n)$ is said to be an L^2-*kernel* if it is measurable in (t,s) on $[0,1] \times [0,1]$ and fulfils (4,1). Given an L^2-kernel $\mathbf{P}$, $\|\!|\mathbf{P}|\!\|$ is defined by (4,1).

Let us recall some basic properties of L^2-kernels and of Fredholm integral equations for $\mathbf{u} \in L^2_n$

$$\mathbf{u}(t) - \int_0^1 \mathbf{P}(t,s) \, \mathbf{u}(s) \, ds = \mathbf{g}(t) \tag{4,2}$$

with an L^2-kernel $\mathbf{P}$. (For the proofs see e.g. Dunford, Schwartz [1] or Smithies [1].)

Let $\mathbf{P}: [0,1] \times [0,1] \to L(R_n)$ be an L^2-kernel. Then for any $\mathbf{u} \in L^2_n$, the n-vector valued function

$$\mathbf{g}(t) = \int_0^1 \mathbf{P}(t,s) \, \mathbf{u}(s) \, ds , \qquad t \in [0,1]$$

is L^2-integrable on $[0,1]$ and the mapping $\mathbf{u} \in L^2_n \to \mathbf{g} \in L^2_n$ is linear and bounded. (This may be shown easily by making use of the Cauchy inequality and the Tonelli-Hobson Theorem I.4.36.) Moreover, a linear operator $\Theta: L^2_n \to L^2_n$ is compact if and only if there exists an L^2-kernel $\mathbf{T}: [0,1] \times [0,1] \to L(R_n)$ such that

$$\Theta: \mathbf{u} \in L^2_n \to \int_0^1 \mathbf{T}(t,s) \, \mathbf{u}(s) \, ds \in L^2_n .$$

If $\|\!|\mathbf{P}|\!\| < 1$, then the equation (4,2) possesses for any $\mathbf{g} \in L^2_n$ a unique solution $\mathbf{u}$ in L^2_n and there exists an L^2-kernel $\mathbf{R}: [0,1] \times [0,1] \to L(R_n)$ such that for any $\mathbf{g} \in L^2_n$ the unique solution $\mathbf{u} \in L^2_n$ of (4,2) is given by

$$\mathbf{u}(t) = \mathbf{g}(t) + \int_0^1 \mathbf{R}(t,s) \, \mathbf{g}(s) \, ds , \qquad t \in [0,1] .$$

$\mathbf{R}$ is called the *resolvent kernel* corresponding to $\mathbf{P}$.

Finally, given an L^2-kernel $\mathbf{P}$, there exist a natural number n', functions $\mathbf{P}_1: [0,1] \to L(R_{n'}, R_n)$ and $\mathbf{P}_2: [0,1] \to L(R_n, R_{n'})$ L^2-integrable on $[0,1]$ and an L^2-kernel $\mathbf{P}_0: [0,1] \times [0,1] \to L(R_n)$ such that

$$\|\!|\mathbf{P}_0|\!\| < 1 \quad \text{and} \quad \mathbf{P}(t,s) = \mathbf{P}_0(t,s) + \mathbf{P}_1(t) \, \mathbf{P}_2(s) \quad \text{on } [0,1] \times [0,1] . \tag{4,3}$$

Let us turn our attention to BVP (2,1), (2,2) fulfilling 2.1 with $p=q=2$ and $\mathbf{P}(t,1)=\mathbf{0}$ a.e. on $[0,1]$. ($\mathbf{P}(t,s)$ is an $L^2[BV]$-kernel, $\mathbf{K}$ is L^2-integrable on $[0,1]$ and $\mathbf{f}\in L^2_{n^{\cdot}}$.)

A function $\mathbf{x}\in W^2_n$ is a solution to BVP (2,1), (2,2) if and only if

$$\mathbf{x} = \boldsymbol{\Phi}\mathbf{c} + \boldsymbol{\Psi}\mathbf{u} + \boldsymbol{\Psi}\mathbf{f},$$

where

(4,4) $$\boldsymbol{\Phi}\colon \mathbf{c}\in R_n \to \mathbf{z}(t)\equiv \mathbf{c}\in W^2_n\,, \qquad \boldsymbol{\Psi}\colon \mathbf{u}\in L^2_n \to \int_0^t \mathbf{u}(\tau)\,\mathrm{d}\tau \in W^2_n$$

and the couple $\begin{pmatrix}\mathbf{u}\\ \mathbf{c}\end{pmatrix}\in L^2_n\times R_n$ verifies the system

(4,5) $$\mathbf{u} - \mathbf{P}\boldsymbol{\Phi}\mathbf{c} - \mathbf{P}\boldsymbol{\Psi}\mathbf{u} = \mathbf{P}\boldsymbol{\Psi}\mathbf{f},$$

(4,6) $$\mathbf{S}\boldsymbol{\Phi}\mathbf{c} + \mathbf{S}\boldsymbol{\Psi}\mathbf{u} = \mathbf{r} - \mathbf{S}\boldsymbol{\Psi}\mathbf{f}.$$

In fact, if $\mathbf{x}\in W^2_n$ is a solution to BVP (2,1), (2,2), then $\mathbf{x} = \boldsymbol{\Phi}\,\mathbf{x}(0) + \boldsymbol{\Psi}\mathbf{P}\mathbf{x} + \boldsymbol{\Psi}\mathbf{f}$ and $\mathbf{S}\mathbf{x} = \mathbf{S}\boldsymbol{\Phi}\,\mathbf{x}(0) + \mathbf{S}\boldsymbol{\Psi}\mathbf{P}\mathbf{x} + \mathbf{S}\boldsymbol{\Psi}\mathbf{f} = \mathbf{r}$. Consequently, $\mathbf{u}=\mathbf{P}\mathbf{x}$ and $\mathbf{c}=\mathbf{x}(0)$ satisfy (4,5) and (4,6). (Clearly $\mathbf{u}\in L^2_{n^{\cdot}}$.) On the other hand, if $\begin{pmatrix}\mathbf{u}\\ \mathbf{c}\end{pmatrix}\in L^2_n\times R_n$ is a solution to the system (4,5), (4,6) and $\mathbf{x} = \boldsymbol{\Phi}\mathbf{c} + \boldsymbol{\Psi}\mathbf{u} + \boldsymbol{\Psi}\mathbf{f}$, then $\mathbf{x}(0)=\mathbf{c}$, $\mathbf{P}\mathbf{x} = \mathbf{P}\boldsymbol{\Phi}\mathbf{c} + \mathbf{P}\boldsymbol{\Psi}\mathbf{u} + \mathbf{P}\boldsymbol{\Psi}\mathbf{f} = \mathbf{u}$ and hence $\mathbf{x} - \boldsymbol{\Phi}\,\mathbf{x}(0) - \boldsymbol{\Psi}\mathbf{P}\mathbf{x} = \boldsymbol{\Psi}\mathbf{f}$ and $\mathbf{S}\mathbf{x}=\mathbf{r}$.

Let us mention that in virtue of I.4.33, the composed operator $\mathbf{P}\boldsymbol{\Psi}\colon L^2_n\to L^2_n$ is given by

(4,7) $$\mathbf{P}\boldsymbol{\Psi}\colon \mathbf{u}\in L^2_n \to -\int_0^1 \mathbf{P}(t,s)\,\mathbf{u}(s)\,\mathrm{d}s\in L^2_n\,.$$

Now, let a natural number n', an L^2-kernel $\mathbf{P}_0\colon [0,1]\times[0,1]\to L(R_n)$ and L^2-integrable functions $\mathbf{P}_1\colon [0,1]\to L(R_{n'},R_n)$ and $\mathbf{P}_2\colon [0,1]\to L(R_n,R_{n'})$ be such that (4,3) holds. Furthermore, let $\mathbf{R}_0\colon [0,1]\times[0,1]\to L(R_n)$ be the resolvent kernel corresponding to $\mathbf{P}_0$. The symbols $\mathbf{P}_0$, $\mathbf{P}_1$, $\mathbf{P}_2$ and $\mathbf{R}_0$ will denote the linear operators

(4,8) $$\begin{aligned}
\mathbf{P}_0\colon\ \mathbf{u}\in L^2_n &\to -\int_0^1 \mathbf{P}_0(t,s)\,\mathbf{u}(s)\,\mathrm{d}s\in L^2_n\,,\\
\mathbf{P}_1\colon\ \mathbf{d}\in R_{n'} &\to -\mathbf{P}_1(t)\,\mathbf{d}\in L^2_n\,,\\
\mathbf{P}_2\colon\ \mathbf{u}\in L^2_n &\to \int_0^1 \mathbf{P}_2(s)\,\mathbf{u}(s)\,\mathrm{d}s\in R_{n'}\,,\\
\mathbf{R}_0\colon\ \mathbf{u}\in L^2_n &\to -\int_0^1 \mathbf{R}_0(t,s)\,\mathbf{u}(s)\,\mathrm{d}s\in L^2_n\,,
\end{aligned}$$

as well. All of them are obviously compact.

By (4,3) and (4,8) we may write

$$\mathbf{P}\boldsymbol{\Psi} = \mathbf{P}_0 + \mathbf{P}_1\mathbf{P}_2$$

and the equation (4,5) becomes

$$\mathbf{u} - \mathbf{P}_0\mathbf{u} = \mathbf{P}\boldsymbol{\Phi}\mathbf{c} + \mathbf{P}_1\mathbf{P}_2\mathbf{u} + \mathbf{P}\boldsymbol{\Psi}\mathbf{f}.$$

Accordingly

(4,9) $$\mathbf{u} - [\mathbf{I} + \mathbf{R}_0](\mathbf{P}\boldsymbol{\Phi}\mathbf{c} + \mathbf{P}_1\mathbf{P}_2\mathbf{u}) = [\mathbf{I} + \mathbf{R}_0]\,\mathbf{P}\boldsymbol{\Psi}\mathbf{f}.$$

Let us denote

$$\mathbf{d} = \mathbf{P}_2\mathbf{u}.$$

Then the equation (4,9) reduces to

(4,10) $$\mathbf{u} = [\mathbf{I} + \mathbf{R}_0]\,\mathbf{P}\boldsymbol{\Phi}\mathbf{c} + [\mathbf{I} + \mathbf{R}_0]\,\mathbf{P}_1\mathbf{d} + [\mathbf{I} + \mathbf{R}_0]\,\mathbf{P}\boldsymbol{\Psi}\mathbf{f}.$$

Applying $\mathbf{P}_2$ to (4,10) and inserting (4,10) into (4,6) we reduce the system (4,5), (4,6) to the system of equations for $\mathbf{c} \in R_n$ and $\mathbf{d} \in R_{n'}$

(4,11) $$\mathbf{B}\begin{pmatrix}\mathbf{c}\\ \mathbf{d}\end{pmatrix} = \begin{pmatrix}\mathbf{F}_1\mathbf{f}\\ \mathbf{r} - \mathbf{F}_2\mathbf{f}\end{pmatrix}$$

where

(4,12) $$\mathbf{B}\colon \begin{pmatrix}\mathbf{c}\\ \mathbf{d}\end{pmatrix} \in R_{n+n'} \to \begin{pmatrix}-\mathbf{P}_2[\mathbf{I} + \mathbf{R}_0]\,\mathbf{P}\boldsymbol{\Phi}\mathbf{c} + (\mathbf{I} - \mathbf{P}_2[\mathbf{I} + \mathbf{R}_0]\,\mathbf{P}_1)\,\mathbf{d}\\ \mathbf{S}(\mathbf{I} - \boldsymbol{\Psi}[\mathbf{I} + \mathbf{R}_0]\,\mathbf{P})\,\boldsymbol{\Phi}\mathbf{c} + \mathbf{S}\boldsymbol{\Psi}[\mathbf{I} + \mathbf{R}_0]\,\mathbf{P}_1\mathbf{d}\end{pmatrix} \in R_{m+n'}$$

and

(4,13) $$\mathbf{F}_1\colon \mathbf{f} \in L_n^2 \to \mathbf{P}_2[\mathbf{I} + \mathbf{R}_0]\,\mathbf{P}\boldsymbol{\Psi}\mathbf{f} \in R_{n'},$$
$$\mathbf{F}_2\colon \mathbf{f} \in L_n^2 \to \mathbf{S}\boldsymbol{\Psi}(\mathbf{I} + [\mathbf{I} + \mathbf{R}_0]\,\mathbf{P}\boldsymbol{\Psi})\,\mathbf{f} \in R_m.$$

The operator $\mathbf{B}$ may be represented by a uniquely determined $(m+n') \times (n+n')$-matrix. Let us denote this matrix again by $\mathbf{B}$.

Thus BVP (2,1), (2,2) possesses a solution $\mathbf{x} \in W_n^2$ if and only if the system (4,11) possesses a solution $\begin{pmatrix}\mathbf{c}\\ \mathbf{d}\end{pmatrix} \in R_{n+n'}$ and $\mathbf{x}$ is then given by

(4,14) $$\mathbf{x} = (\boldsymbol{\Phi} + \boldsymbol{\Psi}[\mathbf{I} + \mathbf{R}_0]\,\mathbf{P}\boldsymbol{\Phi})\,\mathbf{c} + \boldsymbol{\Psi}[\mathbf{I} + \mathbf{R}_0]\,\mathbf{P}_0\mathbf{d} + \boldsymbol{\Psi}[\mathbf{I} + \mathbf{R}_0]\,\mathbf{P}\boldsymbol{\Psi}\mathbf{f} + \boldsymbol{\Psi}\mathbf{f}.$$

Let $\Delta_{1,1} \in L(R_{n'}, R_n)$, $\Delta_{1,2} \in L(R_m, R_n)$, $\Delta_{2,1} \in L(R_{n'})$ and $\Delta_{2,2} \in L(R_m, R_{n'})$ be chosen in such a way that

$$\mathbf{B}^+ = \begin{bmatrix}\Delta_{1,1}, & \Delta_{1,2}\\ \Delta_{2,1}, & \Delta_{2,2}\end{bmatrix} \in L(R_{m+n'}, R_{n+n'})$$

fulfils $\mathbf{B}\mathbf{B}^+\mathbf{B} = \mathbf{B}$ (e.g. $\mathbf{B}^+ = \mathbf{B}^\#$). Then if (4,11) has a solution, the couple

(4,15) $$\mathbf{c} = [\Delta_{1,1}\mathbf{F}_1 - \Delta_{1,2}\mathbf{F}_2]\,\mathbf{f} + \Delta_{1,2}\mathbf{r} \in R_n,$$
$$\mathbf{d} = [\Delta_{2,1}\mathbf{F}_1 - \Delta_{2,2}\mathbf{F}_2]\,\mathbf{f} + \Delta_{2,2}\mathbf{r} \in R_{n'},$$

is also its solution.

Inserting (4,15) into (4,14) we obtain that if BVP (2,1), (2,2) has a solution, then

(4,16) $$\mathbf{x} = \boldsymbol{\Phi}[\mathbf{G}_1\mathbf{f} + \mathbf{H}_1\mathbf{r}] + \boldsymbol{\Psi}[\mathbf{I} + \mathbf{R}_0](\mathbf{G}_2\mathbf{f} + \mathbf{H}_2\mathbf{r}) + \boldsymbol{\Psi}\mathbf{f}$$

with

(4,17) $$\mathbf{G}_1 = \boldsymbol{\Delta}_{1,1}\mathbf{F}_1 - \boldsymbol{\Delta}_{1,2}\mathbf{F}_2, \qquad \mathbf{H}_1 = \boldsymbol{\Delta}_{1,2},$$
$$\mathbf{G}_2 = \mathbf{P}\boldsymbol{\Phi}(\boldsymbol{\Delta}_{1,1}\mathbf{F}_1 - \boldsymbol{\Delta}_{1,2}\mathbf{F}_2) + \mathbf{P}_1(\boldsymbol{\Delta}_{2,1}\mathbf{F}_1 - \boldsymbol{\Delta}_{2,2}\mathbf{F}_2) + \mathbf{P}\boldsymbol{\Psi},$$
$$\mathbf{H}_2 = \mathbf{P}\boldsymbol{\Phi}\boldsymbol{\Delta}_{1,2} + \mathbf{P}_1\boldsymbol{\Delta}_{2,2}$$

is also its solution. As $\mathbf{G}_1: L_n^2 \to R_n$ is a linear bounded n-vector valued functional on L_n^2 and $[\mathbf{I} + \mathbf{R}_0]\,\mathbf{G}_2 \in K(L_n^2)$, there exist an L^2-integrable function $\mathbf{G}_1: [0,1] \to L(R_n)$ and an L^2-kernel $\mathbf{G}_2: [0,1] \times [0,1] \to L(R_n)$ such that

(4,18) $$\mathbf{G}_1: \mathbf{f} \in L_n^2 \to \int_0^1 \mathbf{G}_1(s)\,\mathbf{f}(s)\,\mathrm{d}s \in R_n,$$
$$[\mathbf{I} + \mathbf{R}_0]\,\mathbf{G}_2: \mathbf{f} \in L_n^2 \to \int_0^1 \mathbf{G}_2(t,s)\,\mathbf{f}(s)\,\mathrm{d}s \in L_n^2.$$

Applying the Tonelli-Hobson Theorem I.4.36 we may show that

$$\int_0^t \left(\int_0^1 \mathbf{G}_2(\tau,s)\,\mathbf{f}(s)\,\mathrm{d}s\right)\mathrm{d}\tau = \int_0^1 \left(\int_0^t \mathbf{G}_2(\tau,s)\,\mathrm{d}\tau\right)\mathbf{f}(s)\,\mathrm{d}s$$

for any $\mathbf{f} \in L_n^2$ and $t \in [0,1]$, i.e.

(4,20) $$\boldsymbol{\Psi}[\mathbf{I} + \mathbf{R}_0]\,\mathbf{G}_2: \mathbf{f} \in L_n^2 \to \int_0^1 \left(\int_0^t \mathbf{G}_2(\tau,s)\,\mathrm{d}\tau\right)\mathbf{f}(s)\,\mathrm{d}s \in W_n^2.$$

Furthermore, by (4,3), (4,4) and (4,8) there exist an L^2-integrable function $\tilde{\mathbf{H}}_2: [0,1] \to L(R_m, R_n)$ such that

$$\mathbf{H}_2 = \mathbf{P}\boldsymbol{\Phi}\boldsymbol{\Delta}_{1,2} + \mathbf{P}_1\boldsymbol{\Delta}_{2,2}: \mathbf{r} \in R_m \to \tilde{\mathbf{H}}_2(t)\,\mathbf{r} \in L_n^2.$$

Consequently,

(4,21) $$\boldsymbol{\Psi}[\mathbf{I} + \mathbf{R}_0]\,\mathbf{H}_2: \mathbf{r} \in R_m \to \left(\int_0^t \mathbf{H}_2(\tau)\,\mathrm{d}\tau\right)\mathbf{r},$$

where

$$\mathbf{H}_2(t) = \tilde{\mathbf{H}}_2(t) + \int_0^1 \mathbf{R}_0(t,\tau)\,\tilde{\mathbf{H}}_2(\tau)\,\mathrm{d}\tau, \qquad t \in [0,1]$$

is also L^2-integrable on $[0,1]$. Inserting (4,18), (4,20) and (4,21) into (4,16) we obtain that if BVP (2,1), (2,2) has a solution, then also

(4,22) $$\mathbf{x}(t) = \int_0^1 \mathbf{G}_0(t,s)\,\mathbf{f}(s)\,\mathrm{d}s + \mathbf{H}_0(t)\,\mathbf{r}, \qquad t \in [0,1],$$

with

$$(4{,}23)\qquad \mathbf{G}_0(t,s) = \mathbf{G}_1(s) + \int_0^t \mathbf{G}_2(\tau,s)\,\mathrm{d}\tau + \Delta(t,s) \qquad \text{on } [0,1]\times[0,1],$$

$$\Delta(t,s) = \mathbf{0} \quad \text{if} \quad t < s, \qquad \Delta(t,s) = \mathbf{I} \quad \text{if} \quad t \geq s,$$

$$\mathbf{H}_0(t) = \mathbf{H}_1 + \int_0^t \mathbf{H}_2(\tau)\,\mathrm{d}\tau \qquad \text{on } [0,1]$$

is a solution to BVP (2,1), (2,2). It follows from the definition of the functions $\mathbf{G}_0(t,s)$ and $\mathbf{H}_0(t)$, that the linear operator

$$(4{,}24)\qquad \mathscr{L}^+ : \begin{pmatrix}\mathbf{f}\\ \mathbf{r}\end{pmatrix} \in L_n^2 \times R_m \to \int_0^1 \mathbf{G}_0(t,s)\,\mathbf{f}(s)\,\mathrm{d}s + \mathbf{H}_0(t)\,\mathbf{r} \in W_n^2$$

is bounded. The results obtained are summarized in the following theorem.

4.2. Theorem. *Let the assumptions* 2.1 *with* $p = q = 2$ *be fulfilled and, moreover, let* $\mathbf{P}(t,s)$ *be measurable in* (t,s) *on* $[0,1]\times[0,1]$. *Then there exist functions* $\mathbf{G}_0\colon [0,1]\times[0,1] \to L(R_n)$ *and* $\mathbf{H}_0\colon [0,1] \to L(R_m, R_n)$ *such that for any* $\mathbf{f} \in L_n^2$ *and* $\mathbf{r} \in R_m$ *the function* $\mathbf{x}(t)$ *given by* (4,22) *belongs to* W_n^2 *and the linear operator* $\mathscr{L}^+$ *given by* (4,24) *is bounded. Furthermore, if BVP* (2,1), (2,2) *possesses a solution, then* (4,22) (*i.e.* $\mathbf{x} = \mathscr{L}^+\begin{pmatrix}\mathbf{f}\\ \mathbf{r}\end{pmatrix}$) *is also its solution.*

4.3. Remark. According to the definition IV.3.10 we may say that $\mathbf{G}_0(t,s)$, $\mathbf{H}_0(t)$ is a generalized Green's couple of BVP (2,1), (2,2). The operator $\mathscr{L}^+$ given by (4,24) fulfils the relation $\mathscr{L}\mathscr{L}^+\mathscr{L} = \mathscr{L}$.

4.4. Proposition. *The functions* $\mathbf{G}_0(t,s)$ *and* $\mathbf{H}_0(t)$ *defined by* (4,23) *have the following properties*

- (i) $\mathbf{H}_0$ *possesses a.e. on* $[0,1]$ *a derivative which is* L^2*-integrable on* $[0,1]$,
- (ii) $\mathbf{G}_0$ *is an* L^2*-kernel,* $\mathbf{G}_0(.,s)$ *is of bounded variation on* $[0,1]$ *for a.e.* $s \in [0,1]$,
- (iii) $\gamma(s) = |\mathbf{G}_0(0,s)| + \mathrm{var}_0^1\, \mathbf{G}_0(.,s) \in L^2$,
- (iv) *for almost every* $s \in [0,1]$ *the columns of* $\mathbf{G}_0(.,s) - \Delta(.,s)$ *belong to the space* W_n^2.

Proof follows from the construction of the functions $\mathbf{G}_0(t,s)$ and $\mathbf{H}_0(t)$ ($\mathbf{G}_0(0,s) = \mathbf{G}_1(s)$, $\mathrm{var}_0^1\,\Delta(.,s) \leq 1$ and hence

$$\gamma(s) \leq |\mathbf{G}_1(s)| + \int_0^1 |\mathbf{G}_2(\tau,s)|\,\mathrm{d}\tau + 1 \qquad \text{a.e. on } [0,1].)$$

4.5. Remark. If $k = \dim N(\mathscr{L}) > 0$, let $\mathbf{X}_0$ denote the $n \times k$-matrix function whose columns form a basis in $N(\mathscr{L})$. If $k^* = \dim N(\mathscr{L}^*) > 0$, let $\mathbf{Y}_0\colon [0,1] \to L(R_n, R_{k^*})$

and $\Lambda_0 \in L(R_m, R_{k^*})$ be such that the couples $(\mathbf{y}_j^*, \lambda_j^*)$ $(j = 1, 2, ..., k^*)$ of their rows form a basis in $N(\mathscr{L}^*)$. Then evidently for any L^2-integrable function $\Theta_1: [0, 1] \to L(R_n, R_k)$, any matrix $\Theta_2 \in L(R_m, R_k)$ and any function $\Sigma: [0, 1] \to L(R_{k^*}, R_n)$ of bounded variation on $[0, 1]$

(4,25) $$\mathbf{G}(t, s) = \mathbf{G}_0(t, s) + \mathbf{X}_0(t)\,\Theta_1(s) + \Sigma(t)\,\mathbf{Y}_0(s), \qquad t, s \in [0, 1],$$
$$\mathbf{H}(t) = \mathbf{H}_0(t) + \mathbf{X}_0(t)\,\Theta_2 + \Sigma(t)\,\Lambda_0$$

is also a generalized Green's couple of BVP (2,1), (2,2) and fulfils (i)–(iv) from 4.4 in place of $\mathbf{G}_0(t, s)$ $\mathbf{H}_0(t)$.

4.6. Definition. Generalized Green's couples of the form (4,25) will be called *standard generalized Green's couples.*

4.7. Remark. It is easy to verify that given a standard generalized Green couple $\mathbf{G}(t, s)$, $\mathbf{H}(t)$, the operator

(4,26) $$\mathscr{L}^+ : \begin{pmatrix} \mathbf{f} \\ \mathbf{r} \end{pmatrix} \in L_n^2 \times R_m \to \int_0^1 \mathbf{G}(t, s)\,\mathbf{f}(s)\,\mathrm{d}s + \mathbf{H}(t)\,\mathbf{r}$$

is bounded and fulfils the relation $\mathscr{L}\mathscr{L}^+\mathscr{L} = \mathscr{L}$. *)

4.8. Remark. Making use of the equivalence between BVP (2,1), (2,2) and the linear algebraic equation (4,11) we could obtain (under the assumptions 2.1 with $p = q = 2$) the basic results of the Section V.2 in a more elementary way. An analogous procedure can be applied also to BVP

(4,27) $$\mathbf{x}'(t) - \mathbf{A}(t)\,\mathbf{x}(t) - \mathbf{C}(t)\,\mathbf{x}(0) - \mathbf{D}(t)\,\mathbf{x}(1) - \int_0^1 \mathrm{d}_s[\mathbf{R}(t, s)]\,\mathbf{x}(s) = \mathbf{f}(t)$$
$$\text{a.e. on } [0, 1],$$

(4,28) $$\mathbf{M}\,\mathbf{x}(0) + \int_0^1 \mathbf{K}(t)\,\mathbf{x}'(t)\,\mathrm{d}t = \mathbf{r},$$

where $\mathbf{A}$ is supposed to be only L-integrable on $[0, 1]$ and $\mathbf{K}$ is measurable and essentially bounded on $[0, 1]$. (In general BVP (4,27), (4,28) cannot be rewritten as the system of the form (2,1), (2,2) fulfilling the assumptions of this section.) If $\mathbf{X}(t)$ denotes the fundamental matrix solution of the equation $\mathbf{x}'(t) - \mathbf{A}(t)\,\mathbf{x}(t) = \mathbf{0}$, then BVP (4,27), (4,28) will be transferred to a system of integro-algebraical equations

*) Since in general we may not assume that $\mathbf{X}_0(t)$ has a full rank on $[0, 1]$ (cf. 1.10), we may not apply the procedure from IV.3.12 to show that $\mathscr{L}^+ \in B(L_n^2 \times R_m, W_n^2)$ fulfils $\mathscr{L}\mathscr{L}^+\mathscr{L} = \mathscr{L}$ if and only if $\mathscr{L}^+$ is given by (4,26), where $\mathbf{G}(t, s)$, $\mathbf{H}(t)$ is a standard generalized Green's couple.

for $\mathbf{u} \in L_n^2$ and $\mathbf{c} \in R_n$ of the form (4,5), (4,6) (with an L^2-kernel) by means of the substitution

$$\mathbf{u}(t) = \mathbf{C}(t)\,\mathbf{x}(0) + \mathbf{D}(t)\,\mathbf{x}(1) + \int_0^1 \mathrm{d}_s[\mathbf{R}(t,s)]\,\mathbf{x}(s)$$

$$\mathbf{c} = \mathbf{x}(0)\,.$$

On the other hand,

$$\mathbf{x}(t) = \mathbf{X}(t)\,\mathbf{c} + \mathbf{X}(t)\int_0^t \mathbf{X}^{-1}(s)\,\mathbf{u}(s)\,\mathrm{d}s + \mathbf{X}(t)\int_0^t \mathbf{X}^{-1}(s)\,\mathbf{f}(s)\,\mathrm{d}s\,,$$

i.e. $\mathbf{x} = \mathbf{U}\mathbf{c} + \mathbf{V}\mathbf{u} + \mathbf{V}\mathbf{f}$.

5. Best approximate solutions

We still assume that $\mathbf{P}$: $[0,1] \times [0,1] \to L(R_n)$ *is a measurable* $L^2[BV]$*-kernel,* $\mathbf{P}(t,1) = \mathbf{0}$ *a.e. on* $[0,1]$, *the columns of* $\mathbf{K}$: $[0,1] \to L(R_n, R_m)$ *belong to* L_n^2, $\mathbf{f} \in L_n^2$ *and* $\mathbf{r} \in R_m$. Given $\mathbf{x}, \mathbf{u} \in W_n^2$, let us put

(5,1)
$$(\mathbf{x}, \mathbf{u})_X = \int_0^1 \mathbf{u}^*(t)\,\mathbf{x}(t)\,\mathrm{d}t \in R\,.$$

Clearly, $\mathbf{x}, \mathbf{u} \in W_n^2 \to (\mathbf{x}, \mathbf{u})_X \in R$ is a bilinear form on $W_n^2 \times W_n^2$, while $(\mathbf{x}, \mathbf{u})_X = (\mathbf{u}, \mathbf{x})_X$ for all $\mathbf{x}, \mathbf{u} \in W_n^2$ and $(\mathbf{x}, \mathbf{x})_X = 0$ if and only if $\mathbf{x}(t) \equiv \mathbf{0}$ on $[0,1]$. It means that $(.,.)_X$ is an inner product and $\mathbf{x} \in W_n^2 \to \|\mathbf{x}\|_X = (\mathbf{x}, \mathbf{x})_X^{1/2}$ is a norm on W_n^2.

Analogously,

(5,2)
$$\varphi = \begin{pmatrix} \mathbf{f} \\ \mathbf{r} \end{pmatrix},\ \psi = \begin{pmatrix} \mathbf{g} \\ \mathbf{q} \end{pmatrix} \in L_n^2 \times R_m \to (\varphi, \psi)_Y = \langle \varphi, \psi^* \rangle_{L^2 \times R}$$
$$= \int_0^1 \mathbf{g}^*(t)\,\mathbf{f}(t)\,\mathrm{d}t + \mathbf{q}^*\mathbf{r} \in R$$

is an inner product on $L_n^2 \times R_m$ and $\varphi \in L_n^2 \times R_m \to \|\varphi\|_Y = (\varphi, \varphi)_Y^{1/2}$ is a norm on $L_n^2 \times R_m$. Moreover, as $|\mathbf{c}| \le |\mathbf{c}|_e = (\mathbf{c}^*\mathbf{c})^{1/2} \le n|\mathbf{c}|$ for any $\mathbf{c} \in R_n$,

$$\left\|\begin{pmatrix} \mathbf{f} \\ \mathbf{r} \end{pmatrix}\right\|_{L^2 \times R}^2 = \left(\left(\int_0^1 |\mathbf{f}(t)|^2\,\mathrm{d}t\right)^{1/2} + |\mathbf{r}|\right)^2 \ge \frac{1}{n^2}\left(\left(\int_0^1 |\mathbf{f}(t)|_e^2\,\mathrm{d}t\right)^{1/2} + |\mathbf{r}|_e\right)^2 \ge \left(\frac{1}{n^2}\right)\left\|\begin{pmatrix} \mathbf{f} \\ \mathbf{r} \end{pmatrix}\right\|_Y^2$$

for all $\mathbf{f} \in L_n^2$ and $\mathbf{r} \in R_m$.

On the other hand,

$$\int_0^1 |\mathbf{f}(t)|^2\,\mathrm{d}t + |\mathbf{r}|^2 - 2\left(\int_0^1 |\mathbf{f}(t)|^2\,\mathrm{d}t\right)^{1/2} |\mathbf{r}| \ge 0$$

and hence

$$2\left\|\begin{pmatrix}\mathbf{f}\\ \mathbf{r}\end{pmatrix}\right\|_Y^2 \geq 2\left(\int_0^1 |\mathbf{f}(t)|^2\,\mathrm{d}t + |\mathbf{r}|^2\right) \geq \left(\left(\int_0^1 |\mathbf{f}(t)|^2\,\mathrm{d}t\right)^{1/2} + |\mathbf{r}|\right)^2 = \left\|\begin{pmatrix}\mathbf{f}\\ \mathbf{r}\end{pmatrix}\right\|_{L^2\times R}^2,$$

i.e.

$$(5,3)\qquad \frac{1}{n}\|\boldsymbol{\varphi}\|_Y \leq \|\boldsymbol{\varphi}\|_{L^2\times R} \leq \sqrt{(2)}\,\|\boldsymbol{\varphi}\|_Y \qquad \text{for each} \quad \boldsymbol{\varphi}\in L_n^2 \times R_m.$$

It follows immediately that the space $L_n^2 \times R_m$ endowed with the norm $\|.\|_Y$ is complete, i.e. it is a Hilbert space.

5.1. Notation. In the subsequent text X stands for the inner product space of elements of W_n^2 with the inner product (5,1) and the corresponding norm $\|.\|_X$. Y denotes the Hilbert space of elements of $L_n^2 \times R_m$ equipped with the inner product (5,2) and the corresponding norm $\|.\|_Y$. The operator $\mathbf{x}\in X \to \mathscr{L}\mathbf{x}\in Y$ (cf. (2,3)) is denoted by $\mathscr{A}$.

5.2. Remark. Evidently $\mathscr{A}\in L(X, Y)$, $R(\mathscr{A}) = R(\mathscr{L})$ and $N(\mathscr{A}) = N(\mathscr{L})$. It follows easily from (5,3) and 2.9 that $R(\mathscr{A})$ is closed in Y.

5.3. Remark. Let us notice that in general $\mathscr{A}$ is unbounded.

5.4. Notation. If $k = \dim N(\mathscr{L}) > 0$, then $\mathbf{X}_0$ denotes the $n\times k$-matrix valued function whose columns form a basis in $N(\mathscr{L})$. If $k^* = \dim N(\mathscr{L}^*) > 0$ and $(\mathbf{y}_j, \boldsymbol{\lambda}_j^*) \in L_n^2 \times R_m^*$ $(j = 1, 2, \ldots, k^*)$ is a basis in $N(\mathscr{L}^*)$, let us put $\mathbf{Y}_0^*(t) = [\mathbf{y}_1(t), \mathbf{y}_2(t), \ldots, \mathbf{y}_{k^*}(t)]$ on $[0, 1]$ and $\boldsymbol{\Lambda}_0^* = [\boldsymbol{\lambda}_1, \boldsymbol{\lambda}_2, \ldots, \boldsymbol{\lambda}_{k^*}]$.

5.5. Lemma. *If $k^* > 0$, then the $k^*\times k^*$-matrix*

$$\mathbf{C} = \int_0^1 \mathbf{Y}_0(t)\,\mathbf{Y}_0^*(t)\,\mathrm{d}t + \boldsymbol{\Lambda}_0\boldsymbol{\Lambda}_0^*$$

is regular. If we put

$$(5,4)\qquad \boldsymbol{\Pi}_1\colon \begin{pmatrix}\mathbf{f}\\ \mathbf{r}\end{pmatrix}\in Y \to \begin{pmatrix}\mathbf{f}(t)\\ \mathbf{r}\end{pmatrix} - \begin{bmatrix}\mathbf{Y}_0^*(t)\\ \boldsymbol{\Lambda}_0^*\end{bmatrix}\mathbf{C}^{-1}\left[\int_0^1 \mathbf{Y}_0(s)\,\mathbf{f}(s)\,\mathrm{d}s + \boldsymbol{\Lambda}_0\mathbf{r}\right]\in Y \qquad \text{if} \quad k^* > 0,$$

$$\boldsymbol{\Pi}_1 = \mathbf{I} \qquad \text{if} \quad k^* = 0,$$

then $\boldsymbol{\Pi}_1$ is an orthogonal bounded projection of Y onto $R(\mathscr{A})$.

Proof. If there were $\boldsymbol{\delta}^*\mathbf{C} = \mathbf{0}$ for some $\boldsymbol{\delta}\in R_{k^*}$, then it would be also $0 = \boldsymbol{\delta}^*\mathbf{C}\boldsymbol{\delta}$, i.e.

$$0 = \int_0^1 (\boldsymbol{\delta}^*\,\mathbf{Y}_0(t))\,(\mathbf{Y}_0^*(t)\,\boldsymbol{\delta})\,\mathrm{d}t + (\boldsymbol{\delta}^*\boldsymbol{\Lambda}_0)\,(\boldsymbol{\Lambda}_0^*\boldsymbol{\delta}) = \|(\mathbf{Y}_0^*(t)\,\boldsymbol{\delta},\, \boldsymbol{\Lambda}_0^*\boldsymbol{\delta})\|_Y^2.$$

This may hold if and only if $\boldsymbol{\delta}^*[\mathbf{Y}_0(t), \boldsymbol{\Lambda}_0] = \mathbf{0}$ a.e. on $[0, 1]$. Hence $\boldsymbol{\delta}^*\mathbf{C} = \mathbf{0}$ implies $\boldsymbol{\delta}^* = \mathbf{0}$.

Furthermore, it follows easily from 2.12 that $\boldsymbol{\Pi}_1\boldsymbol{\varphi} \in R(\mathscr{A})$ for any $\boldsymbol{\varphi} \in Y$ and $\boldsymbol{\Pi}_1\boldsymbol{\varphi} = \boldsymbol{\varphi}$ if $\boldsymbol{\varphi} \in R(\mathscr{A})$. Finally, given $\boldsymbol{\varphi} \in \begin{pmatrix} \mathbf{f} \\ \mathbf{r} \end{pmatrix} \in Y$ and $\boldsymbol{\psi} = \begin{pmatrix} \mathbf{g} \\ \mathbf{q} \end{pmatrix} \in R(\mathscr{A})$, we have by 2.12

$$(\boldsymbol{\varphi} - \boldsymbol{\Pi}_1\boldsymbol{\varphi}, \boldsymbol{\psi})_Y = \left[\int_0^1 \mathbf{g}^*(t)\, \mathbf{Y}_0^*(t)\, \mathrm{d}t + \mathbf{q}^*\boldsymbol{\Lambda}_0^*\right]\mathbf{C}^{-1}\left[\int_0^1 \mathbf{Y}_0(s)\, \mathbf{f}(s)\, \mathrm{d}s + \boldsymbol{\Lambda}_0\mathbf{r}\right] = 0\,.$$

The boundedness of $\boldsymbol{\Pi}_1$ is obvious.

5.6. Lemma. *If $k > 0$, then the $k \times k$-matrix*

$$\mathbf{D} = \int_0^1 \mathbf{X}_0^*(t)\, \mathbf{X}_0(t)\, \mathrm{d}t$$

is regular. The mapping

$$\text{(5,5)} \qquad \boldsymbol{\Pi}_2\colon \mathbf{x} \in X \to \mathbf{X}_0(t)\, \mathbf{D}^{-1}\left(\int_0^1 \mathbf{X}_0^*(s)\, \mathbf{x}(s)\, \mathrm{d}s\right) \in X \qquad \textit{if} \;\; k > 0\,,$$

$$\boldsymbol{\Pi}_2 = \mathbf{0} \qquad \textit{if} \;\; k = 0$$

is an orthogonal bounded projection of X onto $N(\mathscr{A})$.

Proof. The regularity of $\mathbf{D}$ follows analogously as the regularity of $\mathbf{C}$. Obviously $R(\boldsymbol{\Pi}_2) \subset N(\mathscr{A})$. Furthermore, if $\boldsymbol{\delta} \in R_k$ and $\mathbf{x}(t) \equiv \mathbf{X}_0(t)\,\boldsymbol{\delta}$ on $[0, 1]$ (i.e. $\mathbf{x} \in N(\mathscr{A})$), then

$$(\boldsymbol{\Pi}_2\mathbf{x})(t) = \mathbf{X}_0(t)\, \mathbf{D}^{-1}\left(\int_0^1 \mathbf{X}_0^*(s)\, \mathbf{X}_0(s)\, \mathrm{d}s\right)\boldsymbol{\delta} = \mathbf{X}_0(t)\,\boldsymbol{\delta} = \mathbf{x}(t)\,.$$

Consequently $R(\boldsymbol{\Pi}_2) = N(\mathscr{A})$ and $\boldsymbol{\Pi}_2^2 = \boldsymbol{\Pi}_2$. Finally, given $\mathbf{x} \in X$, $\boldsymbol{\delta} \in R_k$ and $\mathbf{u}(t) = \mathbf{X}_0(t)\,\boldsymbol{\delta}$,

$$(\mathbf{x} - \boldsymbol{\Pi}_2\mathbf{x}, \mathbf{u})_X$$
$$= \boldsymbol{\delta}^*\left(\int_0^1 \mathbf{X}_0^*(t)\, \mathbf{x}(t)\, \mathrm{d}t\right) - \boldsymbol{\delta}^*\left(\int_0^1 \mathbf{X}_0^*(t)\, \mathbf{X}(t)\, \mathrm{d}t\right)\mathbf{D}^{-1}\left(\int_0^1 \mathbf{X}_0^*(t)\, \mathbf{x}(t)\, \mathrm{d}t\right) = 0\,.$$

5.7. Definition. A function $\mathbf{u}_0 \in W_n^2$ is said to be a *least square solution* or a *best approximate solution of BVP* (2,1), (2,2), if it is a least square solution or a best approximate solution of the operator equation

$$\text{(5,6)} \qquad \mathscr{A}\mathbf{x} = \begin{pmatrix} \mathbf{f} \\ \mathbf{r} \end{pmatrix},$$

respectively.

Let us assume 2.1 *with* $p = q = 2$ *and let* $\mathbf{P}(t, s)$ *be measurable in* (t, s) *on* $[0, 1] \times [0, 1]$. *Then there exist a standard generalized Green's couple* $\mathbf{G}(t, s)$,

$\mathbf{H}(t)$ of BVP (2,1), (2,2) such that for any $\mathbf{f} \in L_n^2$ and $\mathbf{r} \in R_m$

$$(5,7) \qquad \mathbf{u}_0(t) = \int_0^1 \mathbf{G}(t,s)\,\mathbf{f}(s)\,\mathrm{d}s + \mathbf{H}(t)\,\mathbf{r} \qquad \text{on } [0,1]$$

is the unique best approximate solution of BVP (2,1), (2,2).

Proof. Let $\mathbf{G}_0(t,s)$, $\mathbf{H}_0(t)$ be the generalized Green's couple of BVP (2,1), (2,2) given by 4.2 and let $\mathscr{L}^+\colon L_n^2 \times R_m \to W_n^2$ be the corresponding generalized inverse operator to $\mathscr{L}$ given by (4,24). Let us define $\mathscr{A}^+\colon \begin{pmatrix}\mathbf{f}\\ \mathbf{r}\end{pmatrix} \in Y \to \mathscr{L}^+\begin{pmatrix}\mathbf{f}\\ \mathbf{r}\end{pmatrix} \in X$ and

$$(5,8) \qquad \mathscr{A}^{\#}\colon \begin{pmatrix}\mathbf{f}\\ \mathbf{r}\end{pmatrix} \in Y \to (\mathbf{I} - \boldsymbol{\Pi}_2)\,\mathscr{A}^+\boldsymbol{\Pi}_1\begin{pmatrix}\mathbf{f}\\ \mathbf{r}\end{pmatrix} \in X\,,$$

where $\boldsymbol{\Pi}_1 \in B(Y)$ and $\boldsymbol{\Pi}_2 \in B(X)$ are given by (5,4) and (5,5), respectively. Then $\mathscr{A}^+ \in L(Y,X)$, $\mathscr{A}^{\#} \in L(Y,X)$, $\mathscr{A}\mathscr{A}^+\mathscr{A} = \mathscr{A}$ and according to I.3.28 and I.3.29 $\mathbf{u}_0 = \mathscr{A}^{\#}\begin{pmatrix}\mathbf{f}\\ \mathbf{r}\end{pmatrix}$ is the unique best approximate solution of (5,6) for every $\mathbf{f} \in L_n^2$ and $\mathbf{r} \in R_m$. Taking into account 4.4, (5,4), (4,24) and making use of I.4.36 we obtain that for any $\mathbf{f} \in L_n^2$ and $\mathbf{r} \in R_m$

$$(5,9) \qquad \mathscr{A}^+\boldsymbol{\Pi}_1\begin{pmatrix}\mathbf{f}\\ \mathbf{r}\end{pmatrix}(t) = \int_0^1 \tilde{\mathbf{G}}(t,s)\,\mathbf{f}(s)\,\mathrm{d}s + \tilde{\mathbf{H}}(t)\,\mathbf{r} \qquad \text{on } [0,1]\,,$$

where

$$(5,10) \qquad \tilde{\mathbf{G}}(t,s) = \mathbf{G}_0(t,s) - \left(\int_0^1 \mathbf{G}_0(t,\sigma)\,\mathbf{Y}_0^*(\sigma)\,\mathrm{d}\sigma\right)\mathbf{C}^{-1}\,\mathbf{Y}_0(s) \qquad \text{on } [0,1]\times[0,1]\,,$$

$$\tilde{\mathbf{H}}(t) = \mathbf{H}_0(t) - \mathbf{H}_0(t)\,\boldsymbol{\Lambda}_0^*\mathbf{C}^{-1}\boldsymbol{\Lambda}_0 \qquad \text{on } [0,1]\,.$$

Obviously, $\tilde{\mathbf{G}}(t,s)$ is an L^2-kernel and $\tilde{\mathbf{G}}(t,s)$, $\tilde{\mathbf{H}}(t)$ is a standard generalized Green's couple of BVP (2,1), (2,2). By 4.2 and I.4.36 we have

$$\int_0^1 \mathbf{X}_0^*(\tau)\left(\int_0^1 \tilde{\mathbf{G}}(\tau,s)\,\mathbf{f}(s)\,\mathrm{d}s\right)\mathrm{d}\tau = \int_0^1\left(\int_0^1 \mathbf{X}_0^*(\tau)\,\tilde{\mathbf{G}}(\tau,s)\,\mathrm{d}\tau\right)\mathbf{f}(s)\,\mathrm{d}s\,.$$

Consequently, putting

$$(5,11) \qquad \mathbf{G}(t,s) = \tilde{\mathbf{G}}(t,s) - \mathbf{X}_0(t)\,\mathbf{D}^{-1}\int_0^1 \mathbf{X}_0^*(\tau)\,\tilde{\mathbf{G}}(\tau,s)\,\mathrm{d}\tau \qquad \text{on } [0,1]\times[0,1]\,,$$

$$\mathbf{H}(t) = \tilde{\mathbf{H}}(t) - \mathbf{X}_0(t)\,\mathbf{D}^{-1}\int_0^1 \mathbf{X}_0^*(\tau)\,\tilde{\mathbf{H}}(\tau)\,\mathrm{d}\tau \qquad \text{on } [0,1]\,,$$

we obtain

$$\mathbf{u}_0(t) = \int_0^1 \mathbf{G}(t,s)\,\mathbf{f}(s)\,\mathrm{d}s + \mathbf{H}(t)\,\mathbf{r} \qquad \text{on } [0,1]\,.$$

5.9. Remark. Let us notice that $\mathbf{v} \in X$ is a least square solution to BVP (2,1), (2,2) if and only if

$$0 = \left(\mathscr{A}\mathbf{x}, \mathscr{A}\mathbf{v} - \begin{pmatrix} \mathbf{f} \\ \mathbf{r} \end{pmatrix} \right)_Y \quad \text{for any} \quad \mathbf{x} \in X. \tag{5,12}$$

Since by the definition (5,2)

$$\left(\mathscr{A}\mathbf{x}, \mathscr{A}\mathbf{v} - \begin{pmatrix} \mathbf{f} \\ \mathbf{r} \end{pmatrix} \right)_Y = \left\langle \mathscr{L}\mathbf{x}, \left(\mathscr{L}\mathbf{v} - \begin{pmatrix} \mathbf{f} \\ \mathbf{r} \end{pmatrix} \right)^* \right\rangle_{L \times R}$$
$$= \langle \mathbf{x}, \mathscr{L}^*(\mathscr{L}\mathbf{v})^* - \mathscr{L}^*(\mathbf{f}^*, \mathbf{r}^*) \rangle_{L \times R} \quad \text{for any} \quad \mathbf{x} \in W_n^2,$$

the condition (5,12) is equivalent to

$$\mathscr{L}^*(\mathscr{L}\mathbf{v})^* = \mathscr{L}^*(\mathbf{f}^*, \mathbf{r}^*)$$

or

$$\mathbf{v}'(t) + \left[\mathbf{P}(t, 0) + \int_0^1 \mathbf{P}^*(\sigma, t)\, \mathbf{P}(\sigma, 0)\, \mathrm{d}\sigma + \mathbf{K}^*(t)\, \mathbf{M} \right] \mathbf{v}(0) \tag{5,13}$$
$$+ \int_0^1 \left[\mathbf{P}(t, s) + \mathbf{P}^*(s, t) + \int_0^1 \mathbf{P}^*(\sigma, t)\, \mathbf{P}(\sigma, s)\, \mathrm{d}\sigma + \mathbf{K}^*(t)\, \mathbf{K}(s) \right] \mathbf{v}'(s)\, \mathrm{d}s$$
$$= \mathbf{f}(t) + \int_0^1 \mathbf{P}^*(s, t)\, \mathbf{f}(s)\, \mathrm{d}s + \mathbf{K}^*(t)\, \mathbf{r} \quad \text{a.e. on } [0, 1],$$
$$\left[\mathbf{M}^*\mathbf{M} + \int_0^1 \mathbf{P}^*(\sigma, 0)\, \mathbf{P}(\sigma, 0)\, \mathrm{d}\sigma \right] \mathbf{v}(0)$$
$$+ \int_0^1 \left[\mathbf{M}^*\, \mathbf{K}(s) + \mathbf{P}^*(s, 0) + \int_0^1 \mathbf{P}^*(\sigma, 0)\, \mathbf{P}(\sigma, s)\, \mathrm{d}\sigma \right] \mathbf{v}'(s)\, \mathrm{d}s$$
$$= \mathbf{M}^*\mathbf{r} + \int_0^1 \mathbf{P}^*(s, 0)\, \mathbf{f}(s)\, \mathrm{d}s.$$

Let us notice that the system (5,13) of equations for $\mathbf{u} = \mathbf{v}' \in L_n^2$ and $\mathbf{c} = \mathbf{v}(0) \in R_n$ may be treated in the same way as the system (4,5), (4,6) (cf. also Lemma 3.1 in Tvrdý, Vejvoda [1]). If $\mathbf{P}(.,s)$ and $\mathbf{K}$ are of bounded variation on $[0, 1]$, then the system (5,13) may be reduced to the form (2,1), (2,2).

5.10. Remark. Let $\mathbf{r} \in R_m$ be fixed and let us define

$$D_r = \{ \mathbf{x} \in W_n^2;\ \mathbf{S}\mathbf{x} = \mathbf{r} \} \quad \text{and} \quad \mathscr{L}_r\colon \mathbf{x} \in D_r \to \mathbf{D}\mathbf{x} - \mathbf{P}\mathbf{x} \in L_n^2.$$

Then $R(\mathscr{L}_r)$ is closed in L_n^2 (cf. 2.16). Hence if $D_r \neq \emptyset$, then by the Classical Projection Theorem (Luenberger [1], p. 64) $R(\mathscr{L}_r)$ contains a unique element $\mathbf{y}$ of minimum L^2-norm and $\mathbf{y} \in R(\mathscr{L}_0)^\perp$. It follows from 2.17 that $\|\mathbf{y}\|_{L^2} \leq \|\mathbf{f}\|_{L^2}$ for all $\mathbf{f} \in R(\mathscr{L}_r)$

if and only if there exists $\lambda^* \in R_m^*$ such that $(\mathbf{y}^*, \lambda^*) \in N(\mathscr{L}^*)$. Thus $\mathbf{u} \in D_r$ fulfils $\|\mathbf{Du} - \mathbf{Pu}\|_{L^2} \le \|\mathbf{Dx} - \mathbf{Px}\|_{L^2}$ for all $\mathbf{x} \in D_r$ if and only if there exists $\lambda^* \in R_m^*$ such that

$$\mathscr{L}^*((\mathbf{Du} - \mathbf{Pu})^*, \lambda^*) = \mathbf{0}.$$

6. Volterra-Stieltjes integro-differential operator

Let $\mathbf{P}: [0,1] \times [0,1] \to L(R_n)$ be an $L^p[BV]$-kernel and let for a.e. $t \in [0,1]$, $\mathbf{P}(t,s) = \mathbf{P}(t,t)$ if $0 \le t \le s \le 1$. Then

$$\mathbf{P}: \mathbf{x} \in W_n^p \to \int_0^1 \mathrm{d}_s[\mathbf{P}(t,s)]\,\mathbf{x}(s) = \int_0^t \mathrm{d}_s[\mathbf{P}(t,s)]\,\mathbf{x}(s) \in L_n^p$$

and the Fredholm-Stieltjes integro-differential operator $\mathscr{L} = \mathbf{D} - \mathbf{P}$ defined in 1.5 reduces to a *Volterra-Stieltjes integro-differential operator*

$$\mathscr{L} = \mathbf{D} - \mathbf{P}: \mathbf{x} \in W_n^p \to \mathbf{x}'(t) - \int_0^t \mathrm{d}_s[\mathbf{P}(t,s)]\,\mathbf{x}(s) \in L_n^p. \tag{6,1}$$

If $\mathbf{P}(t,s) = \mathbf{P}(t,t) = \mathbf{0}$ for $0 \le t \le s \le 1$, then by I.4.38

$$\int_0^t \left(\int_0^\tau \mathrm{d}_s[\mathbf{P}(\tau,s)]\,\mathbf{x}(s) \right) \mathrm{d}\tau = \int_0^t \left(\int_0^t \mathrm{d}_s[\mathbf{P}(\tau,s)]\,\mathbf{x}(s) \right) \mathrm{d}\tau$$

$$= \int_0^t \mathrm{d}_s \left[\int_0^t \mathbf{P}(\tau,s)\,\mathrm{d}\tau \right] \mathbf{x}(s) = \int_0^t \mathrm{d}_s \left[\int_s^t \mathbf{P}(\tau,s)\,\mathrm{d}\tau \right] \mathbf{x}(s).$$

Thus, if $\mathbf{f} \in L_n^p$, then by integrating the Volterra-Stieltjes integro-differential equation for $\mathbf{x} \in W_n^p$

$$\mathbf{x}'(t) - \int_0^t \mathrm{d}_s[\mathbf{P}(t,s)]\,\mathbf{x}(s) = \mathbf{f}(t) \quad \text{a.e. on } [0,1] \tag{6,2}$$

we obtain

6.1. Proposition. *If* $\mathbf{P}(t,s) = \mathbf{0}$ *for* $0 \le t \le s \le 1$, *then a function* $\mathbf{x} \in BV_n$ *is a solution to* (6,2) *if and only if*

$$\mathbf{x}(t) - \int_0^t \mathrm{d}_s[\mathbf{Q}(t,s)]\,\mathbf{x}(s) = \mathbf{x}(0) + \int_0^t \mathbf{f}(\tau)\,\mathrm{d}\tau \quad \text{on } [0,1], \tag{6,3}$$

where

$$\mathbf{Q}(t,s) = \int_s^t \mathbf{P}(\tau,s)\,\mathrm{d}\tau \quad \text{if } 0 \le s \le t \le 1, \qquad \mathbf{Q}(t,s) = \mathbf{0} \quad \text{if } 0 \le t \le s \le 1. \tag{6,4}$$

(Obviously, if $\mathbf{x} \in BV_n$ fulfils (6,3), then $\mathbf{x} \in W_n^p$.)

6.2. Remark. Let us notice that if $\mathbf{P}_0(t,s) = \mathbf{P}(t,s) - \mathbf{P}(t,t)$ on $[0,1] \times [0,1]$, then $\mathbf{P}_0(t,s) = \mathbf{0}$ for $0 \le t \le s \le 1$ and

$$\int_0^t \mathrm{d}_s[\mathbf{P}_0(t,s)]\,\mathbf{x}(s) = \int_0^t \mathrm{d}_s[\mathbf{P}(t,s)]\,\mathbf{x}(s) \qquad \text{for any} \quad t \in [0,1] \quad \text{and} \quad \mathbf{x} \in C_n .$$

This means that the assumption $\mathbf{P}(t,t) = \mathbf{0}$ for every $t \in [0,1]$ does not cause any loss of generality.

6.3. Proposition. $v_{[0,1]\times[0,1]}(\mathbf{Q}) < \infty$, $\mathbf{Q}(0,s) = \mathbf{0}$ *on* $[0,1]$ *and* $\mathbf{Q}(t,t-) = \mathbf{Q}(t,t) = \mathbf{0}$ *for any* $t \in (0,1]$.

Proof. Let a net-type subdivision $\{0 = t_0 < t_1 < \ldots < t_k = 1;\ 0 = s_0 < s_1 < \ldots < s_k = 1\}$ be given. Then

$$m_{i,j}(\mathbf{Q}) = \left| \int_{t_{i-1}}^{t_i} (\mathbf{P}(\tau, s_j) - \mathbf{P}(\tau, s_{j-1}))\,\mathrm{d}\tau \right| \le \int_{t_{i-1}}^{t_i} |\mathbf{P}(\tau, s_j) - \mathbf{P}(\tau, s_{j-1})|\,\mathrm{d}\tau .$$

Hence

$$\sum_{i=1}^{k} \sum_{j=1}^{k} m_{i,j}(\mathbf{Q}) \le \int_0^1 \sum_{j=1}^{k} |\mathbf{P}(\tau, s_j) - \mathbf{P}(\tau, s_{j-1})|\,\mathrm{d}\tau \le \int_0^1 \varrho(\tau)\,\mathrm{d}\tau$$

and consequently $v_{[0,1]\times[0,1]}(\mathbf{Q}) < \infty$. The other assertions of the lemma follow immediately from (6,4).

Making use of the results obtained for Volterra-Stieltjes integral equations in the Section II.3 we can deduce the variation-of-constants formula for Volterra-Stieltjes integro-differential equations.

6.4. Theorem. *Let* $\mathbf{P}\colon [0,1] \times [0,1] \to L(R_n)$ *be an* $L^p[BV]$*-kernel such that for a.e.* $t \in [0,1]$ $\mathbf{P}(t,s) = \mathbf{0}$ *if* $0 \le t \le s \le 1$. *Then for any* $\mathbf{c} \in R_n$ *and* $\mathbf{f} \in L_n^p$ *there exists a unique solution* $\mathbf{x}$ *of the equation* (6,2) *in* W_n^p *such that* $\mathbf{x}(0) = \mathbf{c}$.

Furthermore, there exists a uniquely determined function $\mathbf{U}\colon [0,1] \times [0,1] \to L(R_n)$ *such that for any* $\mathbf{f} \in L_n^1$ *and* $\mathbf{c} \in R_n$ *this solution is given by*

$$\text{(6,5)} \qquad \mathbf{x}(t) = \mathbf{U}(t,0)\,\mathbf{c} + \int_0^t \mathbf{U}(t,s)\,\mathbf{f}(s)\,\mathrm{d}s , \qquad t \in [0,1] .$$

The function $\mathbf{U}$ *satisfies the equation*

$$\text{(6,6)} \quad \frac{\partial}{\partial t}\mathbf{U}(t,s) = \int_s^t \mathrm{d}_r[\mathbf{P}(t,r)]\,\mathbf{U}(r,s) \qquad \textit{for any} \quad s \in [0,1] \quad \textit{and a.e.} \quad t \in [s,1] .$$

Moreover, $v_{[0,1]\times[0,1]}(\mathbf{U}) + \mathrm{var}_0^1\, \mathbf{U}(0,.) < \infty$, $\mathbf{U}(.,s)$ *is absolutely continuous on* $[0,1]$ *for any* $s \in [0,1]$ *and* $\mathbf{U}(t,s) = \mathbf{I}$ *if* $0 \le t \le s \le 1$.

Proof. Let $\boldsymbol{\Gamma}: [0,1] \times [0,1] \to L(R_n)$ correspond to $\mathbf{Q}$ by II.3.10. In particular, the function $\boldsymbol{x}: [0,1] \to R_n$ given by

$$\boldsymbol{x}(t) = \boldsymbol{c} + \int_0^t \boldsymbol{f}(\tau)\,\mathrm{d}\tau + \int_0^t \mathrm{d}_s[\boldsymbol{\Gamma}(t,s)]\left(\boldsymbol{c} + \int_0^s \boldsymbol{f}(\tau)\,\mathrm{d}\tau\right)$$

is for any $\boldsymbol{c} \in R_n$ and $\boldsymbol{f} \in L_n^p$ a unique solution to (6,3) such that $\boldsymbol{x}(0) = \boldsymbol{c}$. Integration by parts yields

$$\boldsymbol{x}(t) = [\boldsymbol{I} + \boldsymbol{\Gamma}(t,t)]\,\boldsymbol{c} + \int_0^t [\boldsymbol{I} + \boldsymbol{\Gamma}(t,t) - \boldsymbol{\Gamma}(t,s)]\,\boldsymbol{f}(s)\,\mathrm{d}s \qquad \text{on } [0,1]. \tag{6,7}$$

Denoting

$$\boldsymbol{U}(t,s) = \begin{cases} \boldsymbol{I} + \boldsymbol{\Gamma}(t,t) - \boldsymbol{\Gamma}(t,s) & \text{if } 0 \le s \le t \le 1\,, \\ \boldsymbol{I} & \text{if } 0 \le t \le s \le 1\,, \end{cases} \tag{6,8}$$

the expression (6,7) reduces to (6,5). (Recall that $\boldsymbol{\Gamma}(t,0) = \boldsymbol{0}$ for every $t \in [0,1]$.) In our case the function $\boldsymbol{\Gamma}$ satisfies for $0 \le s \le t \le 1$ the relation (cf. (II.3.29))

$$\boldsymbol{\Gamma}(t,s) = \int_s^t \mathbf{P}(\tau,s)\,\mathrm{d}\tau - \int_0^t \mathbf{P}(\tau,0)\,\mathrm{d}\tau + \int_0^t \mathrm{d}_r\left[\int_r^t \mathbf{P}(\tau,r)\,\mathrm{d}\tau\right]\boldsymbol{\Gamma}(r,s)\,. \tag{6,9}$$

Taking into account that $\mathbf{P}(\tau,r) = \boldsymbol{0}$ if $0 \le \tau \le r \le 1$ and $\boldsymbol{\Gamma}(r,s) = \boldsymbol{\Gamma}(r,r)$ if $0 \le r \le s \le 1$ and employing I.4.38 we obtain for $0 \le s \le t \le 1$

$$\begin{aligned} \int_0^t \mathrm{d}_r\left[\int_r^t \mathbf{P}(\tau,r)\,\mathrm{d}\tau\right]\boldsymbol{\Gamma}(r,s) &= \int_0^t \mathrm{d}_r\left[\int_0^t \mathbf{P}(\tau,r)\,\mathrm{d}\tau\right]\boldsymbol{\Gamma}(r,s) \\ &= \int_0^t\left(\int_0^t \mathrm{d}_r[\mathbf{P}(\tau,r)]\,\boldsymbol{\Gamma}(r,s)\right)\mathrm{d}\tau = \int_0^t\left(\int_0^\tau \mathrm{d}_r[\mathbf{P}(\tau,r)]\,\boldsymbol{\Gamma}(r,s)\right)\mathrm{d}\tau \\ &= \int_s^t\left(\int_s^t \mathrm{d}_r[\mathbf{P}(\tau,r)]\,\boldsymbol{\Gamma}(r,s)\right)\mathrm{d}\tau + \int_0^t\left(\int_0^s \mathrm{d}_r[\mathbf{P}(\tau,r)]\,\boldsymbol{\Gamma}(r,r)\right)\mathrm{d}\tau\,. \end{aligned}$$

It is easy to verify (cf. also (6,8) and (6,9)) that

$$\boldsymbol{U}(t,s) = \boldsymbol{I} - \int_s^t \mathbf{P}(\tau,s)\,\mathrm{d}\tau - \int_s^t\left(\int_s^\tau \mathrm{d}_r[\mathbf{P}(\tau,r)]\,(\boldsymbol{\Gamma}(r,s) - \boldsymbol{\Gamma}(r,r))\right)\mathrm{d}\tau$$

$$\text{for } 0 \le s \le t \le 1\,.$$

On the other hand, it follows from (6,8) that

$$\int_s^\tau \mathrm{d}_r[\mathbf{P}(\tau,r)]\,\boldsymbol{U}(r,s) = -\mathbf{P}(\tau,s) - \int_s^\tau \mathrm{d}_r[\mathbf{P}(\tau,r)]\,(\boldsymbol{\Gamma}(r,r) - \boldsymbol{\Gamma}(r,s))$$

$$\text{for } 0 \le s \le \tau \le 1\,.$$

Thus ($\boldsymbol{U}(s,s) = \boldsymbol{I}$)

$$\boldsymbol{U}(t,s) = \boldsymbol{U}(s,s) + \int_s^t\left(\int_s^\tau \mathrm{d}_r[\mathbf{P}(\tau,r)]\,\boldsymbol{U}(r,s)\right)\mathrm{d}\tau \qquad \text{if } 0 \le s \le t \le 1\,,$$

which yields (6,6) immediately.

As $\mathrm{v}_{[0,1]\times[0,1]}(\Gamma) < \infty$ (cf. II.3.10), also $\mathrm{v}_{[0,1]\times[0,1]}(\mathbf{U}) < \infty$. The other assertions of the theorem are evident.

6.5. Remark. Denoting for $\mathbf{c} \in R_n$, $\mathbf{f} \in L_n^p$ and $t \in [0,1]$

$$(\boldsymbol{\Phi}\mathbf{c})(t) = \mathbf{U}(t,0)\,\mathbf{c} \quad \text{and} \quad (\boldsymbol{\Psi}\mathbf{f})(t) = \int_0^t \mathbf{U}(t,s)\,\mathbf{f}(s)\,\mathrm{d}s\,, \tag{6,10}$$

the variation of constants formula (6,5) for solutions of (6,2) becomes

$$\mathbf{x}(t) = (\boldsymbol{\Phi}\mathbf{c})(t) + (\boldsymbol{\Psi}\mathbf{f})(t) \quad \text{on } [0,1] \qquad (\mathbf{x} = \boldsymbol{\Phi}\mathbf{c} + \boldsymbol{\Psi}\mathbf{f})\,. \tag{6,11}$$

By 6.4 the functions $\boldsymbol{\Phi}\mathbf{c}$ and $\boldsymbol{\Psi}\mathbf{f}$ belong to W_n^p for every $\mathbf{c} \in R_n$ and $\mathbf{f} \in L_n^p$. Moreover, the linear operators $\boldsymbol{\Phi}$: $\mathbf{c} \in R_n \to \boldsymbol{\Phi}\mathbf{c} \in W_n^p$ and $\boldsymbol{\Psi}$: $\mathbf{f} \in L_n^p \to \boldsymbol{\Psi}\mathbf{f} \in W_n^p$ are bounded. Indeed, if $\mathbf{f} \in L_n^p$ and $\boldsymbol{\psi} = \boldsymbol{\Psi}\mathbf{f}$, then in virtue of I.4.27, I.6.6 and 6.4 we have for a.e. $t \in [0,1]$

$$\begin{aligned} |\boldsymbol{\psi}'(t)| &= \left| \int_0^t \mathrm{d}_s[\mathbf{P}(t,s)]\,\boldsymbol{\psi}(s) + \mathbf{f}(t) \right| \le \left| \int_0^1 \mathrm{d}_\tau[\mathbf{P}(t,\tau)] \left(\int_0^\tau \mathbf{U}(\tau,s)\,\mathbf{f}(s)\,\mathrm{d}s \right) \right| + |\mathbf{f}(t)| \\ &\le \varrho(t) \sup_{t,s\in[0,1]} |\mathbf{U}(t,s)|\, \|\mathbf{f}\|_{L^1} + |\mathbf{f}(t)|\,. \end{aligned}$$

Consequently

$$\|\boldsymbol{\Psi}\mathbf{f}\|_{W^p} = \|\boldsymbol{\psi}'\|_{L^p} \le (1 + \|\varrho\|_{L^p}(\sup_{t,s\in[0,1]} |\mathbf{U}(t,s)|))\, \|\mathbf{f}\|_{L^p}\,,$$

i.e. $\boldsymbol{\Psi} \in B(L_n^p, W_n^p)$. Analogously we could obtain $\boldsymbol{\Phi} \in B(R_n, W_n^p)$.

6.6. Corollary. *Let $\mathscr{U}$ be a linear normed space and let $\boldsymbol{\Theta} \in B(\mathscr{U}, L_n^p)$. If $\mathbf{P}$: $[0,1] \times [0,1] \to L(R_n)$ is an $L^p[BV]$-kernel such that for a.e. $t \in [0,1]$, $\mathbf{P}(t,s) = \mathbf{0}$ if $s \in [t,1]$, then for any $\mathbf{u} \in \mathscr{U}$, $\mathbf{f} \in L_n^p$ and $\mathbf{c} \in R_n$ there exists a unique solution $\mathbf{x} \in W_n^p$ of*

$$\mathbf{D}\mathbf{x} - \mathbf{P}\mathbf{x} = \boldsymbol{\Theta}\mathbf{u} + \mathbf{f}\,, \qquad \mathbf{x}(0) = \mathbf{c}\,.$$

This solution is given by $\mathbf{x} = \boldsymbol{\Phi}\mathbf{c} + \boldsymbol{\Psi}\boldsymbol{\Theta}\mathbf{u} + \boldsymbol{\Psi}\mathbf{f}$.

6.7. Remark. Let $r > 0$ and let $\mathbf{P}$: $[0,1] \times [-r,1] \to L(R_n)$ be an $L^p[BV]$-kernel on $[0,1] \times [-r,1]$ such that $\mathbf{P}(t,s) = \mathbf{0}$ if $t \le s$ and $\mathbf{P}(t,s) = \mathbf{P}(t,t-r)$ if $s \le t-r$. Let $\mathbf{u} \in BV_n[-r,0]$ and $\mathbf{f} \in L_n^p$ be given and let us look for a function $\mathbf{x} \in BV_n[-r,1]$ absolutely continuous on $[0,1]$ and such that $\mathbf{x}'$ is L^p-integrable on $[0,1]$ and

$$\begin{aligned} \mathbf{x}'(t) - \int_{t-r}^t \mathrm{d}_s[\mathbf{P}(t,s)]\,\mathbf{x}(s) &= \mathbf{f}(t) \quad \text{a.e. on } [0,1]\,, \\ \mathbf{x}(t) &= \mathbf{u}(t) \quad \text{on } [-r,0]\,. \end{aligned} \tag{6,12}$$

If we put

$$\boldsymbol{\Theta}\text{: } \mathbf{u} \in' BV_n[-r,0] \to \int_{-r}^0 \mathrm{d}_s[\mathbf{P}(t,s)]\,\mathbf{u}(s) \in L_n^p\,,$$

then $\boldsymbol{\Theta}$ is a linear compact operator (cf. 1.4). For any $\mathbf{x} \in BV_n[-r, 1]$ and $t \in [0, 1]$ we have

$$\int_{t-r}^{t} \mathrm{d}_s[\mathbf{P}(t, s)]\, \mathbf{x}(s) = \int_{-r}^{0} \mathrm{d}_s[\mathbf{P}(t, s)]\, \mathbf{x}(s) + \int_{0}^{t} \mathrm{d}_s[\mathbf{P}(t, s)]\, \mathbf{x}(s).$$

Thus our problem may be formulated in the form of the operator equation $\mathbf{D}\mathbf{x} - \mathbf{P}\mathbf{x} = \boldsymbol{\Theta}\mathbf{u} + \mathbf{f}$ and according to 6.6 (with $\mathscr{U} = BV_n[-r, 0]$) the equation (6,12) has for any $\mathbf{u} \in BV_n[-r, 0]$ and $\mathbf{x} \in L_n^p$ a unique solution $\mathbf{x} \in W_n^p$ such that $\mathbf{x}(t) = \mathbf{u}(t)$ on $[-r, 0]$. This solution is of the form $\mathbf{x} = \boldsymbol{\Phi}_0\mathbf{u} + \boldsymbol{\Psi}\mathbf{f}$, where $\boldsymbol{\Phi}_0\colon \mathbf{u} \in BV_n[-r, 0] \to \boldsymbol{\Phi}\, \mathbf{u}(0) + \boldsymbol{\Psi}\boldsymbol{\Theta}\mathbf{u} \in W_n^p$ is a linear compact operator. (Let us notice that in virtue of I.4.38

$$(\boldsymbol{\Phi}_0\mathbf{u})(t) = \mathbf{U}(t, 0)\, \mathbf{u}(0) + \int_{-r}^{0} \mathrm{d}_s\left[\int_0^t \mathbf{U}(t, \tau)\, \mathbf{P}(\tau, s)\, \mathrm{d}\tau\right] \mathbf{u}(s) \quad \text{on } [0, 1]$$

for any $\mathbf{u} \in BV_n[-r, 0]$.) Thus, the variation-of-constants formula for functional-differential equations of the retarded type (cf. Banks [1] or Hale [1]) is a consequence of Theorem 6.2.

Analogously we may show that if $0 < r_i \le r$ $(i = 1, 2, \ldots, k)$, $\mathbf{A}_i\colon [0, 1] \to L(R_n)$ $i = 1, 2, \ldots, k)$ are measurable and essentially bounded on $[0, 1]$ and $\mathbf{A}_0\colon [0, 1] \times [-r, 0] \to L(R_n)$ is measurable and essentially bounded on $[0, 1] \times [-r, 0]$, then the system

(6,13)
$$\mathbf{x}'(t) - \sum_{i=1}^{k} \mathbf{A}_i(t) \begin{Bmatrix} \mathbf{0} & \text{if } t - r_i > 0 \\ \mathbf{u}(t - r_i) & \text{if } t - r_i < 0 \end{Bmatrix}$$
$$- \int_{-r}^{0} \mathbf{A}_0(t, s) \begin{Bmatrix} \mathbf{0} & \text{if } t + s > 0 \\ \mathbf{u}(t + s) & \text{if } t + s < 0 \end{Bmatrix} \mathrm{d}s$$
$$- \int_0^t \mathrm{d}_s[\mathbf{P}(t, s)]\, \mathbf{x}(s) = \mathbf{f}(t) \quad \text{a.e. on } [0, 1]$$

has for any $\mathbf{f} \in L_n^p[0, 1]$, $\mathbf{u} \in L_n^p[-r, 0]$ and $\mathbf{c} \in R_n$ a unique solution $\mathbf{x} \in L_n^p[-r, 1]$ such that $\mathbf{x}(t) = \mathbf{u}(t)$ a.e. on $[-r, 0]$, $\mathbf{u}(0) = \mathbf{c}$ and $\mathbf{x}|_{[0,1]} \in W_n^p$. This solution is of the form $\mathbf{x} = \boldsymbol{\Phi}\mathbf{c} + \boldsymbol{\Psi}\boldsymbol{\Theta}\mathbf{u} + \boldsymbol{\Psi}\mathbf{f}$, where

$$\boldsymbol{\Theta}\colon \mathbf{u} \in L_n^p[-r, 0] \to \begin{Bmatrix} \sum_{i=1}^{k} \mathbf{A}_i(t)\, \mathbf{u}(t - r_i) & \text{if } t - r_i < 0 \\ \mathbf{0} & \text{if } t - r_i > 0 \end{Bmatrix}$$
$$+ \int_{-r}^{0} \mathbf{A}_0(t, s) \begin{Bmatrix} \mathbf{u}(t + s) & \text{if } t + s < 0 \\ \mathbf{0} & \text{if } t + s > 0 \end{Bmatrix} \mathrm{d}s \in L_n^p.$$

(Functional-differential equations of the type (6,13) were studied in detail in Delfour-Mitter [1] and [2].)

6.8. Theorem. *Let Λ be a Banach space, $\mathbf{S} \in B(W_n^p, \Lambda)$ and let $\mathbf{P}$: $[0,1] \times [0,1] \to L(R_n)$ be an $L^p[BV]$-kernel such that for a.e. $t \in [0,1]$, $\mathbf{P}(t,s) = \mathbf{0}$ if $s \in [t,1]$. Then the linear bounded operator*

$$\mathscr{L}: \mathbf{x} \in W_n^p \to \begin{bmatrix} \mathbf{Dx} - \mathbf{Px} \\ \mathbf{Sx} \end{bmatrix} \in L_n^p \times \Lambda$$

has a closed range.

Proof. By 6.5, $\begin{pmatrix} \mathbf{f} \\ \mathbf{r} \end{pmatrix} \in L_n^p \times \Lambda$ belongs to $R(\mathscr{L})$ if and only if $\mathbf{r} - \mathbf{S}\boldsymbol{\Psi}\mathbf{f} \in R(\mathbf{S}\boldsymbol{\Phi})$. As $\mathbf{W}$: $\begin{pmatrix} \mathbf{f} \\ \mathbf{r} \end{pmatrix} \in L_n^p \times \Lambda \to \mathbf{r} - \mathbf{S}\boldsymbol{\Psi}\mathbf{f} \in \Lambda$ is bounded and $R(\mathbf{S}\boldsymbol{\Phi})$ is a finite dimensional linear subspace in Λ ($\boldsymbol{\Phi} \in B(R_n, W_n^p)$), it follows that $R(\mathscr{L})$ is closed.

7. Fredholm-Stieltjes integral equations with linear constraints

This section is devoted to the system of equations for $\mathbf{x} \in BV_n$

$$\mathbf{x}(t) - \mathbf{x}(0) - \int_0^1 \mathrm{d}_s[\mathbf{P}(t,s) - \mathbf{P}(0,s)]\, \mathbf{x}(s) = \mathbf{f}(t) - \mathbf{f}(0) \quad \text{on } [0,1], \tag{7,1}$$

$$\int_0^1 \mathrm{d}[\mathbf{K}(s)]\, \mathbf{x}(s) = \mathbf{r}. \tag{7,2}$$

The following hypotheses are pertinent.

7.1. Assumptions. $\mathbf{P}$: $[0,1] \times [0,1] \to L(R_n)$ *and there are* $t_0, s_0 \in [0,1]$ *such that*

$$\mathrm{v}_{[0,1]\times[0,1]}(\mathbf{P}) + \mathrm{var}_0^1\, \mathbf{P}(t_0, .) + \mathrm{var}_0^1\, \mathbf{P}(., s_0) < \infty, \tag{7,3}$$

$\mathbf{K}$: $[0,1] \to L(R_n, R_m)$ *is of bounded variation on* $[0,1]$, $\mathbf{f} \in BV_n$ *and* $\mathbf{r} \in R_m$.

7.2. Definition. Any function $\mathbf{P}$: $[0,1] \times [0,1] \to L(R_n)$ fulfilling (7,3) is called an *SBV-kernel.*

7.3. Remark. If $\mathbf{P}$: $[0,1] \times [0,1] \to L(R_n)$ is an *SBV*-kernel and

$$\mathbf{Q}(t,s) = \begin{cases} \mathbf{P}(t,s) - \mathbf{P}(0,s) & \text{for } t \in [0,1] \text{ and } s \in (0,1], \\ \mathbf{P}(t,0) - \mathbf{P}(0,0) - \mathbf{I} & \text{for } t \in [0,1] \text{ and } s = 0, \end{cases} \tag{7,4}$$

then obviously $\mathbf{Q}(t,s)$ is an *SBV*-kernel and

$$\mathbf{x}(0) + \int_0^1 \mathrm{d}_s[\mathbf{P}(t,s) - \mathbf{P}(0,s)]\, \mathbf{x}(s) = \int_0^1 \mathrm{d}_s[\mathbf{Q}(t,s)]\, \mathbf{x}(s) \quad \text{for any } \mathbf{x} \in BV_n \tag{7,5}$$

(cf. I.4.23). It means that the equation (7,1) is a special case of Fredholm-Stieltjes integral equations studied in Chapter II. Let us denote by $\mathbf{Q}$ the linear operator

$$(7,6) \qquad \mathbf{Q}\colon \mathbf{x} \in BV_n \to \mathbf{x}(0) + \int_0^1 \mathrm{d}_s[\mathbf{P}(t,s) - \mathbf{P}(0,s)]\,\mathbf{x}(s)\,.$$

By (7,5) and II.1.5 $R(\mathbf{Q}) \subset BV_n$ and $\mathbf{Q} \in L(BV_n)$ is compact.

The following assertion follows analogously as 1.8 from I.3.20 and 1.9.

7.4. Proposition. *If* $\mathbf{P}\colon [0,1] \times [0,1] \to L(R_n)$ *is an SBV-kernel and the operator* $\mathbf{Q}$ *is given by* (7,6), *then* $n \leq \dim N(\mathbf{I} - \mathbf{Q}) < \infty$, *while* $\dim N(\mathbf{I} - \mathbf{Q}) = n$ *if and only if the equation* (7,1) *has a solution* $\mathbf{x} \in BV_n$ *for any* $\mathbf{f} \in BV_n$.

Let us mention that the following additional hypotheses do not mean any loss of generality (cf. II.1.4).

7.5. Assumptions. $\mathbf{P}(t,.)$ *is right-continuous on* $(0,1)$ *and* $\mathbf{P}(t,1) = \mathbf{0}$ *for any* $t \in [0,1]$ *and* $\mathbf{P}(0,s) = \mathbf{0}$ *for any* $s \in [0,1]$; $\mathbf{K}$ *is right-continuous on* $(0,1)$ *and* $\mathbf{K}(1) = \mathbf{0}$.

Analogously as in the case of BVP (2,1), (2,2) for Fredholm-Stieltjes integro-differential operators we rewrite the system (7,1), (7,2) of equations for $\mathbf{x} \in BV_n$ as the system of operator equations for $\xi = \begin{pmatrix} \mathbf{x} \\ \mathbf{d} \end{pmatrix} \in BV_n \times R_m$

$$(7,7) \qquad (\mathbf{I} - \mathbf{T})\,\xi = \begin{pmatrix} \mathbf{x} \\ \mathbf{d} \end{pmatrix} - \begin{pmatrix} \mathbf{Qx} \\ \mathbf{d} - \mathbf{Sx} \end{pmatrix} = \begin{pmatrix} \mathbf{\Psi f} \\ \mathbf{r} \end{pmatrix},$$

where $\mathbf{Q} \in K(BV_n)$ is defined by (7,6),

$$(7,8) \qquad \mathbf{S}\colon \mathbf{x} \in BV_n \to \int_0^1 \mathrm{d}[\mathbf{K}(s)]\,\mathbf{x}(s) \in R_m\,,$$

$$(7,9) \qquad \mathbf{T}\colon \begin{pmatrix} \mathbf{x} \\ \mathbf{d} \end{pmatrix} \in BV_n \times R_m \to \begin{pmatrix} \mathbf{Qx} \\ \mathbf{d} - \mathbf{Sx} \end{pmatrix} \in BV_n \times R_m$$

and $\mathbf{\Psi}$ is now given by

$$(7,10) \qquad \mathbf{\Psi}\colon \mathbf{f} \in BV_n \to \mathbf{f}(t) - \mathbf{f}(0) \in BV_n\,.$$

7.6. Proposition. *If* $\mathbf{x} \in BV_n$ *is a solution to* (7,1), (7,2), *then* $\xi = \begin{pmatrix} \mathbf{x} \\ \mathbf{d} \end{pmatrix}$ *is a solution to* (7,7) *for any* $\mathbf{d} \in R_m$. *If* $\mathbf{x} \in BV_n$ *and there exists* $\mathbf{d} \in R_m$ *such that* $\xi = \begin{pmatrix} \mathbf{x} \\ \mathbf{d} \end{pmatrix}$ *verifies* (7,7), *then* $\mathbf{x}$ *is a solution of* (7,1), (7,2).

7.7. Proposition. *Under the assumptions* 7.1 *the operator* $\mathbf{T} \in L(BV_n \times R_m)$ *defined by* (7,6), (7,8) *and* (7,9) *is compact.*

Proof. As obviously $\mathbf{S} \in B(BV_n, R_m) = K(BV_n, R_m)$ (cf. I.3.21) and $\mathbf{Q} \in K(BV_n)$, it is easy to see that $\mathbf{T} \in K(BV_n \times R_m)$.

Our wish is now to establish the duality theory for problems of the form (7,1), (7,2). To this end it is necessary to choose a space $BV_n^{`}$ of functions $[0,1] \to R_n^*$ and an operator $\mathbf{T}^{`} \in L(BV_n^{`} \times R_m^*)$ in such a way that $(BV_n \times R_m, BV_n^{`} \times R_m^*)$ is a dual pair with respect to some bilinear form $[.,.]$ (cf. I.3.1) and

$$(7,11)\qquad \left[\mathbf{T}\begin{pmatrix}\mathbf{x}\\\mathbf{d}\end{pmatrix}, (\mathbf{z}^*, \lambda^*)\right] = \left[\begin{pmatrix}\mathbf{x}\\\mathbf{d}\end{pmatrix}, \mathbf{T}^{`}(\mathbf{z}^*, \lambda^*)\right]$$

$$\text{for all} \quad \begin{pmatrix}\mathbf{x}\\\mathbf{d}\end{pmatrix} \in BV_n \times R_m \quad \text{and} \quad (\mathbf{z}^*, \lambda^*) \in BV_n^{`} \times R_m^* .$$

According to I.5.9 the spaces BV_n and NBV_n form a dual pair with respect to the bilinear form

$$\mathbf{x} \in BV_n,\ \varphi \in NBV_n \to \int_0^1 \mathrm{d}[\varphi^*(t)]\, \mathbf{x}(t) \in R .$$

For the purposes of this section a slightly different choice of the space $BV_n^{`}$ is more suitable.

7.8. Definition. $BV_n^{`}$ denotes the space of all functions $\mathbf{z}^*\colon [0,1] \to R_n^*$ of bounded variation on $[0,1]$, right-continuous on $(0,1)$ and such that $\mathbf{z}^*(1) = \mathbf{0}$.

7.9. Proposition. *The space $BV_n^{`}$ defined in 7.8 becomes a Banach space if it is endowed with the norm $\mathbf{z}^* \in BV_n^{`} \to \|\mathbf{z}^*\|_{BV^{`}} = |\mathbf{z}^*(0)| + \mathrm{var}_0^1\, \mathbf{z}^*$. Moreover, $(BV_n \times R_m, BV_n^{`} \times R_m^*)$ is a dual pair with respect to the bilinear form*

$$(7,12)\qquad \begin{pmatrix}\mathbf{x}\\\mathbf{d}\end{pmatrix} \in BV_n \times R_m ,\ (\mathbf{z}^*, \lambda^*) \in BV_n^{`} \times R_m^*$$

$$\to \left[\begin{pmatrix}\mathbf{x}\\\mathbf{d}\end{pmatrix}, (\mathbf{z}^*, \lambda^*)\right] = \int_0^1 \mathrm{d}[\mathbf{z}^*(t)]\, \mathbf{x}(t) + \lambda^* \mathbf{d} \in R .$$

(For the proofs of analogous assertions for NBV_n see I.5.2 and I.5.9.)

In the following the bilinear form $[.,.]$ is defined by (7,12).

7.10. Proposition. *If the hypotheses 7.1 are fulfilled, $\mathbf{Q}\colon [0,1] \times [0,1] \to L(R_n)$ is defined by (7,4) and*

$$(7,13)\qquad \mathbf{T}^{`}\colon (\mathbf{z}^*, \lambda^*) \in BV_n^{`} \times R_m^* \to \begin{pmatrix} \displaystyle\int_0^1 \mathrm{d}[\mathbf{z}^*(t)]\, \mathbf{Q}(t,s) - \lambda^* \mathbf{K}(s) \\ \lambda^* \end{pmatrix},$$

then (7,11) holds. If 7.5 is also assumed, then $R(\mathbf{T}^{`}) \subset BV_n^{`} \times R_m^*$ and $\mathbf{T}^{`} \in K(BV_n^{`} \times R_m)$.

Proof. Let us denote

$$\mathbf{Q}^{\backprime}: \mathbf{z} \in BV_n \rightarrow \int_0^1 \mathbf{Q}(t,s)\,\mathrm{d}[\mathbf{z}(t)]\,.$$

As $\mathbf{Q}(t,s)$ is an SBV-kernel, $\mathbf{Q}^{\backprime} \in K(BV_n)$ (cf. II.1.9). Moreover, by I.6.20

$$\int_0^1 \mathrm{d}[\mathbf{z}^*(t)] \left(\int_0^1 \mathrm{d}_s[\mathbf{Q}(t,s)]\,\mathbf{x}(s)\right) + \boldsymbol{\lambda}^* \left(\mathbf{d} - \int_0^1 \mathrm{d}[\mathbf{K}(s)]\,\mathbf{x}(s)\right)$$
$$= \int_0^1 \mathrm{d}_s \left[\int_0^1 \mathrm{d}[\mathbf{z}^*(t)]\,\mathbf{Q}(t,s) - \boldsymbol{\lambda}^*\,\mathbf{K}(s)\right] \mathbf{x}(s) + \boldsymbol{\lambda}^* \mathbf{d}$$

for any $\mathbf{x} \in BV_n$, $\mathbf{d} \in R_m$, $\mathbf{z}^* \in BV_n^{\backprime}$ and $\boldsymbol{\lambda}^* \in R_m^*$. If $\mathbf{P}(t,.)$ is right-continuous on $(0,1)$, then according to I.6.16 and I.4.17 also $\mathbf{Q}^{\backprime}\mathbf{z} \in BV_n$ is right-continuous on $(0,1)$ for any $\mathbf{z} \in BV_n$. Consequently, $R(\mathbf{T}^{\backprime}) \subset BV_n^{\backprime} \times R_m^*$ provided that 7.5 is satisfied. The compactness of $\mathbf{T}^{\backprime} \in L(BV_n^{\backprime} \times R_m^*)$ follows readily from the compactness of $\mathbf{Q}^{\backprime}$.

The operators $\mathbf{T}$ and $\mathbf{T}^{\backprime}$ being compact,

$$\text{(7,14)} \qquad \operatorname{ind}(\mathbf{I} - \mathbf{T}) = \operatorname{ind}(\mathbf{I} - \mathbf{T}^{\backprime}) = 0$$

(cf. I.3.20) and we may apply Theorem I.3.2.

7.11. Theorem. *If the hypotheses* 7.1 *and* 7.5 *are satisfied, then the system* (7,1), (7,2) *has a solution* $\mathbf{x} \in BV_n$ *if and only if*

$$\text{(7,15)} \qquad \int_0^1 \mathrm{d}[\mathbf{z}^*(s)]\,(\mathbf{f}(s) - \mathbf{f}(0)) + \boldsymbol{\lambda}^* \mathbf{r} = 0$$

for any $\mathbf{z}^* \in BV_n^{\backprime}$ *and* $\boldsymbol{\lambda}^* \in R_m^*$ *such that*

$$\text{(7,16)} \qquad \mathbf{z}^*(s) - \int_0^1 \mathrm{d}[\mathbf{z}^*(t)]\,\mathbf{P}(t,s) + \boldsymbol{\lambda}^*\,\mathbf{K}(s) = \mathbf{0} \quad \textit{on } [0,1], \quad \mathbf{z}^*(0) = \mathbf{0}\,.$$

Proof. By I.3.2 the system (7,1), (7,2) has a solution if and only if (7,15) holds for any $\mathbf{z}^* \in BV_n^{\backprime}$ and $\boldsymbol{\lambda}^* \in R_m^*$ fulfilling the equation

$$\text{(7,17)} \qquad \mathbf{z}^*(s) - \int_0^1 \mathrm{d}[\mathbf{z}^*(t)]\,\mathbf{Q}(t,s) + \boldsymbol{\lambda}^*\,\mathbf{K}(s) = \mathbf{0} \qquad \text{on } [0,1],$$

i.e. $(\mathbf{I} - \mathbf{T}^{\backprime})(\mathbf{z}^*, \boldsymbol{\lambda}^*) = \mathbf{0}$ (cf. 7.9, 7.10 and (7,14)). Given $\mathbf{z}^* \in BV_n^{\backprime}$,

$$\text{(7,18)} \quad \int_0^1 \mathrm{d}[\mathbf{z}^*(t)]\,\mathbf{Q}(t,s) = \int_0^1 \mathrm{d}[\mathbf{z}^*(t)]\,\mathbf{P}(t,s) - \begin{cases} \mathbf{z}^*(1) - \mathbf{z}^*(0) & \text{if } s = 0 \\ \mathbf{0} & \text{if } s > 0 \end{cases}$$

(cf. (7,4) and I.4.23). After the substitution (7,18), the equation (7,17) becomes

$$(7,19)\qquad \mathbf{z}^*(s) - \int_0^1 \mathrm{d}[\mathbf{z}^*(t)]\,\mathbf{P}(t,s) + \boldsymbol{\lambda}^*\,\mathbf{K}(0) = \mathbf{0} \qquad \text{on } (0,1],$$

$$-\int_0^1 \mathrm{d}[\mathbf{z}^*(t)]\,\mathbf{P}(t,0) + \boldsymbol{\lambda}^*\,\mathbf{K}(s) = \mathbf{0}.$$

According to 7.5 $\mathbf{P}(0,s) = \mathbf{0}$ on $[0,1]$. Thus the value of each of the integrals

$$\int_0^1 \mathrm{d}[\mathbf{z}^*(t)]\,\mathbf{P}(t,s) \quad (s \in [0,1]), \qquad \int_0^1 \mathrm{d}[\mathbf{z}^*(t)]\,(\mathbf{f}(t) - \mathbf{f}(0))$$

does not depend on the value $\mathbf{z}^*(0)$ (cf. I.4.23). Consequently $(\mathbf{z}^*, \boldsymbol{\lambda}^*) \in BV_n^{\backprime} \times R_m^*$ is a solution to (7,19) if and only if $(\mathbf{z}_0^*, \boldsymbol{\lambda}^*)$ with $\mathbf{z}_0^*(s) = \mathbf{z}^*(s)$ on $(0,1]$ and $\mathbf{z}_0^*(0) = \mathbf{0}$ is also its solution. The proof is complete.

The following assertion is also a consequence of I.3.2.

7.12. Proposition. *Let* 7.1 *and* 7.5 *be satisfied and let* $\mathbf{h} \in BV_n^{\backprime}$. *Then there exist* $\mathbf{z}^* \in BV_n^{\backprime}$ *and* $\boldsymbol{\lambda}^* \in R_m^*$ *such that*

$$(7,20)\qquad \mathbf{z}^*(s) - \int_0^1 \mathrm{d}[\mathbf{z}^*(t)]\,\mathbf{Q}(t,s) + \boldsymbol{\lambda}^*\,\mathbf{K}(s) = \mathbf{h}^*(s) \qquad \text{on } [0,1]$$

$((\mathbf{I} - \mathbf{T}^{\backprime})(\mathbf{z}^*, \boldsymbol{\lambda}^*) = (\mathbf{h}^*, \mathbf{0}))$ *if and only if*

$$\int_0^1 \mathrm{d}[\mathbf{h}^*(t)]\,\mathbf{x}(t) = 0$$

holds for every $\mathbf{x} \in N(\mathscr{L})$, *where*

$$(7,21)\qquad \mathscr{L}\colon \mathbf{x} \in BV_n \to \begin{pmatrix} \mathbf{x} - \mathbf{Q}\mathbf{x} \\ \mathbf{S}\mathbf{x} \end{pmatrix} \in BV_n \times R_m .$$

7.13. Theorem. *Let us assume* 7.1 *and* 7.5 *and let* $\mathscr{L} \in B(BV_n, BV_n \times R_m)$ *be given by* (7,6), (7,8) *and* (7,21). *Then* $k = \dim N(\mathscr{L}) < \infty$ *and the system* (7,16) *has exactly* $k^* = k + m - n$ *linearly independent solutions in* $BV_n^{\backprime} \times R_m^*$.

Proof. By 7.4 $k = \dim N(\mathscr{L}) < \infty$. Obviously $\dim N(\mathbf{I} - \mathbf{T}) = k + m$. Since (7,14), it is by I.3.2 $\dim N(\mathbf{I} - \mathbf{T}^{\backprime}) = \dim N(\mathbf{I} - \mathbf{T}) = k + m$. The set $N^{\backprime}$ of all solutions to (7,16) consists of all $(\mathbf{z}^*, \boldsymbol{\lambda}^*) \in N(\mathbf{I} - \mathbf{T}^{\backprime})$ for which $\mathbf{z}^*(0) = \mathbf{0}$. So $\dim N^{\backprime} = \dim N(\mathbf{I} - \mathbf{T}^{\backprime}) - n = k + m - n$. The proof is complete.

In addition to 7.1 *and* 7.5 *we shall assume henceforth that*

$$(7,22)\qquad \mathbf{P}(t-, s) = \mathbf{P}(t,s) \quad \text{for all} \quad (t,s) \in (0,1] \times [0,1],$$

$$\mathbf{P}(0+, s) = \mathbf{P}(0,s) \quad \text{for all} \quad s \in [0,1].$$

In this case we may formulate the adjoint problem to (7,1), (7,2) in a form more similar to (7,1), (7,2).

Integrating by parts (I.4.33) we transfer the system (7,16) of equations for $(\mathbf{z}^*, \boldsymbol{\lambda}^*) \in BV_n^{\backprime} \times R_m^*$ to the form

$$\mathbf{z}^*(s) + \int_0^1 \mathbf{z}^*(t)\, \mathrm{d}_t[\mathbf{P}(t,s)] + \boldsymbol{\lambda}^* \mathbf{K}(s) = \mathbf{0} \quad \text{on } [0,1], \tag{7,23}$$
$$\mathbf{z}^*(0) = \mathbf{z}^*(1) = \mathbf{0}.$$

As by (7,22) $\mathbf{P}(0+,s) = \mathbf{P}(0,s)$ and $\mathbf{P}(1-,s) = \mathbf{P}(1,s)$ for every $s \in [0,1]$, the value of each of the integrals

$$\int_0^1 \mathbf{z}^*(t)\, \mathrm{d}_t[\mathbf{P}(t,s)], \qquad s \in [0,1]$$

does not depend on the value $\mathbf{z}^*(0)$ and $\mathbf{z}^*(1)$. In particular, if $\mathbf{z}^* \in BV_n^{\backprime}$, $\mathbf{z}^*(0) = \mathbf{0}$, $\boldsymbol{\lambda}^* \in R_m^*$ and

$$\mathbf{y}^*(s) = \mathbf{z}^*(s) \ \text{ on } (0,1), \qquad \mathbf{y}^*(0) = \mathbf{z}^*(0+), \qquad \mathbf{y}^*(1) = \mathbf{z}^*(1-), \tag{7,24}$$

then the couple $(\mathbf{z}^*, \boldsymbol{\lambda}^*)$ solves (7,23) (i.e. (7,16)) if and only if

$$\mathbf{y}^*(s) + \int_0^1 \mathbf{y}^*(t)\, \mathrm{d}_t[\mathbf{P}(t,s)] + \boldsymbol{\lambda}^* \mathbf{K}(s) = \mathbf{0} \quad \text{on } (0,1), \tag{7,25}$$
$$\mathbf{0} = \int_0^1 \mathbf{y}^*(t)\, \mathrm{d}[\mathbf{P}(t,0)] + \boldsymbol{\lambda}^* \mathbf{K}(0) \qquad (= \mathbf{z}^*(0)).$$

Applying I.6.16 and I.4.17 we obtain

$$\mathbf{y}^*(0) = \mathbf{y}^*(0+) = -\int_0^1 \mathbf{y}^*(t)\, \mathrm{d}_t[\mathbf{P}(t,0+) - \mathbf{P}(t,0)] - \boldsymbol{\lambda}^*[\mathbf{K}(0+) - \mathbf{K}(0)]$$

and

$$\mathbf{y}^*(1) = \mathbf{y}^*(1-) = -\int_0^1 \mathbf{y}^*(t)\, \mathrm{d}_t[\mathbf{P}(t,1-)] - \boldsymbol{\lambda}^* \mathbf{K}(1-)$$

for every $\mathbf{y} \in BV_n$ and $\boldsymbol{\lambda} \in R_m$ fulfilling (7,25). If for $t \in [0,1]$ we put

$$\mathbf{P}_0(t,s) = \begin{cases} \mathbf{P}(t,0+) & \text{if } s = 0, \\ \mathbf{P}(t,s) & \text{if } 0 < s < 1, \\ \mathbf{P}(t,1-) & \text{if } s = 1, \end{cases} \qquad \mathbf{K}_0(s) = \begin{cases} \mathbf{K}(0+) & \text{if } s = 0, \\ \mathbf{K}(s) & \text{if } 0 < s < 1, \\ \mathbf{K}(1-) & \text{if } s = 1, \end{cases} \tag{7,26}$$
$$\mathbf{C}(t) = \mathbf{P}(t,0+) - \mathbf{P}(t,0), \qquad \mathbf{D}(t) = -\mathbf{P}(t,1-),$$
$$\mathbf{M} = \mathbf{K}(0+) - \mathbf{K}(0), \qquad \mathbf{N} = -\mathbf{K}(1-),$$

then system (7,25) becomes

(7,27)
$$\mathbf{y}^*(s) = \mathbf{y}^*(1) - \int_0^1 \mathbf{y}^*(t)\, \mathrm{d}_t[\mathbf{P}_0(t,s) - \mathbf{P}_0(t,1)] - \boldsymbol{\lambda}^*[\mathbf{K}_0(s) - \mathbf{K}_0(1)] \quad \text{on } [0,1].$$

$$(7,28) \qquad \mathbf{y}^*(0) + \boldsymbol{\lambda}^*\mathbf{M} + \int_0^1 \mathbf{y}^*(t)\,\mathrm{d}[\mathbf{C}(t)] = \mathbf{0},$$

$$(7,29) \qquad \mathbf{y}^*(1) - \boldsymbol{\lambda}^*\mathbf{N} - \int_0^1 \mathbf{y}^*(t)\,\mathrm{d}[\mathbf{D}(t)] = \mathbf{0}.$$

Given $\mathbf{z} \in BV_n$ with $\mathbf{z}(0) = \mathbf{z}(1) = \mathbf{0}$ and $\mathbf{y} \in BV_n$ such that (7,24) holds, we have in virtue of I.4.23

$$\int_0^1 \mathrm{d}[\mathbf{z}^*(s)]\,(\mathbf{f}(s) - \mathbf{f}(0)) = \int_0^1 \mathrm{d}[\mathbf{y}^*(s)]\,\mathbf{f}(s) - \mathbf{y}^*(1)\,\mathbf{f}(1) - \mathbf{y}^*(0)\,\mathbf{f}(0).$$

This completes the proof of the following

7.14. Theorem. *If the hypotheses* 7.1, 7.5 *and* (7,22) *are satisfied, then the problem* (7,1), (7,2) *possesses a solution* $\mathbf{x} \in BV_n$ *if and only if*

$$(7,30) \qquad \mathbf{y}^*(1)\,\mathbf{f}(1) - \mathbf{y}^*(0)\,\mathbf{f}(0) - \int_0^1 \mathrm{d}[\mathbf{y}^*(s)]\,\mathbf{f}(s) = \boldsymbol{\lambda}^*\mathbf{r}$$

for any solution $\mathbf{y} \in BV_n$, $\boldsymbol{\lambda} \in R_m$ *of* (7,27)–(7,29), *where* $\mathbf{P}_0$, $\mathbf{C}$, $\mathbf{D}$, $\mathbf{K}_0$, $\mathbf{M}$ *and* $\mathbf{N}$ *are defined in* (7,26).

7.15. Remark. If (7,22) holds and $\mathbf{f}(t-) = \mathbf{f}(t)$ on $(0, 1]$, $\mathbf{f}(0+) = \mathbf{f}(0)$, then by I.6.16 and I.4.17 any solution $\mathbf{x} \in BV_n$ of (7,1), (7,2) is left-continuous on $(0, 1]$ and right-continuous at 0. On the other hand, if $\mathbf{y} \in BV_n$ and $\boldsymbol{\lambda} \in R_m$ satisfy (7,27)–(7,29), then provided that 7.5 holds, $\mathbf{y}$ is right-continuous on $[0, 1)$ and left-continuous at 1 (cf. 7.24).

7.16. Remark. Let $\mathbf{g} \in BV_n$ be right-continuous on $(0, 1)$, $\mathbf{p}, \mathbf{q} \in R_n$. It is easy to see that $\mathbf{y} \in BV_n$ and $\boldsymbol{\lambda} \in R_m$ satisfy (7,27), (7,28), (7,29) with the right-hand sides $\mathbf{g}^*(s) - \mathbf{g}^*(1)$, $\mathbf{p}^*$ and $\mathbf{q}^*$, respectively, if and only if $\mathbf{y}$ is right-continuous on $(0, 1)$ and the couple $(\mathbf{z}^*, \boldsymbol{\lambda}^*)$, $\mathbf{z}^*(s) = \mathbf{y}^*(s)$ on $(0, 1)$, $\mathbf{z}^*(0) = \mathbf{z}^*(1) = \mathbf{0}$, fulfils (7,20), where $\mathbf{h}^*(s) = \mathbf{g}^*(s) - \mathbf{g}^*(1) + \boldsymbol{\chi}^*(s)$ on $[0, 1]$, $\boldsymbol{\chi}^*(0) = \mathbf{q}^* - \mathbf{p}^*$, $\boldsymbol{\chi}^*(s) = \mathbf{q}^*$ on $(0, 1)$ and $\boldsymbol{\chi}^*(1) = \mathbf{0}$. It follows immediately from 7.12 that the system (7,27), (7,28), (7,29) with the right-hand sides $\mathbf{g}^*(s) - \mathbf{g}^*(1)$, $\mathbf{p}^*$ and $\mathbf{q}^*$, respectively, has a solution $\mathbf{y} \in BV_n$, $\boldsymbol{\lambda} \in R_m$ if and only if (cf. (7,21))

$$\int_0^1 \mathrm{d}[\mathbf{g}^*(t)]\,\mathbf{x}(t) = \mathbf{q}^*\,\mathbf{x}(1) - \mathbf{p}^*\,\mathbf{x}(0) \qquad \text{for each} \quad \mathbf{x} \in N(\mathscr{L}).$$

7.17. Remark. If $\mathbf{P}$: $[0, 1] \times [0, 1] \to L(R_n)$ is an $L^1[BV]$-kernel ($|\mathbf{P}(t, 0)| + \mathrm{var}_0^1\,\mathbf{P}(t, .) = \varrho(t) < \infty$ a.e. on $[0, 1]$ and $\varrho \in L^1$) and $\mathbf{f} \in L^1_n$, then $\mathbf{x}$: $[0, 1] \to R_n$ is a solution to (2,1) on $[0, 1]$ if and only if

$$\mathbf{x}(t) - \mathbf{x}(0) - \int_0^1 \mathrm{d}_s[\mathbf{R}(t, s)]\,\mathbf{x}(s) = \int_0^t \mathbf{f}(\tau)\,\mathrm{d}\tau \qquad \text{on } [0, 1],$$

where

$$\mathbf{R}(t,s) = \int_0^t \mathbf{P}(\tau,s)\,d\tau \qquad \text{on } [0,1]\times[0,1].$$

Given a subdivision $\{0 = t_0 < t_1 < \ldots < t_k = 1;\ 0 = s_0 < s_1 < \ldots < s_k = 1\}$ of $[0,1]\times[0,1]$, we have

$$\sum_{i=1}^{k}\sum_{j=1}^{k} |\mathbf{R}(t_i,s_j) - \mathbf{R}(t_{i-1},s_j) - \mathbf{R}(t_i,s_{j-1}) + \mathbf{R}(t_{i-1},s_{j-1})|$$

$$= \sum_{i=1}^{k}\sum_{j=1}^{k} \left| \int_{t_{i-1}}^{t_i} (\mathbf{P}(\tau,s_j) - \mathbf{P}(\tau,s_{j-1}))\,d\tau \right| \le \int_0^1 \left(\sum_{j=1}^{k} |\mathbf{P}(\tau,s_j) - \mathbf{P}(\tau,s_{j-1})| \right) d\tau$$

$$= \|\varrho\|_{L_1} < \infty .$$

Consequently $v_{[0,1]\times[0,1]}(\mathbf{R}) < \infty$. Clearly $\text{var}_0^1\,\mathbf{R}(.,1) < \infty$. (We may assume $\mathbf{P}(t,1) = \mathbf{0}$ a.e. on $[0,1]$.) As $\mathbf{R}(0,.) = \mathbf{0}$ on $[0,1]$, this implies that $\mathbf{R}$ is an SBV-kernel and the Fredholm-Stieltjes integro-differential equation (2,1) is a special case of the equation (7,1).

7.18. Remark. Let $\mathbf{A}\colon [0,1] \to L(R_n)$, $\text{var}_0^1\,\mathbf{A} < \infty$, $\mathbf{M}$ and $\mathbf{N} \in L(R_n, R_m)$ and

$$\mathbf{P}(t,s) = \begin{cases} \mathbf{A}(0) \quad - \mathbf{A}(t) & \text{if } 0 = s < t \le 1, \\ \mathbf{A}(s+) - \mathbf{A}(t) & \text{if } 0 < s < t \le 1, \\ \mathbf{0} & \text{if } 0 \le t \le s \le 1, \end{cases} \qquad \mathbf{K}(s) = \begin{cases} -\mathbf{M} - \mathbf{N} & \text{if } s = 0, \\ -\mathbf{N} & \text{if } 0 < s < 1, \\ \mathbf{0} & \text{if } s = 1. \end{cases}$$

It can be shown that $v_{[0,1]\times[0,1]}(\mathbf{P}) \le \text{var}_0^1\,\mathbf{A}$. Furthermore, $\mathbf{P}(0,.) = \mathbf{0}$ on $[0,1]$, $\text{var}_0^1\,\mathbf{P}(.,0) = \text{var}_0^1\,\mathbf{A}$ and $\text{var}_0^1\,\mathbf{K} = |\mathbf{M}| + |\mathbf{N}|$. Since for any $t \in [0,1]$ $\mathbf{P}(t,.)$ and $\mathbf{K}$ are right-continuous on $(0,1)$, $\mathbf{K}(1) = \mathbf{0}$ and $\mathbf{P}(t,1) = \mathbf{0}$, the assumptions 7.1 and 7.5 are satisfied in this case. If, moreover, $\mathbf{A}$ is left-continuous on $(0,1]$ and right-continuous at 0, then $\mathbf{P}(t-,0) = \mathbf{A}(0) - \mathbf{A}(t-) = \mathbf{A}(0) - \mathbf{A}(t)$ for $0 < t \le 1$, $\mathbf{P}(t-,s) = \mathbf{A}(s+) - \mathbf{A}(t-) = \mathbf{A}(s+) - \mathbf{A}(t)$ for $0 < s < t \le 1$ and $\mathbf{P}(t-,s) = \mathbf{0}$ for $0 < t \le s \le 1$. Finally, $\mathbf{P}(0+,s) = \mathbf{0}$ for any $s \in [0,1]$. Thus $\mathbf{P}$ fulfils also (7,22). By 7.14 the system (7,1), (7,2) which is now reduced to BVP $d\mathbf{x} = d[\mathbf{A}]\,\mathbf{x} + d\mathbf{f}$, $\mathbf{M}\,\mathbf{x}(0) + \mathbf{N}\,\mathbf{x}(1) = \mathbf{r}$ has a solution if and only if (7,30) holds for all $\mathbf{y} \in BV_n$ and $\boldsymbol{\lambda} \in R_m$ satisfying (7,27), (7,29). In our case $\mathbf{P}_0(t,s) = \mathbf{P}(t,s)$, $\mathbf{C}(t) = \mathbf{D}(t) = \mathbf{0}$ and $\mathbf{K}_0(s) = -\mathbf{N}$. Moreover,

$$\int_0^1 \mathbf{y}^*(t)\,d_t[\mathbf{P}(t,s)] = \int_s^1 \mathbf{y}^*(t)\,d[\mathbf{B}(t)] \qquad \text{for any } \mathbf{y} \in BV_n \text{ and } s \in [0,1],$$

where $\mathbf{B}(s) = \mathbf{A}(s+)$ on $(0,1)$, $\mathbf{B}(0) = \mathbf{A}(0)$ and $\mathbf{B}(1) = \mathbf{A}(1)$. It follows that under the assumptions of this remark the adjoint system (7,27)–(7,29) to (7,1), (7,2) reduces to BVP (III.5,12), (III.5,13). Let us notice that now no assumptions on the regularity of the matrices $(\mathbf{I} + \Delta^+\mathbf{A}(t))$ are needed.

7.19. Remark. Let the matrix valued functions $\mathbf{A}: [0,1] \to L(R_n)$, $\mathbf{P}_1: [0,1] \to L(R_p, R_n)$, $\mathbf{P}_2: [0,1] \to L(R_n, R_p)$, $\mathbf{C}: [0,1] \to L(R_n)$, $\mathbf{D}: [0,1] \to L(R_n)$ and $\mathbf{K}: [0,1] \to L(R_n, R_m)$ be of bounded variation on $[0,1]$, $\mathbf{M}, \mathbf{N} \in L(R_n, R_m)$, $\mathbf{f} \in BV_n$ and $\mathbf{r} \in R_m$ and let us consider the system of equations for $\mathbf{x} \in BV_n$

$$(7,31)\quad \mathbf{x}(t) = \mathbf{x}(0) + \int_0^t \mathrm{d}[\mathbf{A}(s)]\,\mathbf{x}(s) + (\mathbf{C}(t) - \mathbf{C}(0))\,\mathbf{x}(0) + (\mathbf{D}(t) - \mathbf{D}(0))\,\mathbf{x}(1)$$

$$+ (\mathbf{P}_1(t) - \mathbf{P}_1(0)) \int_0^1 \mathrm{d}[\mathbf{P}_2(s)]\,\mathbf{x}(s) + \mathbf{f}(t) - \mathbf{f}(0) \quad \text{on } [0,1],$$

$$(7,32)\qquad \mathbf{M}\,\mathbf{x}(0) + \mathbf{N}\,\mathbf{x}(1) + \int_0^1 \mathrm{d}[\mathbf{K}(s)]\,\mathbf{x}(s) = \mathbf{r}.$$

Introducing new unknowns $\boldsymbol{\alpha}$, $\boldsymbol{\beta}$, $\boldsymbol{\gamma}$, $\boldsymbol{\delta}$, $\boldsymbol{\chi}$ by the relations

$$\boldsymbol{\alpha}(t) = \int_0^t \mathrm{d}[\mathbf{K}(s)]\,\mathbf{x}(s), \qquad \boldsymbol{\beta}(t) = \int_0^t \mathrm{d}[\mathbf{P}_2(s)]\,\mathbf{x}(s),$$

$$\boldsymbol{\gamma}(t) = \mathbf{x}(0), \qquad \boldsymbol{\delta}(t) = \mathbf{x}(1), \qquad \boldsymbol{\chi}(t) = \boldsymbol{\beta}(1),$$

we reduce the given problem to the form

$$\mathrm{d}\mathbf{x} = \mathrm{d}[\mathbf{A}]\,\mathbf{x} + \mathrm{d}[\mathbf{P}_1]\,\boldsymbol{\beta} + \mathrm{d}[\mathbf{C}]\,\boldsymbol{\gamma} + \mathrm{d}[\mathbf{D}]\,\boldsymbol{\delta} + \mathrm{d}\mathbf{f},$$

$$\mathrm{d}\boldsymbol{\alpha} = \mathrm{d}[\mathbf{K}]\,\mathbf{x}, \qquad \mathrm{d}\boldsymbol{\beta} = \mathrm{d}[\mathbf{P}_2]\,\mathbf{x}, \qquad \mathrm{d}\boldsymbol{\gamma} = \mathbf{0}, \qquad \mathrm{d}\boldsymbol{\delta} = \mathbf{0}, \qquad \mathrm{d}\boldsymbol{\chi} = \mathbf{0},$$

$$\mathbf{M}\,\mathbf{x}(0) + \mathbf{N}\,\mathbf{x}(1) + \boldsymbol{\alpha}(1) = \mathbf{r}, \qquad \boldsymbol{\alpha}(0) = \mathbf{0}, \qquad \mathbf{x}(0) - \boldsymbol{\gamma}(0) = \mathbf{0},$$

$$\mathbf{x}(1) - \boldsymbol{\delta}(0) = \mathbf{0}, \qquad \boldsymbol{\beta}(0) = \mathbf{0}, \qquad \boldsymbol{\beta}(1) - \boldsymbol{\chi}(0) = \mathbf{0}$$

which may be expressed in the matrix version

$$\mathrm{d}\xi = \mathrm{d}[\mathfrak{A}]\,\xi + \mathrm{d}\varphi, \qquad \mathfrak{M}\,\varphi(0) + \mathfrak{N}\,\xi(1) = \varrho,$$

where $\xi^* = (\mathbf{x}^*, \boldsymbol{\alpha}^*, \boldsymbol{\beta}^*, \boldsymbol{\gamma}^*, \boldsymbol{\delta}^*, \boldsymbol{\chi}^*)$ and $\mathfrak{A}: [0,1] \to L(R_\nu)$ and $\mathfrak{M}. \mathfrak{N} \in L(R_\nu, R_\mu)$ are appropriately defined matrices, $\mu = 2m + 2n + 2p$, $\nu = m + 3n + 2p$, $\varphi = \begin{pmatrix} \mathbf{f} \\ \mathbf{0}_{\nu-n} \end{pmatrix}$ and $\varrho = \begin{pmatrix} \mathbf{r} \\ \mathbf{0}_{\mu-m} \end{pmatrix}$. By this $\operatorname{var}_0^1 \mathfrak{A} < \infty$. The complicated problem (7,31), (7,32) was transferred to the two-point boundary value problem for a linear generalized differential equation.

Notes

In the case $p = 1$ the compactness of the operator $\mathbf{P}$ and hence also the closedness of $R(\mathbf{L})$ (V.1.4 and V.1.7) were proved by Maksimov [1] and independently by Tvrdý [4]. Theorem V.1.8 is due to Maksimov and Rahmatullina [2]. Our proof follows a different idea. The proofs of the main theorems of Section V.2 (V.2.5, V.2.6 and V.2.12) are carried out in a similar way as the proofs of analogous results for ordinary differential operators in Wexler [1] (cf. also Tvrdý, Vejvoda [1], Tvrdý [3], Maksimov [1]).

For more detail concerning Green's couples see Tvrdý [6]. Systems of the form (4,27), (4,28) were treated in Tvrdý, Vejvoda [1]. Theorem V.6.4 follows also from the variation of constants formula for functional differential equations of the retarded type due to Banks [1]. Equations of the form (V.6,13) were introduced in Delfour, Mitter [1], [2]. Section V.7 is based on the paper Tvrdý [5]. The transformation similar to (7,33) was for the first time used in a simpler situation by Jones [1] and Taufer [1]. For more detail concerning the systems of the form (7,31), (7,32) (Green's function, Jones transformation, selfadjoint problems etc.) see Vejvoda, Tvrdý [1], Tvrdý [1] and Zimmerberg [1], [2].

The oldest papers on the subject seem to be Duhamel [1], Lichtenstein [1] and Tamarkin [1]. Further related references to particular sections are

V.1: Catchpole [1], [2];
V.2: Parhimovič [1]–[3], Lando [1]–[4], Krall [2], [5], Tvrdý [1];
V.3: Maksimov, Rahmatullina [1], [2];
V.6: Hale [1], Maksimov, Rahmatullina [1], Rahmatullina [1], Tvrdý [4];
V.7: Krall [6]–[8], Hönig [1], Tvrdý [2].

Related results may be found also in the papers by N. V. Azbelev and the members of his group (L. F. Rahmatullina, V. P. Maksimov, A. G. Terent'ev, T. S. Sulavko, S. M. Labovskij, G. G. Islamov a.o.) which have appeared mainly in Differencial'nye uravnenija and in the collections of papers published by the Moscow and Tambov institutes of the chemical machines construction.

In Lando [3], [4] and Kultyšev [1] the controllability of integro-differential operators is studied.

VI. *Nonlinear boundary value problems (perturbation theory)*

1. Preliminaries

In this chapter we shall prove some theorems on the existence of solutions to nonlinear boundary value problems for nonlinear ordinary differential equations of the form

$$\mathbf{x}' = \mathbf{f}(t, \mathbf{x}) + \varepsilon\, \mathbf{g}(t, \mathbf{x}, \varepsilon), \qquad \mathbf{S}(\mathbf{x}) + \varepsilon\, \mathbf{R}(\mathbf{x}, \varepsilon) = \mathbf{0}$$

under the assumption that the existence of a solution to the corresponding shortened boundary value problem

$$\mathbf{x}' = \mathbf{f}(t, \mathbf{x}), \qquad \mathbf{S}(\mathbf{x}) = \mathbf{0}$$

is guaranteed. ($\mathbf{S}$ and $\mathbf{R}$ are n-vector valued functionals; $\mathbf{x} \in R_n$, $\mathbf{f}\colon \mathscr{D} \subset R \times R_n \to R_n$, $\mathbf{g}\colon \mathfrak{D} \subset R \times R_n \times R \to R_n$ and $\varepsilon > 0$ is a small parameter.)

The present section provides the survey of the basic theory for the equation

(1,1) $$\mathbf{x}' = \mathbf{f}(t,\mathbf{x}).$$

The proofs may be found in many textbooks on ordinary differential equations (e.g. Coddington, Levinson [1] or Reid [1]).

1.1. Notation. Let $\mathscr{D} \subset R_{p+q}$, $\mathbf{u}_0 \in R_p$ and $\mathbf{v}_0 \in R_q$. Then

$$\mathscr{D}_{(\mathbf{u}_0,.)} = \{\mathbf{v} \in R_q;\ (\mathbf{u}_0, \mathbf{v}) \in \mathscr{D}\} \quad \text{and} \quad \mathscr{D}_{(.,\mathbf{v}_0)} = \{\mathbf{u} \in R_p;\ (\mathbf{u}, \mathbf{v}_0) \in \mathscr{D}\}.$$

If $\mathbf{f}$ maps $\mathscr{D}$ into R_n, then $\mathbf{f}(.,\mathbf{v}_0)$ and $\mathbf{f}(\mathbf{u}_0, .)$ denote the mappings given by

$$\mathbf{f}(.,\mathbf{v}_0)\colon \mathbf{u} \in \mathscr{D}_{(.,\mathbf{v}_0)} \to \mathbf{f}(\mathbf{u}, \mathbf{v}_0) \in R_n$$

and

$$\mathbf{f}(\mathbf{u}_0, .)\colon \mathbf{v} \in \mathscr{D}_{(\mathbf{u}_0,.)} \to \mathbf{f}(\mathbf{u}_0, \mathbf{v}) \in R_n.$$

1.2. Definition. Let $\mathscr{D} \subset R_{n+1}$ be open and let the n-vector valued function $\mathbf{f}(t, \mathbf{x})$ be defined for $(t, \mathbf{x}) \in \mathscr{D}$.

(a) We shall say that $\mathbf{f}$ fulfils the *Carathéodory conditions* on $\mathscr{D}$ and write $\mathbf{f} \in \mathrm{Car}(\mathscr{D})$ if

(i) for a.e. $t \in R$ such that $\mathscr{D}_{(t,\cdot)} \neq \emptyset$, $\mathbf{f}(t,.)$ is continuous;
(ii) given $\mathbf{x} \in R_n$ such that $\mathscr{D}_{(\cdot,x)} \neq \emptyset$, $\mathbf{f}(.,\mathbf{x})$ is measurable;
(iii) given $(t_0, \mathbf{x}_0) \in \mathscr{D}$, there exist $\delta_1 > 0$, $\delta_2 > 0$ and $m \in L^1[t_0 - \delta_1, t_0 + \delta_1]$ such that $[t_0 - \delta_1, t_0 + \delta_1] \times \mathfrak{B}(\mathbf{x}_0, \delta_2; R_n) \subset \mathscr{D}$ and $|\mathbf{f}(t, \mathbf{x})| \leq m(t)$ for a.e. $t \in [t_0 - \delta_1, t_0 + \delta_1]$ and any $\mathbf{x} \in \mathfrak{B}(\mathbf{x}_0, \delta_2; R_n)$;

(b) We shall write $\mathbf{f} \in \mathrm{Lip}(\mathscr{D})$ if

(iv) given $(t_0, \mathbf{x}_0) \in \mathscr{D}$, there exist $\delta_1 > 0$, $\delta_2 > 0$ and $\omega \in L^1[t_0 - \delta_1, t_0 + \delta_1]$ such that $[t_0 - \delta_1, t_0 + \delta_1] \times \mathfrak{B}(\mathbf{x}_0, \delta_2; R_n) \subset \mathscr{D}$ and $|\mathbf{f}(t, \mathbf{x}_1) - \mathbf{f}(t, \mathbf{x}_2)| \leq \omega(t)\,|\mathbf{x}_1 - \mathbf{x}_2|$ for a.e. $t \in [t_0 - \delta_1, t_0 + \delta_1]$ and all $\mathbf{x}_1, \mathbf{x}_2 \in \mathfrak{B}(\mathbf{x}_0, \delta_2; R_n)$.

1.3. Definition. An n-vector function $\mathbf{x}(t)$ is said to be a *solution to the equation* (1,1) *on the interval* $\Delta \subset R$ if it is absolutely continuous on Δ and such that $(t, \mathbf{x}(t)) \in \mathscr{D}$ for a.e. $t \in \Delta$ and

$$\mathbf{x}'(t) = \mathbf{f}(t, \mathbf{x}(t)) \qquad \text{a.e. on } \Delta\,.$$

1.4. Theorem (Carathéodory). *Let* $\mathscr{D} \subset R_{n+1}$ *be open and* $\mathbf{f} \in \mathrm{Car}(\mathscr{D})$. *Given* $(t_0, \mathbf{c}) \in \mathscr{D}$, *there exists* $\delta > 0$ *such that the equation* (1,1) *possesses a solution* $\mathbf{x}(t)$ *on* $(t_0 - \delta, t_0 + \delta)$ *such that* $\mathbf{x}(t_0) = \mathbf{c}$.

1.5. Remark. Obviously, if $\mathbf{f} \in C(\mathscr{D})$, then $\mathbf{f} \in \mathrm{Car}(\mathscr{D})$ and the equation (3,1) possesses for any $(t_0, \mathbf{c}_0) \in \mathscr{D}$ a solution $\mathbf{x}(t)$ on a neighbourhood Δ of t_0 such that $\mathbf{x}(t_0) = \mathbf{c}_0$. Since the function $t \in \Delta \to \mathbf{f}(t, \mathbf{x}(t)) \in R_n$ is continuous on Δ, it follows immediately that $\mathbf{x}'$ is continuous on Δ $(\mathbf{x} \in C^1_n(\Delta))$.

1.6. Definition. The equation (1,1) has the *property* $(\mathscr{U})$ (*local uniqueness*) on $\mathscr{D} \in R_{n+1}$, if for any couple of its solutions $\mathbf{x}_1(t)$ on Δ_1 and $\mathbf{x}_2(t)$ on Δ_2 such that $\mathbf{x}_1(t_0) = \mathbf{x}_2(t_0)$ for some $t_0 \in \Delta_1 \cap \Delta_2$, $\mathbf{x}_1(t) \equiv \mathbf{x}_2(t)$ *on* $\Delta_1 \cap \Delta_2$.

1.7. Theorem. *Let* $\mathscr{D} \subset R_{n+1}$ *and* $\mathbf{f} \in \mathrm{Lip}(\mathscr{D})$. *Then the equation* (1,1) *has the property* $(\mathscr{U})$ *on* $\mathscr{D}$.

1.8. Definition. The *solution* $\mathbf{x}(t)$ of (1,1) on Δ is said to be *maximal* if for any solution $\mathbf{x}_1(t)$ of (1,1) on Δ_1 such that $\Delta \subset \Delta_1$ and $\mathbf{x}(t) = \mathbf{x}_1(t)$ on Δ we have $\Delta = \Delta_1$.

1.9. Lemma. *If the definition domain* $\mathscr{D}$ *of* $\mathbf{f}(t, \mathbf{x})$ *is open and the solution* $\mathbf{x}(t)$ *of* (1,1) *on* Δ *is maximal, then* Δ *is open.*

1.10. Notation. Given $(t_0, \mathbf{c}) \in \mathscr{D}$, $\varphi(.;\, t_0, \mathbf{c})$ denotes the corresponding maximal solution of (1,1), $\Delta(t_0, \mathbf{c})$ its definition interval and

$$\Omega = \{(t, t_0, \mathbf{c}) \in R \times R \times R_n;\ (t_0, \mathbf{c}) \in \mathscr{D},\ t \in \Delta(t_0, \mathbf{c})\}.$$

1.11. Theorem. *Let $\mathscr{D} \subset R_{n+1}$ be open, $\mathbf{f} \in \mathrm{Car}(\mathscr{D})$ and let the equation (1,1) have the property $(\mathscr{U})$. Then for any $(t_0, \mathbf{c}) \in \mathscr{D}$ there exists a unique maximal solution $\mathbf{x}(t) = \varphi(t;\, t_0, \mathbf{c})$ of (1,1) on $\Delta = \Delta(t_0, \mathbf{c}) \subset R$ such that $\mathbf{x}(t_0) = \mathbf{c}$. The set Ω (cf. 1.10) is open and the mapping φ: $(t, t_0, \mathbf{c}) \in \Omega \to \varphi(t;\, t_0, \mathbf{c}) \in R_n$ is continuous ($\varphi \in C(\Omega)$).*

1.12. Corollary. *Let $\mathscr{D} \subset R_{n+1}$, $\mathbf{f} \in \mathrm{Car}(\mathscr{D})$ and (1,1) have the property $(\mathscr{U})$. Let $(t_0, \mathbf{c}_0) \in \mathscr{D}$, $-\infty < a < b < \infty$ and let $[a, b] \subset \Delta(t_0, \mathbf{c}_0)$. Then there exists $\delta > 0$ such that $|\mathbf{c} - \mathbf{c}_0| \le \delta$ implies $(t_0, \mathbf{c}) \in \mathscr{D}$ and $\Delta(t_0, \mathbf{c}) \supset [a, b]$, i.e. for any $\mathbf{c} \in \mathfrak{B}(\mathbf{c}_0, \delta;\, R_n)$ the corresponding maximal solution $\varphi(t, t_0, \mathbf{c})$ of (1,1) is defined on $[a, b]$.*

1.13. Remark. Let us recall that if $\mathbf{f}$: $\mathscr{D} \to R_n$ possesses on $\mathscr{D}$ partial derivatives with respect to the components x_j of $\mathbf{x}$, then $\partial \mathbf{f}/\partial \mathbf{x}$ denotes the Jacobi matrix of $\mathbf{f}$ with respect to $\mathbf{x}$ which is formed by the rows $(\partial \mathbf{f}/\partial x_j)$ $(j = 1, 2, \dots, n)$. If the $n \times n$-matrix valued function $(t, \mathbf{x}) \in \mathscr{D} \to (\partial \mathbf{f}/\partial \mathbf{x})(t, \mathbf{x}) \in L(R_n)$ fulfils the Carathéodory condition (iii) in 1.2, then making use of the Mean Value Theorem I.7.4 we obtain easily that $\mathbf{f} \in \mathrm{Lip}(\mathscr{D})$.

1.14. Theorem. *Let $\mathscr{D} \subset R_{n+1}$, $\mathbf{f} \in \mathrm{Car}(\mathscr{D})$ and $(\partial \mathbf{f}/\partial \mathbf{x}) \in \mathrm{Car}(\mathscr{D})$. Then the equation (1,1) has the property $(\mathscr{U})$ and hence there exist $\Omega \subset R_{n+2}$ and the continuous mapping φ: $\Omega \to R_n$ defined in 1.11. Furthermore $(\partial \varphi/\partial \mathbf{c})(t, t_0, \mathbf{c})$ exists and is continuous in $(t, t_0, \mathbf{c})$ on Ω. For any $(t_0, \mathbf{c}) \in \mathscr{D}$ the $n \times n$-matrix valued function $\mathbf{A}(t) = (\partial \mathbf{f}/\partial \mathbf{x})(t, \varphi(t, t_0, \mathbf{c}))$ is L-integrable on each compact subinterval of $\Omega_{(.,t_0,\mathbf{c})} = \Delta(t_0, \mathbf{c})$ and $\mathbf{U}(t) = (\partial \varphi/\partial \mathbf{c})(t, t_0, \mathbf{c})$ is the maximal solution of the linear matrix differential equation $\mathbf{U}' = \mathbf{A}(t)\,\mathbf{U}$ such that $\mathbf{U}(t_0) = \mathbf{I}_n$.*

1.15. Remark. It follows from 1.14 that $(\partial \varphi/\partial \mathbf{c})(t, t_0, \mathbf{c})$ is for any $(t_0, \mathbf{c}) \in \mathscr{D}$ the fundamental matrix solution of the variational equation

$$\mathbf{u}' = \left(\frac{\partial \mathbf{f}}{\partial \mathbf{x}}(t, \varphi(t, t_0, \mathbf{c}))\right)\mathbf{u}$$

on $\Delta(t_0, \mathbf{c})$. Consequently for any $(t, t_0, \mathbf{c}) \in \Omega$ it possesses an inverse matrix $((\partial \varphi/\partial t)(t, t_0, \mathbf{c}))^{-1}$.

1.16. Theorem. *Let $\mathscr{D} \subset R_{n+1}$, $\mathbf{f} \in \mathrm{Car}(\mathscr{D})$, $(\partial \mathbf{f}/\partial \mathbf{x}) \in \mathrm{Car}(\mathscr{D})$ and $\partial^2 \mathbf{f}/(\partial x_i\, \partial x_j) \in \mathrm{Car}(\mathscr{D})$ for any $i, j = 1, 2, \dots, n$. Then the n-vector valued function φ from 1.11* *possesses on* Ω *all the partial derivatives* $\partial^2 \varphi / (\partial c_i \, \partial c_j)$ $(i, j = 1, 2, \ldots, n)$ *and they are continuous in* $(t, t_0, \mathbf{c})$ *on* Ω $(\varphi \in C^2(\Omega))$.

1.17. Remark. Let $\mathfrak{D} \subset R_1 \times R_n \times R_p$ be open and let the n-vector valued function $\mathbf{h}(t, \mathbf{u}, \mathbf{v})$ map $\mathfrak{D}$ into R_n. The differential equation

$$(1,2) \qquad \mathbf{x}' = \mathbf{h}(t, \mathbf{x}, \mathbf{v})$$

is said to be an equation with a parameter $\mathbf{v} \in R_p$. Let us put

$$\xi = (\mathbf{x}, \mathbf{v}) \quad \text{for} \quad \mathbf{x} \in R_n \quad \text{and} \quad \mathbf{v} \in R_p,$$

$$\tilde{\mathbf{h}}(t, \xi) = \mathbf{h}(t, \mathbf{x}, \mathbf{v}) \quad \text{for} \quad (t, \xi) = (t, (\mathbf{x}, \mathbf{v})) \in \mathfrak{D}$$

and

$$\tilde{\mathbf{f}}(t, \xi) = \begin{pmatrix} \tilde{\mathbf{h}}(t, \xi) \\ \mathbf{0}_p \end{pmatrix} \in R_{n+p} \quad \text{for} \quad (t, \xi) \in \mathfrak{D}.$$

Now, applying the above theorems to the equation

$$\xi' = \tilde{\mathbf{f}}(t, \xi) \qquad \begin{pmatrix} \mathbf{x}' = \mathbf{h}(t, \mathbf{x}, \mathbf{v}) \\ \mathbf{v}' = \mathbf{0} \end{pmatrix}$$

we can easily obtain theorems on the existence, uniqueness, continuous dependence of a solution $\mathbf{x}(t) = \varphi(t; t_0, \mathbf{c}, \mathbf{v})$ of (1,2) on the initial data $(t_0, \mathbf{c})$ and on the parameter $\mathbf{v}$ as well as theorems on the differentiability of φ with respect to t, $\mathbf{c}$ and $\mathbf{v}$. The formulation of the general statements may be left to the reader. For our purposes only the following lemma is needed.

1.18. Lemma. *Let* $\mathscr{D} \subset R_{n+1}$ *and* $\mathfrak{D} \subset R_{n+2}$ *be open,* $\varkappa > 0$, $\mathscr{D} \times [0, \varkappa] \subset \mathfrak{D}$, $\mathbf{f}\colon \mathscr{D} \to R_n$ *and* $\mathbf{g}\colon \mathfrak{D} \to R_n$. *Let us put* $\tilde{\mathbf{g}}(t, \mathbf{y}) = \mathbf{g}(t, \mathbf{x}, \varepsilon)$ *for* $(t, \mathbf{x}, \varepsilon) \in \mathfrak{D}$ *and* $\mathbf{y} = (\mathbf{x}, \varepsilon)$. *Let* $\mathbf{f} \in \mathrm{Car}(\mathscr{D})$, $\tilde{\mathbf{g}} \in \mathrm{Car}(\mathfrak{D})$ *and let for any* $\varepsilon \in [0, \varkappa]$ *the equation*

$$(1,3) \qquad \mathbf{x}' = \mathbf{f}(t, \mathbf{x}) + \varepsilon \, \mathbf{g}(t, \mathbf{x}, \varepsilon)$$

possess the property $(\mathscr{U})$ *on* $\mathfrak{D}$. *Then*

(i) *given* $(t_0, \mathbf{c}, \varepsilon) \in \mathscr{D} \times [0, \varkappa]$, *there exists a unique maximal solution* $\mathbf{x}(t) = \boldsymbol{\psi}(t; t_0, \mathbf{c}, \varepsilon)$ *of* (1,3) *on the interval* $\Delta = \Delta(t_0, \mathbf{c}, \varepsilon)$ *such that* $\mathbf{x}(t_0) = \mathbf{c}$;

(ii) *the set* $\Omega = \{(t, t_0, \mathbf{c}, \varepsilon); (t_0, \mathbf{c}, \varepsilon) \in \mathscr{D} \times [0, \varkappa],\ t \in \Delta(t_0, \mathbf{c}, \varepsilon)\} \subset R_{n+3}$ *is open and the mapping* $\boldsymbol{\psi}\colon \Omega \to R_n$ *is continuous;*

(iii) *if* $-\infty < a < b < \infty$, $(a, \mathbf{c}_0) \in \mathscr{D}$ *and* $[a, b] \subset \Delta(a, \mathbf{c}_0, 0)$, *then there exist* $\varrho_0 > 0$ *and* $\varkappa_0 > 0$, $\varkappa_0 \le \varkappa$ *such that* $[a, b] \subset \Delta(a, \mathbf{c}, \varepsilon)$ *for any* $\mathbf{c} \in \mathfrak{B}(\mathbf{c}_0, \varrho_0; R_n)$ *and* $\varepsilon \in [0, \varkappa_0]$.

The following theorem provides an example of conditions which assure the existence of a solution to the equation on the given compact interval $[a, b] \subset R$.

1.19. Theorem. *Let* $-\infty < a < b < \infty$, $[a,b] \times R_n \subset \mathscr{D} \subset R_{n+1}$, $\mathscr{D}$ *open and let the n-vector valued function* $\mathbf{f}: \mathscr{D} \to R_n$ *fulfil the assumptions*

(i) $\mathbf{f}(t, .)$ *is continuous on* R_n *for a.e.* $t \in [a,b]$;
(ii) $\mathbf{f}(., \mathbf{x})$ *is measurable on* $[a,b]$ *for any* $\mathbf{x} \in R_n$;
(iii) *there exist* $\alpha \in R$, $0 \leq \alpha \leq 1$, *and L-integrable on* $[a,b]$ *scalar functions* $p(t)$ *and* $q(t)$ *such that*

$$|\mathbf{f}(t, \mathbf{x})| \leq p(t) + q(t)\,|\mathbf{x}|^\alpha \qquad \textit{for any} \quad \mathbf{x} \in R_n \quad \textit{and a.e.} \quad t \in [a,b].$$

Let the $n \times n$*-matrix valued function* $\mathbf{A}: [a,b] \to L(R_n)$ *be L-integrable on* $[a,b]$. *Then for any* $t_0 \in [a,b]$ *and* $\mathbf{c} \in R_n$ *there exists a solution* $\mathbf{x}(t)$ *of the equation*

$$\mathbf{x}' = \mathbf{A}(t)\,\mathbf{x} + \mathbf{f}(t, \mathbf{x})$$

on $[a,b]$ *such that* $\mathbf{x}(t_0) = \mathbf{c}$.

This auxiliary section will be completed by proving the following lemmas which illustrate the assumptions on the functions $\mathbf{f}$ and $\mathbf{g}$ employed in this chapter.

1.20. Lemma. *Let* $\mathscr{D} \subset R_{n+1}$ *and* $\mathfrak{D} \subset R_{n+2}$ *be open,* $\varkappa > 0$, $[0,1] \times R_n \subset \mathscr{D}$ *and* $\mathscr{D} \times [0, \varkappa] \subset \mathfrak{D}$. *Furthermore, let us assume that the functions* $\mathbf{f}: \mathscr{D} \to R_n$ *and* $\mathbf{g}: \mathfrak{D} \to R_n$ *are such that* $\mathbf{f} \in \mathrm{Car}(\mathscr{D})$ *and* $\tilde{\mathbf{g}} \in \mathrm{Car}(\mathfrak{D})$, *where* $\tilde{\mathbf{g}}(t, \mathbf{y}) = \mathbf{g}(t, \mathbf{x}, \varepsilon)$ *for* $(t, \mathbf{x}, \varepsilon) \in \mathfrak{D}$ *and* $\mathbf{y} \in (\mathbf{x}, \varepsilon)$. *Let us put*

$$(\mathbf{F}(\mathbf{x}))(t) = \mathbf{f}(t, \mathbf{x}(t)) \quad \textit{and} \quad (\mathbf{G}(\mathbf{x}, \varepsilon))(t) = \mathbf{g}(t, \mathbf{x}(t), \varepsilon)$$

for $\mathbf{x} \in C_n$, $\varepsilon \in [0, \varkappa]$ *and* $t \in [0,1]$. *Then* $\mathbf{F}(\mathbf{x}) \in L^1_n$ *and* $\mathbf{G}(\mathbf{x}, \varepsilon) \in L^1_n$ *for any* $\mathbf{x} \in C_n$ *and* $\varepsilon \in [0, \varkappa]$. *The operators* $\mathbf{F}: \mathbf{x} \in C_n \to \mathbf{F}(\mathbf{x}) \in L^1_n$ *and* $\mathbf{G}: (\mathbf{x}, \varepsilon) \in C_n \times [0, \varkappa] \to \mathbf{G}(\mathbf{x}, \varepsilon) \in L^1_n$ *are continuous.*

Proof. It is sufficient to show only the assertions concerning $\mathbf{G}$.

(a) Let $\varrho > 0$. Since $\tilde{\mathbf{g}} \in \mathrm{Car}(\mathfrak{D})$ ($\tilde{\mathbf{g}}(t, \mathbf{y}) = \mathbf{g}(t, \mathbf{x}, \varepsilon)$, where $\mathbf{y} = (\mathbf{x}, \varepsilon)$), applying the Borel Covering Theorem it is easy to find a function $m \in L^1$ such that

$$|\mathbf{g}(t, \mathbf{x}, \varepsilon)| \leq m(t) \qquad \text{for any} \quad \mathbf{x} \in \mathfrak{B}(\mathbf{0}, \varrho; R_n), \quad \varepsilon \in [0, \varkappa] \quad \text{and a.e.} \quad t \in [0,1]. \tag{1,4}$$

Let the functions $\mathbf{x}_k: [0,1] \to R_n$ and the numbers $\varepsilon_k \in [0, \varkappa]$ $(k = 0, 1, 2, \ldots)$ be such that $\lim_{k \to \infty} \mathbf{x}_k(t) = \mathbf{x}_0(t)$ on $[0,1]$ and $\lim_{k \to \infty} \varepsilon_k = \varepsilon_0$. Under our assumptions on $\mathbf{g}$ this implies that

$$\lim_{k \to \infty} \mathbf{g}(t, \mathbf{x}_k(t), \varepsilon_k) = \mathbf{g}(t, \mathbf{x}_0(t), \varepsilon_0) \qquad \text{a.e. on} \quad [0,1]. \tag{1,5}$$

If each of the functions $\chi_k(t) = \mathbf{g}(t, \mathbf{x}_k(t), \varepsilon_k)$ $(k = 0, 1, 2, \ldots)$ is measurable on $[0,1]$ and $|\mathbf{x}_k(t)| \leq \varrho$ on $[0,1]$ for any $k = 0, 1, 2, \ldots$, then by the Lebesgue Dominated

Convergence Theorem

$$\lim_{k\to\infty}\int_0^1 |\mathbf{g}(t, \mathbf{x}_k(t), \varepsilon_k) - \mathbf{g}(t, \mathbf{x}_0(t), \varepsilon_0)|\,\mathrm{d}t = 0\,.$$

(b) Let $\mathbf{x}_0 \in C_n$ and $\varrho = \|\mathbf{x}_0\|_C + 1$. It is well-known that there exist functions $\mathbf{x}_k\colon [0,1] \to R_n$ $(k = 1, 2, \ldots)$ piecewise constant on $[0,1]$ and such that $|\mathbf{x}_k(t)| \le \varrho$ $(k = 1, 2, \ldots)$ and $\lim_{k\to\infty} \mathbf{x}_k(t) = \mathbf{x}_0(t)$ on $[0,1]$. In particular, (1,5) with $\varepsilon_k = \varepsilon$ $(k = 0, 1, \ldots)$ holds and since any function $\gamma_k\colon t \in [0,1] \to \mathbf{g}(t, \mathbf{x}_k(t), \varepsilon)$ $(\varepsilon \in [0, \varkappa]$, $k = 1, 2, \ldots)$ is obviously measurable, $\gamma_0\colon t \in [0,1] \to \mathbf{g}(t, \mathbf{x}_0(t), \varepsilon)$ is measurable for any $\varepsilon \in [0, \varkappa]$ and hence according to (1,4) $\gamma_0 \in L^1_n$.

The continuity of the operator $\mathbf{G}$ follows easily from the first part of the proof.

1.21. Lemma. *Let $\mathscr{D} \subset R_{n+1}$ and $\mathbf{f}\colon \mathscr{D} \to R_n$ satisfy the corresponding assumptions of 1.20. In addition, let $\partial\mathbf{f}/\partial\mathbf{x} \in \mathrm{Car}(\mathscr{D})$. Then $\mathbf{F}$ defined in 1.20 possesses on C_n the Gâteaux derivative $\mathbf{F}'(\mathbf{x})$ continuous in $\mathbf{x}$ on C_n. Given $\mathbf{x}, \mathbf{u} \in C_n$,*

$$([\mathbf{F}'(\mathbf{x})]\,\mathbf{u})(t) = \left[\frac{\partial\mathbf{f}}{\partial\mathbf{x}}(t, \mathbf{x}(t))\right]\mathbf{u}(t) \qquad \textit{for a.e.} \quad t \in [0,1]\,.$$

Proof. (a) Let us put for $\mathbf{x} \in C_n$ and $t \in [0,1]$

$$[\mathbf{A}(\mathbf{x})](t) = \frac{\partial\mathbf{f}}{\partial\mathbf{x}}(t, \mathbf{x}(t))\,.$$

By 1.14 the $n \times n$-matrix valued function $\mathbf{A}(\mathbf{x})$ is L-integrable on $[0,1]$ for any $\mathbf{x} \in C_n$. If f_j $(j = 1, 2, \ldots, n)$ are the components of $\mathbf{f}$, then

$$[\mathbf{A}_j(\mathbf{x})](t) = \frac{\partial f_j}{\partial\mathbf{x}}(t, \mathbf{x}(t)) \qquad (j = 1, 2, \ldots, n)$$

are columns of $[\mathbf{A}(\mathbf{x})](t)$. By 1.20 the mappings

$$\mathbf{x} \in C_n \to \mathbf{A}_j(\mathbf{x}) \in L^1_n \qquad (j = 1, 2, \ldots, n) \tag{1,6}$$

are continuous. Obviously, for any $\mathbf{x} \in C_n$

$$\mathbf{J}(\mathbf{x})\colon\ \mathbf{u} \in C_n \to [\mathbf{A}(\mathbf{x})](t)\,\mathbf{u}(t) \in L^1_n$$

is a linear bounded operator. Moreover,

$$\|\mathbf{J}(\mathbf{x})\| = \sup_{\|\mathbf{u}\|_C \le 1} \|\mathbf{J}(\mathbf{x})\,\mathbf{u}\|_{L^1} \le \|\mathbf{A}(\mathbf{x})\|_{L^1} = \max_{j=1,2,\ldots,n} \|\mathbf{A}_j(\mathbf{x})\|_{L^1}$$

and consequently the operator $\mathbf{x} \in C_n \to \mathbf{J}(\mathbf{x}) \in B(C_n, L^1_n)$ is continuous.

(b) By the Mean Value Theorem I.7.4

$$\frac{(\mathbf{F}(\mathbf{x}_0 + \vartheta \mathbf{u}))(t) - (\mathbf{F}(\mathbf{x}_0))(t)}{\vartheta} = \frac{\mathbf{f}(t, \mathbf{x}_0(t) + \vartheta\, \mathbf{u}(t)) - \mathbf{f}(t, \mathbf{x}_0(t))}{\vartheta}$$

$$= \left(\int_0^1 \frac{\partial \mathbf{f}}{\partial \mathbf{x}}(t, \mathbf{x}_0(t) + \lambda\vartheta\, \mathbf{u}(t))\, \mathrm{d}\lambda\right) \mathbf{u}(t)$$

and

$$\left\| \frac{\mathbf{F}(\mathbf{x}_0 + \vartheta \mathbf{u}) - \mathbf{F}(\mathbf{x}_0)}{\vartheta} - \mathbf{J}(\mathbf{x}_0)\, \mathbf{u} \right\|_{L^1}$$

$$\leq \int_0^1 \left(\int_0^1 \left| \frac{\partial \mathbf{f}}{\partial \mathbf{x}}(t, \mathbf{x}_0(t) + \lambda\vartheta\, \mathbf{u}(t)) - \frac{\partial \mathbf{f}}{\partial \mathbf{x}}(t, \mathbf{x}_0(t)) \right| \mathrm{d}\lambda \right) \mathrm{d}t\, \|\mathbf{u}\|_C .$$

By the Tonelli-Hobson Theorem I.4.36 we may change the order of the integration in the last integral. The continuity of the mappings (1,6) yields

$$\lim_{\vartheta \to 0+} \int_0^1 \left| \frac{\partial \mathbf{f}}{\partial \mathbf{x}}(t, \mathbf{x}_0(t) + \lambda\vartheta\, \mathbf{u}(t)) - \frac{\partial \mathbf{f}}{\partial \mathbf{x}}(t, \mathbf{x}_0(t)) \right| \mathrm{d}t = 0$$

uniformly with respect to $\lambda \in [0, 1]$. Consequently,

$$\lim_{\vartheta \to 0+} \left\| \frac{\mathbf{F}(\mathbf{x}_0 + \vartheta \mathbf{u}) - \mathbf{F}(\mathbf{x}_0)}{\vartheta} - \mathbf{J}(\mathbf{x})\, \mathbf{u} \right\|_{L^1} = 0$$

for any $\mathbf{x}_0 \in C_n$ and $\mathbf{u} \in C_n$. This completes the proof.

1.22. Remark. Given $\mathbf{x} \in AC_n$ and $\mathbf{L} \in B(C_n, L^1_n)$, $\|\mathbf{x}\|_C \leq \|\mathbf{x}\|_{AC}$, $\mathbf{L} \in B(AC_n, L^1_n)$ and

$$\|\mathbf{L}\|_{B(AC_n, L^1_n)} = \sup_{\|u\|_{AC} \leq 1} \|\mathbf{L}\mathbf{u}\|_{L^1} \leq \sup_{\|u\|_C \leq 1} \|\mathbf{L}\mathbf{u}\|_{L^1} = \|\mathbf{L}\|_{B(C_n, L^1_n)}.$$

It follows readily that 1.20 and 1.21 remain valid also if in their formulations C_n is replaced everywhere by AC_n.

1.23. Remark. If moreover $\partial^2 \mathbf{f}/(\partial x_i\, \partial x_j) \in \mathrm{Car}(\mathscr{D})$ $(i, j = 1, 2, \ldots, n)$, it may be shown that for any $\mathbf{x} \in C_n$, $\mathbf{F}$ possesses the second order Gâteaux derivative $\mathbf{F}''(\mathbf{x})$ such that the mapping $\mathbf{x} \in C_n \to \mathbf{F}''(\mathbf{x}) \in B(C_n, B(C_n, L^1_n))$ is continuous. Given $\mathbf{x}, \mathbf{u}, \mathbf{v} \in C_n$, the components of the n-vector $([\mathbf{F}''(\mathbf{x})\, \mathbf{u}]\, \mathbf{v})(t)$ are given by

$$\sum_{i=1}^n \left(\sum_{j=1}^n \frac{\partial^2 f_k}{\partial x_i\, \partial x_j}(t, \mathbf{x}(t))\, u_i(t) \right) v_j(t), \qquad k = 1, 2, \ldots, n.$$

Let $[0,1] \times \{0\} \times R_n \subset \Omega$. Let us put for $\mathbf{y} \in C_n$

$$\Phi(\mathbf{y})(t) = \varphi(t,0,\mathbf{y}(t)) \quad \text{on } [0,1], \tag{1,7}$$

$$\Phi_t(\mathbf{y})(t) = \frac{\mathrm{d}\varphi}{\mathrm{d}t}(t,0,\mathbf{y}(t)) \quad \text{a.e. on } [0,1],$$

$$\Phi_{\mathbf{c}}(\mathbf{y})(t) = \frac{\partial \varphi}{\partial \mathbf{c}}(t,0,\mathbf{y}(t)) \quad \text{on } [0,1].$$

It is easy to verify that Φ and $\mathbf{F}\Phi$ are continuous mappings of C_n into C_n and L^1_n, respectively, and $\Phi_{\mathbf{c}}$ is a continuous mapping of C_n into the space of $n \times n$-matrix valued function which are continuous on $[0,1]$ (cf. 1.14 and 1.20). Since $\|\mathbf{y}\|_{AC} = |\mathbf{y}(0)| + \|\mathbf{y}'\|_{L^1}$ for any $\mathbf{y} \in AC_n$, it follows readily that Φ is a continuous mapping of AC_n into AC_n. Analogously $\Phi_{\mathbf{c}}$ is a continuous mapping of AC_n into the space of $n \times n$-matrix valued functions absolutely continuous on $[0,1]$, i.e. if $\Phi_{\mathbf{c}}(\mathbf{y})$ denotes also the linear operator $\mathbf{h} \in AC_n \to \Phi_{\mathbf{c}}(\mathbf{y})(t)\,\mathbf{h}(t)$, then $\mathbf{y} \in AC_n \to \Phi_{\mathbf{c}}(\mathbf{y})$ is a continuous mapping of AC_n into $B(AC_n)$. Let us notice that for any $\mathbf{y} \in C_n$

$$\frac{\partial \varphi}{\partial t}(t,0,\mathbf{y}(t)) = \mathbf{f}(t,\varphi(t,0,\mathbf{y}(t))) = \mathbf{F}(\Phi(\mathbf{y}))(t) \qquad \text{a.e. on } [0,1] \tag{1,8}$$

and by 1.14

$$\frac{\partial}{\partial t}\left(\frac{\partial \varphi}{\partial \mathbf{c}}(t,0,\mathbf{y}(t))\right) = \left[\frac{\partial \mathbf{f}}{\partial \mathbf{x}}(t,\varphi(t,0,\mathbf{y}(t))\right]\frac{\partial \varphi}{\partial \mathbf{c}}(t,0,\mathbf{y}(t)) \tag{1,9}$$

$$= ([\mathbf{F}'(\Phi(\mathbf{y}))]\,\Phi_{\mathbf{c}}(\mathbf{y}))(t) \qquad \text{a.e. on } [0,1].$$

Moreover, for any $\mathbf{y} \in AC_n$

$$\Phi_t(\mathbf{y})(t) = \mathbf{F}(\Phi(\mathbf{y}))(t) + \Phi_{\mathbf{c}}(\mathbf{y})(t)\,\mathbf{y}'(t)$$

and thus Φ_t is a continuous operator $AC_n \to L^1_n$.

Let $\mathbf{y}, \mathbf{h} \in AC_n$ and $\vartheta \in (0,1)$. Then

$$\left\|\frac{\Phi(\mathbf{y}+\vartheta\mathbf{h}) - \Phi(\mathbf{y})}{\vartheta} - \Phi_{\mathbf{c}}(\mathbf{y})\,\mathbf{h}\right\|_{AC} \tag{1,10}$$

$$\leq \left|\frac{\varphi(0,0,\mathbf{y}(0) + \vartheta\,\mathbf{h}(0)) - \varphi(0,0,\mathbf{y}(0))}{\vartheta} - \frac{\partial \varphi}{\partial \mathbf{c}}(0,0,\mathbf{y}(0))\,\mathbf{h}(0)\right|$$

$$+ \int_0^1 \left|\frac{\frac{\partial \varphi}{\partial t}(t,0,\mathbf{y}(t) + \vartheta\,\mathbf{h}(t)) - \frac{\partial \varphi}{\partial t}(t,0,\mathbf{y}(t))}{\vartheta} - \frac{\partial^2 \varphi}{\partial t\,\partial \mathbf{c}}(t,0,\mathbf{y}(t))\,\mathbf{h}(t)\right| \mathrm{d}t$$

$$+\int_0^1 \left| \frac{\dfrac{\partial \boldsymbol{\varphi}}{\partial \mathbf{c}}(t, 0, \mathbf{y}(t) + \vartheta\, \mathbf{h}(t))\, \vartheta\, \mathbf{h}'(t)}{\vartheta} - \frac{\partial \boldsymbol{\varphi}}{\partial \mathbf{c}}(t, 0, \mathbf{y}(t))\, \mathbf{h}'(t) \right| \mathrm{d}t$$

$$+\int_0^1 \left| \frac{\left(\dfrac{\partial \boldsymbol{\varphi}}{\partial \mathbf{c}}(t, 0, \mathbf{y}(t) + \vartheta\, \mathbf{h}(t)) - \dfrac{\partial \boldsymbol{\varphi}}{\partial \mathbf{c}}(t, 0, \mathbf{y}(t))\right) \mathbf{y}'(t)}{\vartheta} - \frac{\partial^2 \boldsymbol{\varphi}}{\partial \mathbf{c}^2}(t, 0, \mathbf{y}(t))\, \mathbf{y}'(t)\, \mathbf{h}(t) \right| \mathrm{d}t\,.$$

Obviously, the first and the third terms on the right-hand side of (1,10) tend to 0 as $\vartheta \to 0+$. Furthermore, by (1,8), (1,9) and the Mean Value Theorem the second one becomes

$$\int_0^1 \left| \frac{\mathbf{f}(t, \boldsymbol{\varphi}(t, 0, \mathbf{y}(t) + \vartheta\, \mathbf{h}(t))) - \mathbf{f}(t, \boldsymbol{\varphi}(t, 0, \mathbf{y}(t)))}{\vartheta} - \left[\frac{\partial \mathbf{f}}{\partial \mathbf{x}}(t, \boldsymbol{\varphi}(t, 0, \mathbf{y}(t)))\right] \left(\frac{\partial \boldsymbol{\varphi}}{\partial \mathbf{c}}(t, 0, \mathbf{y}(t))\right) \mathbf{h}(t) \right| \mathrm{d}t$$

$$\leq \int_0^1 \left(\int_0^1 \left\| \left[\frac{\partial \mathbf{f}}{\partial \mathbf{x}}(t, \boldsymbol{\varphi}(t, 0, \mathbf{y}(t) + \vartheta\lambda\, \mathbf{h}(t)))\right] \left(\frac{\partial \boldsymbol{\varphi}}{\partial \mathbf{c}}(t, 0, \mathbf{y}(t) + \vartheta\lambda\, \mathbf{h}(t))\right) - \left[\frac{\partial \mathbf{f}}{\partial \mathbf{x}}(t, \boldsymbol{\varphi}(t, 0, \mathbf{y}(t)))\right] \left(\frac{\partial \boldsymbol{\varphi}}{\partial \mathbf{c}}(t, 0, \mathbf{y}(t))\right) \right\| |\mathbf{h}(t)|\, \mathrm{d}\lambda \right) \mathrm{d}t\,.$$

It is easy to verify that this last expression tends to 0 as $\vartheta \to 0+$. (Obviously, $\mathbf{F}'\boldsymbol{\Phi}_{\mathbf{c}}$ is a continuous operator $C_n \to B(C_n, L_n^1)$.) Analogously, the Mean Value Theorem yields that also the fourth term the right-hand side of (1,10) tends to 0 as $\vartheta \to 0+$.

1.24. Lemma. *Under the assumptions of* 1.16, *the operator* $\boldsymbol{\Phi}$ *given by* (1,7) *is a continuous mapping of* AC_n *into* AC_n *which is Gâteaux differentiable at any* $\mathbf{x} \in AC_n$. *Given* $\mathbf{y}, \mathbf{h} \in AC_n$,

$$([\boldsymbol{\Phi}'(\mathbf{y})]\, \mathbf{h})(t) = \left[\frac{\partial \boldsymbol{\varphi}}{\partial \mathbf{c}}(t, 0, \mathbf{y}(t))\right] \mathbf{h}(t)\,.$$

The mapping $\mathbf{y} \in AC_n \to \boldsymbol{\Phi}'(\mathbf{y}) \in B(AC_n)$ *is continuous.*

1.25. Definition. Let $\mathfrak{D} \subset R_{n+2}$ be open, $\varkappa > 0$, $[0, 1] \times R_n \times [0, \varkappa] \subset \mathfrak{D}$ and $\mathbf{g}\colon \mathfrak{D} \to R_n$. Let $\varepsilon_0 \in [0, \varkappa]$ and let for given $t \in [0, 1]$ and $\mathbf{x}_0 \in R_n$ there exist $\delta_0 = \delta_0(t, \mathbf{x}_0) > 0$, $\varrho_0 = \varrho_0(t, \mathbf{x}_0) > 0$, $\varkappa_0 = \varkappa_0(t, \mathbf{x}_0) > 0$ and $\omega \in L^1(t - \delta_0, t + \delta_0)$ such that $|\tau - t| < \delta_0$, $|\mathbf{x}_1 - \mathbf{x}_0| < \varrho_0$, $|\mathbf{x}_2 - \mathbf{x}_0| < \varrho_0$, $\varepsilon \geq 0$ and $|\varepsilon - \varepsilon_0| < \varkappa_0$ implies $(\tau, \mathbf{x}_1, \varepsilon) \in \mathfrak{D}$, $(\tau, \mathbf{x}_2, \varepsilon) \in \mathfrak{D}$ and

$$|\mathbf{g}(\tau, \mathbf{x}_2, \varepsilon) - \mathbf{g}(\tau, \mathbf{x}_1, \varepsilon)| \leq \omega(\tau)\, |\mathbf{x}_2 - \mathbf{x}_1|\,.$$

Then $\mathbf{g}$ is said to be *locally lipschitzian in* $\mathbf{x}$ *near* $\varepsilon = \varepsilon_0$ and we shall write $\mathbf{g} \in \mathrm{Lip}\,(\mathfrak{D}, \varepsilon_0)$.

1.26. Lemma. *Let* $\mathfrak{D} \subset R_{n+2}$ *and* $\mathbf{g}: \mathfrak{D} \to R_n$ *satisfy the corresponding assumptions of* 1.20. *In addition, let* $\mathbf{g} \in \mathrm{Lip}(\mathfrak{D}, \varepsilon_0)$. *Then* $\mathbf{G}$ *defined in* 1.20 *is locally lipschitzian in* $\mathbf{x}$ *near* $\varepsilon = \varepsilon_0$.

Proof follows from Definition 1.25 applying the Borel Covering Theorem.

1.27. Remark. In order that the operators $\mathbf{F}$ and $\mathbf{G}$ might possess the properties from 1.20 – 1.26 locally, it is sufficient to require that the assumptions of the corresponding lemmas are fulfilled only locally.

2. Nonlinear boundary value problems for functional-differential equations

Let $\varkappa > 0$ and let $\mathbf{F}: C_n \to L^1_n$, $\mathbf{G}: AC_n \times [0, \varkappa] \to L^1_n$, $\mathbf{S}: C_n \to R_n$ and $\mathbf{R}: AC_n \times [0, \varkappa] \to R_n$ be continuous operators. To a given $\varepsilon \in [0, \varkappa]$ we want to find a solution $\mathbf{x}$ of the functional-differential equation

$$\mathbf{x}' = \mathbf{F}(\mathbf{x}) + \varepsilon\, \mathbf{G}(\mathbf{x}, \varepsilon) \tag{2,1}$$

on the interval $[0, 1]$ which verifies the side condition

$$\mathbf{S}(\mathbf{x}) + \varepsilon\, \mathbf{R}(\mathbf{x}, \varepsilon) = \mathbf{0}\,. \tag{2,2}$$

This boundary value problem will be referred to as BVP $(\mathscr{P}_\varepsilon)$. The limit problem for $\varepsilon = 0$

$$\mathbf{x}' = \mathbf{F}(\mathbf{x})\,, \tag{2,3}$$

$$\mathbf{S}(\mathbf{x}) = \mathbf{0} \tag{2,4}$$

is denoted by $(\mathscr{P}_0)$.

2.1. Definition. Let $\varepsilon \in [0, \varkappa]$. An n-vector valued function $\mathbf{x}$ *is a solution to* (2,1) *on* $[0, 1]$ if $\mathbf{x} \in AC_n$ and

$$\mathbf{x}'(t) = (\mathbf{F}(\mathbf{x}))(t) + \varepsilon(\mathbf{G}(\mathbf{x}, \varepsilon))(t) \qquad \text{a.e. on } [0, 1]\,.$$

2.2. Remark. Let $\mathbf{x}_0 \in C_n$, $\omega \in L^1$, $\varrho > 0$ and

$$|(\mathbf{F}(\mathbf{x}_2))(t) - (\mathbf{F}(\mathbf{x}_1))(t)| \leq \omega(t)\, \|\mathbf{x}_2 - \mathbf{x}_1\|_C \tag{2,5}$$

for any $\mathbf{x}_1, \mathbf{x}_2 \in \mathfrak{B}(\mathbf{x}_0, \varrho; C_n)$ and a.e. $t \in [0, 1]$. Then

$$\left|\frac{\mathbf{F}(\mathbf{x}_0 + \vartheta \mathbf{u})(t) - \mathbf{F}(\mathbf{x}_0)(t)}{\vartheta}\right| \leq \omega(t)\, \|\mathbf{u}\|_C$$

for any $\vartheta > 0$, $\mathbf{u} \in C_n$ and a.e. $t \in [0,1]$. If $\mathbf{F}$ possesses the Gâteaux derivative $\mathbf{F}'(\mathbf{x}_0)$ at $\mathbf{x}_0$, then

$$\lim_{\vartheta \to 0+} \left| \frac{\mathbf{F}(\mathbf{x}_0 + \vartheta \mathbf{u})(t) - \mathbf{F}(\mathbf{x}_0)(t)}{\vartheta} - ([\mathbf{F}'(\mathbf{x}_0)]\,\mathbf{u})(t) \right| = 0 \qquad \text{a.e. on } [0,1].$$

It follows easily that

$$|([\mathbf{F}'(\mathbf{x}_0)]\,\mathbf{u})(t)| \le \omega(t)\,\|\mathbf{u}\|_C \qquad \text{for any} \quad \mathbf{u} \in C_n \quad \text{and a.e.} \quad t \in [0,1].$$

In particular, there exists a function $\mathbf{P}\colon [0,1] \times [0,1] \to L(R_n)$ such that $\mathbf{P}(.,s)$ is measurable on $[0,1]$ for any $s \in [0,1]$, $\varrho(t) = |\mathbf{P}(t,0)| + \operatorname{var}_0^1 \mathbf{P}(t,.) < \infty$ for a.e. $t \in [0,1]$, $\varrho \in L^1$ ($\mathbf{P}$ is an $L^1[BV]$-kernel) and

$$([\mathbf{F}'(\mathbf{x}_0)]\,\mathbf{u})(t) = \int_0^1 \mathrm{d}_s[\mathbf{P}(t,s)]\,\mathbf{u}(s) \qquad \text{for any} \quad \mathbf{u} \in C_n \quad \text{and a.e.} \quad t \in [0,1]$$

(cf. Kantorovič, Pinsker, Vulich [1]).

2.3. Theorem. *Let $\mathbf{x}_0 \in AC_n$ be a solution to BVP $(\mathscr{P}_0)$, where $\mathbf{F}\colon C_n \to L_n^1$ and $\mathbf{S}\colon C_n \to R_n$ are continuous operators. Furthermore, let us assume that (2,5) holds and $\mathbf{F}, \mathbf{S} \in C^1(\mathfrak{B}(\mathbf{x}_0, \varrho; C_n))$ for some $\varrho > 0$. If the linear BVP for $\mathbf{u} \in AC_n$*

$$\mathbf{u}' = [\mathbf{F}'(\mathbf{x}_0)]\,\mathbf{u}, \tag{2,6}$$

$$[\mathbf{S}'(\mathbf{x}_0)]\,\mathbf{u} = \mathbf{0} \tag{2,7}$$

possesses only the trivial solution, then there exists $\varrho_0 > 0$ such that there is no other solution $\mathbf{x}$ of BVP $(\mathscr{P}_0)$ such that $\|\mathbf{x} - \mathbf{x}_0\|_{AC} \le \varrho_0$.

Proof. Let us put

$$\mathscr{F}\colon \mathbf{x} \in AC_n \to \begin{pmatrix} \mathbf{x}' - \mathbf{F}(\mathbf{x}) \\ \mathbf{S}(\mathbf{x}) \end{pmatrix} \in L_n^1 \times R_n. \tag{2,8}$$

By the assumption $\mathscr{F}(\mathbf{x}_0) = \mathbf{0}$ and $\mathscr{F} \in C^1(\mathfrak{B}(\mathbf{x}_0, \varrho; AC_n))$,

$$\mathscr{F}'(\mathbf{x})\colon \mathbf{u} \in AC_n \to \begin{pmatrix} \mathbf{u}' - \mathbf{F}'(\mathbf{x})\,\mathbf{u} \\ \mathbf{S}'(\mathbf{x})\,\mathbf{u} \end{pmatrix} \in L_n^1 \times R_n \tag{2,9}$$

for any $\mathbf{x} \in \mathfrak{B}(\mathbf{x}_0, \varrho; AC_n)$.

Let $\mathscr{F}(\mathbf{x}) = \mathbf{0}$ for some $\mathbf{x} \in \mathfrak{B}(\mathbf{x}_0, \varrho; AC_n)$, $\mathbf{x} \neq \mathbf{x}_0$. By the Mean Value Theorem I.7.4 we have

$$\mathbf{0} = \mathscr{F}(\mathbf{x}) - \mathscr{F}(\mathbf{x}_0) = \int_0^1 [\mathscr{F}'(\mathbf{x}_0 + \vartheta(\mathbf{x} - \mathbf{x}_0))]\,(\mathbf{x} - \mathbf{x}_0)\,\mathrm{d}\vartheta.$$

By 2.2 and V.3.12 $\mathscr{F}'(\mathbf{x}_0)$ possesses a bounded inverse

$$\boldsymbol{\Gamma} = [\mathscr{F}'(\mathbf{x}_0)]^{-1}\colon L_n^1 \times R_n \to AC_n.$$

Hence

$$\mathbf{x} - \mathbf{x}_0 = \int_0^1 \Gamma[\mathscr{F}'(\mathbf{x}_0) - \mathscr{F}'(\mathbf{x}_0 + \vartheta(\mathbf{x} - \mathbf{x}_0))] (\mathbf{x} - \mathbf{x}_0)\, d\vartheta$$

and

$$(2,10) \quad \|\mathbf{x} - \mathbf{x}_0\|_{AC} \le \|\Gamma\| \,(\sup_{\vartheta \in [0,1]} \|\mathscr{F}'(\mathbf{x}_0) - \mathscr{F}'(\mathbf{x}_0 + \vartheta(\mathbf{x} - \mathbf{x}_0))\|)\, \|\mathbf{x} - \mathbf{x}_0\|_{AC} .$$

Since the mapping

$$\mathbf{x} \in \mathfrak{B}(\mathbf{x}_0, \varrho;\, AC_n) \to \mathscr{F}'(\mathbf{x}) \in B(AC_n, L_n^1)$$

is continuous, there is $\varrho_0 > 0$ such that $\varrho_0 \le \varrho$ and

$$\|\mathscr{F}'(\mathbf{x}_0) - \mathscr{F}'(\mathbf{x}_0 + \vartheta(\mathbf{x} - \mathbf{x}_0))\| \le \|\Gamma\|^{-1}$$

for any $\mathbf{x} \in \mathfrak{B}(\mathbf{x}_0, \varrho_0;\, AC_n)$ and $\vartheta \in [0, 1]$. Consequently for $\mathbf{x} \in \mathfrak{B}(\mathbf{x}_0, \varrho_0;\, AC_n)$, $\mathbf{x} \neq \mathbf{x}_0$ (2,10) becomes a contradiction $\|\mathbf{x} - \mathbf{x}_0\|_{AC} < \|\mathbf{x} - \mathbf{x}_0\|_{AC}$. This proves that $\mathbf{x} = \mathbf{x}_0$ if $\mathscr{F}(\mathbf{x}) = \mathbf{0}$ and $\mathbf{x} \in \mathfrak{B}(\mathbf{x}_0, \varrho_0;\, AC_n)$.

2.4. Definition. Let $\mathbf{x}_0 \in AC_n$ be a solution of BVP $(\mathscr{P}_0)$ and let the operators $\mathbf{F}$ and $\mathbf{S}$ fulfil the assumptions of 2.3. The problem of determining a solution $\mathbf{u} \in AC_n$ of (2,6) which verifies the side condition (2,7) is called the *variational boundary value problem corresponding to* $\mathbf{x}_0$ and is denoted by $(\mathscr{V}_0(\mathbf{x}_0))$.

2.5. Remark. BVP

$$x' = x + 1\,, \qquad \mathbf{S}(x) = (x(0))^2 + (x(1) + 1 - \exp(1))^2 = 0$$

indicates that in general the converse statement to 2.3 is not true. In fact, the solutions to $x' = x + 1$ are of the form $x(t) = c \exp(t) - 1$, where $c \in R$. The only solution to

$$(2,11) \qquad \mathbf{S}(x) = (c - 1)^2 + (c - 1)^2 (\exp(1))^2 = 0$$

is $c = 1$. Hence $x_0(t) = \exp(t) - 1$ is the only solution to (2,11). The corresponding variational BVP is given by

$$(2,12) \qquad u' = u\,, \qquad [x_0(0)]\, u(0) + [x_0(1) + 1 - \exp(1)]\, u(1) = 0\,.$$

Since $x_0(0) = 0$, $x_0(1) = \exp(1) - 1$, $u(t) = d \exp(t)$ is a solution to (2,12) for any $d \in R$.

2.6. Definition. A solution $\mathbf{x}_0$ of BVP $(\mathscr{P}_0)$ is said to be *isolated* if there is $\varrho_0 > 0$ such that there is no solution $\mathbf{x}$ to $(\mathscr{P}_0)$ such that $\mathbf{x} \neq \mathbf{x}_0$ and $\mathbf{x} \in \mathfrak{B}(\mathbf{x}_0, \varrho_0;\, AC_n)$. It is *regular* if the corresponding variational BVP $(\mathscr{V}(\mathbf{x}_0))$ is defined and possesses only the trivial solution.

2.7. Theorem. *Let* $\mathbf{x}_0 \in AC_n$ *be a solution to BVP* $(\mathscr{P}_0)$ *where* $\mathbf{F}\colon C_n \to L_n^1$ *and* $\mathbf{S}\colon C_n \to R_n$ *are continuous operators such that* (2,5) *holds and* $\mathbf{F}, \mathbf{S} \in C^1(\mathfrak{B}(\mathbf{x}_0, \varrho;\, C_n))$

for some $\varrho > 0$. *Furthermore,* $\varkappa > 0$ *and* $\mathbf{G}: AC_n \times [0, \varkappa] \to L_n^1$ *and* $\mathbf{R}: AC_n \times [0, \varkappa] \to R_n$ *are continuous operators which are locally lipschitzian in* $\mathbf{x}$ *near* $\varepsilon = 0$.

If $\mathbf{x}_0$ *is a regular solution of* $(\mathscr{P}_0)$, *then there exist* $\varepsilon_0 > 0$ *and* $\varrho_0 > 0$ *such that for any* $\varepsilon \in [0, \varepsilon_0]$ *BVP* $(\mathscr{P}_\varepsilon)$ *possesses a unique solution* $\mathbf{x}(\varepsilon)$ *in* $\mathfrak{B}(\mathbf{x}_0, \varrho_0; AC_n)$. *The mapping* $\varepsilon \in [0, \varepsilon_0] \to \mathbf{x}(\varepsilon) \in AC_n$ $(\mathbf{x}(0) = \mathbf{x}_0)$ *is continuous.*

Proof follows by applying I.7.8 to the operator equation

$$\mathscr{F}(\mathbf{x}) + \varepsilon \mathscr{G}(\mathbf{x}, \varepsilon) = \mathbf{0},$$

where $\mathscr{F}: AC_n \to L_n^1 \times R_n$ is given by (2,8) and

$$\mathscr{G}: (\mathbf{x}, \varepsilon) \in AC_n \times [0, \varkappa] \to \begin{pmatrix} \mathbf{G}(\mathbf{x}, \varepsilon) \\ \mathbf{R}(\mathbf{x}, \varepsilon) \end{pmatrix} \in L_n^1 \times R_n .$$

(Under our assumptions there exists a bounded inverse of $\mathscr{F}'(\mathbf{x}_0)$, cf. the proof of 2.3.)

2.8. Remark. The conclusion of Theorem 2.7 may be reformulated as follows.

If $\mathbf{x}_0$ is a regular solution of $(\mathscr{P}_0)$, then there exists for any $\varepsilon > 0$ sufficiently small a unique solution $\mathbf{x}(\varepsilon)$ of BVP $(\mathscr{P}_\varepsilon)$ which is continuous in ε and tends to $\mathbf{x}_0$ as $\varepsilon \to 0$.

Theorem 2.7 assures the existence of an isolated solution to BVP $(\mathscr{P}_\varepsilon)$ which is close to the regular solution $\mathbf{x}_0$ of the limit problem $(\mathscr{P}_0)$. If also the perturbations $\mathbf{G}$ and $\mathbf{R}$ are differentiable with respect to $\mathbf{x}$, then we can prove that for any $\varepsilon > 0$ sufficiently small this solution is regular, too.

2.9. Theorem. *Let the assumptions of 2.7 hold. In addition, let us assume that* $\mathbf{G}$ *and* $\mathbf{R}$ *possess the Gâteaux derivatives* $\mathbf{G}'(\mathbf{x}, \varepsilon)$ *and* $\mathbf{R}'(\mathbf{x}, \varepsilon)$ *with respect to* $\mathbf{x}$ *for any* $(\mathbf{x}, \varepsilon) \in \mathfrak{B}(\mathbf{x}_0, \varrho; AC_n) \times [0, \varkappa]$ *continuous in* $(\mathbf{x}, \varepsilon)$ *on* $\mathfrak{B}(\mathbf{x}_0, \varrho; AC_n) \times [0, \varkappa]$.

Then there exists ε_1, $0 < \varepsilon_1 \le \varepsilon_0$ *such that for any* $\varepsilon \in [0, \varepsilon_1]$ *the corresponding solution* $\mathbf{x}(\varepsilon)$ *of BVP* $(\mathscr{P}_\varepsilon)$ *is regular.*

Proof. Given $\varepsilon \in [0, \varepsilon_0]$, the variational BVP $(\mathscr{V}_\varepsilon(\mathbf{x}(\varepsilon)))$ corresponding to the solution $\mathbf{x}(\varepsilon)$ of BVP $(\mathscr{P}_\varepsilon)$ is given by

$$\mathbf{u}' = [\mathbf{F}'(\mathbf{x}(\varepsilon)) + \varepsilon \mathbf{G}'(\mathbf{x}(\varepsilon), \varepsilon)] \mathbf{u},$$
$$[\mathbf{S}'(\mathbf{x}(\varepsilon)) + \varepsilon \mathbf{R}'(\mathbf{x}(\varepsilon), \varepsilon)] \mathbf{u} = \mathbf{0}.$$

Let $\mathbf{u}$ be its solution, i.e.

$$\mathscr{J}(\varepsilon) \mathbf{u} = [\mathscr{F}'(\mathbf{x}(\varepsilon)) + \varepsilon \mathscr{G}'(\mathbf{x}(\varepsilon), \varepsilon)] \mathbf{u} = \mathbf{0}.$$

Let $\Gamma = [\mathscr{F}'(\mathbf{x}_0)]^{-1}$. Then $\mathbf{u} = \Gamma[\mathscr{J}(0) - \mathscr{J}(\varepsilon)] \mathbf{u}$ and

$$\|\mathbf{u}\|_{AC} \le \|\Gamma\| \, \|\mathscr{J}(0) - \mathscr{J}(\varepsilon)\| \, \|\mathbf{u}\|_{AC} .$$

Since the operators $\varepsilon \in [0, \varepsilon_0] \to \mathbf{x}(\varepsilon) \in AC_n$ and $(\mathbf{x}, \varepsilon) \in \mathfrak{B}(\mathbf{x}_0, \varrho; AC_n) \times [0, \varkappa] \to \mathscr{F}'(\mathbf{x}) + \varepsilon\, \mathscr{G}'(\mathbf{x}, \varepsilon) \in B(AC_n, L_n^1 \times R_n)$ are continuous, their composition $\varepsilon \in [0, \varepsilon_0] \to \mathscr{J}(\varepsilon) \in B(AC_n, L_n^1 \times R_n)$ is also continuous.

Choosing ε_1, $0 < \varepsilon_1 \le \varepsilon_0$ in such a way that $\varepsilon \in [0, \varepsilon_1]$ implies $\|\mathscr{J}(0) - \mathscr{J}(\varepsilon)\| \le \|\boldsymbol{\Gamma}\|^{-1}$ we derive a contradiction $\|\mathbf{u}\|_{AC} < \|\mathbf{u}\|_{AC}$ whenever $\mathbf{u} \neq \mathbf{0}$.

2.10. Remark. The case when $\mathbf{x}_0$ is a regular solution of BVP $(\mathscr{P}_0)$ has appeared to be simple. It is said to be noncritical. The case when $\mathbf{x}_0$ is not a regular solution of $(\mathscr{P}_0)$ is more complicated and said to be critical.

2.11. The critical case. Let $\mathbf{x}_0 \in AC_n$ be a solution to BVP $(\mathscr{P}_0)$, where $\mathbf{F}$: $C_n \to L_n^1$ and $\mathbf{S}$: $C_n \to R_n$ are continuous operators such that (2,5) holds and $\mathbf{F}, \mathbf{S} \in C^2(\mathfrak{B}(\mathbf{x}_0, \varrho; C_n)$ for some $\varrho > 0$. Furthermore, $\varkappa > 0$ and $\mathbf{G}$: $AC_n \times [0, \varkappa] \to L_n^1$ and $\mathbf{R}$: $AC_n \times [0, \varkappa] \to R_n$ are continuous operators such that $\mathbf{G}, \mathbf{R} \in C^{1,1}(\mathfrak{B}(\mathbf{x}_0, \varrho; AC_n) \times [0, \varkappa])$. In general, we do not assume that $\mathbf{x}_0$ is a regular solution of BVP $(\mathscr{P}_0)$. Let us try to find a solution to $(\mathscr{P}_\varepsilon)$ in the form

$$\mathbf{x}(t) = \mathbf{x}_0(t) + \varepsilon\, \boldsymbol{\chi}(t). \tag{2,13}$$

Inserting (2,13) into (2,1) we obtain

$$\mathbf{x}_0' + \varepsilon \boldsymbol{\chi}' = \mathbf{F}(\mathbf{x}_0) + (\mathbf{F}(\mathbf{x}_0 + \varepsilon\boldsymbol{\chi}) - \mathbf{F}(\mathbf{x}_0)) + \varepsilon\, \mathbf{G}(\mathbf{x}_0 + \varepsilon\boldsymbol{\chi}, \varepsilon),$$

i.e.

$$\begin{aligned} \boldsymbol{\chi}' &= \int_0^1 [\mathbf{F}'(\mathbf{x}_0) + (\mathbf{F}'(\mathbf{x}_0 + \varepsilon\vartheta\boldsymbol{\chi}) - \mathbf{F}'(\mathbf{x}_0))]\, \boldsymbol{\chi}\, \mathrm{d}\vartheta \\ &\quad + \mathbf{G}(\mathbf{x}_0, 0) + (\mathbf{G}(\mathbf{x}_0 + \varepsilon\boldsymbol{\chi}, \varepsilon) - \mathbf{G}(\mathbf{x}_0, 0)) \\ &= [\mathbf{F}'(\mathbf{x}_0)]\, \boldsymbol{\chi} + \mathbf{G}(\mathbf{x}_0, 0) + \varepsilon\, \mathbf{H}(\boldsymbol{\chi}, \varepsilon), \end{aligned}$$

where

$$\begin{aligned} \mathbf{H}(\boldsymbol{\chi}, \varepsilon) &= \left(\int_0^1 \left(\int_0^\vartheta \mathbf{F}''(\mathbf{x}_0 + \varepsilon\vartheta_1\vartheta\boldsymbol{\chi})\, \mathrm{d}\vartheta_1 \right) \vartheta\, \mathrm{d}\vartheta\boldsymbol{\chi} \right) \boldsymbol{\chi} \\ &\quad + \left(\int_0^1 \mathbf{G}_x'(\mathbf{x}_0 + \varepsilon\vartheta\boldsymbol{\chi}, \vartheta\varepsilon)\, \mathrm{d}\vartheta \right) \boldsymbol{\chi} + \int_0^1 \mathbf{G}_\varepsilon'(\mathbf{x}_0 + \varepsilon\vartheta\boldsymbol{\chi}, \vartheta\varepsilon)\, \mathrm{d}\vartheta. \end{aligned}$$

Thus (2,13) is a solution to (2,1) on $[0, 1]$ if and only if

$$\boldsymbol{\chi}' = [\mathbf{F}'(\mathbf{x}_0)]\, \boldsymbol{\chi} + \mathbf{G}(\mathbf{x}_0, 0) + \varepsilon\, \mathbf{H}(\boldsymbol{\chi}, \varepsilon). \tag{2,14}$$

Analogously, (2,13) verifies (2,2) if and only if

$$[\mathbf{S}'(\mathbf{x}_0)]\, \boldsymbol{\chi} + \mathbf{R}(\mathbf{x}_0, 0) + \varepsilon\, \mathbf{Q}(\boldsymbol{\chi}, \varepsilon) = \mathbf{0}, \tag{2,15}$$

where

$$\mathbf{Q}(\chi, \varepsilon) = \left(\int_0^1 \left(\int_0^\vartheta \mathbf{S}''(\mathbf{x}_0 + \varepsilon\vartheta_1\vartheta\chi)\,\mathrm{d}\vartheta_1\right)\vartheta\,\mathrm{d}\vartheta\chi\right)\chi$$

$$+ \left(\int_0^1 \mathbf{R}'_x(\mathbf{x}_0 + \varepsilon\vartheta\chi, \vartheta\varepsilon)\,\mathrm{d}\vartheta\right)\chi + \int_0^1 \mathbf{R}'_\varepsilon(\mathbf{x}_0 + \varepsilon\vartheta\chi, \vartheta\varepsilon)\,\mathrm{d}\vartheta\,.$$

It follows that the given BVP $(\mathscr{P}_\varepsilon)$ possesses a solution of the form (2,13) for any $\varepsilon > 0$ sufficiently small if and only if the weakly nonlinear problem (2,14), (2,15) possesses a solution for any $\varepsilon > 0$ sufficiently small. In particular, a necessary condition for the existence of a solution of the form (2,13) to BVP $(\mathscr{P}_\varepsilon)$ is that the linear nonhomogeneous problem

$$\chi' = [\mathbf{F}'(\mathbf{x}_0)]\,\chi + \mathbf{G}(\mathbf{x}_0, 0)\,, \qquad [\mathbf{S}'(\mathbf{x}_0)]\,\chi = \mathbf{R}(\mathbf{x}_0, 0)$$

has a solution. Applying the procedure from I.7.10 to BVP (2,14), (2,15) we should obtain furthermore that BVP $(\mathscr{P}_\varepsilon)$ may possess a solution of the form (2,13) for any $\varepsilon > 0$ sufficiently small only if there exists a solution γ_0 of a certain (determining) equation $\mathbf{T}_0(\gamma) = \mathbf{0}$ for a finite dimensional vector γ and if, moreover, $\mathbf{F}$ and $\mathbf{S} \in C^3$, $\mathbf{G}$ and $\mathbf{R} \in C^{2,1}$ and $\det\left((\partial\mathbf{T}_0/\partial\gamma)(\gamma_0)\right) \neq 0$, then such a solution exists (cf. I.7.11).

The critical case will be treated in more detail in the following paragraph concerning ordinary differential equations with arbitrary side conditions.

2.12. Remark. If $\mathbf{P}\colon [0,1] \times [0,1] \to L(R_n)$ is an $L^1[BV]$-kernel, $\mathbf{f} \in L^1_n$, $\mathbf{S} \in B(AC_n, R_m)$,

$$\mathbf{F}\colon \mathbf{x} \in AC_n \to \int_0^1 \mathrm{d}_s[\mathbf{P}(t,s)]\,\mathbf{x}(s) + \mathbf{f}(t)\,,$$

$\mathbf{G}\colon AC_n \times [0, \varkappa] \to L^1_n$, $\mathbf{R}\colon AC_n \times [0, \varkappa] \to R_m$, then the weakly nonlinear BVP (cf. V.2.4)

(2,16)
$$\mathscr{Q}\mathbf{x} = \begin{pmatrix} \mathbf{D}\mathbf{x} - \mathbf{P}\mathbf{x} \\ \mathbf{S}\mathbf{x} \end{pmatrix} - \begin{pmatrix} \mathbf{f} \\ \mathbf{r} \end{pmatrix} = \varepsilon \begin{pmatrix} \mathbf{G}(\mathbf{x}, \varepsilon) \\ \mathbf{R}(\mathbf{x}, \varepsilon) \end{pmatrix}$$

becomes a special case of BVP $(\mathscr{P}_\varepsilon)$ studied in this section. In particular, if $\mathbf{R}$ and $\mathbf{G}$ are sufficiently smooth and the limit problem $(\mathscr{P}_0)$ possesses a unique solution for any $\begin{pmatrix} \mathbf{f} \\ \mathbf{r} \end{pmatrix} \in L^p_n \times R_m$, then by 2.9 BVP (2,16) possesses a unique solution for $\varepsilon > 0$ sufficiently small.

Since according to V.1.8, V.2.5 and V.2.8 $\mathscr{Q}\colon AC_n \to L^1_n \times R_m$ verifies (I.7,5), the procedure from I.7.10 may be applied to BVP (2,16). Let us mention that in the special case when $\mathbf{P}$ is an $L^2[BV]$-kernel, $\mathbf{f} \in L^2_n$ and $R(\mathbf{G}) \subset L^2_n$ the transformation of BVP $(\mathscr{P}_0)$ to an algebraic equation exhibited in section V.4 may also be used (cf. Tvrdý, Vejvoda [1]).

3. Nonlinear boundary value problems for ordinary differential equations

In this section we shall treat special cases of the problems $(\mathscr{P}_\varepsilon)$ from the previous section, namely the problems of the form (Π_ε)

$$(3,1)\qquad \mathbf{x}' = \mathbf{f}(t, \mathbf{x}) + \varepsilon\, \mathbf{g}(t, \mathbf{x}, \varepsilon),$$

$$(3,2)\qquad \mathbf{S}(\mathbf{x}) + \varepsilon\, \mathbf{R}(\mathbf{x}, \varepsilon) = \mathbf{0}$$

and (Π_0)

$$(3,3)\qquad \mathbf{x}' = \mathbf{f}(t, \mathbf{x}),$$

$$(3,4)\qquad \mathbf{S}(\mathbf{x}) = \mathbf{0}.$$

Our aim is again to obtain conditions for the existence of a solution to the perturbed problem (Π_ε) under the assumption that the limit problem (Π_0) possesses a solution. In doing this only such solutions of BVP (Π_ε) are sought which tend to some solution of BVP (Π_0) as $\varepsilon \to 0+$.

The following assumptions are pertinent.

3.1. Assumptions.

(i) *$\mathscr{D} \subset R_{n+1}$ and $\mathfrak{D} \subset R_{n+2}$ are open, $\varkappa > 0$ and $[0, 1] \times R_n \subset \mathscr{D}$, $\mathscr{D} \times [0, \varkappa] \subset \mathfrak{D}$;*
(ii) *$\mathbf{f}$: $\mathscr{D} \to R_n$, $\mathbf{f} \in \mathrm{Car}(\mathscr{D})$, $\partial \mathbf{f}/\partial \mathbf{x}$ exists on $\mathscr{D}$ and $\partial \mathbf{f}/\partial \mathbf{x} \in \mathrm{Car}(\mathscr{D})$ (cf. 1.2);*
(iii) *$\mathbf{g}$: $\mathfrak{D} \to R_n$, $\mathbf{g} \in \mathrm{Lip}(\mathscr{D}; 0)$ (i.e. $\mathbf{g}$ is locally lipschitzian in $\mathbf{x}$ near $\varepsilon = 0$, cf. 1.25) and if $\tilde{\mathbf{g}}(t, \mathbf{y}) = \mathbf{g}(t, \mathbf{x}, \varepsilon)$ for $(t, \mathbf{x}, \varepsilon) \in \mathfrak{D}$ and $\mathbf{y} = (\mathbf{x}, \varepsilon)$, then $\tilde{\mathbf{g}} \in \mathrm{Car}(\mathfrak{D})$;*
(iv) *$\mathbf{S}$ is a continuous mapping of AC_n into R_n, $\mathbf{S} \in C^1(AC_n)$, $\mathbf{R}$ is a continuous mapping of $AC_n \times [0, \varkappa]$ into R_n which is locally lipschitzian in $\mathbf{x}$ near $\varepsilon = 0$ (cf. I.7.1).*

3.2. Remark. Under the assumptions 3.1 for any $(\mathbf{c}, \varepsilon) \in R_n \times [0, \varkappa]$ there exists a unique maximal solution $\mathbf{x}(t) = \boldsymbol{\psi}(t; 0, \mathbf{c}, \varepsilon)$ of (3,1) on $\Delta = \Delta(\mathbf{c}, \varepsilon)$ such that $0 \in \Delta$ and $\mathbf{x}(0) = \mathbf{c}$ (cf. 1.4, 1.7, 1.11 and 1.13). The set

$$\tilde{\Omega}_{(\cdot, 0, \cdot, \cdot)} = \Omega_0 = \{(t, 0, \mathbf{c}, \varepsilon);\ (\mathbf{c}, \varepsilon) \in R_n \times [0, \varkappa],\ t \in \Delta(\mathbf{c}, \varepsilon)\}$$

is open and the mapping

$$\boldsymbol{\xi}\colon (t, \mathbf{c}, \varepsilon) \in \Omega_0 \to \boldsymbol{\psi}(t; 0, \mathbf{c}, \varepsilon) \in R_n$$

is continuous.

3.3. Notation. In the sequel $\boldsymbol{\xi}(t; \mathbf{c}, \varepsilon) = \boldsymbol{\psi}(t; 0, \mathbf{c}, \varepsilon)$ for $(t, \mathbf{c}, \varepsilon) \in \Omega_0$. In particular, $\boldsymbol{\eta}(t; \mathbf{c}) = \boldsymbol{\psi}(t; 0, \mathbf{c}, 0) = \boldsymbol{\varphi}(t; 0, \mathbf{c})$ for $\mathbf{c} \in R_n$ and $t \in \Delta(\mathbf{c}, 0)$.

3.4. Remark. Given $\mathbf{x} \in AC_n$, the corresponding variational BVP $(\mathscr{V}_0(\mathbf{x}))$ to (Π_0) is given by the linear ordinary differential equation

$$(3,5) \qquad \mathbf{u}' - \left[\frac{\partial \mathbf{f}}{\partial \mathbf{x}}(t, \mathbf{x}(t))\right]\mathbf{u} = \mathbf{0}$$

and by the side condition

$$(3,6) \qquad [\mathbf{S}'(\mathbf{x})]\,\mathbf{u} = \mathbf{0}\,.$$

According to 1.14, given a solution $\mathbf{x}(t) = \boldsymbol{\eta}(t;\, \mathbf{c})$ to (3,3) on $[0, 1]$, the $n \times n$-matrix valued function

$$\mathbf{A}(t) = \left[\frac{\partial \mathbf{f}}{\partial \mathbf{x}}(t, \boldsymbol{\eta}(t;\, \mathbf{c}))\right]$$

is L-integrable on $[0, 1]$. Moreover, the $n \times n$-matrix valued function

$$\mathbf{U}(t) = \left[\frac{\partial \boldsymbol{\eta}}{\partial \mathbf{c}}(t;\, \mathbf{c})\right]$$

is the fundamental matrix solution to (3,5) on $[0, 1]$ such that $\mathbf{U}(0) = \mathbf{I}_n$.

3.5. Remark. Let us notice (cf. 1.20–1.27) that under our assumptions 3.1 the operators $\mathbf{F}\colon AC_n \to L_n^1$ and $\mathbf{G}\colon AC_n \times [0, \varkappa] \to L_n^1$ defined as in 1.21 fulfil all the corresponding assumptions of theorems 2.3 and 2.7 (with AC_n in place of C_n). Moreover, if $\mathbf{x}(t) = \boldsymbol{\eta}(t;\, \mathbf{c})$ and the variational BVP $(\mathscr{V}_0(\mathbf{x}))$ given now by (3,5), (3,6) has only the trivial solution, then according to V.3.12 the linear operator

$$\mathscr{F}'(\mathbf{x})\colon\ \mathbf{u} \in AC_n \to \begin{pmatrix} \mathbf{u}' - [(\partial \mathbf{f}/\partial \mathbf{x})(t, \mathbf{x}(t))]\,\mathbf{u} \\ [\mathbf{S}'(\mathbf{x})]\,\mathbf{u} \end{pmatrix} \in L_n^1 \times R_n$$

possesses a bounded inverse. Thus applying the same argument as in the proofs of Theorems 2.3 and 2.7 we can prove the following assertion.

3.6. Theorem. *Let* 3,1 *hold. Let* $\mathbf{x}_0$ *be a solution to BVP* (Π_0) *and let the corresponding variational BVP* $(\mathscr{V}_0(\mathbf{x}_0))$ *possess only the trivial solution. Then* $\mathbf{x}_0$ *is an isolated solution of* (Π_0) *and for* $\varepsilon > 0$ *sufficiently small BVP* (Π_ε) *has a solution* $\mathbf{x}(\varepsilon)$ *which is continuous in* ε *and tends to* $\mathbf{x}_0$ *as* $\varepsilon \to 0+$.

To obtain some results also for the critical case we shall strengthen our hypotheses.

3.7. Assumptions. *For any* $i, j = 1, 2, \ldots, n$ $\mathbf{f}$ *possesses on* $\mathscr{D}$ *the partial derivatives* $\partial^2 \mathbf{f}/(\partial x_i\, \partial x_j)$ *with respect to the components* x_j *of* $\mathbf{x}$ *and* $\partial^2 \mathbf{f}/(\partial x_i\, \partial x_j) \in \mathrm{Car}(\mathscr{D})$ $(i, j = 1, 2, \ldots, n)$. *Furthermore,* $\partial \mathbf{g}/\partial \mathbf{x}$ *exists on* $\mathfrak{D}$ *and if* $\mathbf{h}(t, \mathbf{y}) = (\partial \mathbf{g}/\partial \mathbf{x})(t, \mathbf{x}, \varepsilon)$ *for* $(t, \mathbf{x}, \varepsilon) \in \mathfrak{D}$ *and* $\mathbf{y} = (\mathbf{x}, \varepsilon)$, *then* $\mathbf{h} \in \mathrm{Car}(\mathfrak{D})$.

$\mathbf{S} \in C^2(AC_n)$ *and* $\mathbf{R} \in C^{1,0}(AC_n \times [0, \varkappa])$ *(i.e. given* $(\mathbf{x}, \varepsilon) \in AC_n \times [0, \varkappa]$, $\mathbf{R}'_x(\mathbf{x}, \varepsilon)$ *exists and the mapping* $(\mathbf{x}, \varepsilon) \in AC_n \times [0, \varkappa] \to \mathbf{R}'_x(\mathbf{x}, \varepsilon) \in B(AC_n, R_n)$ *is continuous).*

The following lemma provides the principal tool for proving theorems on the existence of solutions to BVP (Π_ε) in the critical case. It establishes the *variation-of-constants method for nonlinear equations.*

3.8. Lemma. *Let* 3.1 *and* 3.7 *hold. Let the equation* (3,3) *possess a solution* $\mathbf{x}_0(t) = \boldsymbol{\eta}(t;\mathbf{c}_0)$ *on* $[0,1]$. *Then there exist* $\varrho_0 > 0$ *and* $\varkappa_0 > 0$ *such that for any* $(\mathbf{c},\varepsilon) \in \mathfrak{B}(\mathbf{c}_0, \varrho_0; R_n) \times [0,\varkappa_0]$ *the equation* (3,1) *possesses a unique solution* $\mathbf{x}(t)$ *on* $[0,1]$ *such that* $\mathbf{x}(0) = \mathbf{c}$. *This solution is given by*

$$\mathbf{x}(t) = \boldsymbol{\xi}(t;\mathbf{c},\varepsilon) = \boldsymbol{\eta}(t;\boldsymbol{\beta}(t;\mathbf{c},\varepsilon)) \quad \text{on } [0,1], \tag{3,7}$$

where for any $(\mathbf{c},\varepsilon) \in \mathfrak{B}(\mathbf{c}_0,\varrho_0;R_n) \times [0,\varkappa_0]$ $\mathbf{b}(t) = \boldsymbol{\beta}(t,\mathbf{c},\varepsilon)$ *is a unique solution to*

$$\mathbf{b}' = \varepsilon \left[\frac{\partial \boldsymbol{\eta}}{\partial \mathbf{c}}(t;\mathbf{b})\right]^{-1} \mathbf{g}(t,\boldsymbol{\eta}(t;\mathbf{b}),\varepsilon) \tag{3,8}$$

on $[0,1]$ *such that* $\mathbf{b}(0) = \mathbf{c}$. *The mapping* $(t,\mathbf{c},\varepsilon) \in \mathfrak{B} = [0,1] \times \mathfrak{B}(\mathbf{c}_0,\varrho_0;R_n) \times [0,\varkappa_0] \to \boldsymbol{\beta}(t;\mathbf{c},\varepsilon) \in R_n$ *is continuous and possesses the Jacobi matrix* $(\partial\boldsymbol{\beta}/\partial\mathbf{c})(t;\mathbf{c},\varepsilon)$ *continuous in* $(t,\mathbf{c},\varepsilon)$ *on* $\mathfrak{B}$.

Proof. (a) According to 1.12 there exist an open subset $\Omega \in R_{n+1}$ and $\delta > 0$ such that $\boldsymbol{\eta}(t;\mathbf{c})$ is defined for any $(t,\mathbf{c}) \in \Omega$ and $[0,1] \times \mathfrak{B}(\mathbf{c}_0,\delta;R_n) \subset \Omega$. Furthermore, in virtue of 1.16 the Jacobi matrix $\mathbf{U}(t,\mathbf{c}) = (\partial\boldsymbol{\eta}/\partial\mathbf{c})(t;\mathbf{c})$ and its partial derivatives $\partial\mathbf{U}(t,\mathbf{c})/\partial c_j$ $(j = 1,2,\dots,n)$ with respect to the components c_j of $\mathbf{c}$ exist and are continuous on Ω. Since by 1.15 $\mathbf{U}^{-1}(t,\mathbf{c})$ exists on Ω and for any $j = 1,2,\dots,n$ and $(t,\mathbf{c}) \in \Omega$

$$\mathbf{0} = \frac{\partial}{\partial c_j}(\mathbf{U}(t,\mathbf{c})\,\mathbf{U}^{-1}(t,\mathbf{c})) = \left(\frac{\partial}{\partial c_j}\mathbf{U}(t,\mathbf{c})\right)\mathbf{U}^{-1}(t,\mathbf{c}) + \mathbf{U}(t,\mathbf{c})\left(\frac{\partial}{\partial c_j}\mathbf{U}^{-1}(t,\mathbf{c})\right),$$

$\mathbf{U}^{-1}(t,\mathbf{c})$ possesses on Ω all the partial derivatives

$$\frac{\partial}{\partial c_j}\mathbf{U}^{-1}(t,\mathbf{c}) = -\mathbf{U}^{-1}(t,\mathbf{c})\left(\frac{\partial}{\partial c_j}\mathbf{U}(t,\mathbf{c})\right)\mathbf{U}^{-1}(t,\mathbf{c}) \qquad (j = 1,2,\dots,n).$$

It is easy to see now that the right-hand side

$$\mathbf{h}(t,\mathbf{b},\varepsilon) = \varepsilon\,\mathbf{U}^{-1}(t,\mathbf{b})\,\mathbf{g}(t,\boldsymbol{\eta}(t;\mathbf{b}),\varepsilon) \tag{3,9}$$

of (3,8) possesses the Jacobi matrix $(\partial\mathbf{h}/\partial\mathbf{b})(t,\mathbf{b},\varepsilon)$ on some open subset $\tilde{\Omega}$ of R_{n+2} such that $\Omega \times [0,\varkappa] \subset \tilde{\Omega}$ and if we put $\boldsymbol{\chi}(t,\boldsymbol{\mu}) = (\partial\mathbf{h}/\partial\mathbf{b})(t,\mathbf{b},\varepsilon)$ for $\boldsymbol{\mu} = (\mathbf{b},\varepsilon)$ and $(t,\boldsymbol{\mu}) \in \tilde{\Omega}$, then $\boldsymbol{\chi} \in \mathrm{Car}(\tilde{\Omega})$. By 1.14 (cf. also 1.17) this implies that for any $(\mathbf{c},\varepsilon) \in R_n \times [0,\varkappa]$ sufficiently close to $(\mathbf{c}_0, 0)$ the equation (3,8) possesses a unique solution $\mathbf{b}(t) = \boldsymbol{\beta}(t;\mathbf{c},\varepsilon)$ on $[0,1]$ such that $\mathbf{b}(0) = \mathbf{c}$. Moreover, since for $\varepsilon = 0$, $\mathbf{b}(t) \equiv \mathbf{c}_0$ is a solution to (3,8) on $[0,1]$, there exist $\varrho_0 > 0$ and $\varkappa_0 > 0$ such that $\boldsymbol{\beta}(t;\mathbf{c},\varepsilon)$ is defined and possesses the required properties on $\mathfrak{B} = [0,1] \times \mathfrak{B}(\mathbf{c}_0,\varrho_0;R_n) \times [0,\varkappa_0]$ and in addition $|\boldsymbol{\beta}(t,\mathbf{c},\varepsilon)| \le \delta$ for any $(t,\mathbf{c},\varepsilon) \in \mathfrak{B}$.

(b) Let $(\mathbf{c}, \varepsilon) \in \mathfrak{B}(\mathbf{c}_0, \varrho_0; R_n) \times [0, \varkappa]$. By the first part of the proof (3,7) is defined on $[0, 1]$ and according to the definitions of $\boldsymbol{\eta}(t, \mathbf{c})$ and $\boldsymbol{\beta}(t; \mathbf{c}, \varepsilon)$

$$\mathbf{x}'(t) = \frac{\partial \boldsymbol{\eta}}{\partial t}(t; \boldsymbol{\beta}(t; \mathbf{c}, \varepsilon)) + \frac{\partial \boldsymbol{\eta}}{\partial \mathbf{c}}(t; \boldsymbol{\beta}(t; \mathbf{c}, \varepsilon)) \frac{\partial \boldsymbol{\beta}}{\partial t}(t; \mathbf{c}, \varepsilon)$$

$$= \mathbf{f}(t, \boldsymbol{\eta}(t; \boldsymbol{\beta}(t; \mathbf{c}, \varepsilon))) + \varepsilon\, \mathbf{g}(t; \boldsymbol{\beta}(t; \mathbf{c}, \varepsilon)), \varepsilon) \qquad \text{for a.e.} \quad t \in [0, 1],$$

while $\mathbf{x}(0) = \boldsymbol{\eta}(0; \boldsymbol{\beta}(0; \mathbf{c}, \varepsilon)) = \boldsymbol{\eta}(0; \mathbf{c}) = \mathbf{c}$. Since (3,1) possesses obviously the property $(\mathscr{U})$, it means that

$$\mathbf{x}(t) = \boldsymbol{\eta}(t; \boldsymbol{\beta}(t; \mathbf{c}, \varepsilon)) = \boldsymbol{\xi}(t; \mathbf{c}, \varepsilon) \qquad \text{on } [0, 1].$$

3.9. Notation. $\mathscr{N}$ denotes the naturally ordered set $\{1, 2, \ldots, n\}$. If $\mathscr{I}$ is a naturally ordered subset of $\mathscr{N}$, then $\mathscr{N} \backslash \mathscr{I}$ denotes the naturally ordered complement of $\mathscr{I}$ with respect to $\mathscr{N}$. The number of elements of a set $\mathscr{I} \subset \mathscr{N}$ is denoted by $\nu(\mathscr{I})$. Let $\mathbf{C} = (c_{i,j})_{i,j=1,2,\ldots,n} \in L(R_n)$ and let $\mathscr{I} = \{i_1, i_2, \ldots, i_p\}$ and $\mathscr{J} = \{j_1, j_2, \ldots, j_q\}$ be naturally ordered subsets of $\mathscr{N}$, then $\mathbf{C}_{\mathscr{I},\mathscr{J}}$ denotes the $p \times q$-matrix $(d_{k,l})_{k=1,2,\ldots,p;\, l=1,2,\ldots,q}$, where $d_{k,l} = c_{i_k, j_l}$ for $k = 1, 2, \ldots, p$ and $l = 1, 2, \ldots, q$. In particular, if $\mathbf{b} \in R_n$ $(\mathbf{b} = (b_1, b_2, \ldots, b_n)^*)$, then $\mathbf{b}_{\mathscr{I}}$ denotes the p-vector $(d_1, d_2, \ldots, d_p)^*$, where $d_k = b_{i_k}$ for $k = 1, 2, \ldots, p$. (Analogously for matrix or vector valued functions and operators.)

3.10. Remark. Let $\mathbf{x}_0(t) = \boldsymbol{\eta}(t; \mathbf{c}_0)$ be a solution to the limit problem (Π_0) and let the corresponding variational BVP $(\mathscr{V}_0(\mathbf{x}_0))$ possess exactly k linearly independent solutions on $[0, 1]$ $(\dim N(\mathscr{F}'(\mathbf{x}_0)) = k)$. This means that $\operatorname{rank}(\Delta(\mathbf{c}_0)) = n - k$, where

$$\Delta(\mathbf{c}_0) = [\mathbf{S}'(\mathbf{x}_0)] \frac{\partial \boldsymbol{\eta}}{\partial \mathbf{c}}(.; \mathbf{c}_0)$$

denotes the $n \times n$-matrix formed by the columns $[\mathbf{S}'(\mathbf{x}_0)]\, \mathbf{u}_j$ $(\mathbf{u}_j(t) = (\partial \boldsymbol{\eta}/\partial \mathbf{c}_j)(t; \mathbf{c}_0)$ on $[0, 1]$; $j = 1, 2, \ldots, n)$. Hence there exist naturally ordered subsets $\mathscr{I}$, $\mathscr{J}$ of $\mathscr{N} = \{1, 2, \ldots, n\}$ with k elements such that

$$\det (\Delta(\mathbf{c}_0))_{\mathscr{N} \backslash \mathscr{I}, \mathscr{N} \backslash \mathscr{J}} \neq 0.$$

Let us denote $(\mathbf{c}_0)_{\mathscr{J}} = \boldsymbol{\gamma}_0$ and $(\mathbf{c}_0)_{\mathscr{N} \backslash \mathscr{J}} = \boldsymbol{\delta}_0$. Since for any $\mathbf{c} \in R_n$ sufficiently close to $\mathbf{c}_0$ the value of the Jacobi matrix of the function $\mathbf{d} \in R_n \to \mathbf{S}(\boldsymbol{\eta}(.; \mathbf{d})) \in R_n$ is given by $[\mathbf{S}'(\boldsymbol{\eta}(.; \mathbf{c}))]\, (\partial \boldsymbol{\eta}/\partial \mathbf{c})\, (.; \mathbf{c})$, the Implicit Function Theorem yields that there exist $\sigma > 0$ and a function $\mathbf{p}_0 \colon \mathfrak{B}(\boldsymbol{\gamma}_0, \sigma; R_k) = \Gamma \to R_{n-k}$ such that $\mathbf{p}_0(\boldsymbol{\gamma}_0) = \boldsymbol{\delta}_0$, $(\partial \mathbf{p}_0/\partial \boldsymbol{\gamma})(\boldsymbol{\gamma})$ exists and is continuous on Γ $(\mathbf{p}_0 \in C^1(\Gamma))$ and if the function $\mathbf{q}_0 \colon \Gamma \to R_n$ is defined by $(\mathbf{q}_0(\boldsymbol{\gamma}))_{\mathscr{J}} = \boldsymbol{\gamma}$ and $(\mathbf{q}_0(\boldsymbol{\gamma}))_{\mathscr{N} \backslash \mathscr{J}} = \mathbf{p}_0(\boldsymbol{\gamma})$, then

$$\mathbf{S}_{\mathscr{N} \backslash \mathscr{I}}(\boldsymbol{\eta}(.; \mathbf{q}_0(\boldsymbol{\gamma}))) = \mathbf{0} \qquad \text{for any} \quad \boldsymbol{\gamma} \in \Gamma.$$

If also $\mathbf{S}_{\mathscr{I}}(\boldsymbol{\eta}(.; \mathbf{q}_0(\boldsymbol{\gamma}))) = \mathbf{0}$ for any $\boldsymbol{\gamma} \in \Gamma$, then $\mathbf{x}(t) = \boldsymbol{\eta}(t; \mathbf{q}_0(\boldsymbol{\gamma}))$ is a solution to (Π_0) for any $\boldsymbol{\gamma} \in \Gamma$.

3.11. Theorem. *Let the assumptions* 3.1 *and* 3.7 *hold. In addition, let us assume*

(i) *there exist an integer k, $0 < k < n$, a naturally ordered subset $\mathcal{J}$ of $\mathcal{N}$ with k elements ($\nu(\mathcal{J}) = k$), an open set $\Gamma \subset R_k$ and a function $\mathbf{p}_0: \Gamma \to R_{n-k}$ such that $(\partial \mathbf{p}_0/\partial \gamma)(\gamma)$ exists and is continuous on Γ and if $\mathbf{q}_0: \Gamma \to R_n$ is defined by $(\mathbf{q}_0(\gamma))_{\mathcal{J}} = \gamma$ and $(\mathbf{q}_0(\gamma))_{\mathcal{N}\setminus\mathcal{J}} = \mathbf{p}_0(\gamma)$, then the function $t \in [0, 1] \to \boldsymbol{\eta}(t; \mathbf{q}_0(\gamma)) \in R_n$ is a solution to BVP (Π_0) for any $\gamma \in \Gamma$;*

(ii) $\operatorname{rank}([\mathbf{S}'(\boldsymbol{\eta}(.\,; \mathbf{q}_0(\gamma)))]\,(\partial \boldsymbol{\eta}/\partial \mathbf{c})\,(.\,; \mathbf{q}_0(\gamma)) = n - k$ *for any $\gamma \in \Gamma$.*

Let $\mathcal{J}$ be a naturally ordered subset of $\mathcal{N}$ with k elements such that

$$\operatorname{rank}\left([\mathbf{S}'(\boldsymbol{\eta}(.\,; \mathbf{q}_0(\gamma)))]\,\frac{\partial \boldsymbol{\eta}}{\partial \mathbf{c}}(.\,; \mathbf{q}_0(\gamma))\right)_{\mathcal{N}\setminus\mathcal{J},\mathcal{N}} = n - k \tag{3,10}$$

and let $\boldsymbol{\Theta}: \Gamma \to L(R_{n-k}, R_k)$ be a matrix valued function such that

$$\left([\mathbf{S}'(\boldsymbol{\eta}(.\,; \mathbf{q}_0(\gamma)))]\,\frac{\partial \boldsymbol{\eta}}{\partial \mathbf{c}}(.\,; \mathbf{q}_0(\gamma))\right)_{\mathcal{J},\mathcal{N}} = \boldsymbol{\Theta}(\gamma)\left(\mathbf{S}'(\boldsymbol{\eta}(.\,; \mathbf{q}_0(\gamma)))\,\frac{\partial \boldsymbol{\eta}}{\partial \mathbf{c}}(.\,; \mathbf{q}_0(\gamma))\right)_{\mathcal{N}\setminus\mathcal{J},\mathcal{N}} \tag{3,11}$$

for any $\gamma \in \Gamma$.

Then the mapping

$$\begin{aligned}\mathbf{T}_0: \gamma \in \Gamma \subset R_k \to {} & \left([\mathbf{S}'(\boldsymbol{\eta}(.\,; \mathbf{q}_0(\gamma)))]\,\frac{\partial \boldsymbol{\eta}}{\partial \mathbf{c}}(.\,; \mathbf{q}_0(\gamma))\,\boldsymbol{\zeta}_\gamma + \mathbf{R}(\boldsymbol{\eta}(.\,; \mathbf{q}_0(\gamma)), 0)\right)_{\mathcal{J}} \\ & - \boldsymbol{\Theta}(\gamma)\left([\mathbf{S}'(\boldsymbol{\eta}(.\,; \mathbf{q}_0(\gamma)))]\,\frac{\partial \boldsymbol{\eta}}{\partial \mathbf{c}}(.\,; \mathbf{q}_0(\gamma))\,\boldsymbol{\zeta}_\gamma + \mathbf{R}(\boldsymbol{\eta}(.\,; \mathbf{q}_0(\gamma)), 0)\right)_{\mathcal{N}\setminus\mathcal{J}} \in R_k\end{aligned} \tag{3,12}$$

where

$$\boldsymbol{\zeta}_\gamma(t) = \int_0^t \left[\frac{\partial \boldsymbol{\eta}}{\partial \mathbf{c}}(\tau; \mathbf{q}_0(\gamma))\right]^{-1} \mathbf{g}(\tau, \boldsymbol{\eta}(\tau; \mathbf{q}_0(\gamma)), 0)\,\mathrm{d}\tau \qquad \text{on } [0, 1],$$

possesses the Jacobi matrix $(\partial \mathbf{T}_0/\partial \gamma)(\gamma)$ on Γ.

If, moreover, the equation

$$\mathbf{T}_0(\gamma) = \mathbf{0} \tag{3,13}$$

possesses a solution $\gamma_0 \in \Gamma$ such that

$$\det\left(\frac{\partial \mathbf{T}_0}{\partial \gamma}(\gamma_0)\right) \neq 0, \tag{3,14}$$

then there exists for any $\varepsilon > 0$ sufficiently small a unique solution $\mathbf{x}_\varepsilon(t) = \boldsymbol{\xi}(t; \mathbf{c}(\varepsilon), \varepsilon)$ of BVP (Π_ε) which is continuous in ε and tends to $\boldsymbol{\eta}(t; \mathbf{q}_0(\gamma_0))$ uniformly on $[0, 1]$ as $\varepsilon \to 0+$.

Proof. (a) Let us put

$$\Delta_0(\gamma) = [\mathbf{S}'(\boldsymbol{\eta}(.\,;\mathbf{q}_0(\gamma)))]\frac{\partial \boldsymbol{\eta}}{\partial \mathbf{c}}(.\,;\mathbf{q}_0(\gamma)) \qquad \text{for} \quad \gamma \in \Gamma\,.$$

We shall show that

(3,15) $$\det\left((\Delta_0(\gamma))_{\mathcal{N}\setminus\mathcal{I},\mathcal{N}\setminus\mathcal{I}}\right) \neq 0 \qquad \text{for any} \quad \gamma \in \Gamma\,.$$

In fact, if there were $\det(\Delta_0(\gamma_1))_{\mathcal{N}\setminus\mathcal{I},\mathcal{N}\setminus\mathcal{I}} = 0$, then $h \in \mathcal{I}$ and $\boldsymbol{\mu} \in R_{n-k-1}$ should exist such that

(3,16) $$(\Delta_0(\gamma_1))_{h,\mathcal{N}\setminus\mathcal{I}} = \boldsymbol{\mu}^*(\Delta_0(\gamma_1))_{\mathcal{H},\mathcal{N}\setminus\mathcal{I}}$$

where $\mathcal{H} = (\mathcal{N}\setminus\mathcal{I})\setminus\{h\}$. On the other hand, according to our assumptions and the definition of $\boldsymbol{\eta}(t,\mathbf{c})$

(3,17) $$\mathbf{S}(\boldsymbol{\eta}(.\,;\mathbf{q}_0(\gamma)) = \mathbf{0} \qquad \text{for any} \quad \gamma \in \Gamma\,.$$

Differentiating the identity (3,17) with respect to γ, we obtain

$$\Delta_0(\gamma)\frac{\partial \mathbf{q}_0}{\partial \gamma}(\gamma) = (\Delta_0(\gamma))_{\mathcal{N},\mathcal{N}\setminus\mathcal{I}}\frac{\partial \mathbf{p}_0}{\partial \gamma}(\gamma) + (\Delta_0(\gamma))_{\mathcal{N},\mathcal{I}} = \mathbf{0}$$

for any $\gamma \in \Gamma$. By (3,16)

$$(\Delta_0(\gamma_1))_{h,\mathcal{I}} = -(\Delta_0(\gamma_1))_{h,\mathcal{N}\setminus\mathcal{I}}\frac{\partial \mathbf{p}_0}{\partial \gamma}(\gamma_1)$$

$$= -\boldsymbol{\mu}^*(\Delta_0(\gamma_1))_{\mathcal{H},\mathcal{N}\setminus\mathcal{I}}\frac{\partial \mathbf{p}_0}{\partial \gamma}(\gamma_1) = \boldsymbol{\mu}^*(\Delta_0(\gamma_1))_{\mathcal{H},\mathcal{I}}$$

i.e.

$$(\Delta_0(\gamma_1)_{h,\mathcal{N}} = \boldsymbol{\mu}^*(\Delta_0(\gamma_1))_{\mathcal{H},\mathcal{N}}$$

and $\operatorname{rank}(\Delta_0(\gamma_1))_{\mathcal{N}\setminus\mathcal{I},\mathcal{N}} \leq n - k - 1$. This being a contradiction to (3,10), (3,15) has to hold.

(b) Since (3,10) is assumed, for any $\gamma \in \Gamma$ there exist a $k \times (n-k)$-matrix $\boldsymbol{\Theta}(\gamma)$ such that (3,11) holds on Γ, i.e.

$$(\Delta_0(\gamma))_{\mathcal{I},\mathcal{N}} = \boldsymbol{\Theta}(\gamma)\,(\Delta_0(\gamma))_{\mathcal{N}\setminus\mathcal{I},\mathcal{N}} \qquad \text{on } \Gamma\,.$$

In particular,

$$(\Delta_0(\gamma))_{\mathcal{I},\mathcal{N}\setminus\mathcal{I}} = \boldsymbol{\Theta}(\gamma)\,(\Delta_0(\gamma))_{\mathcal{N}\setminus\mathcal{I},\mathcal{N}\setminus\mathcal{I}}$$

and

(3,18) $$\boldsymbol{\Theta}(\gamma) = (\Delta_0(\gamma))_{\mathcal{I},\mathcal{N}\setminus\mathcal{I}}(\Delta_0(\gamma))^{-1}_{\mathcal{N}\setminus\mathcal{I},\mathcal{N}\setminus\mathcal{I}} \qquad \text{on } \Gamma\,.$$

It is easy to verify that under our assumptions all the partial derivatives $(\partial\Delta_0/\partial\gamma_j)(\gamma)$ $(j = 1, 2, \ldots, k)$ exist and are continuous on Γ. Clearly, for any $j = 1, 2, \ldots, n$

$$\frac{\partial}{\partial \gamma_j}(\Delta_0(\gamma))^{-1} = -(\Delta_0(\gamma))^{-1}\left(\frac{\partial}{\partial \gamma_j}(\Delta_0(\gamma))\right)(\Delta_0(\gamma))^{-1} \qquad \text{on } \Gamma$$

and in virtue of (3,18) also the $k \times (n-k)$-matrix function $\boldsymbol{\Theta}(\gamma)$ possesses all the partial derivatives $(\partial \boldsymbol{\Theta}/\partial \gamma_j)(\gamma)$ $(j = 1, 2, \ldots, n)$ on Γ and they are continuous on Γ. This implies that the function $\mathbf{T}_0\colon \Gamma \subset R_k \to R_k$ defined by (3,12) possesses the Jacobi matrix $(\partial \mathbf{T}_0/\partial \gamma)(\gamma)$ on Γ and it is continuous on Γ.

(c) According to the definition of $\xi(t, \mathbf{c}, \varepsilon)$ an n-vector valued function $\mathbf{x}(t)$ is a solution to BVP (Π_ε) if and only if $\mathbf{x}(t) = \xi(t; \mathbf{c}, \varepsilon)$ on $[0, 1]$ and $\mathbf{c} \in R_n$ fulfils the equation

$$\text{(3,19)} \qquad \mathbf{W}(\mathbf{c}, \varepsilon) \equiv \mathbf{S}(\xi(.\,; \mathbf{c}, \varepsilon)) + \varepsilon\, \mathbf{R}(\xi(.\,; \mathbf{c}, \varepsilon), \varepsilon) = \mathbf{0}\,.$$

The mappings $\mathbf{W}\colon R_n \times [0, \varkappa] \to R_n$ and $\partial \mathbf{W}/\partial \mathbf{c}\colon R_n \times [0, \varkappa] \to L(R_n)$ are clearly continuous.

Let $\gamma_0 \in \Gamma$ be such that $\mathbf{T}_0(\gamma_0) = \mathbf{0}$. Then $\mathbf{W}(\mathbf{q}_0(\gamma_0), 0) = \mathbf{0}$. Furthermore, since

$$\frac{\partial \mathbf{W}}{\partial \mathbf{c}}(\mathbf{q}_0(\gamma), 0) = \Delta_0(\gamma) \qquad \text{on } \Gamma\,,$$

(3,15) means

$$\det\left(\frac{\partial \mathbf{W}}{\partial \mathbf{c}}(\mathbf{q}_0(\gamma), 0)\right)_{\mathcal{N}\setminus\mathcal{I},\,\mathcal{N}\setminus\mathcal{I}} \neq 0 \qquad \text{on } \Gamma\,.$$

It follows that there are $\varrho_1 > 0$ and $\varkappa_1 > 0$ such that

$$\det\left(\frac{\partial \mathbf{W}}{\partial \mathbf{c}}(\mathbf{c}, \varepsilon)\right)_{\mathcal{N}\setminus\mathcal{I},\,\mathcal{N}\setminus\mathcal{I}} \neq 0$$

for all $(\mathbf{c}, \varepsilon) \in \mathfrak{B}_1 = \mathfrak{B}(\mathbf{c}_0, \varrho_1; R_n) \times [0, \varkappa_1]$. By the Implicit Function Theorem there exist $\varrho_2 > 0$, $\varkappa_2 > 0$, $\varkappa_2 \le \varkappa_1$, and a unique function $\mathbf{p}\colon \mathfrak{B}_2 = \mathfrak{B}(\gamma_0, \varrho_2; R_k) \times [0, \varkappa_2] \to R_{n-k}$, $\mathbf{p} \in C^{1,0}(\mathfrak{B}_2)$ such that if $(\mathbf{q}(\gamma, \varepsilon))_{\mathcal{I}} = \gamma$ and $(\mathbf{q}(\gamma, \varepsilon))_{\mathcal{N}\setminus\mathcal{I}} = \mathbf{p}(\gamma, \varepsilon)$, then $\mathbf{q}(\gamma, \varepsilon) \in \mathfrak{B}(\mathbf{c}_0, \varrho_1; R_n)$ and

$$\text{(3,20)} \qquad \mathbf{W}_{\mathcal{N}\setminus\mathcal{I}}(\mathbf{q}(\gamma, \varepsilon), \varepsilon) = \mathbf{0} \qquad \text{for any} \quad (\gamma, \varepsilon) \in \mathfrak{B}_2$$

and $\mathbf{q}(\gamma, 0) = \mathbf{q}_0(\gamma)$ on $\mathfrak{B}(\gamma_0, \varrho_2; R_k)$.

(d) By 3.8 for any $t \in [0, 1]$ and $(\mathbf{c}, \varepsilon)$ sufficiently close to $(\mathbf{c}_0, 0)$ the function $\xi(t; \mathbf{c}, \varepsilon) = \boldsymbol{\eta}(t; \boldsymbol{\beta}(t; \mathbf{c}, \varepsilon))$, where $\mathbf{b}(t) = \boldsymbol{\beta}(t; \mathbf{c}, \varepsilon)$, is the solution of (3,8) on $[0, 1]$ such that $\mathbf{b}(0) = \mathbf{c}$. We may assume that this is true for $(\mathbf{c}, \varepsilon) \in \mathfrak{B}_1$. Let us put

$$\zeta(t; \mathbf{c}, \varepsilon) = \int_0^t \left[\frac{\partial \boldsymbol{\eta}}{\partial \mathbf{c}}(\tau; \mathbf{c}, \varepsilon)\right]^{-1} \mathbf{g}(\tau, \boldsymbol{\eta}(\tau; \boldsymbol{\beta}(\tau; \mathbf{c}, \varepsilon)), \varepsilon)\, \mathrm{d}\tau$$

for $(t; \mathbf{c}, \varepsilon) \in [0, 1] \times \mathfrak{B}_1$. Then

$$\boldsymbol{\beta}(t; \mathbf{c}, \varepsilon) = \mathbf{c} + \varepsilon\, \zeta(t; \mathbf{c}, \varepsilon) \qquad \text{on } [0, 1] \times \mathfrak{B}_1\,.$$

and (cf. (3,12))

$$\lim_{\varepsilon \to 0+} \zeta(t; \mathbf{q}(\gamma, \varepsilon), \varepsilon) = \zeta(t; \mathbf{q}_0(\gamma), 0) = \zeta_\gamma(t) \qquad \text{for any} \quad t \in [0, 1] \quad \text{and} \quad \gamma \in \Gamma\,.$$

By (3,17) and I.7.4 we have for any $(\gamma, \varepsilon) \in \mathfrak{B}_2$, $\varepsilon > 0$,

$$\text{(3,21)} \qquad \frac{1}{\varepsilon} \boldsymbol{W}(\boldsymbol{q}(\gamma, \varepsilon), \varepsilon)$$

$$= \frac{1}{\varepsilon} \left[\boldsymbol{S}(\boldsymbol{\eta}(\,.\,; \boldsymbol{q}(\gamma, \varepsilon) + \varepsilon\zeta)) - \boldsymbol{S}(\boldsymbol{\eta}(\,.\,; \boldsymbol{q}(\gamma, \varepsilon))) \right]$$

$$+ \frac{1}{\varepsilon} \left[\boldsymbol{S}(\boldsymbol{\eta}(\,.\,; \boldsymbol{q}(\gamma, \varepsilon))) - \boldsymbol{S}(\boldsymbol{\eta}(\,.\,; \boldsymbol{q}_0(\gamma))) \right] + \boldsymbol{R}(\boldsymbol{\eta}(\,.\,; \boldsymbol{q}(\gamma, \varepsilon) + \varepsilon\zeta), \varepsilon)$$

$$= \int_0^1 \left[\boldsymbol{S}'(\boldsymbol{\eta}(\,.\,; \boldsymbol{q}(\gamma, \varepsilon) + \varepsilon\vartheta\zeta)) \right] \frac{\partial \boldsymbol{\eta}}{\partial \boldsymbol{c}}(\,.\,; \boldsymbol{q}(\gamma, \varepsilon) + \varepsilon\vartheta\zeta) \, \mathrm{d}\vartheta \zeta$$

$$+ \left(\int_0^1 \left[\boldsymbol{S}'(\boldsymbol{\eta}(\,.\,; \boldsymbol{q}_0(\gamma) + \vartheta[\boldsymbol{q}(\gamma, \varepsilon) - \boldsymbol{q}_0(\gamma)])) \right] \frac{\partial \boldsymbol{\eta}}{\partial \boldsymbol{c}}(\,.\,; \boldsymbol{q}_0(\gamma) + \vartheta[\boldsymbol{q}(\gamma, \varepsilon) - \boldsymbol{q}_0(\gamma)]) \, \mathrm{d}\vartheta \right)$$

$$\cdot \frac{\boldsymbol{q}(\gamma, \varepsilon) - \boldsymbol{q}_0(\gamma)}{\varepsilon} + \boldsymbol{R}(\boldsymbol{\eta}(\,.\,; \boldsymbol{q}(\gamma, \varepsilon) + \varepsilon\zeta), \varepsilon)$$

$$= \left[\boldsymbol{S}'(\boldsymbol{\eta}(\,.\,; \boldsymbol{q}(\gamma, \varepsilon))) \right] \frac{\partial \boldsymbol{\eta}}{\partial \boldsymbol{c}}(\,.\,; \boldsymbol{q}(\gamma, \varepsilon)) \, \zeta + \boldsymbol{R}(\boldsymbol{\eta}(\,.\,; \boldsymbol{q}(\gamma, \varepsilon)), \varepsilon)$$

$$+ \left(\int_0^1 \left\{ \left[\boldsymbol{S}'(\boldsymbol{\eta}(\,.\,; \boldsymbol{q}(\gamma, \varepsilon) + \varepsilon\vartheta\zeta)) \right] \frac{\partial \boldsymbol{\eta}}{\partial \boldsymbol{c}}(\,.\,; \boldsymbol{q}(\gamma, \varepsilon) + \varepsilon\vartheta\zeta) \right.\right.$$

$$\left.\left. - \left[\boldsymbol{S}'(\boldsymbol{\eta}(\,.\,; \boldsymbol{q}(\gamma, \varepsilon))) \right] \frac{\partial \boldsymbol{\eta}}{\partial \boldsymbol{c}}(\,.\,; \boldsymbol{q}(\gamma, \varepsilon)) \right\} \mathrm{d}\vartheta \right) \zeta + (\Delta(\gamma, \varepsilon))_{\mathcal{N}, \mathcal{N} \setminus \mathcal{I}} \frac{\boldsymbol{p}(\gamma, \varepsilon) - \boldsymbol{p}_0(\gamma)}{\varepsilon}$$

$$+ \boldsymbol{R}(\boldsymbol{\eta}(\,.\,; \boldsymbol{q}(\gamma, \varepsilon) + \varepsilon\zeta, \varepsilon) - \boldsymbol{R}(\boldsymbol{\eta}(\,.\,; \boldsymbol{q}(\gamma, \varepsilon)), \varepsilon),$$

where

$$\Delta(\gamma, \varepsilon) = \int_0^1 \left[\boldsymbol{S}'(\boldsymbol{\eta}(\,.\,; \boldsymbol{q}_0(\gamma) + \vartheta[\boldsymbol{q}(\gamma, \varepsilon) - \boldsymbol{q}_0(\gamma)])) \right] \frac{\partial \boldsymbol{\eta}}{\partial \boldsymbol{c}}(\,.\,; \boldsymbol{q}_0(\gamma) + \vartheta[\boldsymbol{q}(\gamma, \varepsilon) - \boldsymbol{q}_0(\gamma)]) \, \mathrm{d}\vartheta.$$

Since for any $\gamma \in \mathfrak{B}(\gamma_0, \varrho_2; R_k)$

$$\lim_{\varepsilon \to 0+} \Delta(\gamma, \varepsilon) = \left[\boldsymbol{S}'(\boldsymbol{\eta}(\,.\,; \boldsymbol{q}_0(\gamma))) \right] \frac{\partial \boldsymbol{\eta}}{\partial \boldsymbol{c}}(\,.\,; \boldsymbol{q}_0(\gamma)) = \Delta_0(\gamma),$$

(3,10) implies that also

$$\text{(3,22)} \qquad \det (\Delta(\gamma, \varepsilon))_{\mathcal{N} \setminus \mathcal{I}, \mathcal{N} \setminus \mathcal{I}} \neq 0$$

for all $\varepsilon > 0$ sufficiently small. Without any loss of generality we may assume that (3,22) holds for all $(\gamma, \varepsilon) \in \mathfrak{B}_2$.

By (3,20)–(3,22) we have for any $\gamma \in \mathfrak{B}(\gamma_0, \varrho_2; R_k)$

$$\text{(3,23)} \qquad \lim_{\varepsilon \to 0+} \frac{\boldsymbol{p}(\gamma, \varepsilon) - \boldsymbol{p}_0(\gamma)}{\varepsilon}$$

$$= (\Delta_0(\gamma))^{-1}_{\mathcal{N} \setminus \mathcal{I}, \mathcal{N} \setminus \mathcal{I}} [(\Delta_0(\gamma))_{\mathcal{N} \setminus \mathcal{I}, \mathcal{N}} \zeta_\gamma + R_{\mathcal{N} \setminus \mathcal{I}}(\boldsymbol{\eta}(\,.\,; \boldsymbol{q}_0(\gamma)), 0)].$$

Differentiating (3,21) with respect to γ and making use of (3,22) we may analogously prove that also

$$(3,24)\qquad \lim_{\varepsilon\to 0+} \frac{\dfrac{\partial \mathbf{p}}{\partial \gamma}(\gamma,\varepsilon) - \dfrac{\partial \mathbf{p}_0}{\partial \gamma}(\gamma)}{\varepsilon}$$

exists.

According to (3,20) for $\varepsilon > 0$ the equation (3,19) is near $\mathbf{c} = \mathbf{q}_0(\gamma_0)$ equivalent to

$$(3,25)\qquad \mathbf{T}(\gamma,\varepsilon) = \frac{1}{\varepsilon}\big[\mathbf{W}_{\mathscr{I}}(\mathbf{q}(\gamma,\varepsilon),\varepsilon) - \boldsymbol{\Theta}(\gamma)\,\mathbf{W}_{\mathscr{N}\setminus\mathscr{I}}(\mathbf{q}(\gamma,\varepsilon),\varepsilon)\big] = \mathbf{0}\,.$$

Moreover, if for any $\varepsilon > 0$ sufficiently small $\gamma_\varepsilon \in \Gamma$ is the solution to (3,25) which tend to γ_0 as $\varepsilon \to 0+$, then

$$\mathbf{x}_\varepsilon(t) = \xi(t;\, \mathbf{q}(\gamma_\varepsilon,\varepsilon),\varepsilon)$$

are solutions of BVP (Π_ε) such that

$$\lim_{\varepsilon\to 0+} \|\mathbf{x}_\varepsilon - \boldsymbol{\eta}(\,.\,;\, \mathbf{q}_0(\gamma_0))\|_C = 0\,.$$

Let $\mathbf{r}(\gamma,\varepsilon)$ denote the n-vector

$$\mathbf{r}(\gamma,\varepsilon) = (\Delta(\gamma,\varepsilon))_{\mathscr{N},\mathscr{N}\setminus\mathscr{I}}\, \frac{\mathbf{p}(\gamma,\varepsilon) - \mathbf{p}_0(\gamma)}{\varepsilon}\,.$$

In virtue of (3,11) and (3,23) for any $\gamma \in \mathfrak{B}(\gamma_0, \varrho_2;\, R_k)$

$$\lim_{\varepsilon\to 0+} [\mathbf{r}_{\mathscr{I}}(\gamma,\varepsilon) - \boldsymbol{\Theta}(\gamma)\,\mathbf{r}_{\mathscr{N}\setminus\mathscr{I}}(\gamma,\varepsilon)] = \mathbf{0}\,.$$

Furthermore, (3,11) implies

$$\left(\frac{\partial \Delta_0}{\partial \gamma}(\gamma)\right)_{\mathscr{I},\mathscr{N}} - \boldsymbol{\Theta}(\gamma)\left(\frac{\partial \Delta_0}{\partial \gamma}(\gamma)\right)_{\mathscr{N}\setminus\mathscr{I},\mathscr{N}} - \frac{\partial \boldsymbol{\Theta}}{\partial \gamma}(\gamma)\,(\Delta_0(\gamma))_{\mathscr{N}\setminus\mathscr{I},\mathscr{N}} \equiv \mathbf{0} \qquad \text{on } I$$

and hence

$$\lim_{\varepsilon\to 0+} \frac{\partial}{\partial \gamma}[\mathbf{r}_{\mathscr{I}}(\gamma,\varepsilon) - \boldsymbol{\Theta}(\gamma)\,\mathbf{r}_{\mathscr{N}\setminus\mathscr{I}}(\gamma,\varepsilon)] = \lim_{\varepsilon\to 0+}\Bigg[\left(\frac{\partial \Delta}{\partial \gamma}(\gamma,\varepsilon)\right)_{\mathscr{I},\mathscr{N}\setminus\mathscr{I}}$$

$$- \boldsymbol{\Theta}(\gamma)\left(\frac{\partial \Delta}{\partial \gamma}(\gamma,\varepsilon)\right)_{\mathscr{N}\setminus\mathscr{I},\mathscr{N}\setminus\mathscr{I}} - \frac{\partial \boldsymbol{\Theta}}{\partial \gamma}(\gamma)\,(\Delta(\gamma,\varepsilon))_{\mathscr{N}\setminus\mathscr{I},\mathscr{N}\setminus\mathscr{I}}\Bigg]\frac{\mathbf{p}(\gamma,\varepsilon) - \mathbf{p}_0(\gamma)}{\varepsilon}$$

$$+ [(\Delta(\gamma,\varepsilon))_{\mathscr{I},\mathscr{N}\setminus\mathscr{I}} - \boldsymbol{\Theta}(\gamma)\,(\Delta(\gamma,\varepsilon))_{\mathscr{N}\setminus\mathscr{I},\mathscr{N}\setminus\mathscr{I}}]\,\frac{\dfrac{\partial \mathbf{p}}{\partial \gamma}(\gamma,\varepsilon) - \dfrac{\partial \mathbf{p}_0}{\partial \gamma}(\gamma)}{\varepsilon} = \mathbf{0}$$

Thus if we put for $\gamma \in \mathfrak{B}(\gamma_0, \varrho_2;\, R_k)$ $\mathbf{T}(\gamma, 0) = \mathbf{T}_0(\gamma)$, then $\mathbf{T}\colon \mathfrak{B}_2 \to R_k$ becomes a continuous operator which possesses the Jacobi matrix $(\partial\mathbf{T}/\partial\gamma)(\gamma,\varepsilon)$ for any

$(\gamma, \varepsilon) \in \mathfrak{B}_2$, while the mapping

$$(\gamma, \varepsilon) \in \mathfrak{B}_2 \to \frac{\partial \boldsymbol{T}}{\partial \gamma}(\gamma, \varepsilon) \in L(R_k)$$

is continuous.

Applying the Implicit Function Theorem to (3,25) we complete the proof of the theorem.

Now, let $\gamma_0 \in \Gamma$ and let us assume that given $\varepsilon > 0$ sufficiently small (e.g. $\varepsilon \in (0, \varkappa_0]$), there exists a solution $\mathbf{x}_\varepsilon(t) = \xi(t; \mathbf{c}_\varepsilon, \varepsilon)$ of BVP (Π_ε) such that $\mathbf{x}_\varepsilon(t)$ tends uniformly on $[0, 1]$ to the solution $\mathbf{x}_0(t) = \boldsymbol{\eta}(t; \mathbf{q}_0(\gamma_0))$ of the limit problem (Π_0) as $\varepsilon \to 0+$. Then, in particular, $\mathbf{c}_\varepsilon = \mathbf{x}_\varepsilon(0)$ tends to $\mathbf{q}_0(\gamma_0)$ and $\gamma_\varepsilon = (\mathbf{c}_\varepsilon)_{\mathscr{I}}$ tends to γ_0 as $\varepsilon \to 0+$. Hence $|\gamma_\varepsilon - \gamma_0| < \varrho_2$ for any $\varepsilon > 0$ sufficiently small and analogously as in the proof of Theorem 3.11 we may show that

$$\lim_{\varepsilon \to 0+} \frac{1}{\varepsilon} \left[\mathbf{W}_{\mathscr{I}}(\mathbf{q}(\gamma_\varepsilon, \varepsilon)) - \boldsymbol{\Theta}(\gamma_\varepsilon) \mathbf{W}_{\mathscr{N} \setminus \mathscr{I}}(\mathbf{q}(\gamma_\varepsilon, \varepsilon)) \right] = \mathbf{T}_0(\gamma_0).$$

Since by the assumption $\mathbf{W}(\mathbf{q}(\gamma_\varepsilon, \varepsilon)) = \mathbf{0}$ for all $\varepsilon \in (0, \varkappa_0]$, this completes the proof of the following theorem.

3.12. Theorem. *Let in addition to* 3.1 *and* 3.7 (i) *and* (ii) *from* 3.11 *hold. Then there exists* $\varepsilon_0 > 0$ *such that given* $\varepsilon \in (0, \varepsilon_0]$, *BVP* (Π_ε) *possesses a solution* $\mathbf{x}_\varepsilon(t)$ *tending uniformly on* $[0, 1]$ *to some solution* $\mathbf{x}_0(t) = \boldsymbol{\eta}(t; \mathbf{q}_0(\gamma))$ *of BVP* (Π_0) *as* $\varepsilon \to 0+$ *only if the equation* (3,13) *has a solution* $\gamma_0 \in \Gamma$.

The next theorem supplements the theorems 3.11 and 3.12.

3.13. Theorem. *Let* 3.1 *and* 3.7 *hold and let* $\Gamma \subset R_n$ *be such an open subset that* $\mathbf{x}_\gamma(t) = \boldsymbol{\eta}(t; \gamma)$ *is a solution to BVP* (Π_0) *for any* $\gamma \in \Gamma$.

Let $\gamma_0 \in \Gamma$. *Then a necessary condition for the existence of an* $\varepsilon_0 > 0$ *such that for a given* $\varepsilon \in (0, \varepsilon_0]$ *there exists a solution* $\mathbf{x}_\varepsilon(t)$ *of BVP* (Π_ε) *and* $\mathbf{x}_\varepsilon(t)$ *tends uniformly on* $[0, 1]$ *to* $\mathbf{x}_{\gamma_0}(t)$ *is that* γ_0 *is a solution to*

$$\mathbf{T}_0(\gamma) = [\mathbf{S}'(\boldsymbol{\eta}(.\,; \gamma))] \frac{\partial \boldsymbol{\eta}}{\partial \mathbf{c}}(.\,; \gamma)\, \zeta_\gamma = \mathbf{0}, \tag{3,26}$$

where

$$\zeta_\gamma(t) = \int_0^t \left[\frac{\partial \boldsymbol{\eta}}{\partial \mathbf{c}}(\tau; \gamma) \right]^{-1} \mathbf{g}(\tau, \boldsymbol{\eta}(\tau; \gamma), 0)\, d\tau.$$

If, moreover, $\det((\partial \mathbf{T}_0 / \partial \gamma)(\gamma_0)) \neq 0$, *then such an* $\varepsilon_0 > 0$ *exists.*

Proof follows readily by an appropriate modification of the proofs of 3.11 and 3.12.

3.14. Remark. Let us notice that the condition (3,10) of 3.11 holds if and only if any variational problem $(\mathscr{V}_0(\boldsymbol{\eta}(.\,; \mathbf{q}_0(\gamma))))$ possesses exactly k linearly independent solutions (cf. IV.2.7). In the next lemma we shall show that the determining equation (3,13) may also be expressed by means of the variational problem.

3.15. Lemma. *Let in addition to* 3.1 *and* 3.7 (i) *and* (ii) *from* 3.11 *hold. Given* $\gamma \in \Gamma$, $\mathbf{T}_0(\gamma) = \mathbf{0}$ *if and only if the nonhomogeneous variational BVP*

$$\mathbf{u}' - \left[\frac{\partial \mathbf{f}}{\partial \mathbf{x}}(t, \boldsymbol{\eta}(t; \mathbf{q}_0(\gamma)))\right]\mathbf{u} = \mathbf{g}(t, \boldsymbol{\eta}(t; \mathbf{q}_0(\gamma)), 0), \tag{3,27}$$

$$[\mathbf{S}'(\boldsymbol{\eta}(.\,; \mathbf{q}_0(\gamma)))]\,\mathbf{u} = -\mathbf{R}(\boldsymbol{\eta}(.\,; \mathbf{q}_0(\gamma)), 0) \tag{3,28}$$

possesses a solution.

Proof. Let Ξ be an $n \times n$-matrix such that for a given $\mathbf{r} \in R_n$

$$\Xi \mathbf{r} = \begin{pmatrix} \mathbf{r}_{\mathcal{N}\setminus\mathcal{I}} \\ \mathbf{r}_{\mathcal{I}} \end{pmatrix}.$$

Then the assumption (3,10) means that there exists a $k \times (n-k)$-matrix valued function $\Theta(\gamma)$ defined on Γ and such that

$$\Lambda(\gamma)\,[\mathbf{S}'(\boldsymbol{\eta}(.\,; \mathbf{q}_0(\gamma)))]\,\frac{\partial \boldsymbol{\eta}}{\partial \mathbf{c}}(.\,; \mathbf{q}_0(\gamma)) = \mathbf{0} \qquad \text{for any} \quad \gamma \in \Gamma, \tag{3,29}$$

where

$$\Lambda(\gamma) = -[-\Theta(\gamma), \mathbf{I}_k]\,\Xi. \tag{3,30}$$

Analogously as in IV.2.2, we may show that to a given $\gamma \in \Gamma$ there exists an $n \times n$-matrix valued function $\mathbf{F}(t, \gamma)$ defined on $[0, 1] \times \Gamma$ and such that

$$\mathbf{T}_0(\gamma) = \Lambda(\gamma)\left(\int_0^1 \mathbf{F}(t, \gamma)\,\mathbf{g}(t, \boldsymbol{\eta}(t; \mathbf{q}_0(\gamma)), 0)\,\mathrm{d}t + \mathbf{R}(\boldsymbol{\eta}(.\,; \mathbf{q}_0(\gamma)), 0)\right) \qquad \text{for any} \quad \gamma \in \Gamma$$

and the couple $(\boldsymbol{\delta}^* \Lambda(\gamma)\,\mathbf{F}(t, \gamma), \boldsymbol{\delta}^* \Lambda(\gamma))$ verifies for any $\boldsymbol{\delta} \in R_n$ and $\gamma \in \Gamma$ the adjoint BVP to BVP (3,27), (3,28). Obviously rank $\Lambda(\gamma) = k$ for any $\gamma \in \Gamma$. Thus, given $\gamma \in \Gamma$, the rows of $\Lambda(\gamma)\,\mathbf{F}(t, \gamma)$, $\Lambda(\gamma)$ form a basis in the space of all solutions of the adjoint BVP to BVP (3,27), (3,28) (cf. V.2.9). Hence by V.2.6 and V.2.12 our assertion follows.

3.16. Remark. Let us assume that BVP (Π_ε) has the *property* $(\mathcal{T})$ (*translation*): $\boldsymbol{\xi}(t; \mathbf{c}, \varepsilon)$ being a solution to BVP (Π_ε), $\boldsymbol{\xi}(t + \delta; \mathbf{c}, \varepsilon)$ is also a solution to BVP (Π_ε) for any $\delta \in R$ such that $\boldsymbol{\xi}(t + \delta; \mathbf{c}, \varepsilon)$ is defined on $[0, 1]$.

Then, if BVP (Π_ε) has a nonconstant solution $\boldsymbol{\xi}(t; \mathbf{c}, \varepsilon)$, it has at least a one-parametric family of solutions $\boldsymbol{\xi}(t; \boldsymbol{\xi}(\delta; \mathbf{c}, \varepsilon), \varepsilon)$ for all $\delta \in R$ such that $|\delta|$ is sufficiently small. Consequently, Theorem 3.11 cannot be used for proving the existence of a solution $\mathbf{x}_\varepsilon(t)$ of BVP (Π) which tends to some solution $\mathbf{x}_0(t)$ of the shortened

BVP (Π_0) as $\varepsilon \to 0+$. This is clear from the fact that this theorem ensures the existence of an isolated solution. In some cases one component of the initial vector $\mathbf{c} = \mathbf{c}(\varepsilon)$ of the sought solution $\xi(t; \mathbf{c}, \varepsilon)$ may be chosen arbitrary (in a certain range) and another parameter has to be taken as a new unknown instead. Theorems on the existence of solutions to such problems can be then formulated and proved analogously as Theorem 3.11 (cf. Vejvoda [2]–[4]).

The most important problems with the property $(\mathcal{T})$ are those of determining a periodic solution to the autonomous differential equation $\mathbf{x}' = \mathbf{f}(\mathbf{x}) + \varepsilon\,\mathbf{g}(\mathbf{x}, \varepsilon)$. Solving such problems, the period $T = T(\varepsilon)$ of the sought solution is usually chosen as a new unknown. In general, two principal cases have to be distinguished. Either the limit BVP (Π_0) associated to the given BVP (Π_ε) has a k-parametric family of T-periodic solutions $\boldsymbol{\eta}(t; \mathbf{c}(\gamma))$, $\gamma \in \Gamma$, with T independent of γ or their periods depend on γ. The former case occurs e.g. if the equation $\mathbf{x}' = \mathbf{f}(\mathbf{x})$ may be rewritten as the equation $\mathbf{z}' = i\mathbf{z} + \mathbf{z}^2$ for a complex valued function $\mathbf{z}$. (All the solutions of this equation with the initial value sufficiently close to the origin are 2π-periodic, cf. Vejvoda [1], Lemma 5.1.) An example of the latter case is treated in the following section.

4. Froud-Žukovskij pendulum

Let us consider the second order autonomous differential equation of the *Froud-Žukovskij pendulum*

(4,1) $$x'' + \sin x = \varepsilon\, g(x, x'),$$

where g is a sufficiently smooth scalar function and $\varepsilon > 0$ is a small parameter. Given $\varepsilon > 0$, we are looking for a real number $T > 0$ and for a solution $x(t)$ to (4,1) on R such that

(4,2) $$x(T) = x(0) \quad \text{and} \quad x'(T) = x'(0).$$

The limit equation (for $\varepsilon = 0$)

(4,3) $$y'' + \sin y = 0$$

is known as being equation of the mathematical pendulum. All the solutions $y(t)$ to (4,3) with sufficiently small initial values $y(0)$, $y'(0)$ are defined on the whole real axis R and may be expressed in the form

$$y(t) = \eta(t + h;\, k),$$

where

(4,4) $$\eta(t; k) = 2 \arcsin (k \operatorname{sn}(t; k)), \qquad h \in R \quad \text{and} \quad k \in (0, 1).$$

(cf. Kamke [1], 6.17). Moreover, for any $h \in R$ and $k \in (0, 1)$ the function $y(t)$

$= \eta(t + h;\, k)$ fulfils the periodic boundary conditions (4,2) with $T = 4\,K(k)$, where

$$K(k) = \int_0^{\pi/2} \frac{\mathrm{d}\vartheta}{1 - k^2 \sin^2 \vartheta}.$$

In (4,4) $\mathrm{sn}\,(t;\, k)$ denotes the value of the Jacobi elliptic sine function with the modulus k at the point t. For the definition and basic properties of the Jacobi elliptic functions sn, cn, dn and of the elliptic integrals $K(k)$, $E(k)$ see e.g. Whittaker-Watson [1], Chapter 22. If no misunderstanding may arise, we write sn, cn, dn instead of $\mathrm{sn}\,(t;\, k)$, $\mathrm{cn}\,(t;\, k)$ and $\mathrm{dn}\,(t;\, k)$, respectively.

Solutions of the perturbed equation (4,1) will be sought in the form

$$x(t) = \xi(t;\, h, k, \varepsilon) = \eta(t + \alpha;\, \beta)\,, \tag{4,5}$$

where $\alpha = \alpha(t) = \alpha(t;\, h, k, \varepsilon)$ and $\beta = \beta(t) = \beta(t;\, h, k, \varepsilon)$ are properly chosen scalar functions such that $\alpha(0) = h$ and $\beta(0) = k$ (cf. 3.8). Differentiating (4,5) with respect to t, we obtain

$$x'(t) = \frac{\partial \eta}{\partial t}(t + \alpha(t);\, \beta(t))\,(1 + \alpha'(t)) + \frac{\partial \eta}{\partial k}(t + \alpha(t);\, \beta(t))\,\beta'(t)\,.$$

Hence, if

$$\frac{\partial \eta}{\partial t}(t + \alpha(t);\, \beta(t))\,\alpha'(t) + \frac{\partial \eta}{\partial k}(t + \alpha(t);\, \beta(t))\,\beta'(t) = 0\,, \tag{4,6}$$

$$\frac{\partial^2 \eta}{\partial t^2}(t + \alpha(t);\, \beta(t))\,\alpha'(t) + \frac{\partial^2 \eta}{\partial k\, \partial t}(t + \alpha(t);\, \beta(t))\,\alpha'(t) = \varepsilon\, g(\eta(t + \alpha(t);\, \beta(t)))\,,$$

then

$$x'(t) = \frac{\partial \eta}{\partial t}(t + \alpha(t);\, \beta(t))$$

and

$$x''(t) - \sin\,(x(t)) = \varepsilon\, g(\eta(t + \alpha(t);\, \beta(t)))\,.$$

Since

$$\frac{\partial\, \mathrm{sn}}{\partial k} = -k^2 \,.\, \mathrm{cn} \,.\, \mathrm{dn} \,.\, J \quad \text{and} \quad \frac{\partial\, \mathrm{cn}}{\partial k} = k^2 \,.\, \mathrm{sn} \,.\, \mathrm{dn} \,.\, J\,,$$

where

$$J = J(t, k) = \int_0^t \frac{\mathrm{sn}^2(\tau;\, k)}{\mathrm{dn}^2(\tau;\, k)}\, \mathrm{d}\tau\,,$$

we have

$$H(t, k) = \begin{pmatrix} 2k \,.\, \mathrm{cn}\,, & 2\dfrac{\mathrm{sn}}{\mathrm{dn}} - 2k^2 \,.\, \mathrm{cn} \,.\, J \\ -2k \,.\, \mathrm{cn} \,.\, \mathrm{dn}\,, & 2\, \mathrm{cn} + 2k^2 \,.\, \mathrm{sn} \,.\, \mathrm{dn} \,.\, J \end{pmatrix}$$

for

$$H(t,k) = \begin{pmatrix} \dfrac{\partial \eta}{\partial t}(t;k), & \dfrac{\partial \eta}{\partial k}(t;k) \\ \dfrac{\partial^2 \eta}{\partial t^2}(t;k), & \dfrac{\partial^2 \eta}{\partial k\,\partial t}(t;k) \end{pmatrix}.$$

Consequently

$$\det H(t+\alpha(t);\ \beta(t)) = 4\beta(t).$$

Provided $\beta(t) \neq 0$, the system (4,6) may be written as follows

$$\text{(4,7)} \qquad \alpha' = \varepsilon . \frac{1}{2}\left[\frac{\operatorname{sn}(t+\alpha;\ \beta)}{\operatorname{dn}(t+\alpha;\ \beta)} - \operatorname{cn}(t+\alpha;\ \beta)\right] g(\eta(t+\alpha;\ \beta)),$$

$$\beta' = \varepsilon . \tfrac{1}{2} \operatorname{cn}(t+\alpha;\ \beta)\, g(\eta(t+\alpha;\ \beta)).$$

Since for $\varepsilon = 0$ the couple $(\alpha(t), \beta(t)) \equiv (h, k)$ is the unique solution of the system (4,7) on R such that $\alpha(0) = h$, $\beta(0) = k$, Lemma 1.18 implies that for any $T > 0$ there exists $\varepsilon_T > 0$ such that for any $\varepsilon \in (0, \varepsilon_T)$ and $h \in R$, $k \in (0, 1)$ the system (4,7) possesses a unique solution $(\alpha(t), \beta(t)) = (\alpha(t; h, k, \varepsilon), \beta(t; h, k, \varepsilon))$ on $[0, T]$, continuous on $[0, T] \times R \times (0, 1) \times (0, \varepsilon_T)$ and such that $\alpha(0) = h$, $\beta(0) = k$, while $\beta(t) \in (0, 1)$ for any $t \in [0, T]$. Let us put $\alpha(t; h, k, 0) = h$ and $\beta(t; h, k, 0) = k$.

Given a solution $x(t)$ to BVP (4,1), (4,2) and $h \in R$, the function $z(t) = x(t + h)$ is also a solution to this problem. Hence without any loss of generality we may put

$$\text{(4,8)} \qquad h = 0.$$

Let $T > 0$ and $k \in (0, 1)$ be for a while fixed. Let $\alpha(t) = \alpha(t; 0, k, \varepsilon)$, $\beta(t) = \beta(t; 0, k, \varepsilon)$ be the corresponding solution of (4,7) on $[0, T]$ $(\varepsilon \in (0, \varepsilon_T))$. Then (4,5) becomes

$$\text{(4,9)} \quad x(t) = 2 \arcsin(\beta(t) \operatorname{sn}(t + \alpha(t);\ \beta(t))) \quad \text{for} \quad t \in [0, T] \quad \text{and} \quad \varepsilon \in (0, \varepsilon_T)$$

and $x(T) = x(0)$ if and only if $\beta(T) \operatorname{sn}(T + \alpha(T);\ \beta(T)) = 0$ or equivalently $(\beta(T) \neq 0)$

$$\text{(4,10)} \qquad T + \alpha(T; 0, k, \varepsilon) - 4K(\beta(T; 0, k, \varepsilon)) = 0.$$

According to (4,6) and (4,9)

$$x'(t) = 2\,\beta(t) \operatorname{cn}(t + \alpha(t);\ \beta(t))$$

and $x'(T) = x'(0)$ if and only if

$$\beta(T) \operatorname{cn}(T + \alpha(T);\ \beta(T)) = k \operatorname{cn}(0; k) = k$$

or in virtue of (4,10)

$$\text{(4,11)} \qquad \beta(T) = \beta(T) \operatorname{cn}(4K(\beta(T));\ \beta(T)) = k.$$

By (4,9)

$$\beta(t) = k + \varepsilon . \tfrac{1}{2}\varkappa(t, k, \varepsilon) \qquad \text{for} \quad t \in [0, T] \quad \text{and} \quad \varepsilon \in (0, \varepsilon_T),$$

where
$$\varkappa(t, k, \varepsilon) = \int_0^t \mathrm{cn}\,(\tau + \alpha(\tau);\, \beta(\tau))\, g(\eta(\tau + \alpha(\tau);\, \beta(\tau)))\, \mathrm{d}\tau\,.$$

This together with (4,11) implies that $x'(T) = x'(0)$ if and only if

(4,12) $$\varkappa(T, k, \varepsilon) = 0\,.$$

If $\varepsilon \to 0+$, then the equation (4,10) becomes $T - 4\,K(k) = 0$ and the system (4,10), (4,12) reduces to the equation

(4,13) $$B(k) = 0\,,$$

where
$$B(k) = \int_0^{4\,K(k)} \mathrm{cn}\,(t;\, k)\, g(\eta(t;\, k))\, \mathrm{d}t\,.$$

This means that a necessary condition for the existence of a solution to the given BVP (4,1), (4,2) for any $\varepsilon > 0$ sufficiently small is the existence of a solution $k \in (0, 1)$ of the equation (4,13).

Taking into account the properties of the Jacobi elliptic functions it can be shown that if e.g.
$$g(x, x') = x' - 3(x')^3\,,$$
then the equation (4,13) possesses a solution $k_0 \in (0, 1)$ such that $(\partial B/\partial k)\,(k_0) \neq 0$. By the Implicit Function Theorem there exists $\varepsilon_0 > 0$ such that for any $\varepsilon \in (0, \varepsilon_0]$ the system (4,10), (4,12) possesses a unique solution $T = T_\varepsilon > 0$ and $k = k_\varepsilon \in (0, 1)$ such that $T_\varepsilon \to 4\,K(k_0)$ and $k_\varepsilon \to k_0$ as $\varepsilon \to 0+$. Given $\varepsilon \in [0, \varepsilon_0]$, $\alpha(t) = \alpha(t;\, 0, k_\varepsilon, \varepsilon)$ and $\beta(t) = \beta(t;\, 0, k_\varepsilon, \varepsilon)$ verify the system (4,7) on $[0, T_\varepsilon]$ and hence $x_\varepsilon(t) = \eta(t + \alpha(t);\, \beta(t))$ is a unique T_ε-periodic solution of the equation
$$x'' + \sin x = \varepsilon(x' - 3(x')^3)$$
such that
$$x_\varepsilon(t) \to x_0(t) = \eta(t;\, k_0) \qquad \text{as} \quad \varepsilon \to 0+\,.$$

Notes

Chapter VI is a generalization of the work by Vejvoda ([4]). The main tools are the Implicit Function Theorem (Newton's method) and the nonlinear variation of constants formula VI.3.8 due to Vejvoda ([4]). Theorems VI.2.3, VI.2.7 and VI.2.9 are contained also in Urabe [2], [3].

The method of a small parameter (perturbation theory) originated from the celestial mechanics (Poincaré [1]). Periodic solutions of nonlinear differential equations were dealt with e.g. by Malkin ([1], [2]), Coddington, Levinson ([1]), Hale [1], Loud ([1], [2]) and others. Further related references concerning the application of the Newton method to perturbed nonlinear BVP are e.g. Antosiewicz [1], [2], Bernfeld, Lakshmikantham [1], Candless [1], Locker [1], Kwapisz [1], Tvrdý, Vejvoda [1], Vejvoda [2], [3] and Urabe [1].

Bibliography

ANTOSIEWICZ, H. A.

[1] Linear problems for nonlinear ordinary differential equations, Proc. US–Japan Sem. on Diff. and Func. Eq., University of Minnesota, Minneapolis 1967, W. A. Benjamin (1967), 1–12

[2] Newton's method and boundary value problems, Journal Computer and System Sciences, 2 (1968), 177–202

ATKINSON, F. V.

[1] Discrete and continuous boundary value problems, Academic Press, New York, London 1964

AUMANN, G.

[1] Reelle Funktionen, Springer Verlag, Berlin, Heidelberg, New York 1969

AZBELEV, N. V.

[1] Linear boundary value problems for functional-differential equations (in Russian), Differencial'nye Uravnenija 10 (1974), 579–584

BANKS, H. T.

[1] Representation for solutions of linear functional differential equations, J. Differential Equations 5 (1969), 399–409

BERNFELD, S. R. – LAKSHMIKANTHAM, V.

[1] An introduction to nonlinear boundary value problems, Academic Press, New York, London 1974

BITZER, C. W.

[1] Stieltjes-Volterra integral equations, Illinois J. Math., 14 (1970), 434–451

BRADLEY, J. S.

[1] Generalized Green's matrices for compatible differential systems, Michigan Math. J., 13 (1966), 97–108

BROWN, R. C.

[1] Generalized Green's functions and generalized inverses for linear differential systems with Stieltjes boundary conditions, J. Differential Equations, 16 (1974), 335–351

[2] Adjoint domains and generalized splines, Czechoslovak Math. J. 25 (100) (1975), 134–147

[3] The operator theory of generalized boundary value problems, Canad. J. Math. 28 (1976), 486–512

BROWN, R. C. – GREEN, G. B. – KRALL, A. M.

[1] Eigenfunction expansions under multipoint-integral boundary conditions, Ann. Mat. Pura Appl., 95 (1973), 231–243

BROWN, R. C. – KRALL, A. M.

[1] Ordinary differential operators under Stieltjes boundary conditions, Trans. Amer. Math. Soc., 198 (1974), 73–92

[2] On minimizing the sum of squares of L^2 norms of differential operators under constraints, Czechoslovak Math. J., 27 (102) (1977), 132–143

[3] n-th order differential systems under Stieltjes boundary conditions, Czechoslovak Math. J. 27 (102) (1977), 119–131

BRYAN, R. N.

[1] A nonhomogeneous linear differential system with interface conditions, Proc. Amer. Math. Soc. 22 (1969), 270–276

BURKILL, J. C.

[1] The Lebesgue integral, Cambridge University Press, 1951

CAMERON, R. H. – MARTIN, W. T.

[1] An unsymmetrical Fubini theorem, Bull. Amer. Math. Soc. 47 (1941), 121–126

CANDLESS, W. L.

[1] Newton's method and nonlinear boundary value problems, J. Math. Anal. Appl. 48 (1974), 434–445

CATCHPOLE, E. A.

[1] A Cauchy problem for an ordinary integro-differential equation, Proc. Roy. Soc. Edinburgh Sect. A, 72 (1972/73), 40–55

[2] An integro-differential operator, J. London Math. Soc., 6 (1973), 513–523

CHITWOOD, H.

[1] Generalized Green's matrices for linear differential systems, SIAM J. Math. Anal., 4 (1973), 104–110

CODDINGTON, E. A. – LEVINSON, N.

[1] Theory of ordinary differential equations, McGraw-Hill, New York, Toronto, London 1955

CODDINGTON, E. A. – DIJKSMA, A.

[1] Selfadjoint subspaces and eigenfunction expansions for ordinary differential subspaces, J. Differential Equations, 20 (1976), 473–526

COLE, R. H.

[1] Theory of ordinary differential equations, Appleton–Century–Crofts, New York 1968

[2] General boundary value problems for an ordinary linear differential system, Trans. Amer. Math. Soc., 111 (1964), 521–550

CONTI, R.

[1] Sistemi differenziali ordinari con condizioni lineari, Ann. Mat. Pura Appl., 46 (1958), 109–130

[2] Recent trends in the theory of boundary value problems for ordinary differential equations, Boll. Un. Mat. Ital., 22 (1967), 135–178

[3] On ordinary differential equations with interface conditions, J. Differential Equations, 4 (1968), 4–11

DELFOUR, M. C. – MITTER, S. K.

[1] Hereditary differential systems with constant delays I – General case, J. Differential Equations, 12 (1972), 213–235

[2] Hereditary differential systems with constant delays II – A class of affine systems and the adjoint problem, J. Differential Equations, 18 (1975), 18–28

DUHAMEL

[1] Sur les phénomènes thermo-mécaniques, Journ. de l'Ecole Polytechnique, 15 (1835), 1–57

DUNFORD, N. – SCHWARTZ, J. T.

[1] Linear operators, Part I, Interscience Publishers, New York, London, 1958

GOLDBERG, S.

[1] Unbounded linear operators; Theory and applications, McGraw-Hill, New York 1966

GONELLI, A.

[1] Un teorema di esistenza per un problema di tipo interface. Le Matematiche, 22 (1967), 203–211

GREEN, G. – KRALL, A. M.
[1] Linear differential systems with infinitely many boundary points, Ann. Mat. Pura. Appl., 91 (1972), 55–67
HALANAY, A. – MORO, A.
[1] A boundary value problem and its adjoint, Ann. Mat. Pura. Appl., 79 (1968), 399–412
HALANAY, A.
[1] Optimal control of periodic solutions, Rev. Roumaine Math. Pures Appl., 19 (1974), 3–16
HALE, J. K.
[1] Oscillations in nonlinear systems, McGraw-Hill, New York, 1963
[2] Functional differential equations, Springer-Verlag, New York, 1971
HELTON, B. W.
[1] Integral equations and product integrals, Pacific J. Math., 16 (1966), 297–322
HENSTOCK, R.
[1] Theory of integration, Butterworths, London 1963
HEROD, J. V.
[1] Multiplicative inverses of solutions for Volterra-Stieltjes integral equations, Proc. Amer. Math. Soc., 22 (1969), 650–656
HEUSER, H.
[1] Funktionalanalysis, B. G. Teubner, Stuttgart 1975
HILDEBRANDT, T. H.
[1] Introduction to the theory of integration, Academic Press, New York, London 1963
[2] On systems of linear differentio-Stieltjes integral equations, Illinois J. Math., 3 (1959), 352–373
HINTON, D. B.
[1] A Stieltjes-Volterra integral equation theory, Canad. J. Math., 18 (1966), 314–331
HNILICA, J.
[1] Verallgemeinerte Hill'sche Differentialgleichung, Časopis pěst. mat., 101 (1976), 293–302
HÖNIG, CH. S.
[1] Volterra-Stieltjes integral equations, Mathematics Studies 16, North-Holland, Amsterdam 1975
JONES, W. R.
[1] Differential systems with integral boundary condition, J. Differential Equations, 3 (1967), 191–202
KAMKE, E.
[1] Differentialgleichungen (Lösungsmethoden und Lösungen) I – Gewöhnliche Differentialgleichungen, Akademische Verlagsgesellschaft, Leipzig 1956
KANTOROVIČ, L. V. – AKILOV, G. P.
[1] Functional analysis in normed spaces, Macmillan, New York, 1964
KANTOROVIČ, L. V. – VULICH, B. Z. – PINSKER, A. G.
[1] Functional analysis in semiordered spaces (in Russian), GITTL, Leningrad 1950
KEMP, R. R. D. – LEE, S. J.
[1] Finite dimensional perturbations of differential expressions, Canad. J. Math. (to appear)
KRALL, A. M.
[1] Nonhomogeneous differential operators, Michigan Math. J. 12 (1965), 247–255
[2] Differential-boundary equations and associated boundary value problems, Proc. US–Japan Sem. on Diff. Func. Eq., Univ. of Minnesota, Minneapolis 1967, W. A. Benjamin (1967), 463–471
[3] Differential operators and their adjoints under integral and multiple point boundary conditions, J. Differential Equations, 4 (1968), 327–336
[4] Boundary value problems with interior point boundary conditions, Pacific J. Math., 29 (1969), 161–166
[5] Differential-boundary operators, Trans. Amer. Math. Soc., 154 (1971), 429–458
[6] Stieltjes differential-boundary operators, Proc. Amer. Math. Soc., 41 (1973), 80–86
[7] Stieltjes differential-boundary operators II, Pacific J. Math., 55 (1974), 207–218

[8] Stieltjes differential-boundary operators III – Multivalued operators – linear relations, Pacific J. Math., 59 (1975), 125–134

[9] The development of general differential and general differential-boundary systems, Rocky Mountain J. Math., 5 (1975), 493–542

KULTYŠEV, C. JU.

[1] The controllability of linear functional-differential systems (in Russian), Differencial'nye Uravnenija, 11 (1975), 1355–1360

KURZWEIL, J.

[1] Generalized ordinary differential equations and continuous dependence on a parameter, Czechoslovak Math. J., 7 (82), 418–449

[2] Generalized ordinary differential equations, Czechoslovak Math. J., 8 (83), (1958), 360–388

[3] On integration by parts, Czechoslovak Math. J., 8 (83) (1958), 356–359

[4] On generalized ordinary differential equations possessing discontinuous solutions (in Russian), Prikl. Mat. Meh., 22 (1958), 27–45

KWAPISZ, M.

[1] On quasilinear differential-functional equations with quasilinear conditions, Math. Nachr., 43 (1970), 215–222

LANDO, JU. K.

[1] The index and the normal solvability of integro-differential operators (in Russian), Differencial'nye Uravnenija, 4 (1968), 1112–1126

[2] The F-index and F-normal solvability of integro-differential operators (in Russian), Differencial'nye Uravnenija, 5 (1969), 546–555

[3] Controllable integro-differential operators (in Russian), Differencial'nye Uravnenija, 9 (1973), 2227–2230

[4] Controllable operators (in Russian), Differencial'nye Uravnenija, 10 (1974), 531–536

LICHTENSTEIN, L.

[1] Über eine Integro-Differential Gleichung und die Entwicklung willkürlicher Funktionen nach deren Eigenfunktionen, Schwarz-Festschrift, Berlin 1914, 274–285

LIGĘZA, J.

[1] On generalized solutions of boundary value problem for linear differential equations of order II, Prace naukowe Uniw. Śląskiego v Katowicach, n. 37 (1973), 57–66

[2] On distributional solution of some systems of linear differential equations, Časopis pěst. mat., 102 (1977), 37–41

LOCKER, J.

[1] An existence analysis for nonlinear boundary value problems, SIAM J. Appl. Math., 19 (1970), 199–207

LOUD, W. S.

[1] Periodic solutions of a perturbed autonomous system, Ann. of Math., 70 (1959), 490–529

[2] Periodic solutions of perturbed second-order autonomous equations, Mem. Amer. Math. Soc., 47 (1964)

[3] Generalized inverses and generalized Green's functions, SIAM J. Appl. Math., 14 (1966), 342–369

LOVELADY, D. L.

[1] Perturbations of solutions of Stieltjes integral equations, Trans. Amer. Math. Soc., 155 (1971), 175–187

LUENBERGER, D. G.

[1] Optimization by vector space methods, John Wiley, New York, 1969

MAC NERNEY, J. S.

[1] Stieltjes integrals in linear spaces, Ann. of Math., 61 (1955), 354–367

[2] Integral equations and semigroups, Illinois J. Math., 7 (1963), 148–173

MAKSIMOV, V. P. – RAHMATULLINA, L. F.

[1] The representation of solutions of a linear functional-differential equation (in Russian), Differencial'nye Uravnenija, 9 (1973), 1026–1036

[2] A linear functional-differential equation that is solved with respect to the derivative (in Russian), Differencial'nye Uravnenija, 9 (1973), 2231–2240

MAKSIMOV, V. P.

[1] The property of being Noetherian of the general boundary value problem for a linear functional differential equation (in Russian), Differencial'nye Uravnenija, 10 (1974), 2288–2291

MALKIN, I. G.

[1] The methods of Lyapunov and Poincaré in the theory of nonlinear oscillations (in Russian), GITTL, Moscow, 1949

[2] Some problems in the theory of nonlinear oscillations, GITTL, Moscow, 1956 (in Russian): English transl. U.S. Atomic Energy Commission Translation AEC-tr-3766, Books 1 and 2

MARCHIÒ, C.

[1] (M, N, F)-controlabilità completa, Questioni di Contrabilità, Istituto Ulisse Dini, 1973/2, 14–16 14–26

MARRAH, G. W. – PROCTOR, T. G.

[1] Solutions of some periodic Stieltjes integral equations, Proc. Amer. Math. Soc., 34 (1972), 121–127

MOORE, E. H.

[1] On the reciprocal of the general algebraic matrix, Bull. Amer. Math. Soc., 26 (1919–20), 394–395

NAJMARK (NEUMARK), M. A.

[1] Lineare Differentialoperatoren, Akademie-Verlag, Berlin 1960

NASHED, M. Z.

[1] Generalized inverses and iteration for singular operator equations, Nonlinear Functional Analysis, Academic Press 1971, 311–359

NATANSON, I. P.

[1] Theory of functions of a real variable, vols. 1. and 2., Frederick Ungar, New York 1955 and 1960

NERING, E. D.

[1] Linear algebra and matrix theory, John Wiley, New York 1963

PARHIMOVIČ, J. V.

[1] Multipoint boundary value problems for linear integro-differential equations in the class of smooth functions (in Russian), Differencial'nye Uravnenija, 8 (1972), 549–552

[2] The index and normal solvability of a multipoint boundary value problem for an integro-differential equation (in Russian), Vesci Akad. Navuk BSSR, Ser. Fiz.-Mat. Navuk, 1972, 91–93

[3] The construction of s-adjoints to integro-differential operators (in Russian), Differencial'nye Uravnenija, 8 (1972), 1486–1493

PAGNI, M.

[1] Equazioni differenziali lineari e problemi al contorno con condizioni integrali, Rend. Sem. Mat. Univ. Padova, 24 (1955), 245–264

PENROSE, R.

[1] A generalized inverse for matrices, Proc. Cambridge Phil. Soc., 51 (1955), 406–413

[2] On best approximate solutions of linear matrix equations, Proc. Cambridge Phil. Soc., 52 (1956), 17–19

POINCARÉ, H.

[1] Les méthodes nouvelles de la mécanique céleste I, Gauthier-Villars, Paris, 1892

RAHMATULLINA, L. F.

[1] On the theory of linear equations with functional argument (in Russian), Differencial'nye Uravnenija, 8 (1972), 523–528

RAHMATULLINA, L. F. – TERENT'EV, A. G.

[1] On the question of the regularization of linear boundary value problems (in Russian), Differencial'nye Uravnenija, 9 (1973), 868–873

REID, W. T.

[1] Ordinary differential equations, Wiley–Interscience, New York, London, Sydney, Toronto 1971

[2] Generalized Green's matrices for compatible systems of differential equations, Amer. J. Math., 53 (1931), 443–459

[3] Generalized Green's matrices for two-point boundary value problems, SIAM J. Appl. Math., 15 (1967), 856–870

SAKS, S.

[1] Theory of the integral, Monografie Matematyczne, Warszawa, Lwów 1937

SCHECHTER, M.

[1] Principles of functional analysis, Academic Press, New York, London 1973

SCHWABIK, Š.

[1] Verallgemeinerte lineare Differentialgleichungssysteme, Časopis pěst. mat., 96 (1971), 183–211

[2] On an integral operator in the space of functions of bounded variation, Časopis pěst. mat., 97 (1972), 297–330

[3] On the relation between Young's and Kurzweil's concept of Stieltjes integral, Časopis pěst. mat., 98 (1973), 237–251

[4] Floquet theory for generalized linear differential equations (in Czech), Časopis pěst. mat., 98 (1973), 416–418

[5] On Volterra-Stieltjes integral equations, Časopis pěst. mat., 99 (1974), 255–278

[6] On an integral operator in the space of functions of bounded variation II, Časopis pěst. mat., 102 (1977), 189–202

[7] Note on Volterra-Stieltjes integral equations, Časopis pěst. mat., 102 (1977), 275–279

SCHWABIK, Š – TVRDÝ, M.

[1] On the generalized linear ordinary differential equation, Časopis pěst. mat. 98 (1973), 206–211

SMITHIES, F.

[1] Integral equations, Cambridge University Press 1965

STALLARD, F. W.

[1] Differential systems with interface conditions, Oak Ridge National Laboratory Publication No. 1876 (Physics)

[2] Functions of bounded variation as solutions of differential systems, Proc. Amer. Math. Soc., 13 (1962), 366–373

TAMARKIN, J. D.

[1] The notion of the Green's function in the theory of integro-differential equations, Trans. Amer. Math. Soc., 29 (1927), 755–800

TAUFER, J.

[1] Lösung der Randwertprobleme für Systeme von linearen Differentialgleichungen, Rozpravy ČSAV, Řada mat. a přír. věd, 83 (1973), 5, Academia Praha

THOMAS, J.

[1] Untersuchungen über das Eigenwertproblem $\mathrm{d}/\mathrm{d}x\,[f(x)\,\mathrm{d}y/\mathrm{d}x] + g(x)\,y = 0$, $\int_a^b A(x)\,y\,\mathrm{d}x = \int_a^b B(x)\,y\,\mathrm{d}x = 0$, Math. Nachr., 6 (1951), 229–260

TUCKER, D. H.

[1] Boundary value problems for linear differential systems, SIAM J. Appl. Math., 4 (1969), 769–783

TVRDÝ, M.

[1] Boundary value problems for generalized linear differential equations and their adjoints, Czechoslovak Math. J., 23 (98) (1973), 183–217

[2] Boundary value problems for generalized linear integro-differential equations with left-continuous solutions, Časopis pěst. mat., 99 (1974), 147–157

[3] Linear boundary value type problems for functional-differential equations and their adjoints, Czechoslovak Math. J., 25 (100) (1975), 37–66
[4] Linear functional-differential operators; normal solvability and adjoints, Colloquia Mathematica Soc. János Bolyai, 15. Differential Equations, Keszthely (Hungary), 1975, 379–389
[5] Fredholm-Stieltjes integral equations with linear constraints: duality theory, Czechoslovak Math. J., (to appear)
[6] Green's couple of linear boundary value problem for linear Fredholm-Stiletjes integro-differential equation,

TVRDÝ, M. – VEJVODA, O.
[1] General boundary value problem for an integro-differential system and its adjoint, Časopis pěst. mat., 97 (1972), 399–419 and 98 (1973), 26–42

URABE, M.
[1] An existence theorem for multipoint boundary value problems, Funkcial. Ekvac., 9 (1966), 43–60
[2] The Newton method and its applications to boundary value problems with nonlinear boundary conditions, Proc. US–Japan Sem. on Diff. and Func. Eq., Univ. of Minnesota, Minneapolis, 1967, W. A. Benjamin (1967), 383–409
[3] The degenerate case of boundary value problems associated with weakly nonlinear differential systems, Publ. Res. Inst. for Math. Sc. Kyoto Univ., Ser A, 4 (1968), 545–584

VEJVODA, O. – TVRDÝ, M.
[1] Existence of solutions to a linear integro-boundary-differential equation with additional conditions, Ann. Mat. Pura Appl., 89 (1971), 169–216

VEJVODA, O.
[1] The stability of solutions of a system of differential equations in complex domain (in Czech, English summary), Časopis pěst. mat., 82 (1957), 137–159
[2] Perturbed boundary value problems and their approximate solution, Proc. Rome Symp. num. treatment etc. 1961, 37–41
[3] On the periodic solution of a quasilinear nonautonomous system, Czechoslovak Math. J., 11 (86) (1961), 62–75
[4] On perturbed nonlinear boundary value problems, Czechoslovak Math. J., 11 (86) (1961), 323–364

WALL, H. S.
[1] Concerning harmonic matrices, Arch. Math., 5 (1954), 160–167

WEXLER, D.
[1] On boundary value problems for an ordinary linear differential system, Ann. Mat. Pura Appl., 80 (1968), 123–134

WHITTAKER, E. T. – WATSON, G. N.
[1] A course of modern analysis II, Cambridge University Press, 1927

WHYBURN, W. M.
[1] Differential equations with general boundary conditions, Sem. Reports Math., 2 (1941), 45–61
[2] Differential equations with general boundary conditions, Bull. Amer. Math. Soc., 48 (1942), 692–704
[3] Differential systems with boundary conditions at more than two points, Proceedings of the Conference on Diff. Eq., University of Maryland, 1955, 1–21
[4] On a class of linear differential systems, Revista de Ciencias, 60 (1958), 43–59

WYLER, O.
[1] Green's operator, Ann. Mat. Pura Appl., 66 (1964), 251–263

ZETTL, A.
[1] Adjoint and self-adjoint boundary value problems with interface conditions, SIAM J. Appl. Math., 16 (1968), 851–859
[2] Adjointness in nonadjoint boundary value problems, SIAM J. Appl. Math., 17 (1969), 1268–1279

ZIMMERBERG, H. J.

[1] Symmetric integro-differential-boundary problems, Trans. Amer. Math. Soc., 188 (1974), 407 – 417

[2] Linear integro-differential-boundary-parameter problem, Ann. Mat. Pura Appl. 105 (1972), 241 – 256

ZUBOV, V. M.

[1] On generalized Green's matrices (in Russian), Mat. Zametki, 15 (1974), 113 – 120

[2] Certain properties of the generalized Green's matrices of linear differential operators (in Russian), Differencial'nye Uravnenija, 10 (1974), 996 – 1002

Index